# *CURRENT SIGNATURE ANALYSIS FOR CONDITION MONITORING OF CAGE INDUCTION MOTORS*

**IEEE Press**
445 Hoes Lane
Piscataway, NJ 08854

# *CURRENT SIGNATURE ANALYSIS FOR CONDITION MONITORING OF CAGE INDUCTION MOTORS*

## Industrial Application and Case Histories


**WILLIAM T. THOMSON**
**IAN CULBERT**



IEEE PRESS SERIES ON POWER ENGINEERING

IEEE PRESS

WILEY

For general information on our other products and services or for technical support, please contact our Customer Care Department within the United States at (800) 762-2974, outside the United States at (317) 572-3993 or fax (317) 572-4002.

Wiley also publishes its books in a variety of electronic formats. Some content that appears in print may not be available in electronic formats. For more information about Wiley products, visit our web site at www.wiley.com.

***Library of Congress Cataloging-in-Publication Data is available.***

ISBN: 978-1-119-02959-5

Printed in the United States of America

10 9 8 7 6 5 4 3 2

# CONTENTS

# ABOUT THE AUTHORS

**William T. Thomson** was born in Scotland in 1946 and started his working life in 1961 as a maintenance electrician until 1970. Evening class study provided the vocational qualifications to enter the University of Strathclyde (Glasgow, Scotland) in 1970. In 1973, he graduated with an Honors degree in Electrical and Electronic Engineering specializing in electrical machines. He was an R&D engineer from 1973 to 1977 with Hoover Ltd. in Glasgow, working on the reduction of noise and vibration from electric motors. In 1977, he was awarded a Master's degree from the University of Strathclyde for a research thesis titled "Reduction of Acoustic Noise and Vibration from Small-Power Electric Motors."

From 1977 to 1979, Thomson was a Lecturer in Electrical Power Engineering at the Hong Kong Polytechnic, and from 1979 until 2001 he was a Lecturer (1979–1983), Senior Lecturer (1983–1990), and Full Professor (1990–2001) at the Robert Gordon University in Aberdeen, Scotland. His research focused on the design, operation, and condition monitoring of induction motors, and in 1980 he initiated research into Motor Current Signature Analysis to detect cage winding breaks and abnormal levels of airgap eccentricity in cage induction motors. This led to his first industrial case history in 1982.

He left academia in 2001 to set-up EM Diagnostics Ltd. and the company now has a team of professional electrical engineers and provides consulting services on the design, operation, and condition monitoring of induction motors. He has published 72 papers on condition monitoring of induction motors in a variety of engineering journals such as *IEEE Transactions* (USA), *IEE Proceedings* (UK), and also at numerous International IEEE and IEE conferences. He is a senior member of the IEEE, a fellow of the IEE, and a Chartered Professional Engineer registered in the UK.

Thomson was awarded the Queen's award for technological achievement in 1992 for his input of knowledge to the development of "Motor monitor" for diagnosing broken rotor bars in 3-phase induction motors via current signature analysis, marketed by Entek of Cincinnati, USA.

In 1999, Thomson provided access to his knowledge on MCSA, via a license from the Robert Gordon University, Scotland to Iris Power (now Iris Power-Qualitrol) Canada for the development of the CS meter (CS meaning current signature) and the third generation of this instrument called the MDSP (Motor Digital Signal Processor) was released by Iris Power-Qualitrol in 2014. The Robert Gordon University and EM Diagnostics Ltd. share (50:50) the royalty payments (in perpetuity) from Iris Power-Qualitrol to the Robert Gordon University.

**Ian Culbert** was a rotating machines specialist at Iris Power-Qualitrol since April 2002 until his very untimely death on September 8, 2015. At this company he provided consulting services to customers, assisted in product development, trained sales and field service staff, and reviewed stator winding partial discharge reports. Culbert also advised customers on Motor Current Signature Analysis and analyzed complex case histories. Before joining Iris Power he was a motor and small generator specialist with Ontario Hydro/Ontario Power Generation from 1977 to 2002 and prior to that time, a motor designer with Parsons Peebles, Scotland and Reliance Electric, Canada.

Culbert was a Registered Professional Engineer in the Province of Ontario, Canada and a senior member of IEEE. He has co-authored two books on electrical machine insulation design, evaluation, aging, testing, and repair and was principal author of a number of Electric Power Research Institute reports on motor repair. Culbert also co-authored a number of papers on motor electrical component on-line and off-line motor diagnostics testing and actively participated in IEEE, IEC, and ISO standards development working groups.

# *OBITUARY TO IAN CULBERT (1943–2015)*

**A POEM** by Robert Burns expresses the thoughts of Ian's family, his many friends, and co-workers whom he met during his professional life.

William T. Thomson

Robert Burns (1759–1796), Scotland's National Bard
"Epitaph on my own friend and my Father's friend Wm. Muir in Tarbolton Mill"
(April 1784)

*An honest man here lies at rest,*
*As e'er God with his image blest:*
*The friend of man, the friend of truth,*
*The friend of age, and guide of youth:*
*Few hearts like his – with virtue warm'd,*
*Few heads with knowledge so inform'd:*
*If there's another world, he lives in bliss;*
*If there is none, he made the best of this.*

Ian was born on February 10, 1943 in Perth, Scotland and started his career as an electric motor designer with Parsons Peebles, Scotland. He emigrated to Canada with his family in February 1974 to take up a position as an electrical machines designer with Reliance Electric, Canada. From 1977 to 2002 Ian was an electrical machines specialist with Ontario Hydro/Ontario Power Generation. In April 2002, Ian joined Iris Power as a Rotating Electrical Machines Expert until his very sad and untimely death on September 8, 2015. At Iris Power he provided consulting services to customers, assisted in product development, trained sales and field service staff, and reviewed stator winding partial discharge reports. Ian also advised customers on Motor Current Signature Analysis and analyzed complex case histories.

Ian was a Registered Professional Engineer in the Province of Ontario, Canada and a Senior Member of IEEE. He co-authored two books on electrical machine insulation design, evaluation, aging, testing, and repair and was principal author of a number of Electric Power Research Institute reports on motor repair. Ian also co-authored and presented papers at IEEE international conferences and published papers in IEEE Transactions on the off-line and on-line testing of electrical machines to diagnose problems in stator and rotor windings and stator core problems. He actively participated in IEEE, IEC, and ISO standards development working groups. Ian was an excellent electrical engineer who set very high standards and was also a gentleman of wit and humor and was very proud of his Canadian citizenship and his native country, Scotland and is very sadly missed by all who knew him. It was indeed a great pleasure to be the co-author of this book with Ian and his input is testament to his knowledge and experience.

To quote Ian's daughter, Georganne "My Dad was a man who lived his life authentically. What you saw was what you got. The generous, intelligent and caring man I knew as my father was the same man his co-workers and friends knew."

# ACKNOWLEDGMENTS

**SINCERE THANKS** are expressed to Mr. Archie Low and Mr. Donald Sutherland, formerly of the Robert Gordon University (RGU), Scotland, for their indispensable contributions to the design and construction of fault producing, 3-phase SCIM test rigs for MCSA tests. The authors also acknowledge the kind assistance of companies that provided photographs, which include, Nidec Corporation – US Electric Motors Division, EASA (Electrical Apparatus Service Association, USA), Parsons Peebles (Scotland), Baldor Electric Company (USA), Quartzelec (Scotland), BW Offshore UK Ltd. (Scotland), Iris Power-Qualitrol (Canada), EM Diagnostics Ltd. (Scotland).

The authors acknowledge Mr. Ellis Hood, former Senior Lecturer at RGU for his invaluable input on checking the manuscript and for the many technical suggestions he made to improve its content and style of presentation. Sincere thanks are expressed to Mr. Angus Brown, consultant electrical and software engineer with EM Diagnostics Ltd. for his excellent production of all the final figures. William T. Thomson and Ian Culbert wish to thank their respective employers EM Diagnostics Ltd. (Scotland) and Iris Power-Qualitrol (Canada) for their vital support during the writing of the book. Mr. Austin H. Bonnett (USA), formerly Vice President Technology, Emerson Motor Technology Center, is sincerely thanked for his all-important technical review of the original manuscript prior to final publication and for writing the foreword to this book. The authors acknowledge Mr. John Middlemiss, formerly chief designer of electrical machines and Mr. Len Jones, Head of Engineering both of Parsons Peebles (Scotland) for their technical advice on certain design aspects of HV induction motors. Dr. Greg Stone (of Iris Power-Qualitrol), Dielectrics Engineer, who is an expert on HV stator winding insulation systems and partial discharge monitoring, is sincerely thanked for proofreading Chapters 2 and 3 on behalf of the late Ian Culbert. Mr. Richard Nailen, Engineering Editor of the *Electrical Apparatus Magazine* (USA) is duly acknowledged for his invaluable and independent review of the final manuscript, which was by personal invitation directly from Thomson.

Thomson and Ian dedicate this book to their respective wives, Mary Thomson and Anne Culbert, for their irreplaceable support throughout their careers which made the writing of this book become a reality, without them, it would not have been possible.

William T. Thomson
Ian Culbert

# FOREWORD

**FOR DECADES** the motor industry associated with squirrel cage induction motors (SCIMs) has been developing, as part of condition monitoring, a process called Motor Current Signature Analysis (MCSA) to determine if the rotor cage winding has broken rotor bars or the motor has an abnormal level of operational airgap eccentricity. Since the 1970s, research, testing, evaluation, and technical papers have been published on this form of condition monitoring and how and where to apply it. Unfortunately, few of these studies and resulting papers have had any significant amount of actual case histories containing enough useful data to assist in the conduction of an accurate analysis.

The authors of this unique book, William T. Thomson and Ian Culbert, have presented 50 industrial case histories, 35 of these are on MCSA to detect broken rotor bars in "Cage Induction Motors," which also include what is referred to as "false positives." These are cases where there are no broken rotor bars but the test data indicate that there are. There are also 15 industrial case histories on MCSA for diagnosing abnormal levels of operational airgap eccentricity including successful and "unsuccessful" cases.

On the surface, false positives may not seem a major issue. One may even breathe a sigh of relief that they do not have to rebuild the rotor. This is the issue with this approach that the authors discuss in great detail. First, if there turns out not to be broken rotor bars in a large, high voltage SCIM and the user shuts down the operation, pulls the motor, transports it to a qualified repair facility, disassembles the motor and then finds nothing wrong, the expense to do so and the loss in production can add up to a greater cost than a new motor! The end user is a very dissatisfied customer and inevitably loses all faith in MCSA as a credible condition monitoring technique. In many cases the "false positive" that drove this decision may indicate a serious problem with the driven equipment that still has not been identified and corrected when the motor is re-installed.

The authors do a great service to industry by identifying the source of many of these false positives. If it is wrongly concluded that there is a "false positive" and the motor is kept in operation, then the danger is that in some cases a broken rotor bar may find its way into the stator winding and cause a catastrophic failure or in some cases may actually exit the motor and cause physical harm to operating personnel, or in a hazardous area may lead to an explosion.

Because of these possibilities there has previously been some hesitance to rely on MCSA as an effective tool for condition monitoring and conducting a "Root Cause Failure Analysis." However, the proper use of this methodology proposed by the authors combined with the case studies now greatly minimizes the possibility of an incorrect root cause failure analysis. The final chapter starts with a very useful flow chart taking the user through a step-by-step process of conducting an MCSA test, which is followed by sections on "*strengths, weaknesses, external constraints, and very importantly the lessons learned.*"

For those who choose to use MCSA as a tool there are some considerations as to where best to apply MCSA as pointed out by the authors. A review of broken rotor bar failures reveals that most broken bars occur on fabricated rotors and not on die-cast rotors which are normally used in smaller motors. The motors with the highest probability of having broken bars are those that have been in service for many years, those that are frequently started, or have high inertial loads. The actual loading of the motor or high ambient temperature conditions can also be a factor.

MCSA is proposed as a useful tool to benchmark the motor prior to shipment to the job site or upon start up. Unfortunately, there are some cases where the original rotor design was not acceptable for its application and therefore has a built in propensity for broken rotor bars during the motor's normal life cycle. There are other cases where, in the process of rebuilding the rotor, adequate steps have not been taken to minimize future bar breakage. The authors have woven into this book the basic knowledge to identify and deal with most of these issues.

When purchasing future motors, especially those with fabricated rotors, the authors recommend that the motor manufacturer also supply the actual motor speed at different loads and the correct number of rotor bars. It is also helpful to know the shaft configuration and whether end ring retaining rings cover the extended bars and end rings area. This additional information will help to ensure a more reliable MCSA diagnosis of cage winding problems. The book contains valuable details as to why this additional information is useful. If the motor is dismantled for repair much of this information can be obtained then.

Another feature of the book is that it contains useful information on basic SCIM theory to assist those who may be technical people, such as mechanical, maintenance, and instrumentation engineers and technicians but who may have had no prior training on basic induction motor theory. At the end of each chapter there are 10 questions that make the book useful in the training process.

This book can be very beneficial for those who design, install, operate, maintain, troubleshoot, and repair SCIMs. The authors have many years of experience in all of these areas and have chosen to include information pertaining to this vast array of users. I would also recommend it to university or trade schools as a training tool and a reference book.

It is quite apparent that the time spent to obtain and study the material presented by the authors is more than offset by the costs and time spent to avoid just one catastrophic motor failure. This book will be a valuable asset in the library of those who have to deal with SCIMs in any of the many aspects from design through to successful operation.

Austin Bonnett, *IEEE Life Fellow*

**Formerly:**

Vice President of Engineering, US Motors, USA
Vice President Technology
Emerson Motor Technology Center
St. Louis, MO, USA

**Currently:**

President of Austin Bonnett Engineering LLC

5341 July Avenue

Gallaton, MO, 64460, USA

# PREFACE

**CONDITION MONITORING** of 3-phase squirrel cage induction motors (SCIMs) is now extensively used by end users to prevent catastrophic failures, unscheduled downtime with consequent loss of income, and hazardous conditions that may lead to major accidents. This book is dedicated to one condition monitoring technique, namely, Motor Current Signature Analysis (MCSA) and although its content is focused to suit the needs of industry it should also be of interest to academia.

The content is uniquely different from all other books on condition monitoring of electrical machines and also those with a part content on the use of MCSA, since it documents 50 industrial case histories on the application of MCSA to diagnose broken rotor bars, unacceptable levels of airgap eccentricity, and abnormal drive dynamics downstream of the motor. A key feature of the case histories is that, wherever possible, they seek to close the loop between diagnosing a problem and strip down of the motor, to provide photographic evidence that the diagnosis was correct or incorrect. MCSA case histories of motors with power ratings from 127 kW (170 HP) to 10,000 kW (13,340 HP) and operating voltages from 415 to 13,800 V are presented.

The reason for this book arises from the nature of the existing literature. Since the late 1970s there have been hundreds of research papers on MCSA which have been predominantly published by academia, with the test results obtained from small power SCIMs operating under controlled experimental conditions. These papers justify their research on the basis that MCSA is required by the end users and this is perfectly acceptable; however, the number of papers containing actual industrial case histories amount to only about 2% of the total. There was therefore an overwhelming need for a book on MCSA that focused on industrial case histories. Since 1982 the authors have applied MCSA in industry, William T. Thomson for 34 years and Ian Culbert for 14 years. Further, the authors' have 108 years of combined experience in the installation, maintenance, repair, design, manufacture, operation, and condition monitoring of SCIMs.

The successful application of MCSA requires the engineer to understand the operation of SCIMs, have an appreciation of the design, construction, and manufacture of cage rotors, the causes of breaks in cage windings, and the fundamentals on the use of MCSA to detect broken rotor bars. These topics are covered in Chapters 1 to 4, respectively. Chapters 5 and 6 contain case histories on MCSA used to assess the operational condition of different designs of cage windings when SCIMs are driving steady mechanical loads. Chapter 7 reports on case histories where MCSA attempts to diagnose broken rotor bars when cyclical disturbances from the mechanical loads are reflected back into the motor, which reflection can cause an incorrect diagnosis of broken rotor bars, referred to as a false positive.

Chapter 8 presents case histories, which include false positives of broken rotor bars due to the reflected mechanical dynamics from the combination of low speed gearboxes and fluctuating loads from conveyors and crushers. Chapter 9 presents miscellaneous MCSA case histories on the detection of broken rotor bars. For example, cases in which the number of spider support arms on the shaft (and axial ducts) of a cage rotor is equal to the number of poles, since this design can give a false positive of cage winding defects; the detection of slack and worn belts in belt driven cooling fans and the detection of imperfections in the caisson of a submersible seawater lift pump driven by a SCIM. Chapter 9 also includes case histories on the application of MCSA to inverter-fed SCIMs.

Chapter 10 covers the definitions and practical causes of different types of airgap eccentricity and discusses the resulting unbalanced magnetic pull (UMP) that can cause a rotor to stator rub. The predictor equations required to detect the unique current signature pattern, which is a function of the combination of static and dynamic airgap eccentricity, are presented and the diagnostic strategy for the signal processing is explained. The interpretation of the current spectrum and methodology used to estimate the operational airgap eccentricity are also included in Chapter 10. Chapter 11 presents industrial case histories on the application of MCSA to estimate the operational airgap eccentricity, including successes and failures.

The MCSA case histories are deliberately presented in "*great detail*" since a "*broad brush, superficial presentation*" that leaves the reader wondering how the diagnosis was achieved is meaningless. The inclusion of cases, when MCSA was not successful, is in complete contrast to the hundreds of research papers published on MCSA, which tend to only report on successful laboratory-based experiments. Each case history in this book stands alone so that the reader does not need to scroll backward and forward to find information and inevitably, there is repetition of formulae and other relevant knowledge. It is the authors' opinion that the style of presentation of the case histories is advantageous to the readers and particularly to engineers who apply MCSA technology. Root Cause Failure Analysis (RCFA) investigations are very time consuming, expensive, and normally delay the repair of motors, which the end user wants to get back in service as soon as possible. Consequently, RCFA investigations for each of the 50 industrial case histories were certainly not carried out by the end users but by the authors only when requested to so do. These requests were very infrequent, but any investigations are reported.

Chapter 12 presents an appraisal on the reasons why end users have not been receptive to the application of MCSA to diagnose shorted turns in LV or HV stator windings or faults in roller element bearings in SCIMs. Chapter 13 starts with a flow chart on the application of MCSA, which is formulated in a practical style, directly applicable for industrial engineers. This is followed by an appraisal on the strengths, weaknesses, external constraints, and very importantly the lessons learned by the authors spanning a period of 34 years.

The identities of manufacturers and end users of the motors in the case histories are not given and neither are the motors' serial numbers. At the start of the chapters on case histories (5–9 and 11) a list is presented to assist the reader to select the ones of personal interest contained therein. For completeness, metric and imperial units are included since the latter are used by electrical machine manufacturers, motor repair companies, and end users in the USA and also by the NEMA MG1: "Motors and Generators," 2012, USA, and API 541, USA, 5th edition, December 2014 standards. Immediately after the list of contents there are lists of symbols, abbreviations, and relevant units of equivalence between SI, metric and Imperial units The equations in each chapter are presented after Chapter 13. There are 10 questions at the end of each chapter (except Chapter 13)—the publisher should be contacted directly to obtain access to the answers.

William T. Thomson
Ian Culbert

# NOMENCLATURE

| Quantity | Quantity Symbol | Unit | Unit Symbol |
|---|---|---|---|
| Angular frequency | $\omega$ | Radians per sec | rad/sec |
| Angular position around circumference | $\theta$ | Degrees | deg |
| Airgap—radial design value | $g$ | Millimeters/inches $10^{-3}$ | mm/mils |
| Airgap as a function of time and angle | $g(\theta,t)$ | Millimeters/inches $10^{-3}$ | mm/mils |
| Airgap eccentricity—static | $e_s$ | Millimeters/inches $10^{-3}$ | mm/mils |
| Airgap eccentricity—dynamic | $e_d$ | Millimeters/inches $10^{-3}$ | mm/mils |
| Backward rotating field from the rotor | $N_{sb}$ | Revolutions per minute | r/min |
| Equivalent broken rotor bar factor at any operating slip below full-load slip | $BB_f$ | Number | – |
| Equivalent broken rotor bar factor at operating slip | $BB_{fs}$ | Number | – |
| Broken rotor bar correction factor | $BB_c$ | Number | – |
| Broken rotor bar index at full-load current and slip | $n$ | Number | – |
| Broken rotor bar index at any operational slip below full-load slip and current | $n_{fs}$ | Number | – |
| Ball diameter in roller element bearing | $BD$ | Millimeters | mm |
| Centrifugal force | $CF$ | Newtons/pounds force | N/lbsf |
| Coil distribution factor | $k_d$ | Number | – |
| Coil span factor | $k_s$ | Number | – |
| Current (rms) | $I$ | Ampere | A |
| Current magnitude of supply frequency component | $I_p$ | Ampere | A |
| Current magnitude of $f - 2sf$ at any slip | $I_{LSB}$ | Ampere | A |
| Current magnitude of $f + 2sf$ at any slip | $I_{USB}$ | Ampere | A |

| Quantity | Quantity Symbol | Unit | Unit Symbol |
|---|---|---|---|
| Current magnitude of $f - 2sf$ at any slip but referred to full-load slip | $I_{LSBr}$ | Ampere | A |
| Current components sum of $f \pm 2sf$ at full-load | $I_n$ | Ampere | A |
| Current input per phase | $I_i$ | Ampere | A |
| Current—no-load per phase | $I_o$ | Ampere | A |
| Current in rotor per phase | $I_r$ | Ampere | A |
| Current in rotor bar | $I_{rb}$ | Ampere | A |
| Current per phase due to core losses | $I_c$ | Ampere | A |
| Current per phase—magnetizing | $I_m$ | Ampere | A |
| Current per phase in rotor referred to stator | $I'_r$ | Ampere | A |
| Contact angle on bearing raceways | $\beta$ | Degrees | deg |
| Diameter of rotor core | $D_r$ | Millimeters/inches | mm/inches |
| Decibel difference between $f - 2sf$ and $f$ | $N$ | Decibels | dB |
| Decibel difference (average) between $f \pm 2sf$ and $f$ | $N_{av}$ | Decibels | dB |
| Decibel difference—the corrected average of measured $N_{av}$ at reduced load and slip between $f \pm 2sf$ and $f$ referred to full-load slip | $N_{cav}$ | Decibels | dB |
| Decibel difference (average) between $f_{rs}$ and $f_{rs} \pm f_r$ | $N_{ec}$ | Decibels | dB |
| Electromotive force (instantaneous) | $e$ | Voltage | V |
| Electrical degrees | $\theta_e$ | Degrees | deg |
| Flux per pole | $\phi_p$ | Webers | Wb |
| Frequency of mains supply | $f$ | Hertz | Hz |
| Frequency of rotor emf and current | $f_2$ | Hertz | Hz |
| Frequency of lower sideband | $f_{sb}$ | Hertz | Hz |
| Frequencies of upper and lower sidebands | $f_s$ | Hertz | Hz |
| Frequencies of rotor slotting current components | $f_{rs}$ | Hertz | Hz |
| Frequency of oscillation of mechanical load | $f_c$ | Hertz | Hz |

| Quantity | Quantity Symbol | Unit | Unit Symbol |
|---|---|---|---|
| Frequency of rotational speed of rotor | $f_r$ | Hertz | Hz |
| Frequency of current—faulty bearings | $f_b$ | Hertz | Hz |
| Frequency of vibration—faulty bearings | $f_v$ | Hertz | Hz |
| Frequency of current—inner race defect | $f_{ci}$ | Hertz | Hz |
| Frequency of current—outer race defect | $f_{co}$ | Hertz | Hz |
| Frequency of current—roller element defect | $f_{cb}$ | Hertz | Hz |
| Frequency of current—bearing cage defect | $f_{cc}$ | Hertz | Hz |
| Frequencies of current components due to shorted turns | $f_{st}$ | Hertz | Hz |
| Flux density (peak) | $B_p$ | Tesla | T |
| Force due to unbalanced magnetic pull | $F_{UMP}$ | Newtons/pounds force | N/lbsf |
| Harmonic integer of time domain mmf | $n_{ws}$ | Integer | – |
| Harmonic integer | $n_v$ | Integer | – |
| Length of rotor core | $L$ | Millimeters | mm |
| Leakage reactance referred to stator | $X_{eq}$ | Ohms | Ω |
| Line current | $I_L$ | Amperes | A |
| Line voltage | $V_L$ | Volts | V |
| Magnetic flux | $\phi$ | Weber | Wb |
| Magnetizing reactance | $X_m$ | Ohms | Ω |
| Magneto motive force (mmf) | $F$ | Ampere-turn | A |
| Magnetic pull | $F_m$ | Newtons/pounds force | N/lbsf |
| Moment of inertia | $J$ | Kilogram metre$^2$ Pounds feet$^2$ | kg-m$^2$/lbs-ft$^2$ |
| Moment of inertia of motor | $WK^2M$ | Pounds feet$^2$ | lbs-ft$^2$ |
| Moment of inertia of mechanical load | $WK^2DE$ | Pounds feet$^2$ | lbs-ft$^2$ |
| Number of turns | $N_T$ | Integer | – |
| Number of rollers in a roller element bearing | $n_e$ | Integer | – |
| Pole-pairs | $p$ | Integer | – |
| Poles | $2p$ | Integer | – |
| Power output at shaft of induction motor | $P$ | Horsepower/kilowatts | HP/kW |

| Quantity | Quantity Symbol | Unit | Unit Symbol |
|---|---|---|---|
| Power input to the induction motor | $P_{in}$ | Kilowatts | kW |
| Pole-pair harmonic of rotor slotting flux waves | $m$ | Integer | – |
| Pole-pair odd harmonic stator winding | $m_o$ | Integer | – |
| Pole-pair even harmonic of stator winding | $m_e$ | Integer | – |
| Power factor | $p.f.$ | Number | – |
| Pitch diameter of roller element bearing | $PD$ | Millimeters/inches | mm/inches |
| Pressure | $P_{re}$ | Pounds per inch$^2$/megapascals | psi/MPa |
| Reluctance | $S$ | Ampere-turns/weber | A/Wb |
| Resistor (fictitious) used in equivalent circuit | $R_c$ | Ohms | Ω |
| Rotor speed | $N_r$ | Revolutions per minute | r/min |
| Relative permeability of free space | $\mu_0$ | Henry per meter | H/m |
| Resistance of shunt across a current transformer | $R_{sh}$ | Ohms | Ω |
| Rotor bars | $R$ | Integer | – |
| Stator resistance per phase | $R_s$ | Ohms | Ω |
| Second moment of area of a round shaft | $I_M$ | Meters$^4$ | m$^4$ |
| Synchronous speed | $N_s$ | Revolutions per minute | r/min |
| Slip | $s$ | – | Number or % |
| Slip at full-load speed | $s_{FL}$ | – | Number or % |
| Slip at any speed | $s_o$ | – | Number or % |
| Speed of forward rotating field from rotor | $N_{fr}$ | Revolutions per minute | r/min |
| Spider support arms or axial ducts on rotor | $S_p$ | Integer | – |
| Stator bore diameter | $D_s$ | Millimeters/inches | mm/inches |
| Stator slots | $S$ | Integer | – |
| Static eccentricity integer | $n_s$ | Integer | – |
| Saturation index | $n_{sa}$ | Integer | – |
| Stator space harmonic index | $n\theta$ | Integer | – |
| Time | $t$ | Seconds | s |
| Torque | $T$ | Newton meter/pounds force feet | N·m/lbsf·ft |

| Quantity | Quantity Symbol | Unit | Unit Symbol |
|---|---|---|---|
| Torque supplied by motor | $T_m$ | Newton meter/pounds force feet | N·m/lbsf·ft |
| Torque—accelerating | $T_{ma}$ | Newton meter/pounds force feet | N·m/lbsf·ft |
| Torque required by mechanical load | $T_L$ | Newton meter/pounds force feet | N·m/lbsf·ft |
| Torque—full-load | $T_{FL}$ | Newton meter/pounds force feet | N·m/lbsf·ft |
| Torque—starting | $T_s$ | Newton meter/pounds force feet | N·m/lbsf·ft |
| Turns per phase | $T_{ph}$ | Integer | – |
| Voltage | $V$ | Volts/kilovolts | V/kV |

# ACRONYMS AND ABBREVIATIONS

| | |
|---|---|
| ac | Alternating current |
| API 541, 5th edition, Dec., 2014 | American Petroleum Institute |
| CCGT | Combined cycle gas turbine |
| CT | Current transformer |
| CM | Condition monitoring |
| DOL | Direct on line |
| dc | Direct current |
| DE | Drive end |
| EU | Engineering units |
| ERR | End ring retaining ring |
| FAT | Factory Acceptance Test |
| FD | Forced draft |
| FLC | Full-load current |
| FLT | Full-load torque |
| FFT | Fast Fourier transform |
| FPSO | Floating Production and Oil Offloading ship |
| GR | Gas recirculating |
| GWI | Ground wall insulation |
| HV | High voltage—see Appendix 13.A |
| ID | Induced draft |
| ITI | Inter-turn insulation |
| ITT | Invite to tender |
| LP | Low pressure |
| LNG | Liquid Natural Gas |
| LV | Low voltage—see Appendix 13.A |
| MIG | Metal inert gas |
| MCSA | Motor current signature analysis |
| NEMA MG1: Motors and Generators | National Electrical Machines Association, USA |
| NDE | Non-drive end |
| NDT | Non-destructive testing |
| OEM | Original equipment manufacturer |
| PM | Planned maintenance |
| PTM | Port-side thruster motor |
| PAM | Pole amplitude modulation |
| PWM | Pulse Width Modulation |

| | |
|---|---|
| rms | Root mean square |
| RSPF | Rotor slot passing frequency |
| ROV | Remote operated vehicle |
| SCIM | Squirrel cage induction motor |
| SWLP | Seawater lift pump |
| STM | Starboard side thruster motor |
| UMP | Unbalanced magnetic pull |
| TIG | Tungsten inert gas |
| TIR | Total indicated run out |
| TSBGB | Top-side bevel gearbox |
| VT | Voltage transformer |
| VSA | Vibration spectrum analysis |

# *RELEVANT UNITS OF EQUIVALENCE USEFUL FOR THIS BOOK*

| Metric/SI Units | Imperial |
|---|---|
| 1.0 m | 39.37 inches |
| 25.4 mm | 1.0 inch |
| 1.0 mm | 0.0394 inches/39.4 mils (≅40 thou/mils) |
| 50 μm | 2.0 thou/mils |
| 25 mm/sec | 1.0 inch/sec |
| 1.0 mm/sec | ≅ 0.04 inches/sec |
| 1.0 kg | 2.2046 lbs (≅ 2.2 lbs) |
| 1.0 N | 0.2248 lbsf (≅ 0.225 lbsf) |
| 1.0 N·m | 0.73756 lbsf ft (≅0.738 lbsf ft) |
| 1.0 kg-$m^2$ | 0.042 lbs-$ft^2$ |
| 1.0 N/$m^2$ | $145 \times 10^{-6}$ lbsf/$inch^2$ |
| 745.7 W (≅746 W) | 1.0 HP |

CHAPTER 1

# MOTOR CURRENT SIGNATURE ANALYSIS FOR INDUCTION MOTORS

*William T. Thomson*

## 1.0 INTRODUCTION

In 1824, Francois Arago initially formulated the concept of a rotating magnetic field which was subsequently known as Arago's rotations. On June 28, 1879, Walter Bailey presented to the Royal Society of London, a paper entitled *A Mode of Producing Arago's Rotations,* in which he proposed an early form of the induction motor. On May 16, 1888, Nikola Tesla presented a paper on a "New System of Alternate Current Motors and Transformers" to The American Institute of Electrical Engineers [1.1]. This paper and Tesla's practical demonstrations proved how a rotating magnetic field could be produced and also verified its application to the principles of operation of an induction motor. Also, in April 1888, The Royal Academy of Sciences of Turin published Galileo Ferraris' research findings on his ac polyphase motor in which he presented the basis for the operation of an induction motor.

Nikola Tesla's contributions, which, very importantly, he patented in the United States in 1888 are considered to be the catalyst for the subsequent industrial development of the modern generation and distribution of electricity using the 3-phase alternating current (ac) system, which is universally used throughout the modern world. It is now generally accepted that Nikola Tesla's greatest achievement was his invention of the induction motor and these motors dwarf all other electric motors in industrial importance since they are used in their millions throughout industry around the world. Without this electrical machine, modern society as we know it today would probably not exist. The induction motor, without doubt is one of the greatest inventions of all time but one of which, it is sad to state, the world at large is completely unaware.

Induction motors typically consume 40–50% of the generated electricity in an industrialized country. In 2008, the United States was the largest consumer of electricity at 4,401,698 GWhrs/annum [1.2], hence the vast number of these motors used in industry becomes very clear. Induction motors are the "workhorses" of a modern industrialized country and as such, a condition assessment of their

*Current Signature Analysis for Condition Monitoring of Cage Induction Motors: Industrial Application and Case Histories*, First Edition. William T. Thomson and Ian Culbert.

operational integrity via condition monitoring is normal practice by end users to avoid the following:

**(i)** Unscheduled downtime and lost (delayed) production and income
**(ii)** Catastrophic failures
**(iii)** Hazardous operating conditions that may lead to major accidents

There are a number of key categories that cause failures in induction motors and these are

**(i)** Bearing failures
**(ii)** Stator winding failures
**(iii)** Broken rotor bars or end rings in cage induction motors
**(iv)** High airgap eccentricity and unbalanced magnetic pull that may lead to a consequential rotor to stator rub

Vibration monitoring and analysis to detect bearing faults in rotating plant (which includes induction motors) is well documented via thousands of published papers and also in text books and is therefore not the subject of this book. Stator winding failures and on-line partial discharge monitoring to assess the operational condition of high voltage (HV) stator windings are covered by Stone et al. [1.3].

This book is dedicated to motor current signature analysis (MCSA) for condition monitoring of 3-phase induction motors (SCIMs) and its content is specifically focused to suit the needs of industry. It differs from all other books on condition monitoring of electrical machines, for example, by Tavner et al. [1.4] which has a small part content (pp. 207–212) on current monitoring for rotor faults but does give a broad coverage of all the different condition monitoring technologies applicable in industry to electrical machines. A book by Toliyat et al. [1.5] provides a chapter on MCSA (Chapter 9, pp. 199–219), but there are no industrial case histories on this subject in that chapter and to quote the authors in the preface: "*the book was written to provide a full review of diagnostic technologies and as an application guide for graduate and undergraduate students.*" The book achieves its objectives and includes coverage on theoretical modeling, condition monitoring, and fault diagnosis for different types of electrical machines, but it is essentially a book for academia and students.

In contrast, this book is unique and the first of its kind since its central theme is on the industrial application of MCSA.

1. It contains a "unique data base" of 50 industrial case histories on the application of MCSA to 3-phase SCIMs in the range from 127 kW (170 HP) up to 10,160 kW (13,620 HP) at voltages from 415 V up to 13,800 V, and covers the following:
   **(i)** Successful and unsuccessful diagnosis of broken rotor bars in cage rotors.
   **(ii)** Successful diagnosis of unacceptable levels of airgap eccentricity in cage induction motors and also cases which were very difficult/impossible to analyze via MCSA.
   **(iii)** Samples of abnormal mechanical load dynamics downstream of the motor.

2. The case histories are deliberately presented in great detail so that practitioners of MCSA understand the complexity of making a final decision as to whether an induction motor actually has or has not a cage winding break, or an abnormal level of operational airgap eccentricity or that a problem may exist in the drive train downstream of the motor causing reflections back into the cage rotor.
3. Each case history is different with a variety of new and practical knowledge being presented.
4. As appropriate, the strengths and weaknesses of MCSA are also covered, particularly in the industrial case histories, since failed diagnoses have to be recognized as being as important as successful ones. Failures in diagnosis, when understood and properly interpreted, alert exponents of MCSA to conditions, which render the identification of fault current signatures impossible or at least very difficult.
5. The industrial case histories have been accumulated by the authors during the past 34 years (as of 2016) from their combined knowledge and experience of the installation, maintenance, design, operation, and condition monitoring of induction motors.
   - **(i)** Published papers by academia have indicated that shorted turns in stator windings and bearing defects in roller element bearings in 3-phase SCIMs can be diagnosed via current monitoring and research papers are referenced on these topics in Chapter 12. However, the application of MCSA in industry to detect these faults in 3-phase SCIMs has not come to fruition in comparison to the widespread usage by end users of MCSA for diagnosing cage winding breaks and operational airgap eccentricity.
   - **(ii)** The authors do not apply current monitoring in industry to detect shorted turns in stator windings or bearing faults in roller element bearings in 3-phase SCIMs for very good reasons, which will be explained and justified in Chapter 12.

The manufacturers of induction motors do not normally release design details of their motors to end users or condition monitoring companies since they have to protect their intellectual property (IP) in a highly competitive market place. Without that complete design information, it is therefore impossible to calculate a definitive magnitude of current components, which are a function of the severity of broken rotor bars or level of airgap eccentricity. During MCSA testing of induction motors in industry, the test engineer has no control over the operating conditions and load currents of the motors being tested and has to accept the actual load on the motor at the time of testing, since plant operators are very reluctant indeed to alter their process to provide a different load current (e.g., full-load amperes). This can be a significant constraint on the application of MCSA in industry.

This is distinctly different from MCSA testing of small induction motors in a laboratory, under controlled conditions, for research results. The core of an MCSA diagnostic strategy and the industrial case histories presented in this book are as follows:

**(i)** Identify unique current signature patterns (frequency content) indicative of the problem, using proven theoretical principles and frequency equations.

**(ii)** Estimate the possible seriousness of the problems based on the authors' experience of analyzing MCSA data from thousands of induction motors. This experience is also supported by fundamental research and industrial application papers by the authors, Thomson et al. [1.6 to 1.36] and Culbert et al. [1.37 to 1.42].

Section 6.4 summarizes, with respect to reliable diagnosis of broken rotor bars, all the additional factors which need to be considered before making a final diagnosis and recommendation for the action to be taken by the end user. This information follows on at the end of Chapter 6 on the diagnoses of cage winding breaks, which have been presented in Chapters 5 and 6 and the reader has had the opportunity to consider the practicalities underlying these recommendations. There is also an appraisal in the last chapter that covers successes, external constraints/weaknesses, and lessons learned on the application of MCSA by the author of this chapter, which spans 34 years and also 15 years by Ian Culbert.

## 1.1 HISTORICAL DEVELOPMENT OF MCSA AND GOALS OF THIS BOOK

In the late 1970s to mid-1980s, novel and fundamental research and development work was simultaneously initiated in the United States, United Kingdom, and mainland Europe on the study of current (and spectra) as a function of cage winding breaks in induction motors. In the United States, this work was reported by, for example, Kliman et al. [1.43, 1.44] and in the United Kingdom and mainland Europe various researchers reported on this topic, including, Williamson, [1.45], Vas [1.46], Deleroi [1.47], Hargis et al. [1.48], Tavner [1.49], Filipetti et al. [1.50], and Thomson [1.34]. In 1982, Thomson initiated research into the diagnosis, via MCSA, of unacceptable levels of operational airgap eccentricity in large HV induction motors operating in power stations and offshore oil production platforms and was the first to report an industrial case history in 1986, when an airgap eccentricity problem was diagnosed [1.37].

With advances in digital signal processing, in the late 1970s, it became possible to produce accurate current spectra of the electrical current to the motor and thus diagnose current signatures indicative of cage winding breaks or abnormal levels of airgap eccentricity between the rotor and stator. Both these problems can lead to consequential stator winding and core damage and failures. Although spectrum analyzers and commercially available MCSA instruments can produce current spectra, which present information pertaining to a cage winding break or abnormal airgap eccentricity and are now widely used by industry, it has to be recognized that such instruments are measurement tools to provide current spectra as the initial source of information to be subsequently interpreted as to whether a problem may or may not exist.

It is important to appreciate that MCSA *"cannot distinguish"* between broken rotor bars and a broken end ring and it certainly cannot identify the position in the cage winding where there is a broken bar. In practice this is not required by industrial end users, since they are only interested in the operational integrity of cage windings in induction motors and whether there is truly a cage winding break that can lead to a motor failure and downtime. When an MCSA instrument indicates that broken rotor bars exist, it cannot deliver a decision on the action to be taken by the end user and it is here, via the case histories, that this book provides the knowledge to assist end users in decisions on the action to be taken to prevent a catastrophic failure.

To achieve that goal, a combination of expert abilities is necessary for reliable diagnoses; these range over the acquisition and interpretation of the current spectra, which includes digital signal processing knowledge, estimation of the severity of the problem, technical appraisal of the rotor cage design, appreciation of the operational modes of the motor, and very importantly, previous experience, via case histories, of the detection of cage winding breaks or abnormal airgap eccentricity. Interpretation of that information and determination of the required action can be supported by reference to appropriate case histories in this book. Knowledge is also required to enable a risk assessment of the probability of either a stator winding failure caused by cage winding breaks or a rotor to stator rub caused by an airgap eccentricity problem and whether it is safe to continue running the motor. These are the key aptitudes required by engineers, when advising plant managers, so that correct decisions are made. Some of the possible questions and/or decisions depend on the estimated severity of the problem, such as a cage winding break, and how vital the motor is to the end user's production process, and samples of these are now given.

**(i)** Let it run and minimize the number of direct-on-line (DOL) starts.

**(ii)** Is there a potential safety hazard if the motor is kept running and what consequential damage could be done, due to, for example, broken rotor bars?

**(iii)** Whether to stop a motor to prevent a possible catastrophic failure?

**(iv)** Is it possible to carry out a boroscope inspection while the motor is *in situ*?

**(v)** Plan a shutdown for repairs as soon as possible.

**(vi)** Order spare parts, such as new rotor bars and end rings, or even a new rotor.

Consequently, this book is focused on providing a knowledge source for industrial engineers, who are responsible at various levels for the operation, maintenance, and condition monitoring of induction motors driving strategic mechanical plant. This book should be of interest to motor manufacturers (OEMs) and electric motor repair shops, since certain end users are now requiring MCSA tests on brand new motors and after repairs to a SCIM to provide base-line current spectra for comparisons with future on-site MCSA measurements. The OEMs and motor repair workshops should also find MCSA a useful test for *their own internal QA/QC checks* to estimate the operational condition of a cage winding and operational airgap eccentricity during a full-load heat run. However, this would not be part of a Factory Acceptance Test unless there was a formal contractual agreement to so do.

The application of MCSA as part of an overall condition monitoring methodology for strategic SCIMs, is certainly not simply a matter of pressing buttons on an instrument and hoping for a reliable and definitive diagnosis of the actual problem. It demands more than that and an engineer who intends to apply MCSA should obtain the following competencies:

1. The ability to correctly understand the principles of operation of a 3-phase induction motor during normal operation and to correctly apply the basic equations which are required for MCSA testing.
2. An understanding of the basic construction of large SCIMs.
3. An understanding of the implications of DOL starts, the torque versus speed curves of the SCIM and its mechanical loads and how to calculate, for example, the run-up time of the motor when driving a mechanical load such as a compressor or pump. This initially receives attention in Section 1.3 and in more depth in Chapter 3 because many of the industrial case histories presented in Chapters 5 and 6, in which MCSA detected broken rotor bars, were due to the following:
   - Too many sequential (DOL) starts causing inherent high starting currents, outwith the motor's design capabilities, Richard Nailen [1.51, 1.52], for example, when the end user does not abide by the original manufacturer's (OEM's) specification for the time delays between sequential starts.
   - Incorrect matching of the motor's torque–speed curve to the torque–speed curve of the load, so that there is insufficient accelerating torque to cope with all the starting conditions, to which the motor may be subjected by the end user.

These include the effect on the torque–speed curve of the motor and a consequential reduction in the available accelerating torque, due to voltage dips during a DOL start, for example, on an offshore oil production platform, which has an isolated generating system. Also the effect of starting the motor against an open valve set up in a centrifugal pump which has a greater torque versus speed demand compared to starting against a closed valve system when the SCIM was designed for the latter. This can happen when the end user has not specified all the possible starting conditions (normal and abnormal) to the manufacturer of a SCIM.

Numerous papers by other researchers and industrial engineers, which are relevant to MCSA will be referenced in due course, but only where applicable to the specific context and content of the particular chapter, and furthermore it is impossible to list all the publications on this subject. A Bibliography in an IEEE Transactions paper by Benbouzi [1.53] lists 372 references on *Induction Motor Faults Detection and Diagnosis* covering various diagnostic techniques and is an excellent source for published papers.

## 1.2 BASIC THEORY OF OPERATION OF THE 3-PHASE INDUCTION MOTOR

There are many text books on electrical machines, and a sample of these include, El-Hawary [1.54], Liwschitz and Whipple [1.55], Say [1.56], Sen [1.57], Chapman

[1.58], Hughes [1.59], and Slemon [1.60], which contain the theory of operation of a 3-phase squirrel cage motor. There are also specialist books dedicated solely to induction motors by, for example, Phillip Alger on "The Nature of Induction Machines" [1.61] and an older but excellent book by Herbert Vickers on "The Induction Motor" [1.62]. These particular books have been referenced since they provide an excellent and comprehensive treatment of the induction motor that covers, for example, the theory of operation of an induction motor, starting characteristics, 3-phase winding theory, design and construction of induction motors, unbalanced magnetic pull, slot combinations, magnetic noise and vibration, testing, and the industrial application of induction motors.

The style of presentation in Sections 1.2 and 1.3 is hopefully suited to the needs of electrical, mechanical, maintenance, and condition monitoring engineers and also chief electricians and senior technicians working in industry, who are responsible for the maintenance and condition monitoring of 3-phase induction motors. It is not the intention to merely regurgitate information, on the operation of the induction motor, which is already in the aforementioned books. The content that follows is deliberately written to avoid being over theoretical and contains the minimum of mathematics since this is not a classical style text book for academia and students, but nevertheless, it should be of interest to graduate and inexperienced engineers working with induction motors in a wide range of industries.

The 3-phase induction motor produces torque and power at the output shaft due to the combination of three fundamental phenomena and these are

**(i)** The production of a rotating magnetic field from a 3-phase winding supplied by a 3-phase supply voltage—Tesla's rotating magnetic field principle.

**(ii)** An electromotive force (emf) is induced in a coil placed in a changing, in this case, rotating magnetic field—Faraday's law of electromagnetic induction.

**(iii)** A current carrying coil placed in a magnetic field experiences a force—Oersted and Faraday's experiments.

The mathematical proof that a 3-phase balanced voltage (each phase spaced 120 degrees apart) supplied to a balanced 3-phase stator winding (with each phase spatially distributed 120 degrees apart) produces a rotating magnetic field in only one direction, can be found in numerous text books on electrical machines [1.54–1.62] and will not be repeated in this book. A stator core, into which the 3-phase winding is inserted, is made of high quality, electrical steel laminations to reduce eddy current losses [1.56, 1.61, 1.62], these laminations are insulated from each other and a photo of part of an unwound stator core assembly is shown in Figure 1.1.

Examples of stator core assemblies for large HV SCIMs are shown in Figures 1.2 and 1.3 with the HV windings inserted and Figure 1.3a is a sectioned view of a 3-phase induction motor. The coils are inserted into the stator core slots and the insulation on the coils is designed to suit the applied voltages, full details of insulation types for stator coils and coil shapes can be found in the book by Stone et al. [1.3].

A 3-phase induction motor cannot deliver power and torque solely due to the fact that a 3-phase winding supplied by a 3-phase balanced voltage supply produces a rotating magnetic field, Tesla [1.1]. To achieve output torque and power at the shaft

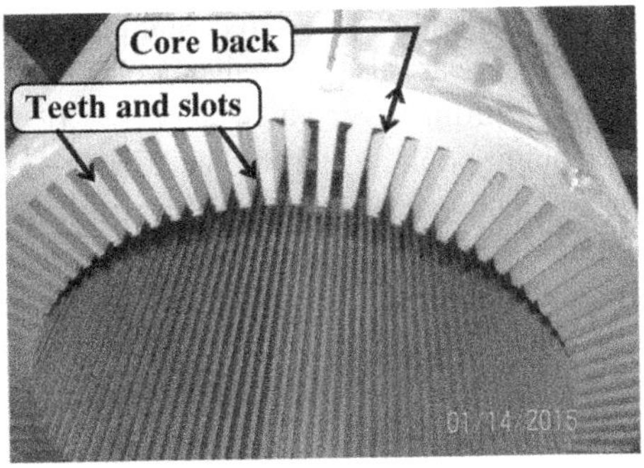


Figure 1.1 Example of an unwound stator core. Reproduced with permission of Nidec Corporation, US Electric Motors Division.

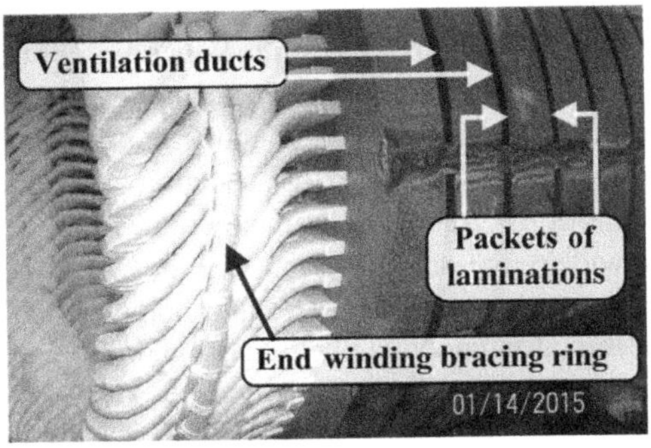


Figure 1.2 HV winding inserted into the stator core. Reproduced with permission of Nidec Corporation, US Electric Motors Division.

requires the application of the second phenomenon, namely, Michael Faraday's discovery of electromagnetic induction which is *"that an electromotive force (emf) will be induced in a coil due to the rate of change of flux linking that coil"* and he demonstrated and presented the phenomenon to the Royal Society in the United Kingdom on November 24, 1831, Dunsheath [1.65]. Faraday carried out a series of experiments

Figure 1.3 Complete HV stator core assembly. Reproduced with permission of Baldor Electric Company, USA, a member of the ABB group.

Figure 1.3a Sectional view of 3-phase SCIM. Reproduced with permission of Baldor Electric Company, USA, a member of the ABB group.

that led to this famous discovery of electromagnetic induction during a very short period of 10 days [1.64] in 1831, but it is quite incorrect to think that this was a spontaneous result, since this was after many years of dedicated research, experimentation on electric currents and magnetism and their interaction. A list of Faraday's diaries, and other references is given in Reference 1.65. Faraday's discovery at that time, was a truly remarkable achievement, by an experimenter, and may be considered to be one of the most important discoveries in electrical engineering, which produced by one individual, shaped the world as we know it today. For example, the generation of electricity is dependent on Faradays' law of electromagnetic induction and for completeness it is presented mathematically.

$$e = -N_T d\phi/dt \text{ volts} \tag{1.1}$$

$e$ = instantaneous magnitude of the electromagnetic force (emf) in volts

$N_T$ = number of turns

$\phi$ = magnetic flux in webers

$t$ = time in seconds

The negative sign indicates that a current flowing in response to the induced voltage ($e$) produces a magnetic field opposing the original field (Lenz's law). When a set of open circuited conductors is placed in a rotating magnetic field and inserted into a laminated, magnetic steel rotor core in the stator bore of a stator winding assembly shown in Figure 1.4, then by Faraday's law, an emf will be induced in these conductors.

The loop has been closed on the second phenomenon that leads to an induction motor being able to turn and produce torque. However, at this stage, the rotor will not turn, since the induction motor is simply a transformer whose primary winding is the stator and its secondary winding is the open circuited conductors in the rotor, inserted into the stator bore.

Now consider the third phenomenon, Hans Oersted, on April 21, 1820 observed that a compass needle moved from its magnetic North, when an electric current (supplied by a battery) passed through a wire close to the compass needle, and crucially, it only moved when the current was switched *"on and off."* Michael Faraday repeated

Figure 1.4 3-phase HV stator winding assembly.

Oersted's experiments but also found that a magnet exerts a force on a wire carrying an electric current. Thus if the conductors in the stator bore carry current in a rotating magnetic field they will experience a force. This can be achieved by simply short circuiting the conductors and Faraday's induced emfs will drive currents through them. The final step has been explained since these conductors will turn and produce torque.

The practical fruition of a set of current carrying conductors is achieved by using a set of copper bars, which are joined (via brazing or induction heating) to short circuiting end rings as illustrated in Figure 1.5. The copper cage winding is installed into a laminated rotor core by first inserting the rotor bars into the core and by then brazing or welding the bars to the end rings. Figures 1.6 and 1.7 give examples of squirrel cage rotors.

Each rotor bar in the cage winding experiences, in the direction of the rotating magnetic field, a force which is transferred to the rotor core structure. The rotor then rotates and produces torque and power output. The principles of operation of a 3-phase induction motor can be summarized as follows:

> *"In the induction motor there is a rotating electromagnetic field from the 3-phase stator winding and by Faraday's law emfs are induced in the rotor conductors, which drive currents through the short-circuited cage winding, and the cage winding experiences a force since it is carrying current in a rotating field."*

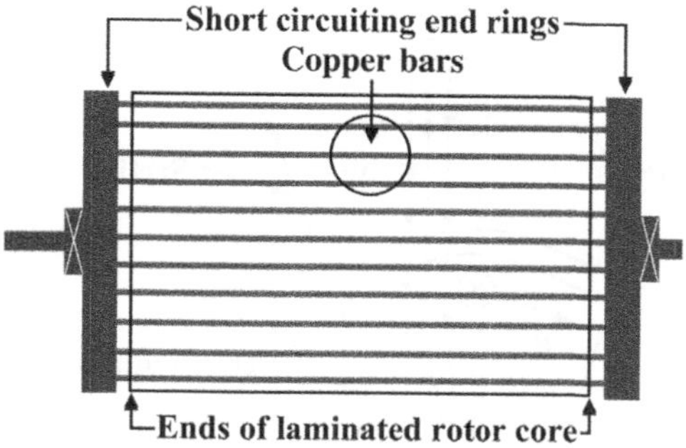


Figure 1.5 Schematic illustration of a squirrel cage copper or aluminum fabricated winding.

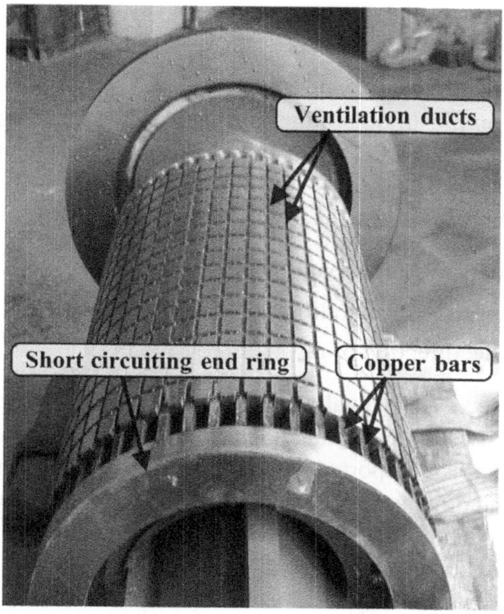


Figure 1.6 Copper fabricated squirrel cage rotor.

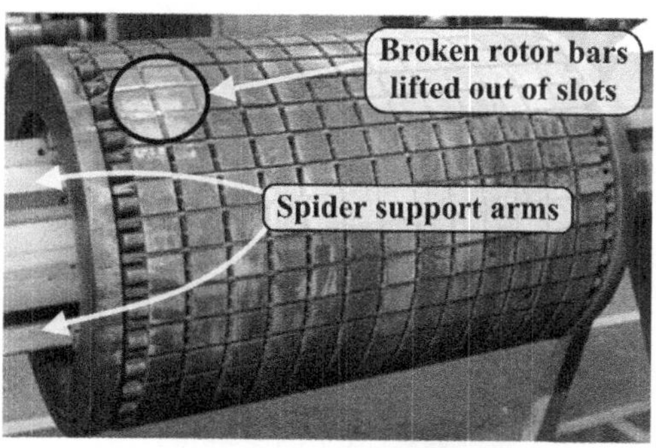


Figure 1.7 Squirrel cage rotor with a broken rotor bar that has lifted. *Source:* Thomson and Fenger [1.13]. Reproduced with permission of IEEE.

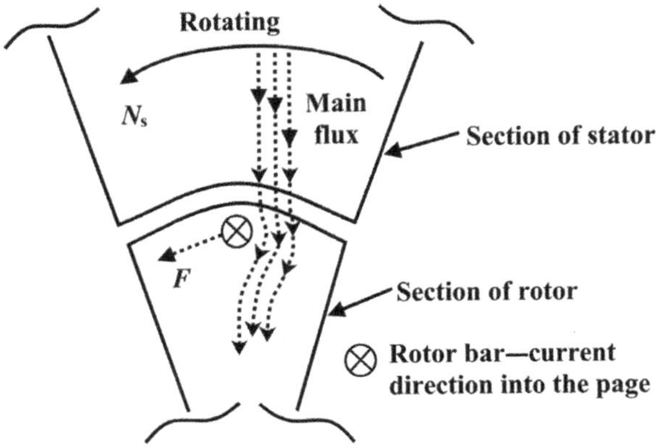


Figure 1.8a Induction motor action.

Figure 1.8a is an illustration of the force and torque on a cage rotor due to the interaction between the main flux (at synchronous speed, $N_S$), which links the induced currents in the rotor bars. The direction of force/motion is in the tangential direction and is obtained via Fleming's left hand rule for motoring action.

- The **F**irst finger is the direction of **F**lux from the stator to the rotor.
- The se**C**ond finger is the direction of **C**urrent.
- The thu**M**b is the direction of the **M**otion.

These three fingers are mutually perpendicular to each other.

Hopefully the style of the explanations will meet its objectives, namely to integrate theoretical principles supported by examples of practical implementation. As a visual aid to understanding, a finite element plot of the distribution of the magnetic field for an 8-pole induction motor is shown in Figure 1.8b and in Figure 1.9 the lines of magnetic flux are clearly shown. These follow the paths of least magnetic

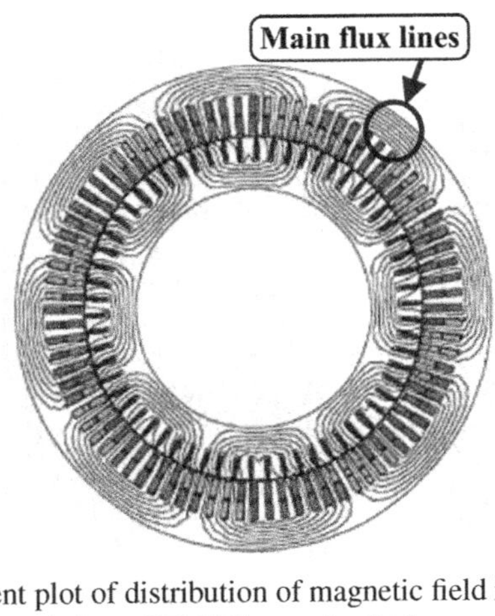


Figure 1.8b Finite element plot of distribution of magnetic field in an 8-pole induction motor. Reproduced with permission of EM Diagnostics Ltd.

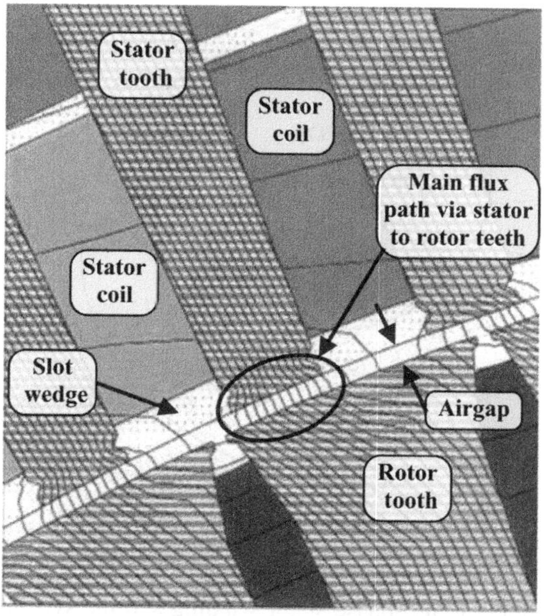


Figure 1.9 Finite element model showing the magnetic flux lines taking the path of least magnetic reluctance through the stator and rotor teeth to cross the airgap. Reproduced with permission of EM Diagnostics Ltd.

reluctance* (opposition to flow of flux*), via the stator and rotor teeth, to cross the air gap and link the rotor cage winding, as opposed to the higher reluctance route via the stator and rotor slots. Leakage flux lines from the main flux paths do of course diverge into the stator and rotor slot regions as specifically shown in Figure 1.9.

The construction and design features of different types of squirrel cage rotor are fully covered in Chapter 2 of this book while the forces and stresses that a cage winding experiences are presented in Chapter 3.

The use of the name squirrel in squirrel cage rotor comes from a rotating cage, which was sometimes used in the United States for squirrels as an external exercise aid and since the animal is fast and agile it was a very appropriate name to give to the cage rotor. The Westinghouse Corporation (who bought the rights from Tesla on his patents of the induction motor) in the United States was the first company to manufacture induction motors on an industrial scale. On a light hearted note, in the United Kingdom the same form of cage was used to provide pet mice with an exercise aid. Clearly the use of squirrel, in squirrel cage rotor, is much more elegant than the use of a "*mouse*" cage rotor.

### 1.2.1 Key Equations for MCSA Based on Operation of a 3-Phase Induction Motor

The 3-phase winding on the stator is supplied with a balanced 3-phase voltage supply at a certain frequency and is wound to produce a forward rotating magnetic field with a specific number of poles. The equation that relates the fundamental supply frequency

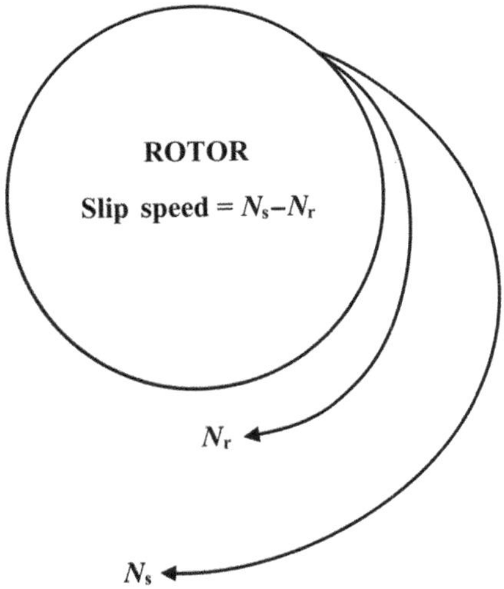


Figure 1.10 Illustration of slip speed in r/min.

$f$, of the voltage supply to a 3-phase stator winding and the synchronous speed $N_S$ (r/min) of the rotating magnetic field produced by the stator winding of pole-pairs $p$, is given by

$$f = N_S p/60 \quad \text{Hz} \tag{1.2}$$

where

$N_S$ is in r/min

$p$ = pole-pairs

The rotor rotates at a speed of $N_r$r/min as illustrated in Figure 1.10 and the diagram shows that, to supply the load torque, the rotor always rotates at a speed less than the synchronous speed. A measure of the slipping back of the rotor is termed the slip and is given by

$$s = (N_S - N_r)/N_S \tag{1.3}$$

This is a per unit term with the synchronous speed being the base reference speed and the slip normally given as a percentage.

The slip speed, which equals $N_S - N_r$ is the actual difference in r/min between the speed of the rotating magnetic field from the stator winding and the actual speed of the rotor but note that the definition of "slip frequency" is not the slip speed $N_S - N_r$ divided by 60 to convert to hertz. The term slip frequency in induction motor theory [1.54–1.62] has a specific electrical meaning, which is the frequency of the rotor currents and is given by

$$f_2 = (N_S - N_r)p/60 = sN_S p/60 = sf \quad \text{Hz} \tag{1.4}$$

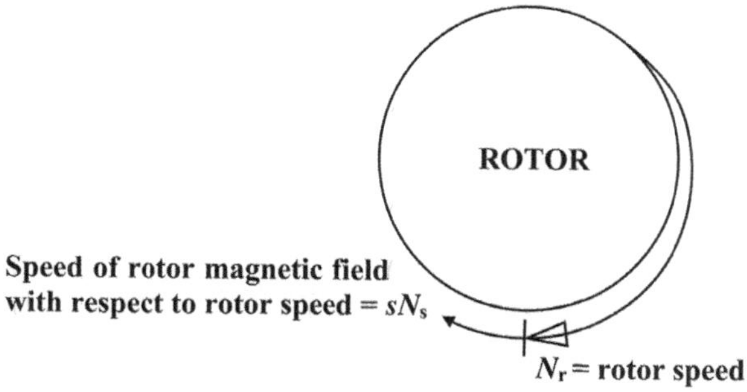


Figure 1.11 Illustration of forward rotating field $N_{fr}$ from rotor currents.

$f_2 = sf$ = slip frequency in hertz of induced emfs and currents in the cage winding.

This is the correct terminology for slip frequency but unfortunately a considerable number of condition monitoring companies who offer MCSA monitoring services, particularly those whose main services are in vibration monitoring, have incorrectly defined slip frequency as being equal to $(N_s - N_r)/60$ Hz. The forward rotating magnetic field, defined by $N_{fr}$ and produced by the rotor currents with respect to a fixed position on the rotating rotor moves faster than the actual rotor speed as illustrated in Figure 1.11 and is at a speed of

$$N_{fr} = N_s - N_r = sN_s \tag{1.5}$$

The speed of the forward rotating magnetic field produced by the current carrying rotor conductors with respect to the stationary stator winding is given by

$$\text{Rotor speed } N_r + \text{Rotor field speed } sN_s = \text{Synchronous speed}$$
$$N_r + sN_s = N_r + N_s - N_r = N_s$$

With respect to a stationary observer on the fixed stator winding, the speed of the rotating magnetic field from the rotor equals the speed of the stator rotating magnetic field, namely, the synchronous speed, $N_s$. This has to be the case and it is an important result but initially it can be somewhat challenging to understand. Both fields must be locked together to give steady torque production from an induction motor.

### 1.2.2 Interpretation of Motor Nameplate and Application of Basic Equations

It is now logical to interpret the nameplate data on a 3-phase induction motor, with particular reference to the application of MCSA, since that is the first step in carrying out the measurement of current and analysis of the current spectrum. Typical information on the nameplates of SCIMs manufactured for the UK and US markets is shown in Figures 1.12 and 1.13, respectively. It is often the case that the relevant nameplate data from SCIMs is not fully recorded in condition monitoring reports and this perhaps suggests that it is not appreciated how much useful information can

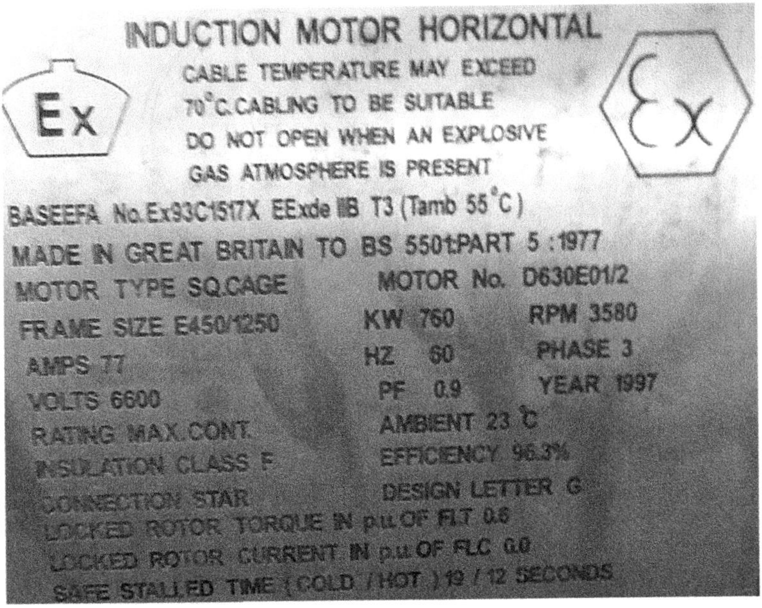


Figure 1.12 Nameplate from a 6.6 kV, 760 kW/1018 HP, 60 Hz SCIM.

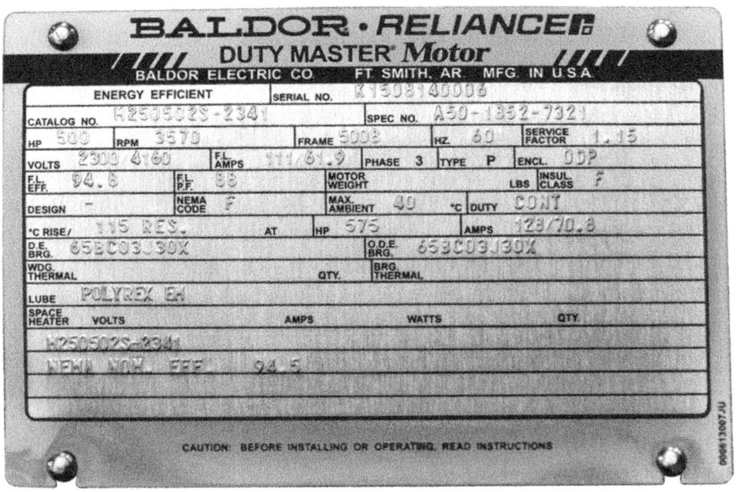


Figure 1.13 Photo of a nameplate from a 4160 V, 500 HP/373 kW, SCIM, reproduced by kind permission of © Baldor Electric Company, USA, a member of the ABB group.

be obtained and calculated from that data. It is the authors' experience that far too little attention is given to the interpretation of the nameplate, particularly by young and recently graduated electrical engineers, who have studied electrical engineering but may well be unsure of the actual meaning of the data thereon. It is important to stress the importance of that data and to comment on what is relevant information for MCSA.

The inclusion of this section on nameplate information is not about the total content of an induction motor's nameplate, since that is the domain of the manufacturers and the various standards, such as NEMA MG1 [1.66]. However, the only comment the authors wish to make to motor manufacturers in a worldwide context is that it would be extremely helpful to the end user and for the application of MCSA to stamp the number of rotor slots on the nameplate.

The information on the nameplates of motors manufactured for the US market is normally governed by the specifications of NEMA MG1 [1.66], and these requirements include

**(a)** Manufacturer's type and frame designation

**(b)** Horsepower output

**(c)** Time rating

**(d)** Temperature rise

**(e)** r/min at rated load

**(f)** Frequency

**(g)** Number of phases

**(h)** Voltage

**(i)** Rated-load amperes

**(j)** Code letter (see Section 20.9 in Reference 1.66)

**(k)** Service factor

Induction motors manufactured in the United Kingdom and on mainland Europe use kW for the power output at the shaft and N·m for Torque, whereas in the United States the units HP and lbsf·ft are used. The information on the nameplate nominally means that, when the motor is supplied at the rated volts and frequency and appropriately loaded it will deliver a full-load rated output power at the shaft in HP or kW and at the speed on the nameplate. The input current and power factor will also be the full-load rated values.

The nameplate of an HV SCIM manufactured in the United Kingdom, is shown in Figure 1.12 and as per normal, it does not give the number of poles, synchronous speed, operating slip $s_{FL}$, or the full-load rated torque. It is often the case that OEMs of large, HV SCIMs in the United Kingdom, stamp on the nameplate, the locked rotor torque and locked rotor current in per unit or percentage quantities of the full-load torque (FLT) and full-load current (FLC). Although the following may well be considered to be obvious to electrical power engineers, recall that MCSA is also used by non-electrical power engineers, thus it is important to explicitly explain the meaning of information on the nameplate which is relevant to MCSA.

With respect to Figure 1.12

$$\text{Power output at the shaft } P = 760\,\text{kW or } 1018\text{ HP}$$

Since the full-load efficiency is given the electrical power input is

$$P_{in} = 760/0.963 = 789\text{ kW}$$

For completeness, recall power input $P_{in} = \sqrt{3}V_L I_L \cos\phi$

The full-load rated speed is 3580 r/min and on a 60 Hz supply the synchronous speed in r/min of the rotating magnetic field from the stator winding of a 2-pole (pole-pairs, $p = 1$) motor is

$$N_s = (3600)/p = 3600\text{ r/min}$$

And since the rated full-load speed is 3580 r/min, which is just below 3600 r/min it is a 2-pole motor. The full-load slip, $s_{FL} = (N_s - N_r)/N_s = 0.0055$ or 0.55% at a nominal full-load rated speed of 3580 r/min.

The full-load slip is the first reference information that is required for MCSA, it is a crucial parameter as verified in Chapter 4. However, SCIMs are often operating below full-load, hence the input current, output speed, and power output will be less than the full-load rated values. At this early stage in the book, it should be mentioned that the nominal full-load speed on the nameplate may not be that which occurs in practice and the implications of this with respect to MCSA testing will be discussed in more detail in the case histories in Chapters 5 and 6.

The frequency of the currents in the squirrel cage rotor at full-load for this motor is $f_2 = s_{FL}f = 0.0055 \times 60 = 0.33\text{Hz}$ (close to dc) which is a very low frequency in comparison to the stator supply frequency. At switch on when the slip is 1.0 the frequency of the rotor currents equals the supply frequency.

The nameplate shown in Figure 1.12 does not give the full-load rated torque but this can be calculated from

$$\text{Power output at the shaft } P = \omega T \quad \text{N}\cdot\text{m}$$

$$\text{Full-load torque } T = 760 \times 10^3/((2\pi 3580/60)) = 2027\ (\cong 2000)\text{N}\cdot\text{m}$$

Due cognizance is given to the units used for power output and torque at the shaft used by industry in the United States, which are HP and lbsf·ft and as stated in NEMA MG1, therefore $T = 1496$ lbsf·ft (often stated in lbs·ft by industry but pounds force (lbsf·ft) is the correct unit since lbs is a mass and not a force).

$$1.0\text{ N}\cdot\text{m} \cong 0.74\text{ lbsf ft}$$

Two useful formulae for calculating torque but recall 746 W = 1.0 HP

$$T = \frac{(9550 \times \text{Power out in kW})}{\text{Rotor speed in r/min}}\ \text{N}\cdot\text{m}$$

$$T = \frac{(5250 \times \text{HP})}{\text{Rotor speed in r/min}}\ \text{lbsf}\cdot\text{ft}$$

**(i)** Figure 1.12 gives the locked rotor current or starting current presented on the nameplate as being FLC 6.0 which is six times the FLC and gives a starting current ($I_s$) of 462 A. Thus the cage winding is subjected to high mechanical

**TABLE 1.1 Relationships Between Frequency, Poles, Rotor Speed, Slip, and Frequency of Rotor Currents**

| Mains Supply Frequency | Poles | Synch. Speed of Stator Rotating Field | Rotor Speed Variable | Slip Depends on Load | Freq. of Rotor Currents |
|---|---|---|---|---|---|
| $f$ (Hz) | Number | $N_s$ (r/min) | $N_r$ (r/min) | $s$ | $f_2$ (Hz) |
| 60 | 2-poles | 3600 | 3564 | 1.0% | 0.6 |
| 50 | 2-poles | 3000 | 2985 | 0.5% | 0.25 |
| 60 | 4-poles | 1800 | 1782 | 1% | 0.6 |
| 50 | 4-poles | 1500 | 1470 | 2% | 1.0 |
| 60 | 6-poles | 1200 | 1194 | 0.5% | 0.3 |
| 50 | 6-poles | 1000 | 952 | 0.8% | 0.4 |
| 60 | 12-poles | 600 | 590 | 1.66% | 1.0 |

and electromagnetic forces and high thermal stresses due to temperature being proportional to current squared ($I^2$). These are discussed in detail in Chapter 3.

**(ii)** The locked rotor or starting torque ($T_s$) given on the nameplate is FLT 0.6, which is 60% of the calculated FLT (≅2000 N·m) and is 1200 N·m or 885 lbsf ft.

**(iii)** Table 1.1 gives sample values of frequency, pole-pairs, synchronous speed, actual rotor speed, slip, and slip frequency of rotor currents for SCIMs.

When an induction motor is supplied at rated volts and frequency and is running on no-load (and uncoupled) the input current consists of two components, which are phasors and therefore cannot be directly added to give a resultant.

**(i)** The magnetizing current, which sets up the rotating magnetic field lags the supply volts by 90°

**(ii)** An active component of the no-load current to the motor which is in phase with the supply voltage and supplies the core losses

There can be a misconception in calculating the percentage load at which the motor is operating, when the input current is incorrectly used to give that percentage loading. The power output and shaft speed are not normally measured on SCIMs operating in industry but the input current to large, HV SCIMs is normally continuously measured. This motor actually takes a current of 14 A on a no-load, uncoupled run and that current is part of the input current at any loading of the motor.

- Consider that this motor, which has a FLC of 77 A, is operating at an input current of 40 A. That does not mean that the motor is operating at 52% of its rated full-load power output at the shaft, because a portion of the 40 A provides the constant 14 A no-load mainly magnetizing current which does not contribute to the output power.
- The motor is taking 52% of the rated full-load amperes but that is different from the output power at the shaft and the operating slip is certainly not 52% of the full-load slip. This is an important fact, when applying MCSA and interpreting the current spectrum, as is discussed and demonstrated in Chapter 4.

## 1.3 STARTING AND RUN-UP CHARACTERISTICS OF SCIMS

There are many publications which cover the reasons for a SCIM taking a large starting current, when started DOL, and these publications also discuss the design features, which can change the starting current and it is not the intention of this book to repeat already published information. The reader is referred to Nailen's papers, [1.51, 1.52], which discuss in detail the starting demands and implications of DOL starting of 3-phase SCIMs for the oil industry but these papers are equally relevant to other industries and are particularly relevant due to the causes of cage winding breaks in a number of the MCSA case histories presented in Chapters 5 and 6. The main focus in this book is on the application of MCSA to motors that are started DOL and Chapter 3, with appropriate references, discusses the forces and stresses on an induction motor during a DOL start.

Suffice to state that a SCIM takes a large starting current, which can be up to six or seven times the rated FLC, when switched DOL, since at the instant of switch on, a SCIM is effectively a short circuited transformer, which inevitably takes a large starting current. The starting current is a function of numerous parameters such as the supply voltage, resistance and reactance of the rotor, the operating flux density and the magnetic circuit, and airgap length in the motor. In addition, the frequency of the voltage and currents in the rotor cage is at the supply frequency at switch on (slip is 1.0 and $f_2 = sf, f_2 = f$). There is the phenomenon known as skin effect [1.55, 1.61, 1.62, 1.63], which causes the current distribution in the rotor bars to be a function of frequency and the ac resistance of the rotor cage winding changes as the frequency changes from the supply frequency at switch on down to nearly dc at steady-state operation. This is fully discussed in Chapter 2. It is the expertise of the electrical machine designer to design the motor to have a specific starting current as specified by the end user, to also suit any limitations of the supply network, to which it will be connected. Also, the motor must have the required torque-speed curve to suit the starting and running requirements from the mechanical load.

Also, a high starting current taken by a large SCIM operating on an offshore oil production platform (an isolated generating system) inevitably causes a voltage dip during start up, which means the available torque from the motor drops since torque is proportional to the voltage squared, $V^2$. As an illustration, consider the following motor, which was specifically designed to have a relatively low starting current. It was used on an offshore oil and gas production platform, which has its own generators and thus there are restrictions on the starting current to prevent an excessive voltage dip on start up. The motor was driving a reciprocating compressor.

**Motor Name Plate Data**

3-phase SCIM, 6600 V, 2900 kW/3887 HP 340 A, 60 Hz, 890 r/min, S1, MCR

Locked rotor current 1275 A.

Locked rotor torque 17,113 N·m. (12,630 lbsf·ft)

Breakdown torque 46,674 N·m. (34,445 lbsf·ft)

The FLT is given by

$$T_{FL} = (9550 \times \text{Power output in kW})/\text{Rotor speed in r/min}$$
$$= 9550 \times 2900/890 = 31{,}116\ \text{N} \cdot \text{m}(22{,}931\ \text{lbsf} \cdot \text{ft})$$

**The following items (i–iv) were reproduced with permission of Parsons Peebles, Scotland:**

**(i)** Run-up time at 100% volts: Motor + Load = 2.6 seconds

**(ii)** Total moment of inertia: 350 kg-m$^2$ (8314 lbs-ft$^2$)

**(iii)** Allowable run-up time at 100% volts: 9 seconds from cold, 6 seconds from hot

**(iv)** Safe stall time at 100% volts: 10 seconds from cold, 7 seconds from hot

The current versus speed curve from switch on is shown in Figure 1.14.

This shows that the current is greater than three times the FLC up to around 90% of synchronous speed. Thus the faster the motor can run up to its steady-state operation the less exposure the rotor bars have to high currents, with their consequential thermal stresses, electromagnetic forces, and mechanical bending stresses on the bars and end rings external to the rotor core ends. To meet the needs of industry, this specially designed motor has a low starting current (3.75 FLC) for a large HV motor. Recall that in Figure 1.12, which gives the nameplate data for a different HV SCIM, the starting current is six times FLC.

The torque versus speed curve for the motor under discussion is shown in Figure 1.15 and all the area under the curve is available for accelerating the motor up to its no-load speed since there is no load on the motor. OEMs normally provide to their customers a torque–speed curve similar to that shown in Figure 1.15. The curve

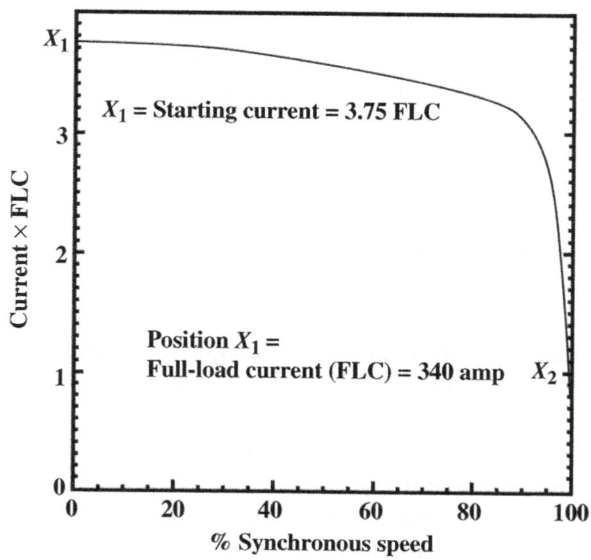


Figure 1.14 Current versus speed. Reproduced with permission of Parsons Peebles, Scotland.

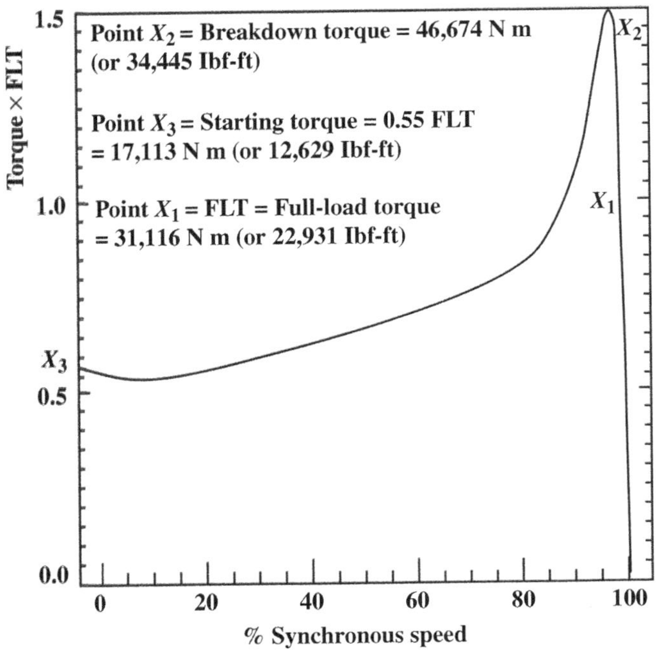


Figure 1.15 Torque versus speed curve for the motor. Reproduced with permission of Parsons Peebles, Scotland.

is non-linear up to a point immediately beyond the breakdown torque ($X_2$) and cannot be expressed by a single algebraic equation over that speed range but it is linear in the straight line portion (negative slope) of the plot down to the no-load speed (virtually synchronous speed, $N_s$).

Due to the speed scale used in Figure 1.15, which covers the full speed range, it is not possible to obtain the torque and corresponding speed directly from the plot between $N_s$ and the full-load speed, $N_{rFL}$ (at $X_1$). It is fully appreciated that the graph in Figure 1.5 between no-load to FLT versus speed can be redrawn to a different scale whereby the torque at different speeds between no-load and full-load can be read directly from a new graph. However, this requires the end user to do so and to avoid this the following analysis shows that it is unnecessary. The objective of the following analysis is to

(i) Provide a simple equation to calculate the torque at any speed ($N_r$) or vice versa from the linear part of the graph between no-load and full-load at position $X_1$.

Figure 1.15a is obtained from Figure 1.15 but with speed on the $y$-axis and torque on the $x$-axis to focus on the speed–torque curve between $N_s$ and the full-load speed and torque. The graph shown in Figure 1.15a can be represented by the very simple equation that describes a straight line curve, namely $y = mx + c$.

In this case the $y$-axis is the speed ($N_r$), $m$ is the slope of the curve, which is negative in this case (speed drops as torque increases), the $x$-axis is torque, and $c$ is

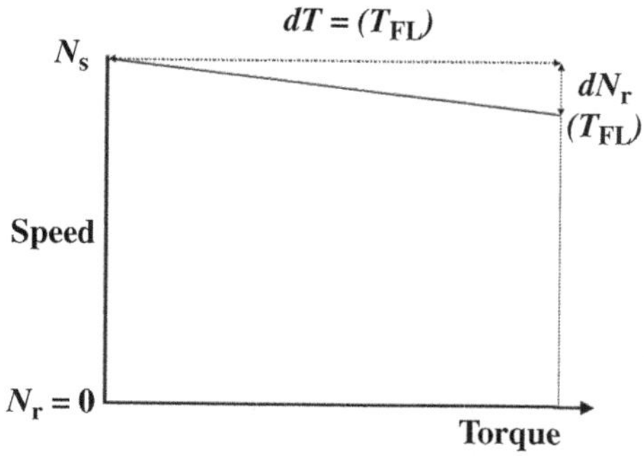


Figure 1.15a Speed versus torque for the operating region between no-load and full-load.

the intercept on the y-axis, which is the synchronous speed ($N_s$). The equation for the graph on Figure 1.15a is therefore

$$N_r = \left[\left(-\frac{dN_r}{dT}\right)T\right] + N_s \tag{1.6}$$

Equation (1.6) can be rearranged into a more convenient form

$$T = \left[\frac{(N_S - N_r)}{(N_S - N_{rFL})}\right] T_{FL} \tag{1.6a}$$

The operational slip $s_o = \dfrac{(N_S - N_r)}{N_S}$

The full-load slip $s_{FL} = \dfrac{(N_S - N_{rFL})}{N_S}$

Rearranging equation (1.6) gives

$$\frac{T_{FL}}{T} = \frac{s_{FL}}{s_o} \tag{1.7}$$

The torque is therefore proportional to slip in the linear region of the torque–speed graph.

Where

$N_s$ = synchronous speed, r/min

$N_r$ = speed at any torque between synchronous speed and the speed at full-load torque, r/min

$N_{rFL}$ = speed at full-load, r/min

$dN_r$ = change in speed between synchronous speed and full-load speed, ($N_s - N_{rFL}$), r/min

$T$ = torque at any speed between $N_s$ and $N_{rFL}$, N·m (lbsf·ft)

$T_{FL}$ = full-load torque, N·m (lbsf·ft)

$dT$ = change in torque between torque (zero) at synchronous speed and $T_{FL}$, N·m (lbsf·ft)

**Example** The torque–speed curve in Figure 1.15 is for a 3-phase, 6600 V, 2900 kW/3887 HP, 340 A, 60 Hz, 890 r/min SCIM. The FLT was previously calculated from the nominal full-load rated speed and power output (nameplate data) and is 31,116 N·m. The synchronous speed is 900 r/min and the full-load speed (nameplate data) is 890 r/min. Calculate the operating torque at a speed of 894 r/min.

$$\text{From equation (1.6a) } T = \left[\frac{(N_S - N_r)}{(N_S - N_{rFL})}\right] T_{FL} = \left[\frac{(900 - 894)}{(900 - 890)}\right] 31{,}116 = 18{,}670 \text{ N m}$$

$$\text{Or from equation (1.7) } T = \left(\frac{s_o}{s_{FL}}\right) T_{FL} = \left(\frac{0.0067}{0.011}\right) 31{,}116 = 18{,}670 \text{ N m}$$

The answers are equal and prove that the torque for a 3-phase SCIM is proportional to slip (equation 1.7) in the steady-state operating region between full-load and no-load.

### 1.3.1 Calculation of Run-Up Time of SCIM Driving a Mechanical Load

It is assumed that the windage and friction torques are negligible. The area between the two curves in Figure 1.16 is the accelerating torque available to accelerate the motor up to the steady-state operating speed demanded by the mechanical load. This occurs at the cross over point in Figure 1.16 between the two curves, provided that

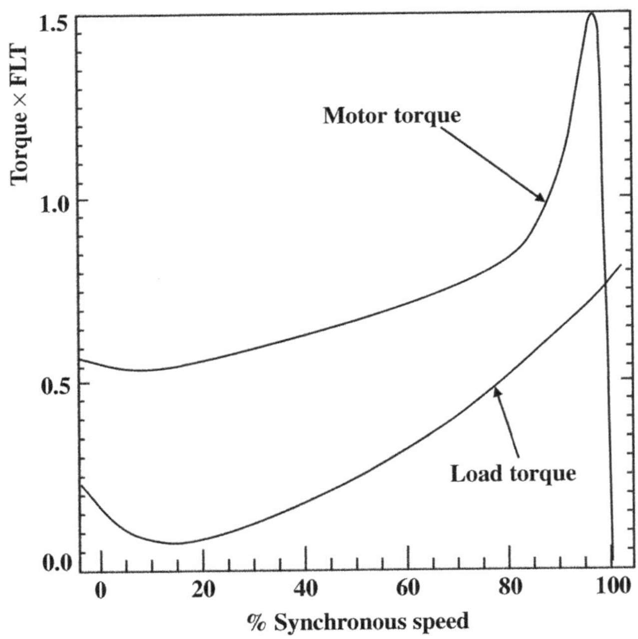


Figure 1.16 Torque versus speed curves for the motor and load. Reproduced with permission of Parsons Peebles, Scotland.

the cross over occurs in the negative slope of the motor's torque curve, in which case steady state will be obtained.

Torque required from the motor to drive the load is given by (and using metric units)

$$T_m = T_{ma} + T_L$$

$T_m$ = total torque from motor, N·m

$T_{ma}$ = accelerating torque to start the motor and overcome its inertia (J), N·m

$T_L$ = load torque, N·m

A step–by-step mathematical integration solution can be used to calculate the run-up time. The accelerating torque is

$$T_{ma} = J(d_{wr}/d_t)$$

$$\int dt = J/T_{ma} \int_{wr_1}^{wr_2} d_{wr}$$

$$t = (J/T_{ma})(w_{r2} - w_{r1}) \quad (1.8)$$

If small steps in $d_{wr} = w_{r2} - w_{r1}$ are taken, it is assumed that the curve is linear during each step, therefore the time for the motor to accelerate in that period can be calculated. The smaller the step in $d_{wr}$ the more accurate is the assumption of linearity. In this illustration only 20 steps are shown in Figure 1.17 but the greater the number of steps of $d_{wr}$, the more accurate is the result for the total run-up time ($t$). The average accelerating torque is taken directly from the curves and the $\Delta t$ time is determined between the two curves as shown in Figure 1.17. All the $\Delta t$ times are then used to plot the run-up time versus the speed from zero up to the steady-state speed. In this illustration, for this 8-pole motor and reciprocating compressor the total inertia is taken to be 350 kg-m$^2$ (8314 lbs-ft$^2$).

**Using Metric Units**

From Figure 1.17, each step is 45 r/min, $d_{wr} = 2\pi 45/60 = 4.7\text{rad/sec}$

The time to accelerate between each step is given by

$$\Delta t = (J/T_{av})(w_{r2} - w_{r1})$$

A sample calculation for $\Delta t_1$ for the first step between zero speed to 45 r/min (i.e., $d_{wr} = 4.71$ rad/sec) the average accelerating torque ($T_{av}$) between the two curves is 12,466 N·m.

$$\Delta t_1 = (350/12{,}466)(4.7) = 0.13 \text{ seconds}$$

**Using Imperial units**, Dymond [1.64]

$\Delta t$ = the time taken to accelerate during the selected step $\Delta S$ in r/min is given by

$$\Delta t = \frac{\left(WK^2M + WK^2DE\right) \times \Delta S}{308 \times (Ta)} \text{ seconds} \quad (1.9)$$

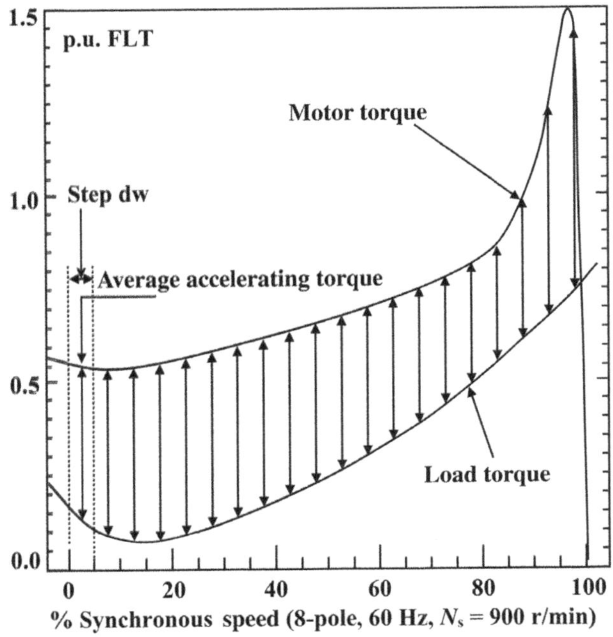


Figure 1.17 Per unit torque versus percentage of synchronous speed—step-by-step illustration to calculate the run-up time.

where

$\Delta t$ = time taken to accelerate during the selected step $\Delta S$ in r/min

$\Delta S$ = step in r/min

$WK^2M$ = inertia of rotor in the motor in lbs-ft$^2$ (lbs in mass)

$WK^2DE$ = inertia of mechanical load downstream of the motor in lbs-ft$^2$

$T_a$ = average accelerating torque in lbsf·ft during the step $\Delta S$ in r/min

Recall 1.0 lbs-ft$^2$ = 1.0 kg-m$^2$/42.1 × 10$^{-3}$ and 1.0 lbsf·ft = 0.738 N·m

For this example

Recall 1 p.u. or 100% FLT = 31,116 N·m (22,964 lbsf·ft)

$WK^2M + WK^2DE$ = 8314 lbs-ft$^2$

$\Delta S$ = 45 r/min

$T_a$ = 9200 lbsf·ft for the first step of $\Delta S$ = 45 r/min

Therefore $\Delta t_1 = \{8314/(308 \times 9200)\}45 = 0.13$ seconds

which is exactly the same as when calculated using metric units. This is repeated for the remaining 19 steps and Table 1.2 gives the results. A simpler method of arriving at all the values of $\Delta t$ for both 100% and 80% volts is suggested.

First calculate $\Delta t_1$ (or any other $\Delta t$) by the method previously used to calculate $\Delta t_1$ = 0.13 seconds.

**TABLE 1.2 100% Volts, Time $\Delta t$ in Seconds for Each Step in $d_{wr}$**

| | | | | | | | | | |
|---|---|---|---|---|---|---|---|---|---|
| $\Delta t_1$ | $\Delta t_2$ | $\Delta t_3$ | $\Delta t_4$ | $\Delta t_5$ | $\Delta t_6$ | $\Delta t_7$ | $\Delta t_8$ | $\Delta t_9$ | $\Delta t_{10}$ |
| 0.132 | 0.12 | 0.12 | 0.11 | 0.12 | 0.12 | 0.12 | 0.12 | 0.12 | 0.14 |
| $\Delta t_{11}$ | $\Delta t_{12}$ | $\Delta t_{13}$ | $\Delta t_{14}$ | $\Delta t_{15}$ | $\Delta t_{16}$ | $\Delta t_{17}$ | $\Delta t_{18}$ | $\Delta t_{19}$ | $\Delta t_{20}$ |
| 0.132 | 0.14 | 0.15 | 0.154 | 0.16 | 0.17 | 0.18 | 0.15 | 0.1 | 0.084 |

But $\Delta t$ is inversely proportional to the length of the line representing average accelerating torque, so if $\Delta t_1$ is calculated via length 1 then $\Delta t_2$ must equal $\Delta t_1$. *length 1/length* 2.

$$\text{That is, } \Delta t_x = \Delta t_1.l_1/l_x$$

The summation of all the times for each step gives an estimate of the run-up time $t = 2.6$ seconds. Which is the same as the OEM's data sheet stated a run-up time of 2.6 seconds on a 100% volts supply. This example provides industrial engineers with

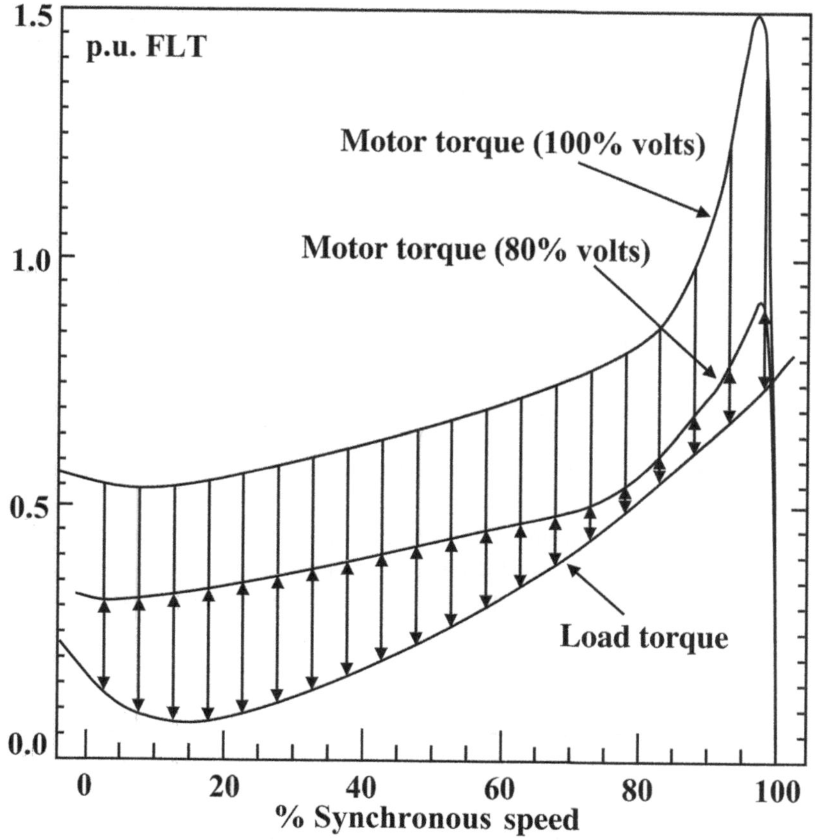


Figure 1.18 Motor torque (100% volts), motor torque (80% volts), and load torque versus synchronous speed—step-by-step calculation of run-up times at different applied volts.

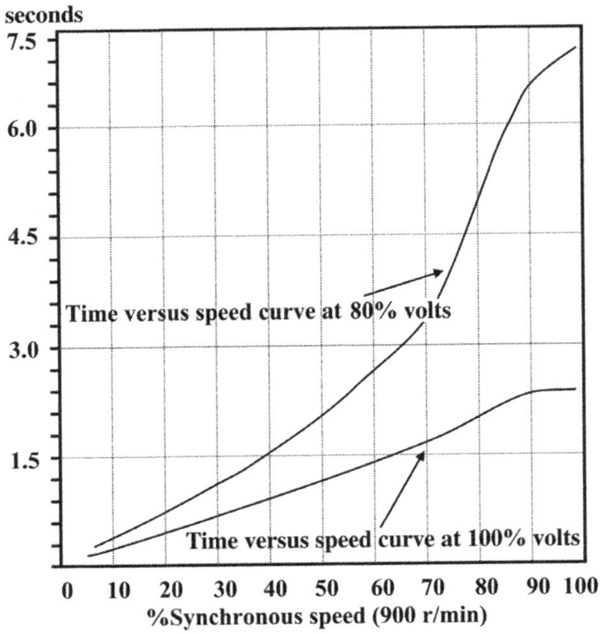


Figure 1.19 Run-up time versus speed for 100% and 80 % volts driving a reciprocating compressor.

knowledge so that they can repeat these calculations, if required to do so, provided they have the torque–speed curves of the motor and load and the total moment of inertia, which should all be available from the OEMs. The client estimated that the voltage would only drop to 95% of its rated value.

As an illustration the torque–speed curve from the motor at 80% volts is shown in Figure 1.18, which shows there is a large drop in the accelerating torque available between 75% and 95 % of the synchronous speed, in visual terms it is "*very thin*" and as Richard Nailen [1.51] very aptly put it in his paper, the available accelerating torque between the load and motor torque–speed curves should be "*as thick as possible.*"

The curves for the run-up time in seconds versus speed for a start up at 100% and 80% volts, respectively, when driving the reciprocating compressor, are shown in Figure 1.19. The results from the step-by-step solution for an applied 80% of

**TABLE 1.3 80% Volts, Time $\Delta t$ in Seconds for Each Step in $d_{wr}$**

| | | | | | | | | | |
|---|---|---|---|---|---|---|---|---|---|
| $\Delta t_1$ | $\Delta t_2$ | $\Delta t_3$ | $\Delta t_4$ | $\Delta t_5$ | $\Delta t_6$ | $\Delta t_7$ | $\Delta t_8$ | $\Delta t_9$ | $\Delta t_{10}$ |
| 0.25 | 0.2 | 0.18 | 0.18 | 0.18 | 0.18 | 0.2 | 0.2 | 0.25 | 0.26 |
| $\Delta t_{11}$ | $\Delta t_{12}$ | $\Delta t_{13}$ | $\Delta t_{14}$ | $\Delta t_{15}$ | $\Delta t_{16}$ | $\Delta t_{17}$ | $\Delta t_{18}$ | $\Delta t_{19}$ | $\Delta t_{20}$ |
| 0.28 | 0.3 | 0.37 | 0.43 | 0.76 | 0.95 | 0.96 | 0.64 | 0.38 | 0.28 |

rated voltage are given in Table 1.3 and overall the result is that the run-up time has increased to approximately 7.5 seconds, an increase by a factor of 2.9. If this drop in volts were to actually occur, the motor would still start but the rotor bars would be exposed to the starting current for a longer time and thus the thermal stresses, electromagnetic forces, mechanical forces, and bending stresses on the rotor bars and end rings external to the rotor core ends would be present for an undesirable length of time.

An in-depth case history on a completely different HV motor driving a centrifugal pump is presented in Section 6.1 of Chapter 6, in which the cage rotor had 10 broken rotor bars. The cause of the breakages was starting the motor against an open valve pump setting, when it should always have been a closed valve setting. This resulted in much longer run-up times and a stall if the volts dropped to 80% of rated value. All the torque–speed curves are presented in that particular case history.

## 1.4 ILLUSTRATIONS OF CONSTRUCTION OF A LARGE HV SCIM

Before moving on to further chapters and to support their content, the following Figures 1.20 to 1.26 illustrate the constructional features of a large HV squirrel cage induction motor.

Figure 1.20 3-phase, 13,800 V, 5720 kW/7668 HP, 60 Hz, 1774 r/min SCIM, ready for delivery to the client. Reproduced with permission of Parsons Peebles, Scotland.

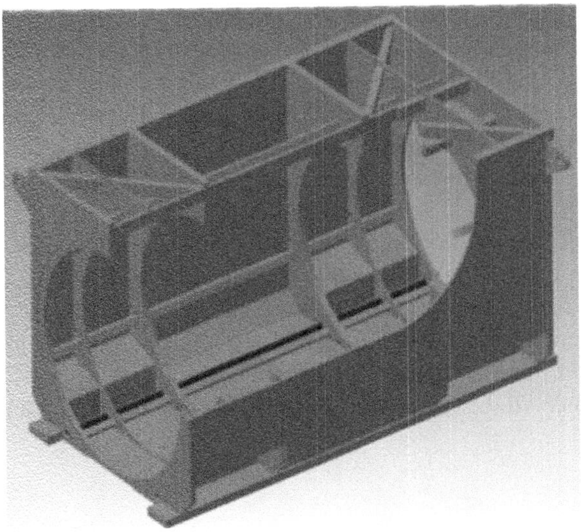

Figure 1.21 Sectional view of frame assembly. Reproduced with permission of Parsons Peebles, Scotland.

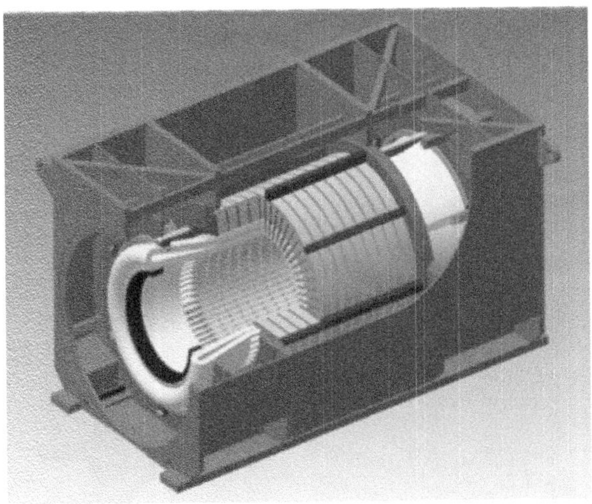

Figure 1.22 Sectional view of frame and stator assembly. Reproduced with permission of Parsons Peebles, Scotland.

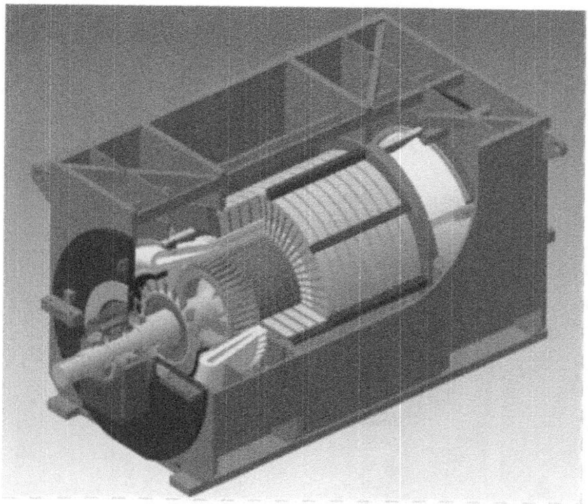

Figure 1.23 Sectional view of frame, stator, and rotor. Reproduced with permission of Parsons Peebles, Scotland.

Figure 1.24 Stator core laminations. Reproduced with permission of Parsons Peebles, Scotland.

Figure 1.25 Squirrel cage rotor. Reproduced with permission of Parsons Peebles, Scotland.

Figure 1.26 HV stator winding and core in process of varnishing. Reproduced with permission of Parsons Peebles, Scotland.

## 1.5 QUESTIONS

**1.5.1** A 3-phase SCIM is connected to a 60 Hz supply and its full-load rated speed is 3580 r/min.

**(a)** Determine the number of poles.

**(b)** Calculate the synchronous speed of the rotating magnetic field from the stator winding.

**(c)** The slip speed in r/min between the stator's rotating field and the actual speed of the rotor is given by ($N_s - N_r$). Is it correct or incorrect to divide this r/min by 60 and call it the slip frequency with reference to terminology used in induction motor theory? Explain your answer.

**(d)** Calculate the operating slip in p.u. and as a percentage.

**(e)** Calculate the slip frequency of the induced emf and current in the rotor conductors.

**(f)** Repeat (a), (b), (d), and (e) with SCIMs supplied at 60 Hz but when operating at speeds of (i) 1782, (ii) 1185, (iii) 885, and (iv)195 r/min.

**(g)** A SCIM is fed from a 50 Hz supply and its full-load rated speed is 1470 r/min repeat the calculations for (a), (b), (d), and (e).

**1.5.2** The frequency of the currents in the cage winding is constant at all times. Is this statement correct or incorrect? Explain the reasons for your answer.

**1.5.3** The speed of the rotating magnetic field produced by the current carrying conductors in the cage winding is at the same speed as the rotor. Is this statement correct or incorrect? Explain the reasons for your answer.

**1.5.4** The speed of the rotating magnetic field produced by the current carrying rotor conductors is not at the same speed as the rotating magnetic field produced by the stator winding with respect to a stationary observer on the stator. Is this statement correct or incorrect? Explain the reasons for your answer.

**1.5.5** For a 3-phase, star connected, 460 V SCIM, 74.6 kW/100 HP, 60 Hz, 1782 r/min, 0.88 p.f. and an efficiency of 96.5%, calculate the following

**(a)** The nominal full-load rated current.

**(b)** The nominal FLT in Nm and in lbsf ft.

**1.5.6** For the motor specified in Question 1.5.5, the mechanical load on the motor is reduced and the speed increases to 1792 r/min.

**(a)** What is the new output power and torque at the shaft in kW and HP and in lbsf ft and in N·m? Hint: Refer to Figures 1.15 and 1.15a and the associated text and worked example.

**(b)** During another change in the mechanical load on this motor the input current drops to 50% of rated FLC, which means that the motor is delivering 50% of its rated full-load power and torque output. Is this statement correct or incorrect? Justify your answer.

**1.5.7** Why does a SCIM take a much larger current during a DOL start compared to its rated FLC?

**1.5.8** What are the various undesirable effects on the copper cage winding caused by a DOL start?

**1.5.9** Why does a SCIM run-up faster on no-load compared to the time taken when connected to a mechanical load?

**1.5.10** For the torque–speed curves of the motor and load shown in Figure 1.17 with the same total inertia determine the run-up time when the supply volts is 90% or 0.9 p.u. of the rated volts.

## REFERENCES


[1.1] N. Tesla, "A new system of alternate current motors and transformers," *Transactions of American Institute of Electrical Engineers*, vol. V, no. 10, 1888, pp. 308–327.

[1.2] Source en.wikipedia.org/wiki/electricity consumption.

[1.3] G. C. Stone, E. A. Butler, I. Culbert, and H. Dhiranai, *Electrical Insulation for Rotating Machines – Design, Evaluation, Testing and Repair*, Wiley-IEEE Press, 2004, ISBN 0-471-44506-1.

[1.4] P. Tavner, L. Ran, J. Penman, and H. Sedding, *Condition Monitoring of Rotating Electrical Machines*, IET Power and Energy Series 56, The Institution of Engineering and Technology, London, UK, 2008, ISBN 978-0-86341-739-9.

[1.5] H. A. Toliyat, S. Nandi, S. Choi, and H. Meshgin-Kilk, *Electric Machines-Modeling, Condition Monitoring and Fault Diagnosis*, CRC Press Taylor & Francis Group, New York, 2013, ISBN 978-0-8493-7027-4.

[1.6] W. T. Thomson, *MCSA Industrial Case Studies Proves the On-Line Detection of Airgap Eccentricity in HV Induction Motors*, PCIC Europe, International European Conference, London, June 2015.

[1.7] W. T. Thomson, *MCSA Case Histories on Detection of Broken Rotor Bars in Induction Motors - Including Influences of Mechanical Load Dynamics on the Analysis*, COMET Conference, 2014, Hosted by Iris Power Qualitrol, Canada, in Austin TX, December 2014.

[1.8] W. T. Thomson and M. Fenger, "Case Histories of Current Signature Analysis to Detect Faults in Induction Motor Drives," Proceedings of IEEE International conference on Electrical Machines and Drives (IEMDC), University of Wisconsin, Madison, WI, June 2003.

[1.9] W. T. Thomson and R. J. Gilmore, "Motor Current Signature Analysis to Detect Faults in Induction Motor Drives - Fundamentals, Data Interpretation and Industrial Case Histories," Proceedings of 32nd Turbomachinery Symposium, A&M University, TX, September 2003.

[1.10] W. T. Thomson and P. Orpin, "Current and Vibration Monitoring for Fault Diagnosis and Root Cause Analysis of Induction Motors," Proceedings of 31st Turbomachinery Symposium, A&M University, TX, September 2002.

[1.11] W. T. Thomson and W. H. Ross, "Application of MCSA to a Diverse Range of Induction Motor Drives in Power Stations," Iris Rotating Machines Conference, IRMC 01, Washington, D.C., June 2001.

[1.12] M. Fenger, M. Susnik, and W. T. Thomson, "Development of a Fully Portable Current Signature Analysis Meter to Detect Electrical and Mechanical Faults in Induction Motor Drives," Iris Rotating Machines Conference, IRMC 01, Washington, D.C., June 2001.

[1.13] W. T. Thomson and M. Fenger, "Current signature analysis to detect induction motor faults", *IEEE Industry Applications Magazine*, vol. 7, no. 4, 2001, pp. 26–34.

[1.14] W. T. Thomson, D. Rankin, and D. G. Dorrell, "On-line current monitoring to diagnose airgap eccentricity in large 3-phase induction motors - an industrial case history verifies the predictions", *IEEE Transactions on Energy Conversion*, vol. 14, no. 4, December, 1999, pp. 1372–1378.

[1.15] W. T. Thomson and A. Barbour, "The On-Line Prediction of Airgap Eccentricity Levels in Large (MW Range) 3-Phase Induction Motors," Proceedings of IEEE, IEMDC Conference, Seattle, WA, May 1999—US$1000 best paper award.

[1.16] W. T. Thomson and A. Barbour, "An Industrial Case Study of On-Line Current Monitoring and Finite Element Analysis to Diagnose Airgap Eccentricity Problems in Large High Voltage 3-phase

Induction Motors," Proceedings of the 9th IEE International Conference on Electrical Machines and Drives (EMD), Canterbury Christ Church College, UK, September 1999.

[1.17] W. T. Thomson, "A Review of On-Line Condition Monitoring Techniques for Three-Phase Squirrel-Cage Induction Motors -Past Present and Future," Keynote address at IEEE Symposium on Diagnostics for Electrical Machines, Power Electronics and Drives, Gijon, Spain, September 1999 pp. 3–18.

[1.18] W. T. Thomson and A. Barbour, "On-line current monitoring and application of a finite element method to predict the level of airgap eccentricity in 3-Phase induction motors", *IEEE Transactions on Energy Conversion*, vol. 13, no. 4, December 1998, pp. 347–357 (includes discussion and closure).

[1.19] W. T. Thomson, A. Barbour, C. Tassoni, and F. Filippetti, "An Appraisal of the MMF-Permeance Method and Finite Element Models to Study Static Airgap Eccentricity and its Diagnosis in Induction Machines," Proceedings of ICEM'98, Istanbul, 1998.

[1.20] W. T Thomson, D. Rankin, and D. G. Dorrell, "On-line Current Monitoring to Diagnose Airgap Eccentricity - An Industrial Case History of Large HV, 3-Phase Induction Motors," Proceedings of IEEE International, Conference, IEMDC'97, Milwaukee, WI, May 1997.

[1.21] D. G. Dorrell, W. T. Thomson, and S. Roach, "Analysis of airgap flux, current and vibration signals as a function of the combination of static and dynamic airgap eccentricity in 3-Phase induction motors", *IEEE Transactions on Industry Applications*, vol. 33, no. 1, January/February 1997, pp. 24–34.

[1.22] A. Barbour and W. T. Thomson, "Finite Element Study of Rotor Slot Designs with respect to Current Monitoring for detecting Static Airgap Eccentricity in Squirrel-cage Induction Motors," Proceedings of IEEE Industrial Applications Conference, New Orleans, LA, October 1997.

[1.23] A. Barbour and W. T. Thomson, "Finite Element Analysis and On-Line Current Monitoring to Diagnose Airgap Eccentricity in 3-Phase Induction Motors," Proceedings of IEE International Conference, EMD, University of Cambridge, September 1997.

[1.24] D. G. Dorrell, W. T. Thomson, and S. Roach, "Combined Effects of Static and Dynamic Eccentricity on Airgap Flux Waves and the Application of Current Monitoring to Detect Dynamic Eccentricity in Three-phase Induction Motors," IEE International Conference on Electrical Machines and Drives, September 1995.

[1.25] D. G. Dorrell, W. T. Thomson, and S. Roach, "Analysis of Airgap Flux, Current and Vibration Signals as a Function of the Combination of Static and Dynamic Airgap Eccentricity in Three-phase Induction Motors," IEEE International Conference on Industry Applications, Orlando, FL, October 1995.

[1.26] W. T. Thomson, "On-Line Current Monitoring - The Influence of Mechanical Loads/Unique Rotor Designs on the Detection of Broken Rotor Bars in Squirrel Cage Induction Motors," ICEM '92, UMIST, September 1992.

[1.27] S. Elder, J. F. Watson, and W. T. Thomson, "The Analysis of Transient Currents in Induction Motors for the Purpose of Detecting Rotor Faults," Proceedings of the 25th Universities Power Engineering Conference, RGIT, Aberdeen, September 1990.

[1.28] W. T. Thomson, J. R. Cameron, and A. B. Dow, "On-Line Diagnostics of Large Induction Motors," NATO ARW (by invitation only), Catholic University of Leuven, Belgium, August 1986, published in the NATO Api Series, Pub Martines Hijhoff, July 1988.

[1.29] W. T. Thomson and I. D. Stewart, "On-Line Current Monitoring for Fault Diagnosis in Inverter Fed Induction Motors," Proceedings of IEE 3rd International Conference in Power Electronics and Variable Speed Drives, Savoy Place, London, July 1988.

[1.30] W. T. Thomson, J. R. Cameron, and A. B. Dow, "On-Line Current Monitoring and Analysis to Quantify the Level of Airgap Eccentricity in Induction Motors," Proceedings of International Conference in Electrical Machines, ICEM '88, Pisa, Italy, September 1988.

[1.31] J. R. Cameron, W. T. Thomson, and A. B. Dow, "On-Line Current Monitoring of Induction Motors - A Method of Calculating The Level of Airgap Eccentricity," Proceedings of IEE EMDA '87 Conference, London, November 1987.

[1.32] W. T. Thomson and D. Rankin, "Case Histories of Rotor Winding Fault Diagnosis in Induction Motors," Proceedings 2nd International Conference on Condition Monitoring, University College of Swansea, March 1987.

[1.33] W. T. Thomson, S. J. Chalmers, and D. Rankin, "On Line Current Monitoring and Fault Diagnosis in HV Induction Motors - Case Histories and Cost Savings in Offshore Installations," SPE Proceedings of Offshore Europe '87 Conference, Aberdeen, September, 1987.

[1.34] J. R. Cameron, W. T. Thomson, and A. B. Dow, "Vibration and current monitoring for detecting airgap eccentricity in large induction motors", *IEE Proceedings*, vol. 133, Part B. no 3, May 1986.

[1.35] J. R. Cameron, W. T. Thomson, and A. B. Dow, "A New On-Line Current and Vibration Monitoring Technique for Detecting Airgap Eccentricity in Large Induction Motors," Proceedings of IEE EMDA '85 Conference, Savoy Place, London, November 1985.

[1.36] W. T. Thomson, N.D. Deans, R. A. Leonard, and A. J. Milne, "Monitoring Strategy for Discriminating Between Different Types of Rotor Cage Faults," Proceedings 18th Universities Power Engineering Conference Proceedings, University of Surrey, Surrey, April, 1983.

[1.37] I. Culbert, H. Jivajee, and P. Laderoute, "Case Studies with New Current Signature Analysis Technology," IRIS Rotating Machines Conference, New Orleans, LA, June 14–17, 2004

[1.38] I. Culbert and H. Jivajee, "New Current Signature Analysis Technology to Reliably Detect Cage Winding Defects and Abnormal Airgap Eccentricity in Induction Motors," EXFOR 2005, Palais des Congress, Montreal, QC, Canada, February 8–10, 2005.

[1.39] M. Fenger, I. Culbert, and B. Lloyd, "Case Histories of Current Signature Analysis to Detect Faults in Induction Motor Drives," IEEE-IAS/PCA Cement Industry Conference, Dallas, TX, May 4–9, 2003.

[1.40] I. M. Culbert, and W. Rhodes, "Using current signature analysis technology to reliably detect cage winding defects in squirrel-cage induction motors", *IEEE-TIA*, vol. 43, no. 2, March/April 2007.

[1.41] I. Culbert, "Motor Maintenance Testing & Diagnostics On-Line, Off-Line," IEEE 55th Annual IEEE Pulp and Paper Conference, Renaissance Ross Bridge Golf Resort & Spa, Birmingham, AL, June 21–26, 2009.

[1.42] G. C. Stone, M. Sasic, D. Dunn, and I. Culbert, "Recent Problems Experienced with Motor and Generator Windings," IEEE PCIC 2009, Anaheim, CA, September 14–16, 2009.

[1.43] G. B. Kliman and J. Stein, *Methods of Motor Current Signature Analysis Electric Machines and Power Systems*, Hemisphere Publishing, New York vol. 20, 1982.

[1.44] G. B. Kliman, R. A. Koegl, J. Stein, R. D. Endicott, and M. W. Madden, "Non-invasive detection of broken rotor bars in operating induction motors", *IEEE Transactions on Energy Conversion*, vol. 3, no. 4, December 1988, pp. 873–879.

[1.45] S. Williamson, and S. C. Smith, "Steady-state analysis of 3-phase cage motors with rotor-bar and end-ring faults", *IEEE Proceedings B-Electric Power Applications*, vol. 129, no. 3, 1982.

[1.46] P. Vas, "Steady state and transient performance of induction motors with rotor asymmetry", *IEEE Transactions on Power Apparatus band Systems*, vol. PAS-101, no. 9, September 1982, pp. 3246–3251.

[1.47] W. Deleroi, "Broken Bar in Squirrel-Cage Rotor of an Induction Motor," Part I, Description by Superimposed Fault Currents, Archiv fur Eleektrotechnik 67, 1984, pp. 91–99

[1.48] C. Hargis, B. Gaydon, and K. Kamish, "The Detection of Rotor Defects in Induction Motors," Proceedings of 1st IEE International Conference on Electrical Machines, Design and Application. London, UK, pp. 216–220.

[1.49] P. Tavner, K. E. Armin, and C. Hargis, "An Electrical Technique for Monitoring Induction Cage Rotors", Proceedings of 3rd IEEE International Conference on Electrical Machines and Drives, London, UK, pp. 43–46.

[1.50] F. Filippetti, G. Franceschini, M. Martelli, and C. Tassoni, "An Approach to a Knowledge Representation about Induction Machine Diagnostics in Expert Systems," International Conference on Electrical Machines ICEM'88, Pisa, Italy, September 1988.

[1.51] R. L. Nailen, "Large motor starting problems in the petroleum industry", *IEEE Transactions on Industry and General Applications*, vol. IGA-5, no. 4, 1969, pp. 422–427.

[1.52] R. L. Nailen, "New rotor design concept solves pipeline motor acceleration" *IEEE Transactions on Industry Applications Problem*, vol. IA-9, no. 2, 1973, pp. 201–205.

[1.53] M. E. H. Benbouzid, "Bibliography on induction motors faults detection and diagnosis" *IEEE Transactions on Energy Conversion*, vol. 14, no. 4, December 1999, pp. 1065–1074.

[1.54] M. E. El-Hawary, *Principles of Electric Machines with Power Electronic Applications*, 2nd edition, Hardcover, Wiley-IEEE Press, July 2002.

[1.55] M. Liwschitz-Garik and C. C Whipple, *Electric Machinery Vol. II, A-C Machines,* Van Nostrand Company, 1st published, September 1946.

[1.56] M. G. Say, *Alternating Current Machines*, 4th edition, ELBS and Pitman Publishing, 1976.

[1.57] P. C. Sen, *Principle of Electrical Machines and Power Electronics*, 2nd edition, John Wiley & Sons, 1997.

[1.58] A. Chapman, *Electric Machinery Fundamentals*, McGraw Hill, 1985.

[1.59] A. Hughes, *Electric Motor and Drives – Fundamentals, Types and Applications*, Butterworth-Heinemann, 1990.

[1.60] G. Slemon, *Electric Machines and Drives*, Addison-Wesley Publishing Company Inc., 1992.

[1.61] P. L. Alger, *Induction Machines-Their Behavior and Uses*, Gordon and Breach Science Publications Inc., 2nd edition, published by OPA Amsterdam, 3rd printing with additions, 1995.

[1.62] H. Vickers, *The Induction Motor*, Sir Isaac Pitman and Sons Ltd., London (1st edition, 1924; 2nd edition 1953).

[1.63] M. M. Liwschitz-Garik, Computation of skin effect in bars of squirrel-cage rotors [includes discussion, *Power Apparatus and Systems, Part III. Transactions of the American Institute of Electrical Engineers*, vol. 74, no. 3, 1955.

[1.64] J. H. Dymond, Stall time, acceleration time, frequency of starting: the myths and the facts [electric motors], *IEEE Transactions on Industry Applications*, vol. 29, no. 1, 1993, pp. 42–51.

[1.65] P. Dunsheath, *A History of Electrical Engineering*, Faber and Faber, 1st published in 1962.

[1.66] NEMA MG1: "Motors and Generators," 2012.

CHAPTER 2

# DESIGN, CONSTRUCTION, AND MANUFACTURE OF SQUIRREL CAGE ROTORS

Ian Culbert

## 2.0 INTRODUCTION

The focus of this chapter is to present the essential features of the design, construction, and manufacture of squirrel cage induction rotors. This is to provide knowledge for industrial engineers, who use Motor Current Signature Analysis (MCSA) to determine if there are any cage winding breaks in their squirrel cage induction motors (SCIMs). The actual cage winding design and manufacture can influence the decision to remove a motor from service or let it continue to run if MCSA suggests that there is a cage winding break such as a broken rotor bar or bars. An end user normally always carries out a risk assessment following a diagnosis of a problem and design issues, which increase the risk of a broken rotor bar lifting and crashing into an HV stator winding, have to be carefully assessed. It is the authors' opinion that an engineer using MCSA should at least have an appreciation of the design features of cage rotors and certainly a good understanding of cage construction and a thorough understanding of how a SCIM actually operates as presented in Section 1.2, Chapter 1.

This chapter is not about the actual design of cage rotors hence a selection of references (not a fully inclusive list) is given in which key elements of the design of induction motors are included and this can provide the reader with an introduction to the design of SCIMs. These include books by, for example, Liwschitz-Garik et al. [2.1], Say [2.2], Alger [2.3], Vickers [2.4] and papers by Lloyd [2.5], Bonnett et al. [2.6], Barr et al. [2.7], and Nailen [2.8]. However, it is really only electrical machine designers employed by electrical machine manufacturers who can truly design SCIMs which can be manufactured such that they perform as per the design and are fit for purpose to suit the clients' specifications. OEMs of electric motors have a vast amount of knowledge and many years of experience of actually producing SCIMs and that knowledge and experience is of course the intellectual property (IP) of the OEM and cannot be released into the public domain.

*Current Signature Analysis for Condition Monitoring of Cage Induction Motors: Industrial Application and Case Histories*, First Edition. William T. Thomson and Ian Culbert.

The predominance of the SCIM is attributed to the simplicity and ruggedness of its rotor winding. The actual cage winding (e.g., copper or aluminum) is not electrically insulated from the laminated rotor core and the rotor can be designed for operating speeds that range from less than 100 to at least 20,000 r/min, when the motor is supplied from a variable voltage and frequency converter. A squirrel cage rotor consists of a number of bars installed in rotor core slots that are connected together to form a winding by short circuiting end rings as shown in Figures 1.5 and 1.6. Such windings can be formed by a die-casting process to form a homogeneous winding or by fabricating rotor bars and end rings and connecting them together by means of a brazing or welding process. Die-cast aluminum windings have been around since the 1930s, while copper-based windings have existed since the 1920s.

There are four types of squirrel cage windings in common use. These are

- Aluminum die-cast
- Copper die-cast
- Copper or copper alloy fabricated
- Aluminum fabricated

## 2.1 ALUMINUM AND COPPER DIE-CAST WINDINGS

Die-cast windings are used because they are much less expensive to manufacture and more intricate and variable bar shapes can be obtained, since the bars take the form of the slots in the rotor core. The reasons for having different bar shapes are discussed in Section 2.2.1. While the lower resistance of copper windings provides higher motor efficiencies as a result of lower $I^2R$ losses, the aluminum die-cast type is popular with motor manufacturers due to its lower material costs. Cast copper rotors also have lower stray losses, when used in motors fed from variable frequency drives, Anderson [2.9]. However, the fact that aluminum has a much lower melting temperature (660°C/1220°F) than copper (1083°C/1981°F) makes it easier to cast and so it is much more commonly used than the copper type. There are two commonly used die-casting processes for both aluminum and copper windings, which are injection casting and centrifugal casting. To illustrate how each is done the following are the procedures for the manufacture of die-cast aluminum types. The first few steps for both die-casting processes are as follows:

(i) Manufacture the rotor core laminations and stack them so that the rotor slots are aligned. If skewed rotor slots are required, the laminations have to be oriented to provide this feature.

(ii) Molds in the form of end rings (which can also incorporate cooling fan blades) are installed on either end of the rotor core and axial pressure is applied to the core/end mold assembly (see Figure 2.1).

(iii) The core and mold assembly is often heated before the copper or aluminum is injected into them to avoid rapid cooling of the molten aluminum. The aluminum is heated to around 843°C/1550°F before it is injected or cast into the core/mold cavity, EASA [2.10].

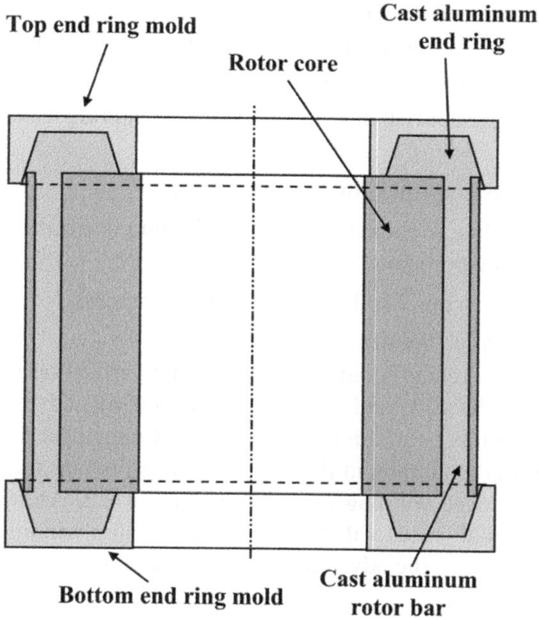


Figure 2.1 Rotor core with molds on either end ready for die-casting.

**(a)** *High Injection Die-Casting Process*

The rotor core and mold are mounted in a vertical orientation as shown in Figure 2.1. Pressurized molten aluminum is injected into the top and the pressure forces it through the core slots into the bottom to form a squirrel cage winding as shown in Figure 2.2. Some manufacturers draw a vacuum to further

Figure 2.2 Die-cast aluminum rotor cage winding without the rotor core.

help the molten metal to completely fill all the cavities. There will always be some voids present, when the molten aluminum cools, since it will shrink. Some of the features of this process are

**(i)** There is a greater risk of voids in the lower end ring as it is the last end to fill.

**(ii)** Bar shape is controlled by slot dimensions.

**(iii)** During cooling to room temperature, aluminum shrinks about 6%, which creates some porosity.

**(iv)** The process is quicker than the centrifugal process.

**(b)** *Centrifugal Casting Process*

The rotor core and mold are mounted in a vertical orientation as shown in Figure 2.1. The core and mold assembly are rotated like a top and the molten aluminum is poured into the top mold. Centrifugal forces create pressure that forces the aluminum through the core slots into the bottom mold to form a squirrel cage winding as shown in Figure 2.2. As with the injection process some small voids will form in the aluminum winding, when it cools and shrinks. Some of the advantages and disadvantages of this process are as follows:

**(i)** Dross and light contaminates gather in the top mold at the bar interfaces.

**(ii)** The process is less likely to create shrink holes in the casting.

(a)

(b)

Figure 2.3 (a) and (b) Examples of aluminum die-cast cage rotors.

**(iii)** Aluminum shrinks about 6% after cooling to room temperature, which creates some porosity.

**(iv)** The process typically has a longer cycle time than injection molding. The surfaces of the finished product are not as smooth as those from injection die-casting.

Figures 2.3a and 2.3b show examples of completed rotors produced by the high pressure injection die-casting process. Die-cast aluminum rotor windings have been used in 2- and 4-pole motors with ratings up to around 2240 kW/3000 HP.

It should be pointed out that the aluminum die-cast rotor in Figure 2.3a has closed slots and is a good quality die-cast rotor, in contrast, Figure 2.3b uses semi-closed slots and the zig zag finish of the bars at the top of the slots indicates a poor quality die-cast rotor. It is now normally the case that aluminum die-cast rotors use closed slots.

## 2.2 FABRICATED SQUIRREL CAGE WINDINGS

### 2.2.1 Rotor Bar Design, Shape, and Installation

For large motors, fabricated copper or copper alloy windings are preferred because they can better withstand the mechanical and thermal stresses imposed during motor starting and running and have lower $I^2R$ losses during motor operation, since copper has a much lower resistivity and higher mechanical stress withstand capability. The primary reason for using fabricated aluminum squirrel cage windings is to reduce the cost of manufacture because of the lower cost of the metal, from which they are made. In addition, further cost reductions can be achieved by using the end rings as end support rings to maintain adequate core pressure (Figure 2.13). These fabricated rotor windings are constructed from individual bars by machining or extruding the material from which they are made, to obtain the desired shape.

The starting torque of a SCIM partly depends on the rotor resistance, and is also a function of the applied voltage squared ($V^2$). So one of the ways to obtain a higher starting torque is to use a higher resistivity bar material in the form of an alloy of copper or aluminum. However, this has the disadvantage of inherently higher rotor $I^2R$ losses during motor operation. Motor designers overcome this problem by using different bar shapes to provide the appropriate starting torque and torque versus speed curve for a particular application. This shape change takes advantage of what is called "skin effect" [2.1–2.3], and Dymond [2.11], which causes a non-uniform distribution of the rotor current during motor starting and particularly just after the motor breaker is first closed when the frequency of the current in the rotor is close to 50/60 Hz. This is illustrated by Figure 2.4 which shows the leakage flux around a rotor slot with a bar in it. If the rotor bar is assumed to be made of a number of layers, connected in parallel, as shown in Figure 2.4, then with close to line frequency current in the rotor winding, the top-most layer 1 is linked with the minimum leakage flux and therefore it has the lowest leakage inductance. On the other hand, the bottom layer 3 links the maximum flux and so it has the highest leakage inductance.

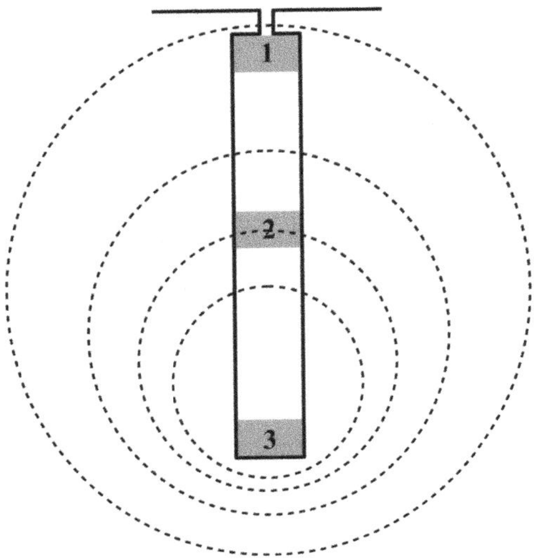


Figure 2.4 Illustration of skin effect in rotor bar.

Therefore, with a common voltage considered across the ends of the bar, more current will flow in the upper low impedance layer than in the relatively higher impedance bottom layer. Associated with this concentration of current in the top of the bar will be an increase in the effective bar resistance and hence starting torque, which is proportional to $\frac{I_r^2 R_r}{s}$, where

$I_r$ = rotor current; $R_r$ = rotor resistance; $s$ = operational slip = 1.0 at starting

The concentration of current in the top of the bar will also result in an overall reduction of the rotor bar leakage reactance compared with its running value.

So if the bar shape is made such that the top section of the bar has a higher resistance, higher motor starting torques can be obtained. Of course this uneven current distribution in the rotor bars results in the top sections of the bars becoming much hotter than the bottom sections and can cause a thermal bow in the bars. Examples of commonly used bar shapes include a rectangular bar (Figure 2.5a), which is

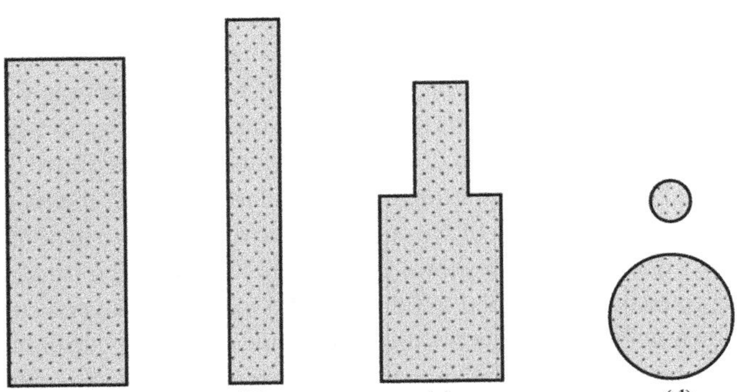

(a) (b) (c) (d)

Figure 2.5 Examples of different bar shapes.

used for centrifugal pump and fan applications, which do not require a high starting torque. If a slightly higher starting torque is required, then a deep narrow bar is used (Figure 2.5b). When the required starting torque is around 100–125% of motor full-load torque then an inverted "T' shaped bar (Figure 2.5c) is often used. To achieve a very high starting torque in the region of 200% or higher, of full-load torque, for applications such as rock crushers, a double cage winding with a small top bar made from a high resistivity material, such as bronze or brass (Figure 2.5d), is used.

After the rotor core has been stacked, pressed, and clamped to maintain lamination tightness, the rotor bars are installed in the core slots. The bars should be a tight fit in the core and so are usually driven into the slots unless, as shown in Figure 2.9, they have key bars to tighten them. For bars that are driven into the slots it is best to install alternate bars from opposite ends of the core. This is because, if all bars are driven from the same end they may all be slightly loose, the complete cage winding tends to ratchet toward the end they were originally driven toward. This cage winding ratcheting migration is due to thermal cycling produced by rotor winding cyclic heat from motor starts and stops. The rotor bars are then brazed or welded to the shorting rings. No matter what construction is used it is important that once installed the rotor bars are held tightly in their slots. Some of the common techniques used by manufacturers to ensure bar tightness are

**(a)** Swaging the tops of all the bars at regular intervals along their length by inserting a round ended tool thorough the narrow opening at the top of the rotor slot and striking this tool with a hammer, Finley et al [2.12] (see Figure 2.6). This method is not recommended for inverted "T" bars since tightening the narrow top section can result in high stresses in the area where the transition occurs from a wide to a narrow section.

**(b)** Using bars with sharp corners and slots with round corners to create an interference fit between the two (Figure 2.7). Swaging is also used with this method to ensure bar tightness in the slots.

**(c)** Installing custom sized steel slot liners between the bars and slots to ensure a tight fit between the two (Figure 2.8). Swaging is used at rotor radial vent ducts to tighten bars and deform liners to lock them in place. However, this method is very rarely used in the USA.

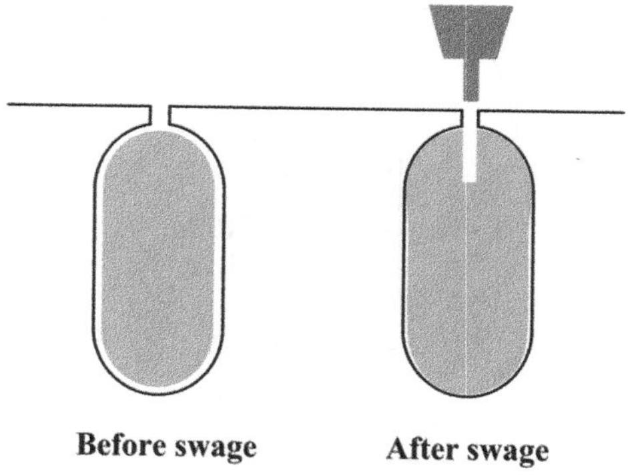


Figure 2.6 Swaging.

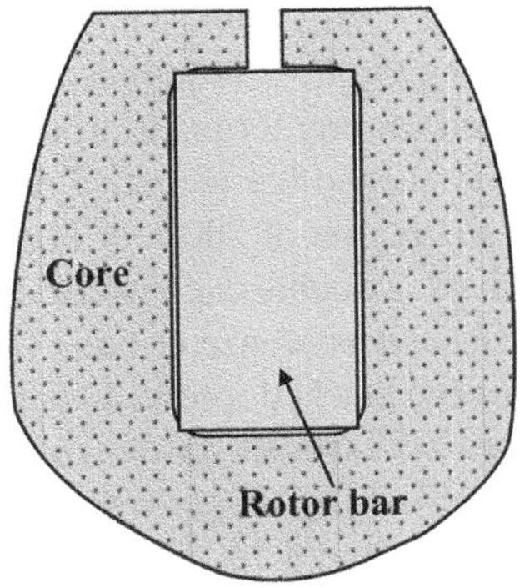


Figure 2.7 Bar with sharp corners installed in slot with round corners.

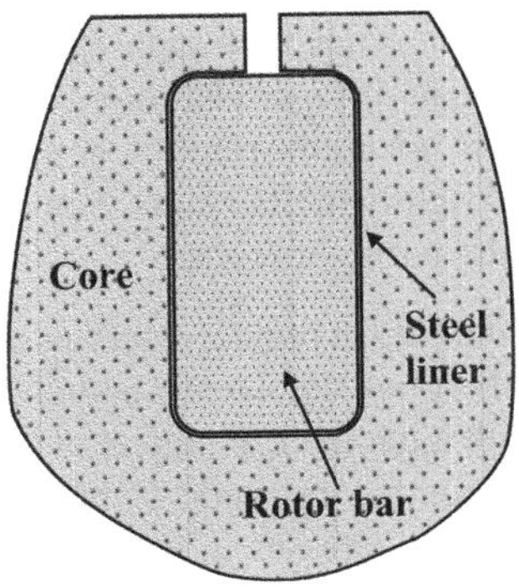


Figure 2.8 Slot liner.

**(d)** Having a keyway below each bar slot into which an interference fit key is driven after bar installation (Figure 2.9a gives a simplified illustration) and this method ensures a tight fit which has proved to be very successful at keeping the bars tight in the slots for the lifetime of the SCIM. Figures 2.9b and 2.9c show a

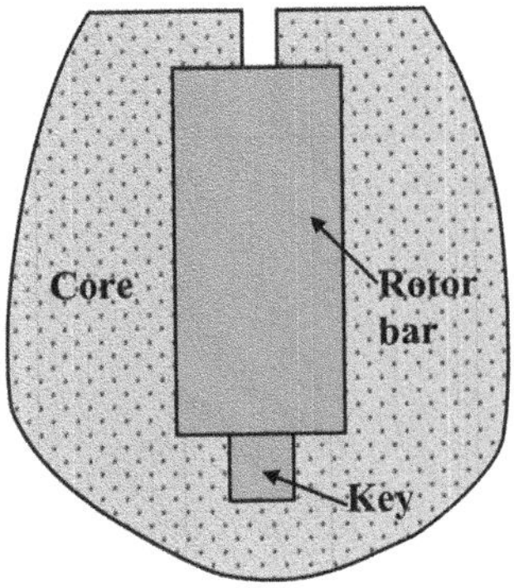


Figure 2.9a Simplified illustration of a key under bar.

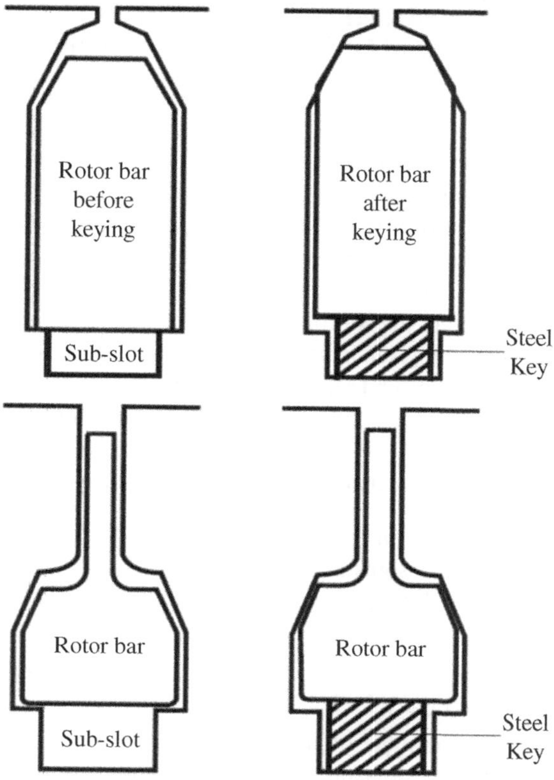


Figure 2.9b Actual designs before and after fitting steel locking keys. Reproduced with permission of Parsons Peebles, Scotland.

schematic before and after the steel keys are fitted and a photo of a cage rotor with the key bars fitted under the copper cage rotor bars.

Once the rotor bars are installed they have to be joined at either end to the shorting rings via brazing or welding.

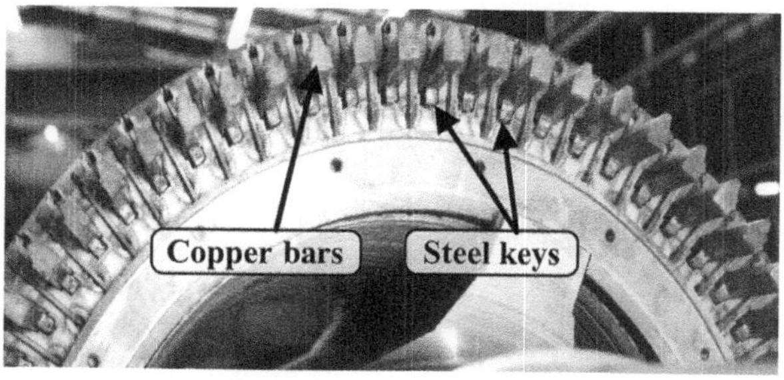


Figure 2.9c Key bars fitted under copper bars. Reproduced with permission of Parsons Peebles, Scotland.

### 2.2.2 Design and Construction of Short Circuiting End Rings, Bar to End Ring Joints, and End Ring Retaining Rings (ERRs)

To ensure their integrity under starting and operating thermal and mechanical stress the shorting rings should be of a single piece construction formed by a casting or forging process. Some manufacturers make shorting rings for slower speed motors by rolling them from rectangular copper or copper alloy strips and joining the two ends by a butt or overlapping brazed joint. With this construction, failure of the brazed connections may occur, due to high thermal and mechanical stresses, in motors used to directly start high inertia devices. The bars in copper or copper alloy windings are most commonly connected to the shorting rings by brazing them together, but some manufacturers have used a tungsten inert gas (TIG) welding process. Since aluminum is difficult to braze, the bars and end rings, which are made from this material, are usually welded together. There are a number of different bar-to-shorting ring connection configurations.

The most common of which are

**(a)** Butt connection between the bar ends and the face of the shorting ring, see Figure 2.10. In some designs a groove, approximately the same depth as the rotor bars, is machined in the ring to give a larger braze connection surface to provide a more robust mechanical connection, see Figure 2.11.

**(b)** Butt connection between the outside diameter of the shorting ring and under side of the bars, see Figure 2.12.

**(c)** Machining the same number of slots in the shorting ring as there are bars, so that the bars fit in these slots and their outer ends are approximately flush with the outer face of the ring (Figure 2.14).

**(d)** Although it is possible to braze aluminum bars and end rings together, it is much easier to weld them. Figure 2.13 [2.10] illustrates the typical construction of the rotor bar to end ring welded joint configuration [2.12]. The end ring is slotted to accommodate each rotor bar. In designs where the end rings are used to maintain core pressure (Figure 2.13), once all the bars have been installed in their core slots the end rings are fitted over them and clamped together to exert pressure on the rotor core. With this pressure maintained the rotor is installed in a lathe and circumferential grooves are machined in the end ring and bar

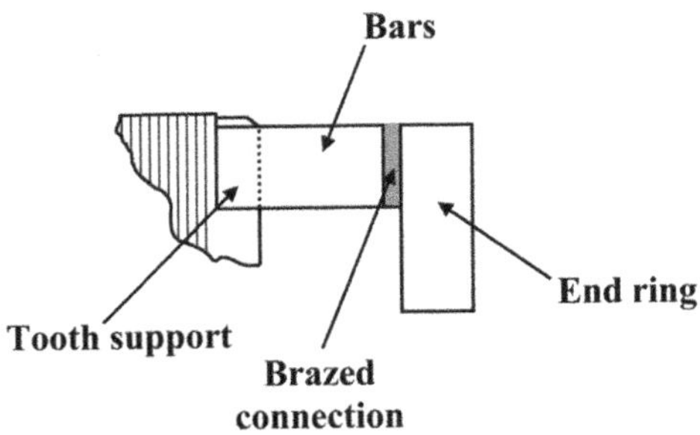


Figure 2.10 Butt connection.

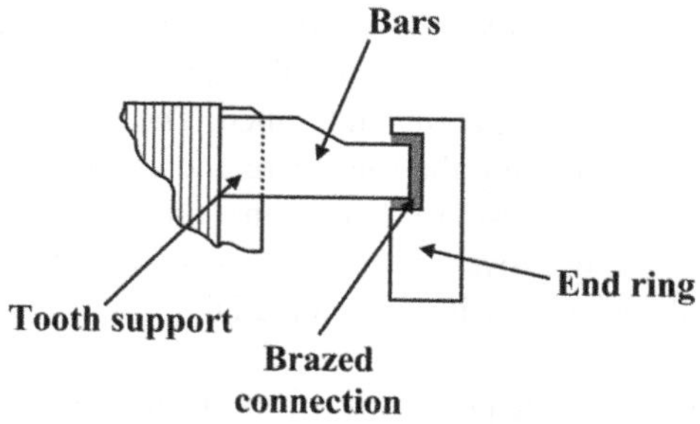


Figure 2.11 Butt connection with groove machined in end ring.

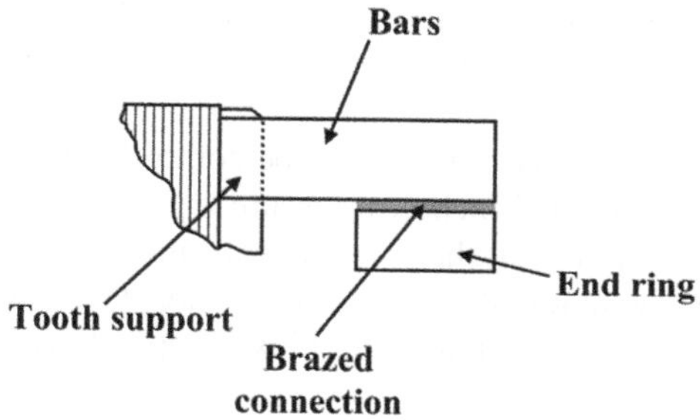


Figure 2.12 End ring under rotor bars.

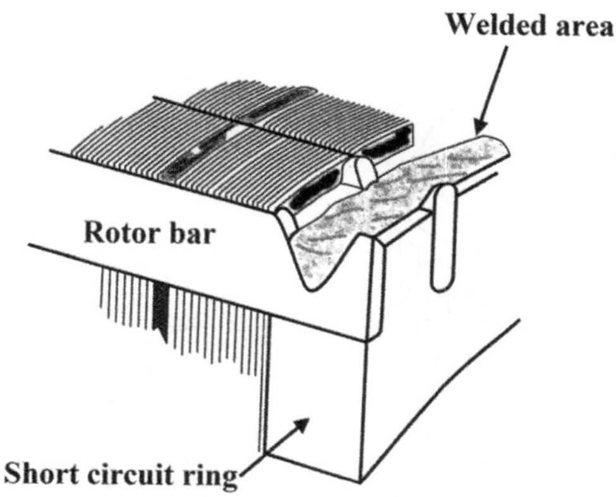


Figure 2.13 Weld joint in fabricated aluminum winding. *Source*: Finley and Hodowanec [2.12]. Reproduced with permission of IEEE.

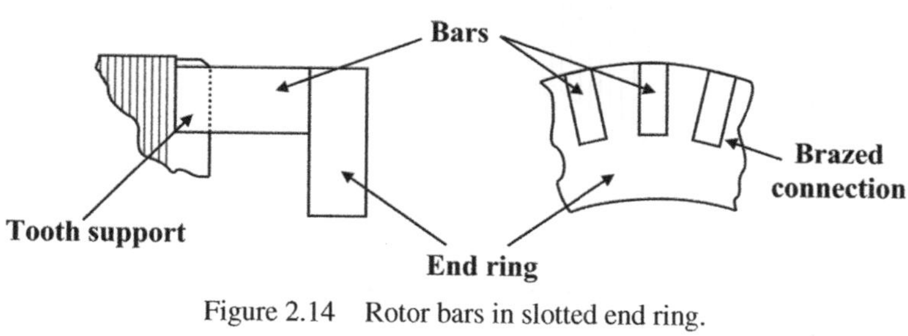


Figure 2.14 Rotor bars in slotted end ring.

assemblies. This groove is then filled with weld material using a TIG or metal inert gas (MIG) welding process [2.10] with aluminum wire. Layers of weld are applied until the grooves are completely filled. The other construction method that is used involves building the rotor core and clamping it with end fingers and core clamping rings, using a bar to end ring configuration as shown in Figure 2.14, machining groves in the end rings and bars as shown in Figure 2.13 and filling the grooves with weld material.

The bar to end ring brazing process involves ensuring the metal surfaces to be brazed together are very clean. Then the rotor is vertically mounted and the end rings installed close to or over the bars. In larger machines the mass of the end ring makes it difficult to support, since mounting it on a large mass will create a heat sink, which can reduce the temperature of the components to be brazed and can affect the quality of the bar to end ring joints.

The space between the end rings has to be controlled to be within 0.05 mm/0.002 inches to 0.127 mm/0.005 inches, to ensure a good brazed connection. Two common methods are used to heat the bars and end rings to raise their temperature to about 425°C/797°F, the required brazing temperature. The first involves using gas burners and the other uses induction heating. Flux is applied to ensure good bonding

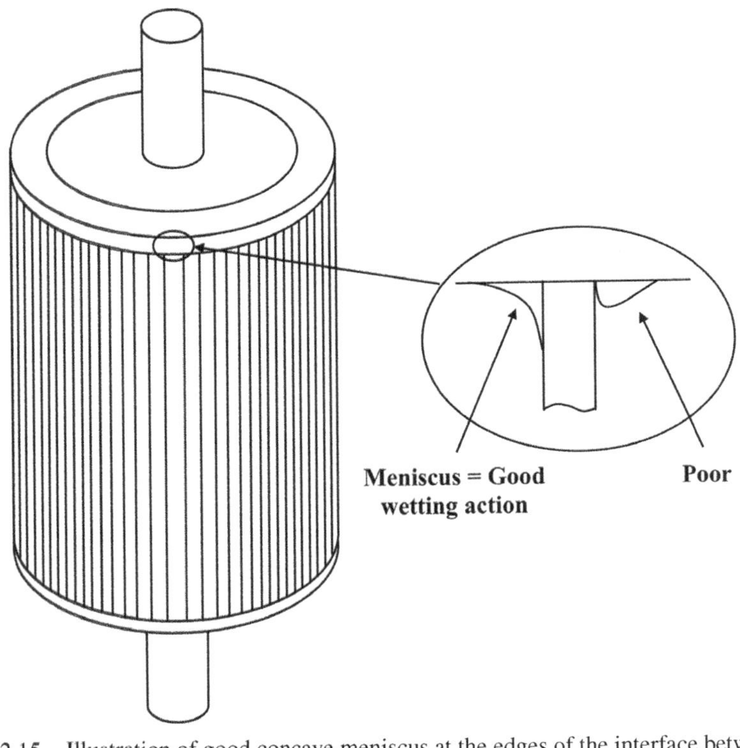


Figure 2.15 Illustration of good concave meniscus at the edges of the interface between bars and end rings and a poor finish.

between the bars and end rings. A gas burner is used to heat the brazing metal, which is usually an alloy containing silver, phosphorus, and copper and in rod form, is fed into the bar end ring connections, so that it flows and fills the small space between the two. This type of brazing requires a *highly skilled* operator, who can control the amount of heat applied to the joint being brazed. High frequency induction heating requires the brazing material to be placed on the end ring before its installation under or on the bars. An induction heater is installed under the end ring and when current is induced in its coil the temperature of the end ring and bar ends is raised to around 425°C/797°F causing the brazing metal to flow and fill the gaps between the two. With this heating process it is much easier to control the temperature of the bar.

For both processes, it is important to ensure that the end ring is maintained at a uniform temperature to avoid creating residual mechanical stresses in the joint area. Once the brazing process is completed the cage winding assembly should be allowed to cool before it is turned over to braze the other end ring in place. For a good brazed joint, the brazing material should form a concave meniscus at the edges of the interface between bars and end rings (see Figure 2.15). If the rotor winding being brazed is to be installed in a motor that will operate in a plant, where hydrogen sulfide gas could be present, a special brazing material that does not contain phosphorus should be used since chemical action between this gas and the normal brazing material can cause the brazed joints to crack and fail.

The end winding stresses in fabricated rotor cage windings, which are used in larger 2- and 4-pole SCIMs can be high enough during starting and running to cause rotor bar or end ring cracking and failure if they are not controlled within certain limits. Such stress control can be achieved by fitting non-magnetic, high strength retaining rings, made from materials such as stainless steel and naval brass, over the end rings (See Figure 2.16) or bar extensions beyond the rotor core and end rings (See Figure 2.17). The latter of these two configurations provides better stress control. The end rings have to be machined to ensure that there is an interference fit between the two, so that when the retaining ring grows under the influence of centrifugal forces during motor operation, there is still an interference fit between the two. This ensures

Figure 2.16 Retaining ring over end ring. Photo taken by Ian Culbert. Reproduced with permission of Iris Power Qualitrol.

Figure 2.17 Retaining ring over end ring and bar extensions outside rotor slots. Photo taken by Ian Culbert. Reproduced with permission of Iris Power Qualitrol.

that the retaining rings do not become loose and migrate axially. Because of this interference fit the retaining rings have to be heated up to around 300°C/572°F before they can be installed.

During operation, as a further precaution against a loose fit causing axial migration, retaining rings are often locked to the endings. This is usually done by means of four set screws in tapped holes circumferentially located in the retaining ring 90° apart. Some manufacturers use retaining rings made from resin bonded glass banding in smaller 2- and 4-pole motors, but this is not as robust a design and such banding has been known to fail, rub on the stator winding, and cause it to fail.

## 2.3 DESIGN AND MANUFACTURING FEATURES OF SQUIRREL CAGE ROTOR WINDINGS TO MINIMIZE FAILURES

The following is a summary of squirrel cage winding design and manufacturing features that should minimize catastrophic failures that can cause consequential damage to other motor parts, including the stator winding.

**(a)** The motor design and manufacturing processes should ensure that the bars are tight in their slots. This should prevent the bars from failing under the influence of the high twice power supply frequency radial and torsional electromagnetic forces imposed on them by the high rotor winding currents present during starting.

**(b)** The brazed or welded connections between the rotor bars and end rings should be of high quality to ensure that they are less likely to fail under the high thermal and mechanical stresses imposed on them during motor operation. If the motor is to operate in an environment containing hydrogen sulfide, brazing or welding materials should not contain phosphorus.

**(c)** During starting, high mechanical and thermal stresses are imposed in the squirrel cage end winding structure of fabricated windings in motors used in high torque and high inertia applications. The magnitudes and nature of these stresses should be determined at the motor design stage by 3D finite analysis modeling. This should ensure that an appropriate squirrel cage rotor winding design can be developed to withstand these stresses, Curiac [2.13]. For example, such analyses may indicate the need for retaining rings to be installed.

**(d)** For high inertia motor applications, the motor manufacturer's limits for consecutive starts from cold and hot and the delay time before attempting further starts should be strictly adhered to because excessive starting can result in winding failures.

**(e)** A torsional natural frequency analysis should be performed on the section of the squirrel cage winding outside the rotor core to ensure that it does not have a natural frequency close to 100 Hz (50 Hz motors) or 120 Hz (60 Hz motors). If such natural frequencies exist they can be excited by the twice power supply frequency forces induced by high motor starting currents.

**(f)** Die-cast windings in motors for critical applications should have quality checks such as X-rays or ultrasonic tests performed to establish whether large voids exist. The presence of voids in a die-cast winding causes the rotor to bend under load, due to non-uniform heating. Such bending causes high vibration levels. This is especially true of 2-pole motors with long and smaller diameter rotors. If this problem is suspected, an effective test to confirm it exists, is to measure the no-load vibration levels of the unloaded motor and to then plug, which is to reverse two of the phases while the motor is running, two or three times to heat up the rotor winding. A repeat vibration test is then performed. If the rotor cage winding does not have any large voids in it, the "hot rotor" vibration levels should be no greater than 120% of the "cool rotor" ones."

**(g)** Fitting end ring retaining rings (ERRs) as shown in Figure 2.17 can prevent a broken rotor bar crashing into the stator winding and a possible catastrophic failure.

## 2.4 QUESTIONS

**2.4.1** Why are aluminum die-cast windings very often used in SCIMs up to 500 HP/373 kW?

**2.4.2** Why are aluminum die-cast cage windings much less likely to have broken rotor bars than copper fabricated cage windings?

**2.4.3** Why are copper fabricated cage windings very often used in SCIMs greater than 1000 HP/746 kW?

**2.4.4** If the starting torque of a SCIM is 7380 lbsf ft/10,000 N·m at its rated volts of 1.0 p.u., what is the starting torque if the voltage drops to 0.8 p.u.?

**2.4.5** Explain what is meant by the term skin effect with respect to the current distribution at start-up of a SCIM.

**2.4.6** How can the phenomenon of skin effect be used to advantage to produce a high starting torque?

**2.4.7** What can cause thermal bow of the rotor bars?

**2.4.8** What is a double cage winding and in what applications are they often used?

**2.4.9** Why must the rotor bars in a fabricated copper cage winding be tight in the rotor slots?

**2.4.10** Why are ERRs often fitted to copper fabricated 2-pole, 60 Hz squirrel cage rotors? What are the advantages of having ERRs that cover the joints between the bars and end rings and the exposed part of the bars between the inner faces of the end rings and the rotor core ends in high speed 2-pole, 60 Hz, HV, SCIMs?

## REFERENCES


[2.1] M. Liwschitz-Garik and C. C. Whipple, *Electric Machinery Vol. II, A-C Machines*, Van Nostrand Company (1st published, September 1946).

[2.2] M. G. Say, *Alternating Current Machines*, 4th edition, ELBS and Pitman Publishing, 1976.

[2.3] P. L. Alger, "Induction Machines-Their Behavior and Uses," Gordon and Breach Science Publications Inc., 2nd edition, published by OPA Amsterdam, 3rd printing with additions, 1995.

[2.4] H. Vickers, *The Induction Motor*, Sir Isaac Pitman and Sons Ltd., London (1st edition, 1924; 2nd edition, 1953).

[2.5] T. C. Lloyd, "Some aspects of electric motor design polyphase induction motor design to meet fixed specifications," *Transactions of the American Institute of Electrical Engineers*, vol. 63, no. 1, 1944, pp. 14–20.

[2.6] A. H. Bonnett and T. Albers, "*Squirrel cage rotor options for AC induction motors*," IEEE Annual Pulp and Paper Industry Technical Conference, 2000, Conference Record, pp. 54–67.

[2.7] H. Barr, A. H. Bonnett, and C. Yung, "*Understanding the design of stators and rotors of squirrel cage induction motors*," 55th IEEE PCIC Conference, USA 22–24 September, 2008, Cincinnati, Ohio, pp. 1–11.

[2.8] R. L. Nailen, "New rotor design concept solves pipeline motor acceleration," *IEEE Transactions on Industry Applications Problem*, vol. IA-9, no. 2, 1973, pp. 201–205.

[2.9] G. R. Anderson, "Copper spun squirrel-cage rotor," *Electrical Engineering*, vol. 66, February 12, 2013, pp. 980–982.

[2.10] Electrical Apparatus Service Organization (EASA), *Mechanical Repair Fundamentals of Electric Motors*, 2nd edition.

[2.11] J. H. Dymond, "Stall time, acceleration time, frequency of starting: the myths and facts," *IEEE Transactions on Industry Applications*, vol. 29, no. 1, January/February 1993, pp. 42–51.

[2.12] W. R. Finley and M. M. Hodowanec, "*Selection of Copper vs. Aluminum Rotors for Induction Motors*," IEEE 2000 PCIC, San Antonio, TX, September 11–13, 2000, PCIC-2000-19.

[2.13] R. S. Curiac, "*Forces and Stresses in Squirrel Cage Motors during Starting*," IEEE PPIC 2008, Seattle, WA, June 22–27, 2008, pp. 7–14.

CHAPTER 3

# CAUSES OF BREAKS IN SQUIRREL CAGE WINDINGS DURING DIRECT-ON-LINE STARTS AND STEADY-STATE OPERATION

*Ian Culbert*

## 3.0 INTRODUCTION

Provided that squirrel cage rotor windings are designed to cope with mechanical and electromagnetic forces and thermal stresses during direct-on-line (DOL) starting and also any abnormal steady-state operating conditions which the motor may experience, the cage rotor should be a reliable part of a Squirrel Cage Induction Motor (SCIM). However, this requires the user to provide, in the first instance, all the necessary information to the motor manufacturer, or as is often the case, to the supplier of the complete pump, compressor, or fan drive train, who then liaises directly with the motor manufacturer (OEM). It is the authors' view that the end user should have an opportunity to deal directly with the OEM in full consultation with the supplier of the complete drive train. The susceptibility to failure of a cage winding, is dependent on the type of winding construction, the motor application, operating duty, winding geometry, the materials of construction, and in particular the actual DOL starting regime, to which the motor will be subjected by the user. To ensure, as much as possible, the long term integrity of the cage winding, a sample of the key facts, which the user should provide to the OEM of the motor is as follows:

- **(i)** Limitations on starting current during a Direct On Line (DOL) start following a system study of the supply network.
- **(ii)** The expected drop in voltage at the motor's terminal during a DOL start.
- **(iii)** The torque–speed curve of the mechanical load to be provided by the supplier of the pumps, compressors, or fans and to include, for example, the curves for

*Current Signature Analysis for Condition Monitoring of Cage Induction Motors: Industrial Application and Case Histories*, First Edition. William T. Thomson and Ian Culbert.

open and closed valves in pumps, open and closed inlets for compressors, open and closed dampers for fans, and inertia downstream of the motor.

**(iv)** The user should confirm the expected DOL starting regime for the mechanical load, for example, could DOL starts be required against an open valve system for a centrifugal pump or a closed valve set-up for an axial flow pump. If yes, the motor has to be designed to start against that torque–speed curve at the specified reduced volts and to accelerate up to steady-state speed, without unacceptable stresses, during the run-up time.

**(v)** The user should discuss with the OEM of the motor the specifications for sequential DOL starts from cold and from hot and the user must appreciate the implications of not abiding by the OEM's specifications.

**(vi)** The type of load characteristics from, for example, large induced draft fans with high inertias, reciprocating compressors, coal crushers, and centrifugal compressors on long term recycling.

This chapter begins with a general review of the failure mechanisms, which lead to actual breaks in the cage winding, such as broken rotor bars or end rings. This is followed by features on the design of cage windings used for different applications, taking into account DOL starts, starting currents, combined inertia of the motor's rotor and load, torque–speed curves of the motor and load, and run-up times, all of which can affect the reliability of a cage rotor and may lead to cage winding failures. A selection of excellent papers by Bonnett et al. [3.1–3.3] and Bonnett [3.4], including a comprehensive coverage of failure mechanisms and consequential cage winding breaks in SCIMs is referenced. It is the authors' opinion that to simply regurgitate already published information is quite unnecessary. The key facts supported by additional relevant references are presented in this chapter.

## 3.1 MECHANICAL STRESSES AND CONSEQUENTIAL FORCES ON ROTOR BARS AND END RINGS

The mechanical stresses in a squirrel cage winding result from centrifugal forces during motor operation and are highest at the largest radius of rotation in a rotor [3.1], Finlay et al. [3.5], Curiac [3.6], and Yabiku et al. [3.7]. The two main forces which develop during motor starting are

**(i)** Centrifugal forces (CF) on the rotor bars and end rings and from the sections of rotor core above the bars.

**(ii)** A reaction force (RF), which holds the bars in place.

Since centrifugal force is proportional to the square of the rotor speed, both CF and RF reach their maximum magnitudes when the rotor reaches operating speed. Figure 3.1 Culbert et al. [3.8] indicates the locations and directions of these mechanical forces on

**(i)** The slot sections of the rotor bars ($CF_1$).

**(ii)** The sections of rotors bars outside the rotor slots ($CF_2$) which are not applicable to die-cast (Figures 2.3a and 2.3b) or some fabricated aluminum windings (Figure 2.13).

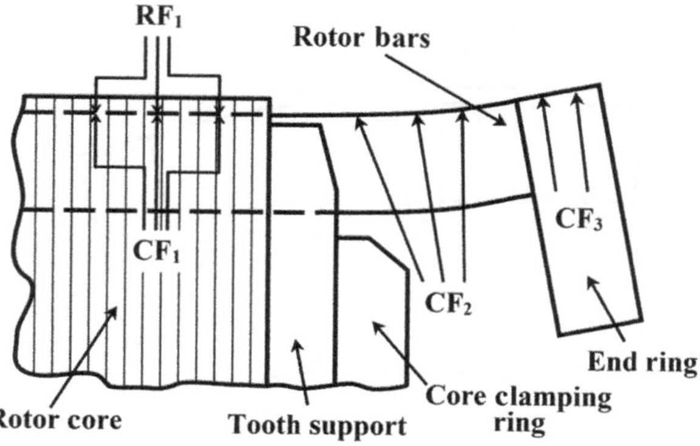


Figure 3.1 Mechanical stresses on squirrel cage winding [3.8].

**(iii)** The end rings ($CF_3$).

**(iv)** The centrifugal stresses on the rotor bars in the slot areas are insignificant, but this is not true for the stresses in the rotor laminations immediately above the rotor bars, since they provide a retaining reaction force to keep the bars in their slots. This core reaction force above the bars is much higher if the bars are made from copper or copper alloy rather than aluminum. This is so because centrifugal forces exerted by the bars on the core are a function of their mass and copper and its alloys are approximately three times heavier than aluminum. Therefore, to prevent the core bridge from fracturing, its compressive strength above the bars must be higher than the stresses imposed on it.

The mechanical stresses from centrifugal forces in the bar sections outside the rotor slots and end rings (hereafter referred to as the end windings) are much higher than those on the bars in the rotor core slots, since these sections of the rotor winding are not restrained by the core (Figure 3.1). Again, because aluminum is much lighter than copper and its alloys, the mechanical stresses on the end winding sections are much lower for the same geometric configuration [3.6] of aluminum compared to copper. As illustrated in Figure 3.1, $CF_3$ enlarges the end rings and this exerts a bending force on the rotor bar sections outside the rotor slots.

If the motor sees frequent starts and stops, the stresses from centrifugal forces will be cyclic and this can lead to winding failures from low cycle fatigue. The end winding components can crack and fail if the stresses in them exceed their fracture and/or fatigue strength. Such failure possibilities should be addressed by the motor designer and, if necessary, retaining rings should be installed over the end rings (Figure 2.16) or over both them and the sections of bar outside the rotor slots (Figure 2.17) to remove the possibility of them crashing into the HV stator winding.

## 3.2 THERMAL STRESSES IN THE ROTOR BARS AND END RINGS

Every time a motor is started directly on-line, a significant amount of heat is generated in its rotor windings. The current in the rotor during starting can be more than five times the full-load rated current. The amount of heat generated and therefore

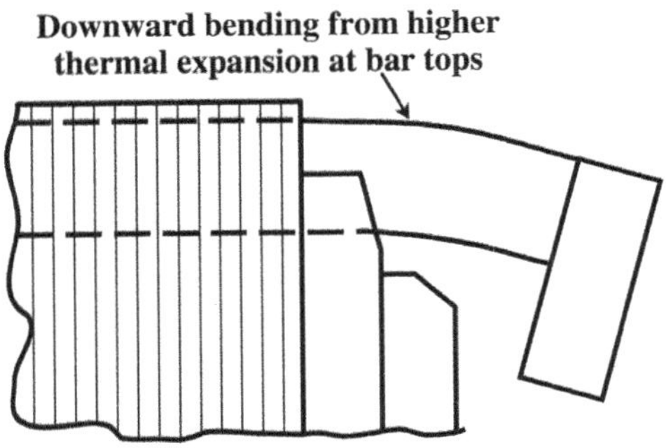


Figure 3.2 Bar bowing outside rotor slots due to differential thermal expansion.

the maximum rotor winding temperature is a function of the torque margin between torque–speed curves for the motor and driven equipment, together with the combined inertia of the motor's rotor and driven equipment. The longer the time that the rotor takes to reach the operational speed, the greater will be the temperature rise, which the rotor winding experiences. Much less rotor winding heating occurs after the rotor is at its steady-state operating speed, since heating is proportional to $I_r^2$ and the current at the motor's operating speed is much less than during starting or run-up currents. Another factor that determines the maximum rotor bar temperature is "skin effect" Alger [3.9], Liwschitz-Garik [3.10], as discussed in Section 2.2.1.

As the motor speeds up the frequency of the rotor bar current falls from supply frequency at standstill to slip frequency (very close to dc) at the steady-state running speeds between no-load and full-load. The high concentration of current, and hence the $I^2 \times R$ loss in the top of bar will therefore be highest at switch on. The non-uniform rotor bar heating loss distribution results in an associated non-uniform temperature rise distribution with the temperature rise at the top of the bar being greater than at the bottom of the bar.

Therefore, during run-up, this non-uniform temperature distribution can give rise to "bar bowing" outside the rotor slots due to differential thermal expansion as shown in Figure 3.2 [3.5]. However, as the rotor speeds up, the slip frequency ($f_2 = sf$) of the rotor current drops and finally the current in the bars becomes uniformly distributed at the normal running speeds.

This temperature increase is more significant in the following designs:

**(i)** In double cage windings the outer cage winding has rotor bars of a smaller cross-sectional area than the inner run winding bars.

**(ii)** Deep, thin bars in single cage designs.

**(iii)** Inverted "T" bars, which are much narrower at the top than the bottom.

Repetitive starts (three or more) over a short period of time (see Section 3.8) can produce excessive rotor winding temperatures in motors used in high inertia drive systems, since not all of the heat created during the first start(s) will have been radiated or conducted from the rotor. Thus, the rotor winding temperature may still be high from the first start, when the second start causes a further increase in temperature (see Section 3.8). Thus if repeated starts are performed, cumulative rotor heating can lead to undesirably high rotor winding temperatures. Such rotor winding overheating problems are much more likely in motors that directly start high inertia driven equipment, such as induced and forced draft fans in fossil power plants. Those problems are accentuated if the wrong bar design is used for a high inertia driven application, for example, the tops of inverted "T" bars have been known to crack, due to excessive temperatures developed during motor starting.

## 3.3 BROKEN BARS AND END RINGS DUE TO COMBINED MECHANICAL AND THERMAL STRESSES WHEN STARTING HIGH INERTIA LOADS

Each time a motor is started, the mechanical and thermal stresses, described in Sections 3.1 and 3.2 , act in combination to produce much higher mechanical stresses than they would individually. The effects of these combined stresses can become more serious, if repetitive starts are performed without allowing sufficient time for the rotor winding to cool between each start. Such stresses can cause both the rotor bars and the end rings to crack and eventually fail, see Section 3.11 for examples of cage winding failures. Also in die-cast and fabricated aluminum rotor windings with no bar extensions outside the rotor slots, the lack of a transition section of bars between the core and end rings produces very high stresses at the interface joint between bars and rings at the end of the core [3.5]. This is illustrated in Figure 3.3, which shows that because of skin effect on the bars there is a consequentially greater expansion at the top sections of the bars relative to the bottom sections, therefore the top of the end ring section is pushed further from the rotor core. This tilting action of the end ring relative to the core induces very high stress in the joints between the bars and end rings. This in conjunction with the poorer mechanical properties of aluminum, especially at higher temperatures, makes the joints and bars in these winding types more susceptible to failure, if used in applications with high inertia driven equipment.

Copper and copper alloy windings, with sections of bar between the core and end rings, provide the best design for high inertia applications, since these materials

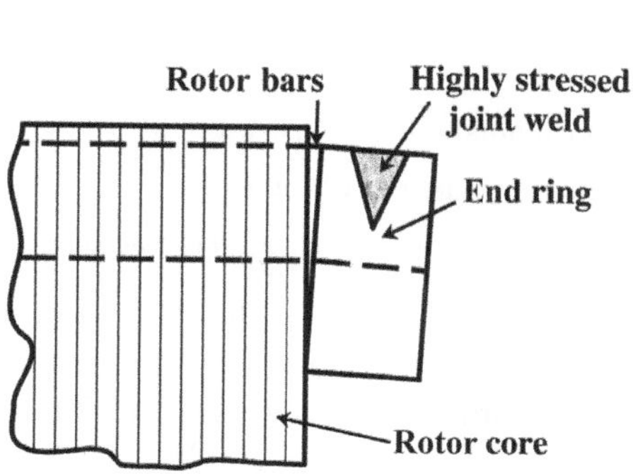


Figure 3.3 High stresses in aluminum squirrel cage winding with no bar extensions beyond core.

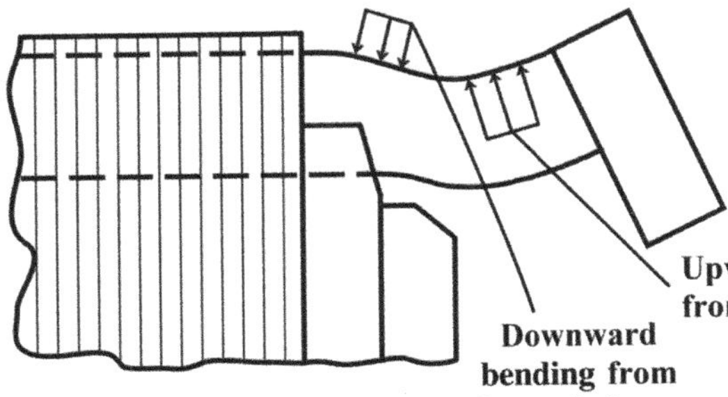


Figure 3.4 Distortion of rotor end winding due to combined mechanical and thermal stresses.

have a much higher thermal capacity than aluminum and the flexibility of the bar extensions reduces the stresses on the brazed joints between the two. The cage windings in motors directly driving high inertia loads are susceptible to failure from low cycle fatigue due to the change in the rotor end winding stresses between stand-still and running conditions. These forces, which mainly affect fabricated rotor windings, are described in more detail as follows. The effects of these stresses are illustrated in Figure 3.4, which exaggerates the rotor winding deformation resulting from thermal bow in the rotor bars and centrifugal forces from rotor rotation. Small cracks in the tops of the bars can be introduced, if the swaging tool used is too sharp. These cracks may grow as a result of combined thermal and mechanical stresses during motor operation, especially during starting.

## 3.4 ROTOR BAR STRESSES RESULTING FROM A LOOSE SLOT FIT

Any design or manufacturing deficiencies, which lead to the loss of tightness of the bars in the slots will lead to the bars becoming even more loose, due to electromagnetic and centrifugal forces, and thus to possible failures from this cause. If only some bars are loose, non-uniform bar expansion can occur during motor operation and this can create sufficient force to bend the rotor and put it out of balance. Two-pole motors are most susceptible to this problem. Examples of issues leading to loose bars are

- **(i)** The use of bars with rounded corners instead of sharp ones.
- **(ii)** Loose slot liners which migrate out of the slots.
- **(iii)** Omitting the bar swaging operation during manufacture, see Chapter 2.

When rotor bars are loose they vibrate in the slots, especially during starting, when large electromagnetic forces act on them. This force is at its highest at switch-on and run-up due to the high starting currents in each rotor bar and for a 60 or 50 Hz supply frequency the bar vibration is at 120 or 100 Hz. For the avoidance of doubt, this electromagnetic force, which vibrates the rotor bars at a frequency of twice the supply frequency ($2f$) is due to the interaction between the slot leakage flux, which

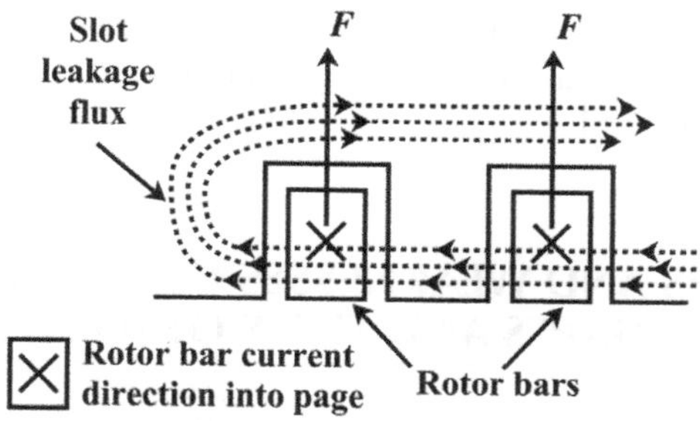


Figure 3.5a Illustration of radial force on the rotor bars, $F \propto I_{rb}^2$ and vibration.

is in the horizontal direction and links adjacent current carrying rotor bars. It is well known that this force on the rotor bars is proportional to the rotor bar current squared ($I_{rb}^2$) [3.1, 3.2, 3.9]. By "Fleming's left hand rule" this force is in the radial direction as shown in Figure 3.5a.

This is in complete contrast to Figure 1.8a, which illustrates the force in the tangential direction which turns the rotor and is due to the interaction between the main rotating flux at synchronous speed ($N_s$), which links the stator and rotor windings and the corresponding current induced in the rotor bars.

These radial forces cause stresses in the bars and the joints between them and the end rings. In the long term, this can lead to bars making and breaking contact with the core causing arcing, since the bars are at some voltage and the core is grounded. Through time, this arcing can cause overheating and mechanical deterioration of the core bridge above the bars, weakening it (see Figure 3.5b) and the bars can begin to

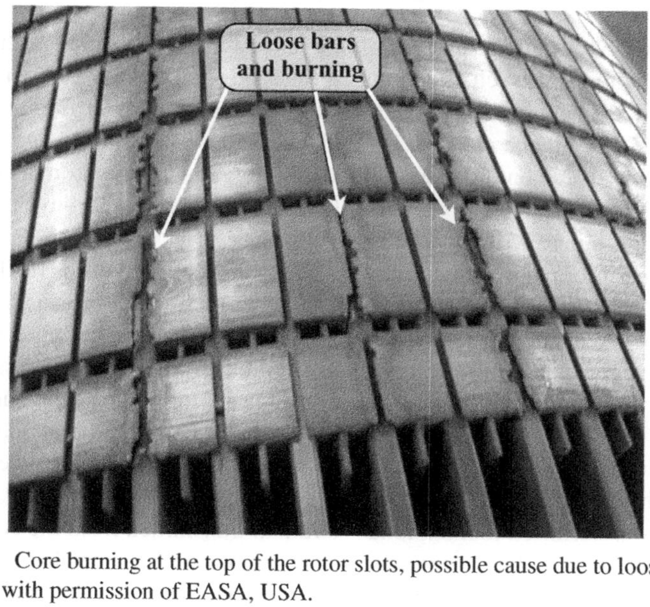


Figure 3.5b Core burning at the top of the rotor slots, possible cause due to loose bars. Reproduced with permission of EASA, USA.

migrate out of the slots since there is a reduced or zero reaction force (see Section 3.1) to retain them. Bar looseness and the resulting vibration can also lead to bar cracking and fracture in the slots [3.6].

## 3.5 STRENGTHS AND WEAKNESSES OF CERTAIN BAR AND END RING SHAPES AND TYPES OF JOINTS

As indicated in Section 3.2, it is important to select the appropriate bar shape to suit the characteristics of the driven load and method of starting. For high inertia drives deep narrow, inverted "T" shaped or a double cage bar configuration should "*not*" be used, since there is a much higher risk of them failing from low cycle fatigue. The best bar shape and type for such applications is a rectangular bar made of a copper alloy, which takes advantage of the higher starting torque and shorter run-up time such higher resistance bars provide. On the other hand, for low inertia driven equipment requiring a high starting torque, the use of a narrow and deep, inverted "T," or double cage bar configurations is quite appropriate since the thermal stresses produced during starting are low. Since significant heating of a rotor winding occurs during a DOL motor start of high inertia equipment, it is beneficial to use larger end rings, which will act as heat sinks to reduce the maximum temperature at the tops of the rotor bars. Also, a larger interface area in the joints between the bars and end rings will promote better heat transfer to the end rings.

## 3.6 PULSATING LOADS DUE TO CRUSHERS AND COMPRESSORS

Pulsating loads on a motor can excite torsional natural frequencies in the motor's rotor and its components such as cooling air fans and squirrel cage end windings and produce speed and supply current fluctuations (see Figure 3.6), Griffith et al. [3.11], Perrin et al. [3.12], Hanna et al. [3.13] and Middlemiss [3.14]. Although there are published papers documenting cases of motor cooling fan failures from torsional resonances [3.12] in such applications, a literature survey did not reveal any on fabricated squirrel winding failures from this cause.

However, informal e-mail discussions with Bill Lockley, Fellow IEEE Calgary, Canada, and also professional networking indicate there have been isolated cases of fixed supply frequency, 6- and 8-pole SCIMs driving reciprocating compressors which have experienced squirrel cage winding failures from torsional stresses, which caused the brazed joints between the bars and end ring to fracture. The bars in these motors were only locked by swaging in the center, so both the radial and torsional stiffness of the cage end winding structures were lower than if the bars were tight in their slots. This reduced stiffness decreases the torsional natural frequency of the end windings. It was surmised that torsional vibration, transmitted from the compressor, excited a cage end winding natural frequency, imposing high mechanical stresses on the brazed joints and since the joints were weaker than the bars and end rings

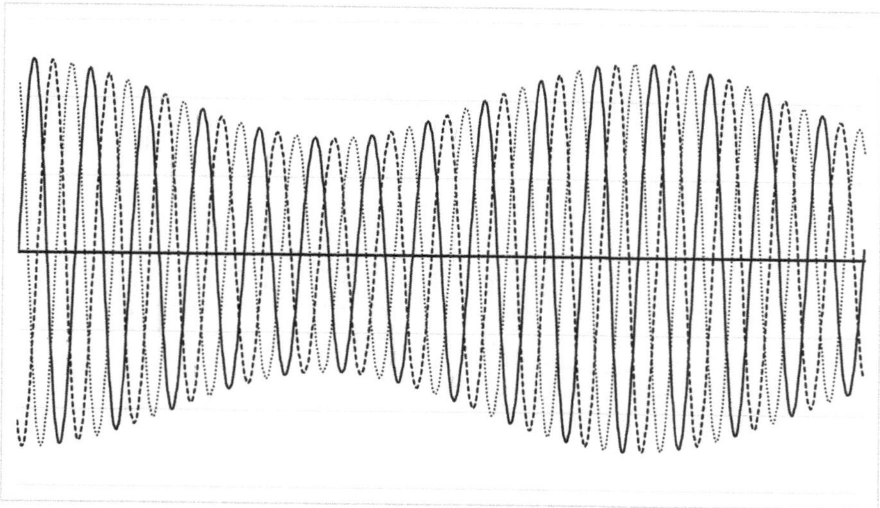

Figure 3.6 Motor driving a reciprocating compressor showing fluctuations in all 3-phase currents. *Source:* Perrin et al. [3.12]. Reproduced with permission of IEEE.

they failed from cyclic fatigue. It was also found that the breaks at the brazed joints were more likely to be at the non-drive end (NDE) of the winding, since the vibration mode shapes usually produced the highest amplitudes furthest from the coupling between the motor and compressor. Informal discussions with OEMs have indicated that there have also been isolated cases of cage winding failures in variable frequency, inverter-fed SCIMs as a result of rotor torsional oscillations induced in the motors by harmonics from the inverter supply.

The speed and load fluctuations imposed on the rotors are most significant in reciprocating compressors and they cause fluctuating motor line currents (see Figure 3.6) and varying rotor speed and hence slip. Fitting a flywheel to a motor-compressor driven train will reduce these fluctuations [3.11], but not sufficiently to prevent significant pulsations of the rotor and thus continual slip variations with time. Severely fluctuating loads such as those found in crushers, ball mills, and reciprocating compressors can impose abrupt transient overloading on the motor, which subjects the rotor cage winding to fast temperature rises. Copper or copper alloy cage windings cope better with these problems since they have a much higher thermal capacity than aluminum types.

## 3.7 DIRECT-ON-LINE STARTING OF LARGE INDUCTION MOTORS DRIVING HIGH INERTIA FANS

During DOL starting of high inertia centrifugal fans such as induced and forced draft types used in coal fired power stations, the temperatures reached in the cage windings are a function of the torque–speed curve of the fan during starting and the combined inertia of the motor's rotor and the fan's rotating parts, Dymond [3.15]. Note that not all induced draft (ID) and forced draft (FD) fans are centrifugal for

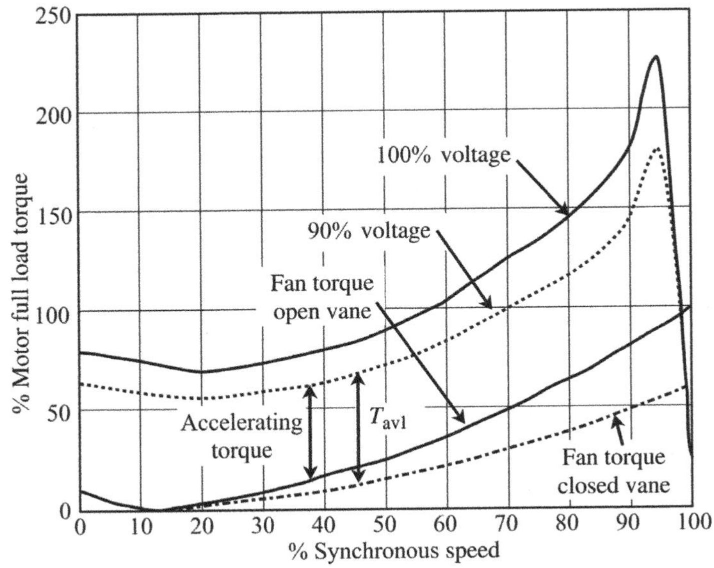


Figure 3.7 Motor and centrifugal fan torque versus speed curves.

example ID fans can be axial flow. One way to reduce the maximum temperature of a squirrel cage winding is to start the fan with its dampers closed rather than open. As can be seen from Figure 3.7, starting with the fan dampers closed decreases the fan torque and increases the motor accelerating torque during the starting period at both 100% and 90% of motor rated voltage, so the starting time and rotor winding heating can be reduced. It is important to stress to the end user that if reliance on keeping the temperature rise on the rotor winding to an acceptable level is dependent on the dampers being closed during starting, the motor must not be started with open dampers. If the motor must be capable of starts with open or closed dampers, the original specification to the OEM must reflect that necessity.

When assessing the starting capability of a squirrel cage induction motor, it is important to take account of the minimum voltage at its terminals as illustrated in Figure 3.7, since the available accelerating torque does decrease significantly with voltage because motor torque is approximately proportional to supply voltage squared. The motor accelerating time can be calculated via a step–by-step solution using equation (3.1) as given in Reference 3.15 in an IEEE publication, which used imperial units, but for metric units equation (3.2) must be used.

**In imperial units**

$\Delta t$ = the time taken to accelerate during the selected step $\Delta S$ in r/min is given by

$$\Delta t = (WK^2M + WK^2DE \times \Delta S)/(308 \times T_a) \tag{3.1}$$

where

$\Delta t$ = the time taken to accelerate during the selected step $\Delta S$ in r/min

$\Delta S$ = step in r/min

$WK^2M$ = inertia of rotor in the motor in lb-ft$^2$

$WK^2DE$ = inertia of mechanical load in lb-ft$^2$

$T_a$ = average accelerating torque in lbf·ft during selected step $\Delta S$ in r/min

**In metric units**

$\Delta t$ = time taken in seconds for each step in *dwr* (rad/sec) is given by

$$\Delta t = J(dwr)/T_{av} \tag{3.2}$$

where

$\Delta t$ = time taken for each step in *dwr* (rad/sec) = $2\pi$ (step in r/min)/60

$J$ = total inertia of complete rotor system (motor + load) in drive train kg-m$^2$

$dwr$ = $2\pi$ (step in r/min)/60 in rad/sec

$T_{av}$ = the average accelerating torque during the step in *dwr*, N·m

For the torque–speed curves presented in Figure 3.7, the total accelerating time of the motor and driven load can be calculated using either equation (3.1) or (3.2) via a step-by-step solution as shown in the following example.

A 3-phase, 6600 V, 2230 kW/3000 HP, 6600 V, 3-phase, 60 Hz SCIM with a full-load speed of 710 r/min directly coupled to a forced draft fan has a rotor inertia of 205 kg-m$^2$ (5000 lb-ft$^2$) while the fan inertia is 4210 kg-m$^2$ (100,000 lb-ft$^2$). This is a 10-pole motor, $N_s$ = 720 r/min.

The motor full-load torque ($T_{FL}$) is 30,085 N m (22,203 lbf·ft) and the torque versus speed curves are as shown in Figure 3.7 and the accelerating torques given as percentages of full-load torque.

**Sample calculation at 90% of rated volts**

Metric Units

From the curves in Figure 3.7, the average accelerating torque ($T_{av1}$) at 45% of $N_s$, that is, between the step in speed between 40% and 50% of $N_s$.

From Figure 3.7,

$$T_{av1} = 0.55 \times 30{,}085 = 16{,}561 \text{ N m}$$

$$J = \text{total inertia of the drive} = 205 + 4210 = 4415 \text{ kg-m}^2$$

$$dwr = 2\pi(\text{step in r/ min})/60 = 2\pi\, 72/60 = 7.54 \text{ rad/ sec}$$

$$\Delta t = J(dwr)/T_{av} = 4415(7.54)/16{,}561 = 2.01 \text{ seconds}$$

Imperial Units

$$\Delta t = ((WK^2M + WK^2DE) \times \Delta S)/(308 \times T_a) = \frac{((5000 + 100{,}000)72)}{308 \times 0.55 \times 22{,}203} = 2.0 \text{ seconds}$$

The values for $\Delta t$ using metric and imperial units are equal at 2.0 seconds.

The calculated steps in time ($\Delta t$) using this step by step method for the motor–fan combination, using equations (3.1) and (3.2), are shown in Table 3.1. The results at 90% motor rated voltage with both closed and open fan outlet vanes are given in

**TABLE 3.1 Calculation of Motor/Fan Starting Times with Closed and Open Fan Outlet Vanes with 90% of Motor Rated Volt**

| Speed Change (r/min) | Accel. Torque Closed Vane (N·m / lbf ft) | Accel. Time Closed Vane (sec) | Accel. Torque Open Vane (N·m/lbf ft) | Accel. Time Open Vane (sec) |
|---|---|---|---|---|
| 0–72 | 16,546/12,211 | 2.0 | 16,546/12,211 | 2.0 |
| 72–144 | 18,051/13,322 | 1.8 | 17,450/12,878 | 2.0 |
| 144–216 | 17,750/13,100 | 1.9 | 14,741/10,879 | 2.4 |
| 216–288 | 16,547/12,212 | 2.0 | 15,042/11,101 | 2.3 |
| 288–360 | 16,561/12,212 | 2.0 | 14,440/10,657 | 2.4 |
| 360–432 | 19,556/14,432 | 1.7 | 14,140/10,435 | 2.5 |
| 432–504 | 21,060/15,542 | 1.6 | 15,042/11,101 | 2.3 |
| 504–576 | 24067/17,762 | 1.4 | 15,042/11,101 | 2.3 |
| 576–648 | 26,174/19,317 | 1.3 | 17,450/12,878 | 2.0 |
| 648–710 | | | 25,512/18,872 | 1.2 |
| 648–714 | 36,103/26,644 | 1.0 | | |
| **Total Accel. Time (sec)** | | **16.7** | | **21.4** |

1.0 N·m = 0.738 lbf·ft

Table 3.1 using nine 10% percent motor synchronous speed (72 r/min) change increments and one 66 r/min (closed vane) or 62 r/min. (open vane) speed increments. A comparison of the open and closed vane starting times at 90% of motor rated voltage shows that with closed fan vanes the motor reaches its steady-state operating speed in 5.0 seconds faster than with open fan vanes. This reduced starting time decreases the maximum rotor winding temperature and thus the thermal stresses on the cage winding.

## 3.8 DIRECT-ON-LINE STARTING OF LARGE INDUCTION MOTORS DRIVING CENTRIFUGAL PUMPS

Many high speed centrifugal pumps have fairly low inertias and therefore do not contribute to excessive rotor winding temperatures during starting. The impact of starting duty from such pumps can be further reduced by starting them with a closed discharge valve (see Figure 3.8), which reduces the pump torque values during motor acceleration up to the operating speed. Some pump-set types and applications such as nuclear power plant reactor cooling pumps with slow speed, circulating water pumps have a high inertia to accelerate Eliasen [3.16]. The pump inertia of reactor cooling pump-sets is fairly low, but a cage rotor with a high inertia is required, so that the pump-set will continue to circulate water through the reactor for a short period of time, after a loss of power, to promote cooling by thermosiphoning (i.e., a physical effect, which refers to a method of passive heat exchange based on natural convection, which circulates a fluid without the necessity of a mechanical pump).

This high inertia is achieved by fitting a flywheel to the motor's shaft or by designing the motor with a large enough rotor diameter to achieve the required inertia. Note that axial flow pumps have a higher torque demand with the discharge

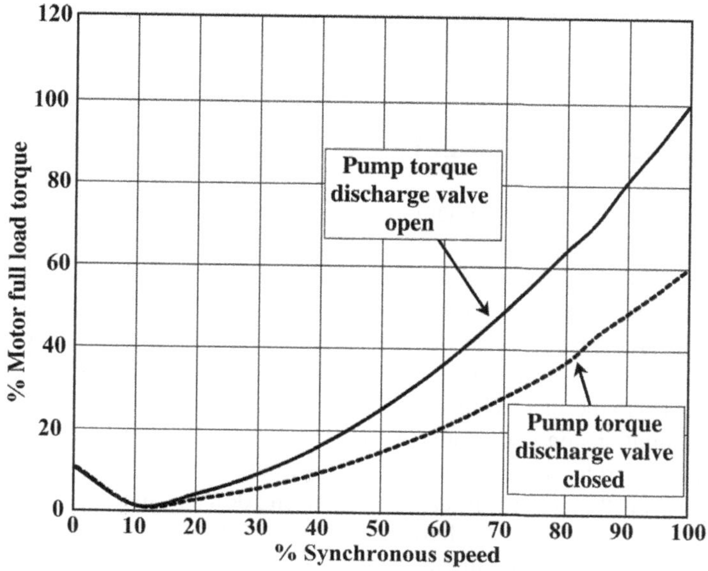


Figure 3.8 Centrifugal pump torque versus speed curves with discharge valve open and closed.

valve closed, which is the opposite effect from that of a centrifugal pump. Also for such applications there are no pump discharge valves to close during motor starting so the same problems of high squirrel cage rotor winding temperatures and the consequential thermal stress during starting discussed in Section 3.8 apply.

Since these are high output motors with ratings in the range of 4167 kW/6000 HP to 9325 kW/12,500 HP the supply voltage during motor starting is often reduced due to their high starting current. This results in significantly longer run-up times which, although the starting current is lower than it would have been had the supply voltage remained at 100%, can cause unacceptably high rotor bar temperatures and increases the risk of cage winding breaks. These motors can be susceptible to rotor winding damage from thermal stresses during commissioning, when trips often occur and repetitive starts are required.

The starting duty for large, slow speed circulating water pumps is less severe than that for reactor cooling pumps, since the pump's torque versus speed curve during pump-set starting can be reduced by closing the discharge valve (see Figure 3.8). However, high maximum rotor bar temperatures can still occur, due to relatively long starting times. Another factor that can affect the maximum temperature of a cage winding is due to DOL starts of SCIMs driving pumps and compressors on offshore oil platforms with relatively small capacity, isolated power supplies derived from gas turbine generators. These starts can cause a considerable drop in voltage at the motor's terminals. Large motors used in such plants may reduce the supply voltage to levels at or below 80% of rated voltage during starting. This requires motors with higher starting torques that often require deep narrow, or inverted "T" shaped bars, which will have higher maximum temperatures at their tops during motor starting. Such motors are often subjected to repetitive starts when problems occur, since oil production loss is very costly. This, of course, makes the rotor cage windings in such motors more susceptible to failure from cyclic thermal and mechanical stresses.

## 3.9 LIMITATIONS ON REPETITIVE MOTOR STARTS

Motor manufacturers can set limitations on the number of permissible consecutive motor starts, if they are given the torque versus speed curve, driven equipment inertia, and minimum motor terminal voltage during starting. From this information the rotor winding temperature rise during a start, the number of permissible consecutive starts and the time between a second start from a hot rotor winding condition and subsequent starts can be calculated. Obviously if the motor is directly started when connected to a high inertia mechanical load, such as an induced draft fan, there will be significant limitations on consecutive starts. This is illustrated in Figure 3.9, which demonstrates that rotor winding heating during starting is further compounded by repetitive starts. If the motor trips and, after coasting down in speed for a short time, is restarted, the winding will not have cooled to ambient temperature and therefore its maximum temperature and thermal stresses will be significantly higher after the second start [3.16].

If a motor is specified to meet NEMA MG1 requirements, then Section 20.12 [3.19] of this standard specifies the following minimum starting requirements.

### 3.9.1 Starting Capability

Squirrel cage induction motors (or induction generators specified to start and accelerate a connected load) shall be capable of making the following starts, providing the

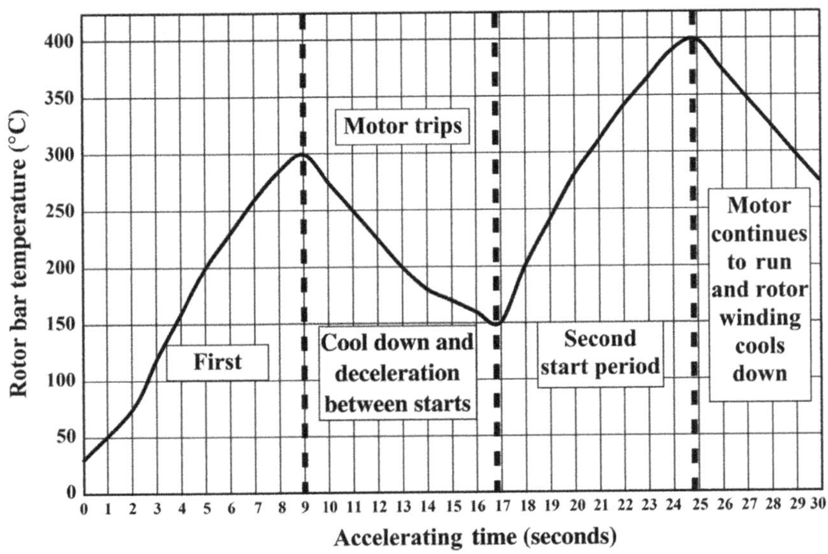


Figure 3.9 Increased squirrel cage winding temperature during direct-on-line starting of high inertia fans due to repeated starts.

$WK^2$ (inertia) of the load, the load torque during acceleration, the applied voltage, and the method of starting are those for which the motor was designed.

**(i)** Two starts in succession, coasting to rest between starts, with the motor initially at ambient temperature.

**(ii)** One start with the motor initially at a temperature not exceeding its rated load operating temperature.

### 3.9.2 Additional Starts

If additional starts are required, it is recommended that none be made until all conditions affecting operation have been thoroughly investigated and the apparatus has been examined for evidence of excessive heating. It should be recognized that the number of starts should be kept to a minimum since the life of the motor is affected by the number of starts.

### 3.9.3 Additional Name Plate Data

When requested by the end user, a separate starting information plate should be supplied on the motor with the starting specification with respect to allowed number of sequential starts from cold, time span between sequential starts when hot, and so on.

## 3.10 CRITERIA FOR DESIGN OF SQUIRREL CAGE ROTOR WINDINGS

The suitability of a particular cage winding design for a specific application depends on

**(i)** The total inertia of the motor and driven equipment rotating parts.

**(ii)** The driven equipment torque versus speed curve during starting.

**(iii)** The motor torque versus speed curve at the minimum supply voltage during motor starting.

**(iv)** Motor consecutive starting requirements.

**(v)** The squirrel cage rotor winding mechanical configuration.

**(vi)** The rotor bar material and shape.

### 3.10.1 Total Motor and Driven Equipment Inertia

As discussed in Section 3.7, a large total inertia to be accelerated by the motor can result in excessive temperatures at the bar tops due to "skin effect" therefore it is important that the motor manufacturer is provided with the driven equipment inertia. Some guidance on the maximum driven inertia, that a standard motor design can accelerate, without its windings reaching an injurious temperature, is given in NEMA MG1, Section 20.11, [3.17]. For standard torque motors driving loads such as centrifugal fans and pumps requiring the following torque characteristics the maximum drive equipment inertia is calculated by formula 3.3.

**(i)** Locked rotor torque: 60% of motor full-load torque

**(ii)** Pull-up torque: 60% of motor full-load torque

**(iii)** Breakdown torque: 175% of motor full-load torque

Maximum driven equipment inertia

$$(WK^2) = A[HP^{0.95}/(RPM/1000)^{2.4}] - 0.0685[HP^{1.5}/(RPM/1000)^{1.8}] \quad (3.3)$$

This formula is from NEMA MG1 motors and generators [3.17], which is a current USA standard used by electrical machine manufacturers and users of 3-phase induction motors, and the inertia $WK^2$ is in lb-ft$^2$.

where

A = 24 for motors with synchronous speeds between 300 and 1800 r/min.

A = 27 for motors with synchronous speeds of 3600 r/min (although not stated, this factor should also apply to motors with synchronous speeds of 3000 r/min).

For high torque motors with the following characteristics, this section of NEMA MG1 indicates that the maximum driven equipment inertia would be 50% of the value obtained from formula 3.3 with

**(i)** Locked rotor torque: 200% of motor full-load torque.

**(ii)** Pull-up torque: 150% of motor full-load torque.

**(iii)** Breakdown torque: 190% of motor full-load torque.

### 3.10.2 Driven Equipment Torque–Speed Curve During Starting

Motor starting time and rotor bar maximum temperature during a start can be reduced if the load torque versus speed curve profile can be decreased by

**(i)** Starting centrifugal pumps and compressors with their discharge valves closed.

**(ii)** Starting centrifugal fans with their discharge dampers closed or axial flow fans with dampers open.

**(iii)** Starting crushers and ball mills, when they are empty.

When purchasing a motor, it is most important that the motor manufacturer be given details of the driven equipment torque versus speed curve during motor starting.

### 3.10.3 Motor Torque–Speed at Minimum Supply Voltage During Starting

The motor manufacturer should be advised of the minimum supply voltage to the motor during its starting period. From this the motor manufacturer can ensure that there is an adequate margin (at least 10% of motor full-load torque) between the motor and driven equipment torque versus speed curves during the starting period.

### 3.10.4 Motor Consecutive Starting Requirements

Unless the inertia to be accelerated is extremely high, all motors should be capable of at least two consecutive starts from cold or one from hot and then subsequent starts after a time interval, which can be calculated by the motor manufacturer. If the requirements stated in NEMA MG1, Section 20.12 are included in the "Invite to Tender" (ITT) technical specification, a motor with this starting capability should be provided. There are some applications, such as power plant service and instrument air compressor drives, which require the motor to be started when the system pressure decreases to a pre-determined level. The starting frequency for such motors needs to be specified when they are being purchased. The limiting factor for motors that are started on a frequent basis is usually the total rotor rotating component inertia of the motor and driven equipment as discussed earlier, for example, high speed service water centrifugal pumps and their drive motors used in power plants have a low inertia, so their start time is of the order of 1–2 seconds and so rotor winding heating during this time is minimal, which means that such motors can be frequently started.

### 3.10.5 Squirrel Cage Rotor Mechanical Configuration

The mechanical configuration of the end winding sections of a squirrel cage winding should be based on the mechanical and thermal stresses imposed on them during starting (see Sections 3.1–3.3) and these can be determined by the motor manufacturer by finite element analysis, which can provide

(i) Suitability of the selected bar, material and shape (see Section 3.10.4).

(ii) Length of bar between the end of the rotor core and end ring.

(iii) End ring size and the need for retaining rings with high yield and tensile strengths (see Table 3.2) over the end rings or complete end winding can be determined.

As discussed in Section 3.3 and illustrated in Figure 3.3, fabricated aluminum winding designs with no bar extensions beyond the end of the stator core should not be used for applications where the thermal and mechanical stresses in the cage winding are high during starting.

**TABLE 3.2 An Indication of Material Properties Used in Cage Windings**

| Rotor Bar Material | Yield Strength (lbf/in$^2$)/(MN/m$^2$) | Tensile Strength (lbf/in$^2$)/(MN/m$^2$) | Melting Point (°F/°C) |
|---|---|---|---|
| Aluminum alloy, e.g. 60161-T6 | 34,809/240* | 42,000/290* | 1150/621 |
| Cupro Nickel | 18,855/130 | 50,763/350 | 2264/1240 |
| Naval Brass UNS C46400 | 46,000/317 | 69,000/476 | 1625/885 |
| Aluminum–Bronze (retaining rings) | 60,000/414 | 110,000/759 | — |
| ASTM A286 stainless steel (retaining rings) | 100,000/690 | 146,000/1007 | — |

*at 100°C/212°F

### 3.10.6 Rotor Bar Material and Shape

Aluminum rotor bars have only about 33.3% of the density weight and 2.5 times the specific heat of copper bars. The coefficient of thermal expansion for a given temperature change is 35% greater in aluminum than that of copper and in addition, as can be seen from Table 3.2, the aluminum alloy 60161 T6 has a tensile strength of 83% of that of Cupro Nickel. These material density, specific heat, and stress withstand capability differences from those of copper result in aluminum bar maximum temperatures and thermal expansion being greater than those of copper. For comparably sized bars, these in turn result in much higher cage winding stress being developed in windings made from aluminum, while accelerating the same rotor system inertia. Also, as indicated in Table 3.2, copper has a much higher melting point than aluminum. For these reasons aluminum cage windings should not be used in motors that directly start rotor systems plus load, with a high inertia.

For some high inertia driven applications, copper alloy bars and end rings are used because their higher resistivity produces high motor torques during starting.

Bar shape selection should also be carefully considered in the determination of a suitable cage winding design for a particular application. For centrifugal pumps and fans, a reasonably wide rectangular bar (see Figure 2.5a) provides adequate accelerating torque and the lowest maximum temperature reached by the rotor winding, during motor starting of low and high inertia rotor systems. Of course, as indicated earlier, aluminum bars should not be used if the inertia to be accelerated is high. Inverted "T" shaped (see Figure 2.5c) and double cage windings (Figure 2.5d) should never be used in applications with high inertia rotating systems. NEMA MG1 gives some guidance on how to assess whether the rotational (motor and driven equipment rotors) have a high inertia. Also, narrow and deep rectangular bars may not be suitable for high inertia applications due to the very high temperatures that are likely to develop at the tops of the bars due to "skin effect."

## 3.11 SAMPLES OF BREAKS IN SQUIRREL CAGE ROTOR WINDINGS

Breaks in squirrel cage windings can occur in the rotor bars, end rings, and joints between the two. The location usually depends on where the highest combined thermal and rotational mechanical stresses are located. Another cause of such breaks is poor manufacturing. For fabricated windings, the most common manufacturing problems that cause breaks are poorly brazed or welded joints between the bars and end rings (Figure 3.10) and loose bars in the rotor core slots. The most common manufacturing defect in die-cast cage windings is voids in the casting, which are most commonly found at the interface between the bars and end rings (Figure 3.13).

The photographs in Figures 3.10–3.21 illustrate breaks in cage windings and, where possible, some of their likely causes. Note that it has to be appreciated that a root cause failure analysis (RCFA) is certainly not carried out in all cases of cage winding faults, in fact RCFAs are in the minority. Also, the case histories presented in Chapters 5–9 on the application of MCSA to diagnose cage winding breaks illustrate numerous and different types of breaks in cage windings.

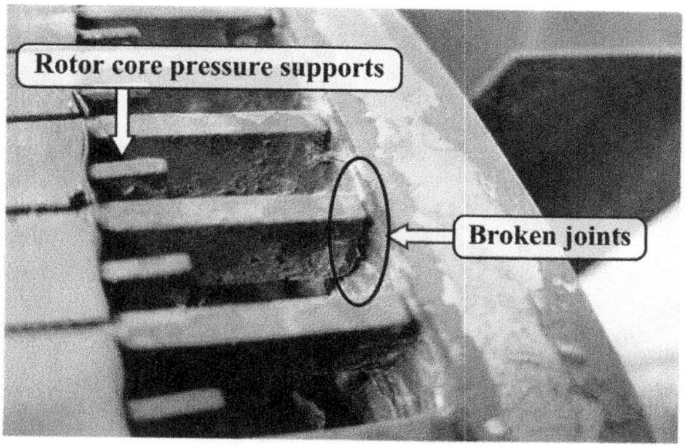


Figure 3.10 Failed joint in a copper cage in a SCIM driving an ID high inertia fan, caused by thermal and mechanical cyclic stresses. Reproduced with permission of EASA, USA.

Figure 3.11 The fractured bars were staked to prevent axial movement. Thermal expansion of the cage exerted a tensile stress, which eventually led to the staked bars fracturing. Reproduced with permission of EASA, USA.

Figure 3.12 Cracked end ring in die-cast aluminum cage winding. Reproduced with permission of EASA, USA.

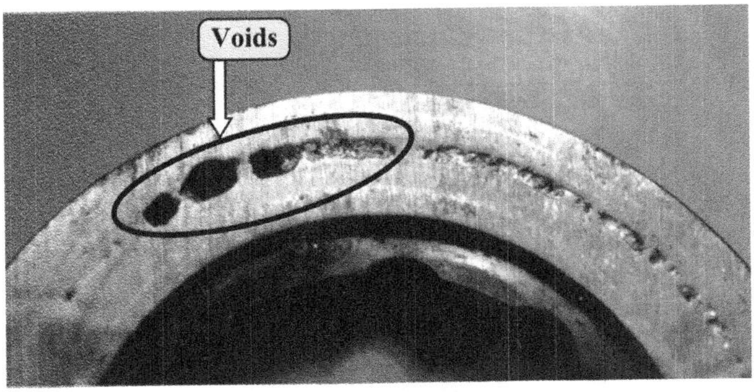


Figure 3.13 Voids in die-cast aluminum cage winding. Reproduced with permission of EASA, USA.

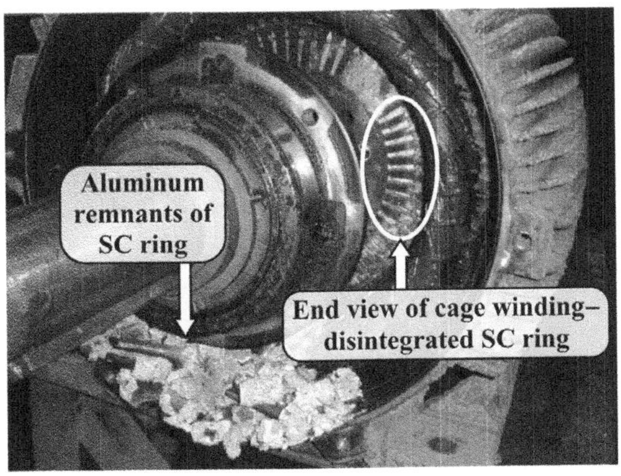


Figure 3.14 Catastrophic failure of die-cast aluminum cage winding, overheated while running caused by overload, too many sequential starts, high ambient temperature. Reproduced with permission of EASA, USA.

Figure 3.15 Fabricated aluminum cage winding bar breaks in slot region, due to absence of airflow through the ducts during sequential starts. Reproduced with permission of EASA, USA.

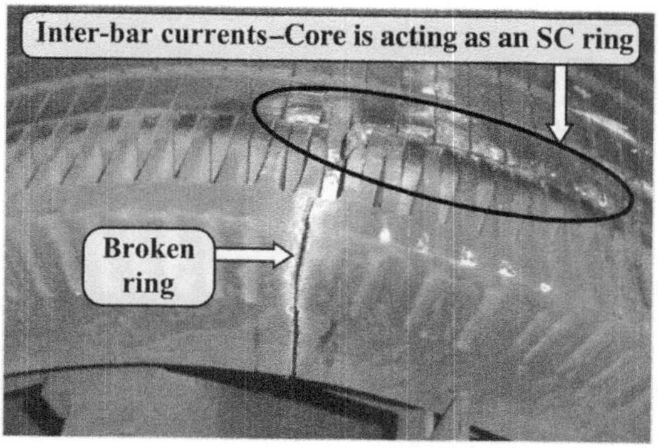


Figure 3.16 Crack in a cage end ring of a SCIM driving an ID fan motor, causes are due to cyclic thermal and mechanical stresses. Photo taken by Ian Culbert. Reproduced with permission of Iris Power Qualitrol.

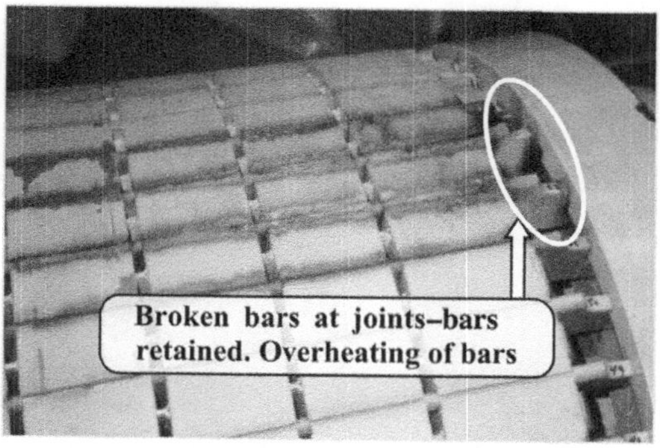


Figure 3.17 A 2-pole boiler feed pump motor cage winding with two broken bars at edge of retaining ring. The core laminations are burnt due to the flow of inter-bar currents through the core from the broken bars to adjacent healthy ones. Reproduced with permission of EASA, USA.

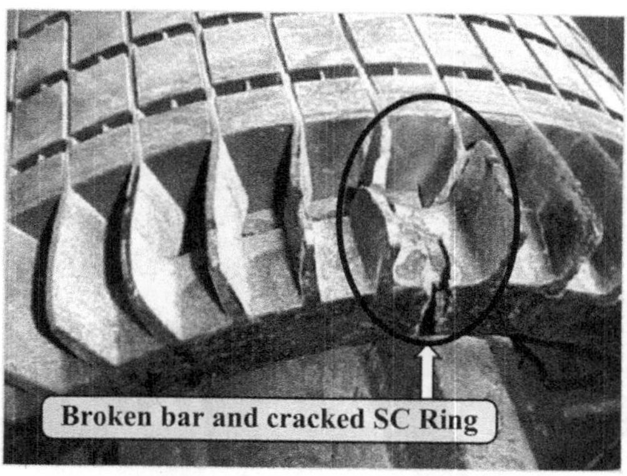


Figure 3.18 Fabricated copper cage winding with breaks in bars and end ring. Reproduced with permission of EASA, USA.

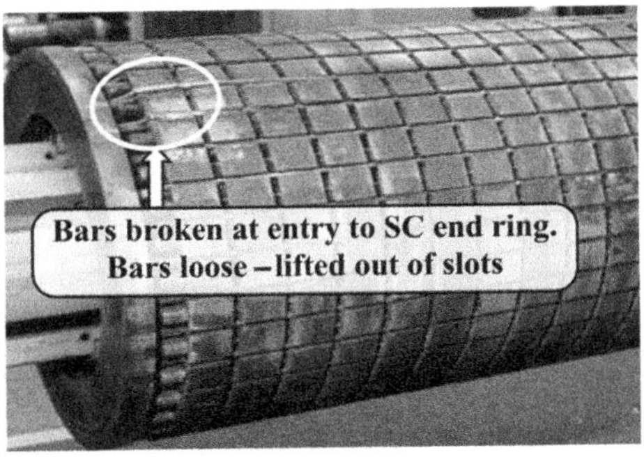


Figure 3.19 Fabricated copper cage winding with two broken bars, which have lifted out of the slots, caused by too many sequential DOL starts, bars were not originally tight enough in the slots. *Source:* Thomson and Fenger [3.18]. Reproduced with permission of IEEE.

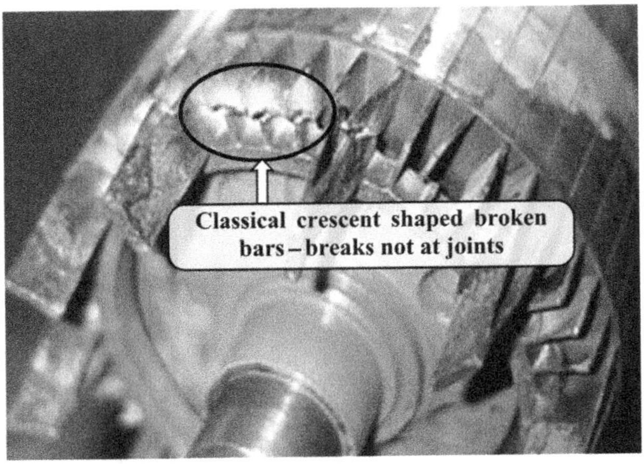


Figure 3.20 Fabricated copper cage winding, which was not fit for purpose due to the deep thin bars experiencing very high temperatures during frequent DOL starts. The motor was driving a crane on an offshore platfrom and during very cold conditions the lubrication system became "*very stiff*" and caused long run-up times due to the increased torque demands during run-up.

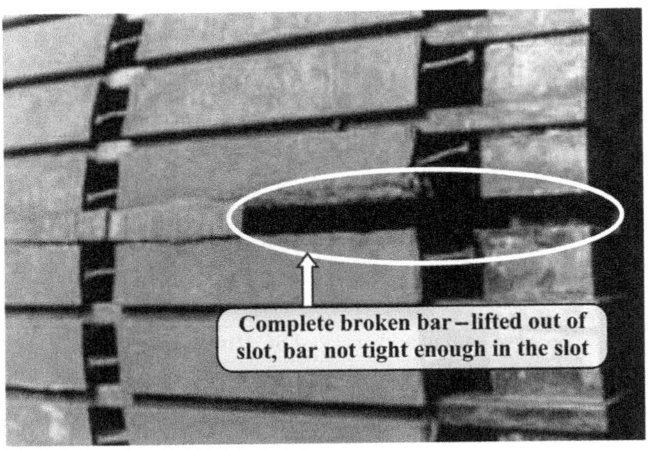


Figure 3.21 Fabricated copper cage winding with one completely broken bar that has lifted out of its slot. The causes were too many DOL starts enhanced by this bar not being tight enough in the slot.

Note that a truly broken rotor bar exists when the bar is mechanically broken through its full depth and there is no electrical conducting path whatsoever between the end of the broken bar and the short circuiting ring.

## 3.12 QUESTIONS

**3.12.1** What additional information should the end user include in an invite to tender (ITT) to the OEM to supply a brand new 3-phase, 13.8 kV, 6800 kW/9115 HP, 326 A, 3570 r/min, 60 Hz, class F, p.f. = 0.89, efficiency = 97.2%, star connected SCIM, which will be used to drive a centrifugal pump on an offshore oil and gas production platform?

**3.12.2** Explain why too many DOL starts can be a major cause of broken rotor bars or cracked end rings in copper fabricated cage windings when the end user does *not adhere* to the starting specifications set by the OEM for a large, HV SCIM.

**3.12.3** Why can repetitive DOL starts of copper fabricated cage rotors in large, HV SCIMs used to drive high inertia loads such as FD fans in fossil fuel power stations be a cause of broken rotor bars in these drive trains?

**3.12.4** What manufacturing procedures can be taken to ensure rotor bars are tight in the slots to prevent unacceptable levels of vibration of the bars?

**3.12.5** What method do you consider to be the most reliable for ensuring rotor bars are tight during all operating conditions?

**3.12.6** What type of load would an inverted T shaped bar in a cage rotor possibly be used for and definitely not be used for?

**3.12.7** Comment on the undesirable effects of oscillating currents (as shown in Figure 3.6) supplied to a SCIM due to the mechanical load characteristics.

**3.12.8** Reciprocating compressors cause fluctuations of the supply currents to a SCIM. Why should these fluctuations be reduced and how is this achieved?

**3.12.9** What undesirable effects do severely and often randomly fluctuating loads, caused by rock or coal crushers have on copper fabricated cage windings? What design features are used to cater for such mechanical load characteristics?

**3.12.10** Explain why longer run-up times at lesser currents, due to a dip in the supply voltage caused by a DOL start, can increase the rotor temperature more than higher currents for a much shorter time.

## REFERENCES

[3.1] A. H. Bonnett and G. C. Soukup, "Rotor failures in squirrel cage induction motors," *IEEE Transactions on Industry Applications*, vol. IA-22, no. 6, 1986, pp. 1165–1173.

[3.2] A. H. Bonnett and G. C. Soukup, "Analysis of rotor failures in squirrel-cage induction motors," *IEEE Transactions on Industry Applications*, vol. 24, no. 6, 1988, pp. 1124–1130.

[3.3] A. H. Bonnett and G. C. Soukup, "Cause and analysis of stator and rotor failures in three-phase squirrel-cage induction motors," *IEEE Transactions on Industry Applications*, vol. 28, no. 4, 1992, pp. 921–937.

[3.4] A. H. Bonnett, "Root Cause Failure Analysis for AC Induction Motors in the Petroleum and Chemical Industry," IEEE PCIC 2010, San Antonio, TX, September 20–22, 2010, pp. 1–13.

[3.5] W. R. Finley and M. M. Hodowanec, "Selection of Copper vs. Aluminum Rotors for Induction Motors," IEEE PCIC 2000, San Antonio, TX, 2000, pp. 187–197.

[3.6] R. S. Curiac, "Forces and Stresses in Squirrel Cage Motors During Starting," IEEE PPIC 2008, Seattle, WA, June 22–27, 2008, pp. 7–14.

[3.7] R. Yabiku, R. Fialho, L. Teran, A. Santos, E. Rangel, and D. Dutra, "A Comparative Study Between Copper and Aluminum Induction Squirrel Cage Constructions," IEEE PCIC 2010, San Antonio, TX, September 20–22, 2010, pp. 1–9.

[3.8] I. M. Culbert and W. Rhodes, "Using current signature analysis technology to reliably detect cage winding defects in squirrel-cage induction motors," *IEEE Transactions on Industry Applications*, vol. 43, no. 2, March/April 2007.

[3.9] P. L. Alger, *Induction Machines-Their Behavior and Uses*, 2nd edition, Gordon and Breach Science Publications, published by OPA Amsterdam, 3rd printing with additions, 1995.

[3.10] M. M. Liwschitz-Garik, "Computation of skin effect in bars of squirrel-cage rotors [includes discussion]," *Power Apparatus and Systems, Part III. Transactions of the American Institute of Electrical Engineers*, vol. 74, no. 3, 1955.

[3.11] T. Griffith, J. Krukowski, "Synchronous motors for compressors," *IEEE Industry Applications Magazine*, vol. 20, no. 6, 2014, pp. 14–23.

[3.12] M. Perrin, G. Kohn, S. Mugford, and G. Seggewiss, "Induction Motors, Reciprocating Compressors and Variable Frequency Drives," IEEE 1997 PCIC, Banff, Alberta, September 15–17, 1997, pp. 1–9.

[3.13] R. A. Hanna, W. Hiscock, and P. Klinowski, "Failure Analysis of Three Slow Speed Induction Motors for Reciprocating Load Applications," IEEE 2005 PCIC, Denver, CO, September 12–14, 2005, pp. 239–245.

[3.14] J. J. Middlemess, "Current Pulsation of Induction Motor Driving a Reciprocating Compressor," IEEE Proceedings of the Institution of Electrical Engineers, vol. 121, no. 11, November 1974, pp. 1399–1403.

[3.15] J. H. Dymond, "Stall time, acceleration time, frequency of starting: the myths and the facts [electric motors]," *IEEE Transactions on Industry Applications*, vol. 29, no. 1, Part 1, 1993, pp. 42–51.

[3.16] A. N. Eliasen, "High-inertia drive motors and their starting characteristics," *IEEE TPAS*, vol. PAS-99, no. 4, July/August 1980, pp. 472–1482.

[3.17] NEMA MG1, Motors and Generators.

[3.18] W. T. Thomson and M. Fenger, "Current signature analysis to detect induction motor faults," *IEEE Industry Applications Magazine*, vol. 7, no. 4, July/August, 2001, pp. 26–34.

## FURTHER READING

[3.19] R. L. Nailen, "Large motor starting problems in the petroleum industry," *IEEE Transactions on Industry and General Applications*, vol. IGA-5, no. 4 1969, pp. 422–427.

[3.20] R. L. Nailen, "New rotor design concept solves pipeline motor acceleration," *IEEE Transactions on Industry Applications Problem*, vol. IA-9, no. 2, 1973, pp. 201–205.

[3.21] M. Thirugnanasambandamoorthy, R. Krusemark, and C. Mouton, "Recommendations for Design and Protection of Induction Motors for Extruder Applications," IEEE 2005 PCIC, Denver, CO, September 12–14, 2005, pp. 233–238.

[3.22] A. H. Bonnett, "The benefits of allowing for increased starting current in AC squirrel cage induction motors," *IEEE Transactions on Industry Applications*, vol. 27, no. 6, 1991, pp. 1169–1174.

[3.23] H. Barr, A. H. Bonnett, and C. Yung, "Understanding the Design of Stators and Rotors of Squirrel Cage Induction Motors," 55th IEEE 2008 PCIC, Cincinnati, OH, September 22–24, 2008, pp. 1–11.

[3.24] A. H. Bonnett and T. Albers, "Squirrel Cage Rotor Options for AC Induction Motors," IEEE Annual Pulp and Paper Industry Technical Conference, Atlanta, GA, June 19–23, 2000, pp. 54–67.

[3.25] J. H. Dymond and R. D. Findlay, "Some commentary on the choice of rotor bar material for induction motors," *IEEE Transactions on Energy Conversion*, vol. 10, no. 3, 1995, pp. 425–430.

[3.26] K. K. Schwarz, "The Design of Reliable Squirrel-Cage Rotors," Publication No 189, Laurence, Scott & Electromotors Ltd., Norwich, England.

CHAPTER 4

# MOTOR CURRENT SIGNATURE ANALYSIS (MCSA) TO DETECT CAGE WINDING DEFECTS

*William T. Thomson*

## 4.0 SUMMARY

The focus of this chapter is the provision of fundamental knowledge on MCSA in a style suitable for electrical, mechanical, and condition monitoring engineers working in industry. Therefore, its content is deliberately biased toward industrial applications and is not highly theoretical. In terms of SCIMs, the effects on the magnetic field of cage winding breaks within the motor, which result in voltages at frequencies of $\{f(1 \pm 2s)\}$ Hz being induced in each phase of the stator winding, are presented. These induced voltages drive currents in the stator, which in turn modulate the supply current at twice slip frequency ($2sf$) around the supply component ($f$) and frequency components at $\pm 2sf$ around $f$ exist in the Fourier spectrum.

A brief overview of signal processing, necessary for spectrum analysis of the supply current and the detection of the twice slip frequency sidebands around the supply frequency component, is presented. A typical specification for a spectrum analyzer of sufficient sensitivity is included, together with practical information to allow the practicing engineer to complete the requisite tests. Using the magnitudes of the $\pm 2sf$ components relative to the magnitude of the supply component and knowing the number of rotor bars, a review of the formulae for estimating the condition of the cage winding is presented. A discussion is also presented on the application of using only the dB amplitude differences between the $\pm 2sf$ sidebands and the supply component to assess the condition of the cage winding, when the number of rotor bars is unknown.

## 4.1 INTRODUCTION

There are now hundreds of published papers on the topics of "*broken rotor bars*" and "*motor current signature analysis*" with respect to SCIMs, for example, a basic

*Current Signature Analysis for Condition Monitoring of Cage Induction Motors: Industrial Application and Case Histories*, First Edition. William T. Thomson and Ian Culbert.

literature search using these two "*generic*" topics via the IEEE Xplore Digital library listed 671 and 488 publications, respectively. Based on the authors' combined knowledge and experience of 108 years, only a selection of what they consider to be key publications relevant to this chapter are referenced. It has been known since the 1920s that an asymmetrical rotor winding, whether it be in a cage or slip-ring 3-phase induction motor, will induce a voltage in the stator winding at a frequency of $f(1-2s)$ Hz and thus drive a current at that frequency in the stator. This was initially verified theoretically in books by, for example, Kron [4.1], Jones [4.2], and Vickers [4.3], using equivalent circuit analysis methods during steady state operation of the motor at a constant speed. Because these were solely electrical circuit models only one current component at $f(1-2s)$ emerged from the analysis. This component is correctly referred to as a twice slip frequency sideband ($-2sf$) below the supply component, $f$. These equivalent circuit models did not predict the absolute magnitude in amperes of this component as a function of the motor's design parameters together with a number of, for example, one or more broken rotor bars in a cage winding. During these early years of the twentieth century, digital signal processors had not been invented and it was impossible to accurately measure the frequency and magnitude of the current component at $f(1-2s)$ Hz.

In 1982, Williamson and Smith [4.4] published a landmark paper, which modeled a 3-phase SCIM using advanced circuit analysis for the rotor cage with one broken rotor bar and a balanced stator winding and 3-phase supply that included the required design parameters. A theoretical study was presented [4.4] using the design specification for a 3-phase, star connected, 2300 V, 186.5 kW (250 HP), 60 Hz, 8-pole SCIM taken, by Reference 4.4 and acknowledged, from a book by Liwschilz-Garick and Whipple [4.5]. The full-load current and corresponding speed were not given in Reference 4.4. The magnitude in amperes of the $f(1-2s)$ component for this particular motor and design was computed with one broken rotor bar, as shown in Figure 4.1.

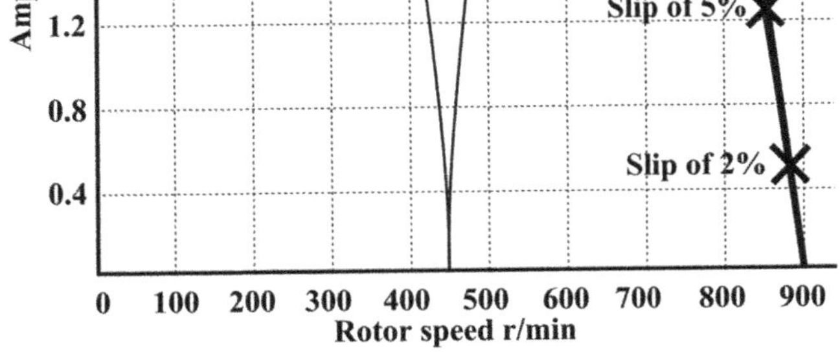


Figure 4.1 Variation of amplitude of $f(1-2s)$ component of stator current against rotor speed. *Source:* Williamson and Smith [4.4]. Reproduced with permission of IET.

The curve was redrawn by Thomson (author of this chapter) with the addition of slip values at 2% and 5% and the line which passes through these points. Examination of Figure 4.1 shows that up to a slip of 5% for this particular design and rating of motor, the $f(1-2s)$ component in amperes versus speed relationship is linear. The positions on the curve corresponding to a slip of 2% (882 r/min) and 5% (855 r/min) have been added to show a typical operating range on load for this motor. The magnitude of the current component at $f(1-2s)$ from Figure 4.1 at a slip of 2% is 0.5 A. By extrapolation, using the linear relationship, the current at a slip of 5% is 0.5(5/2) which equals 1.25 A and from the actual curve the current is 1.3 A. This is obvious but is presented to demonstrate that, when a measurement of the lower sideband $f(1-2s)$ in amperes $\{I_{LSB}\}$ is made at an operating slip ($s_o$), which is below the full-load slip ($s_{FL}$), the magnitude $\{I_{LSBr}\}$ of the lower sideband referred to the full-load slip is given by:

$$I_{LSBr} = I_{LSB}(s_{FL}/s_o)$$

Although simple, this is an obvious and very useful correction factor, as it enables compensation to be made, when the motor is not running at full-load slip and rated current.

Where

$I_{LSB}$ = magnitude of $f(1-2s)$ component at any slip below full-load slip, amperes.

$I_{LSBr}$ = magnitude of $f(1-2s)$ component referred to the full-load slip, amperes.

$s_{FL}$ = full-load slip using the rated speed, and frequency from the motor's nameplate.

$s_o$ = operating slip at which the current was actually measured.

In practice, as demonstrated by the numerous industrial case histories in Chapters 5–9, the motors are very often operating at less than full-load current and slip and the end user will not change the load merely to carry out an MCSA test at full-load. The paper by Williamson and Smith [4.4], did not include a comparison between the predicted current at $f(1-2s)$ from their model, with one broken rotor bar and a measured current spectrum from an actual motor of the same design with one broken rotor bar. Of course it is accepted that they did not have access to carry out experimental tests on a 186.5 kW SCIM, which they had theoretically modeled. Nevertheless, the paper 4.4 provided very important new knowledge, hitherto not available, on the magnitude of the $f(1-2s)$ current component, as a function of the operational speed, in a SCIM with one broken rotor bar. It has to be recognized that this was for one motor, of a given design and the design specifications were required for the analysis. However, in Section 4.6 it will also be proven via experimental tests, not previously published by Thomson (author) that a linear relationship can be used to provide a good estimate of the magnitude of the $f(1-2s)$ sideband as a function of slip with one broken rotor bar. The practical results used to support that theory were taken at different load currents and slips from a commercially produced SCIM but with different rotors having different numbers of broken rotor bars, see Section 4.5.2.

It is impossible to review all the key papers on MCSA used for the detection of broken rotor bars and the reader is referred to publications from the following research work. In the early 1980s, research and development funded by The Electrical Power Research Institute (EPRI) in the United States was being carried out for the detection of broken rotor bars via MCSA, which led to the development of a prototype MCSA instrument for the detection of broken rotor bars as reported by Kliman et al. [4.6,4.7]. In 1980, Thomson (author) commenced his research and development work on MCSA for the detection of broken rotor bars in large induction motors. This research was funded by oil companies and power utilities and consequently the focus was on experimental tests and exhaustive field trials and numerous papers have been published by Thomson et al. as already referenced in Chapter 1, References 1.6–1.35.

## 4.2 DERIVATION OF CURRENT COMPONENT AT $f(1-2s)$

Without using circuit analysis techniques and to help industrial engineers understand the fundamental concepts a derivation is now presented using, as a starting point, the basic equations and induction motor concepts already presented in Section 1.2, to explain why a rotating flux wave with respect to the stationary stator at a frequency of $f(1-2s)$ is produced.

Recall the frequency of the rotor currents is given by

$$f_2 = (N_s - N_r)\,p/60 = sN_s p/60 \text{ Hz}$$

$f_2 = sf$ = slip frequency is the definition for a specific electrical quantity, Hertz.

When rotor cage asymmetry occurs due to a broken rotor bar this produces a backward rotating field defined as $N_{sb}$ at a speed of $sN_s$ in the opposite direction from the forward rotating field produced by the rotor currents with respect to an observer sitting on the rotating rotor as shown in Figure 4.2.

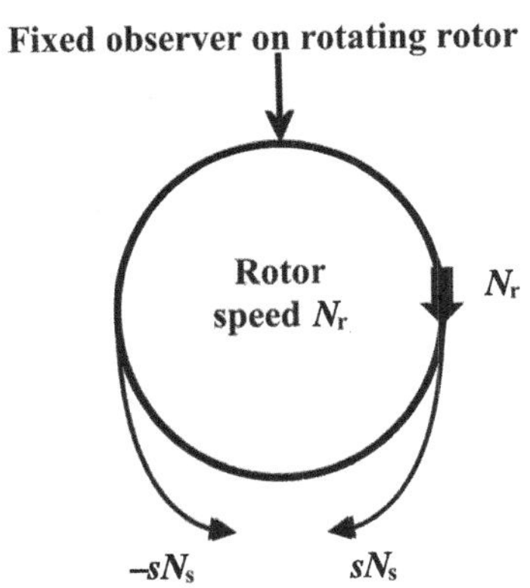


Figure 4.2 Forward and backward rotating fields when a broken bar exists.

With respect to the stationary stator winding, the backward rotating magnetic field speed due to a broken rotor bar is given by

$$N_{sb} = N_r - sN_s \text{ and } N_{sb} = N_s(1 - s) - sN_s = N_s(1 - 2s)$$

The stationary stator winding now sees a rotating field at $N_{sb} = N_s(1 - 2s)$, in terms of frequency, (recall $f = N_s p/60$) speed of rotating magnetic field in r/min, and number of pole-pairs this gives a rotating magnetic field at a frequency given by

$$f_{sb} = (N_s - 2sN_s)\,p/60 = ((60f/p) - 2s(60f/p))p$$

$$f_{sb} = f(1 - 2s) \text{ Hz} \tag{4.1}$$

This means that due to a broken rotor bar, a rotating magnetic field at $f(1 - 2s)$ is produced, which links the stator windings and induces an emf which drives a current through the stator and can therefore be detected via spectrum analysis of the supply current to the motor. This frequency component ($f_{sb}$) is referred to as a twice slip frequency sideband at $2sf$ down from $f$.

## 4.3 REASONS FOR CURRENT COMPONENT AT $f(1 + 2s)$

The cyclic variation of current caused by a broken rotor bar produces a torque variation at twice slip frequency and this produces a speed variation which is a function of the drive train inertia. This normally causes a reduction in magnitude of the $f(1 - 2s)$ current component and a new current component appears at $f(1 + 2s)$ and its magnitude may be enhanced by modulation of the third time-harmonic flux in the stator.

The larger the inertia of the drive train the greater the resistance to the torque and speed oscillation at $2sf$ and thus the smaller the magnitude of the upper sideband at $+2sf$ compared to the lower sideband $-2sf$ around $f$, the supply component. Thus cage winding breaks produce two sidebands at $\pm 2sf$ around $f$, as illustrated in Figure 4.3 and given by equation (4.2)

$$f_s = f \pm 2sf \tag{4.2}$$

At this point, it is relevant to mention that certain suppliers (not all), of MCSA instruments refer to the twice slip frequency sidebands simply as "*pole-pass frequencies*" and make no reference to fundamental induction motor theory, nor justify the use of this terminology. None of the papers referenced in this book use the term pole-pass frequencies to describe twice slip frequency sidebands, nor do the authors of this book.

The term "*pole-pass frequencies*" is only included for completeness, because the reader may see it elsewhere. It conveys no information with respect to the actual current spectrum, whereas the term "*twice slip frequency sidebands*" around the supply component details exactly their position in the frequency spectrum. Every effort should be made to discourage the use of the expression pole-pass frequencies, which seems to be used and promoted by condition monitoring companies.

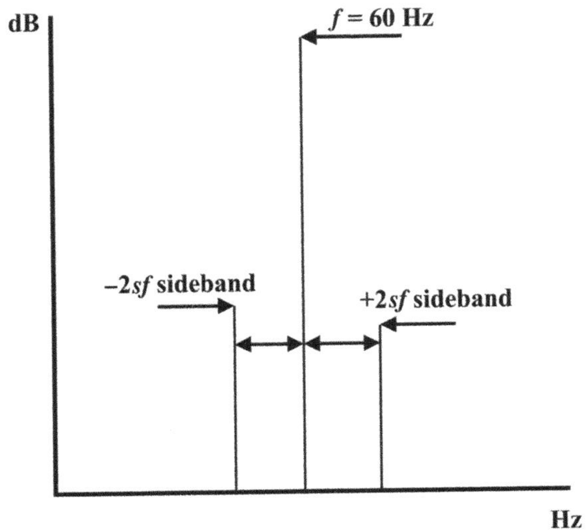


Figure 4.3 Illustration of $\pm 2sf$ sidebands around supply component.

Equivalent circuit models as explained in Section 4.1 have previously not taken account of the rotor torque and speed variation and thus only the $f(1 - 2s)$ component was predicted. Fillipetti et al. [4.8] in an excellent paper also proposed a dynamic model of a 3-phase induction motor that included the torque and speed oscillations and also the rotor's inertia. Paterson [4.9], Watson et al [4.10] and Thomson et al [4.11] also verified the influence of these parameters. They verified the influence of these parameters on the magnitudes of the twice slip frequency sidebands around the supply component.

The majority of classical research papers, which have been published with respect to the measurement of the $\pm 2sf$ sidebands from a motor with one or more broken bars have used small power LV SCIMs coupled to a load dynamometer. The broken bar or bars have been produced by drilling holes in die-cast aluminum rotors, next to an end ring or by machining the bar, next to the inside of the end ring, in a copper fabricated rotor. Comparisons were made between the measured $\pm 2sf$ sidebands and predicted magnitudes but many of the publications only presented results at full-load current and slip. However, in industry it is very often the case that SCIMs are operating at less than full-load output and slip. Also, it has to be recognized, as clearly demonstrated in the industrial case histories in this book, that the type of cage defects, which occur in industrial usage, can be much more complex than the clean breaks of bars which have been used in research studies. The use of small power, LV SCIMs up to a maximum of, for example, 30 kW is perfectly understandable, and not a criticism, but a statement of fact, since there are normally no facilities in university laboratories to carry out tests on medium and large induction motors, for example, between 100 kW/134 HP and 10,000 kW/13,400 HP.

Chapters 5–9 present 35 industrial case histories on the application of MCSA to detect cage winding defects, in a wide variety of rotor cage designs, in induction motors ranging from 127 kW/170 HP up to 10,000 kW/13,400 HP, driving different mechanical loads. The type of cage defects in large industrial SCIMs can be much more varied and subtle than clear cut breaks in the bars. The following conditions will affect the magnitude of the $\pm 2sf$ sidebands.

**(a)** Change in load and slip with a fixed rotor cage defect.

**(b)** Faulty bar to end ring joints creating an asymmetrical cage.

**(c)** Porosity and consequential arcing in aluminum die-cast cage rotors.

**(d)** Partially broken rotor bars, which are still making contact with an end ring via high resistance joints caused by a reduction in the cross section of the effective joint areas between bars and end rings.

**(e)** Actual broken bars still making face-to-face contact with an end ring, before dissection of the faulty cage winding, thus masking the severity of the cage winding defects.

**(f)** When an end ring retaining ring (ERR) covers the bar to end ring joints and a bar breaks at a joint, it can lift due to centrifugal forces. In doing so it can make face-to-face contact with a conducting end ring. This provides an alternative path for current flow, albeit via a high resistance path, which masks the severity of the fault, that is, the actual number of broken bars. Therefore, the magnitude of the $\pm 2sf$ sidebands will be lower than if the rotor did not have an ERR but had the same number of broken bars.

**(g)** There is the possibility of inter-bar currents flowing via the rotor core, when a broken bar or bars occur and this can in some cases severely mask the severity of the broken rotor bar problem by reducing the magnitude of the $\pm 2sf$ sidebands, when applying MCSA, as reported by Landy [4.12]. This can, under unique circumstances, and depending on the conductivity of the rotor core, and damage to the insulation on the laminations caused by overheating, between the broken bars, result in a gross underestimate of the severity of the cage winding breaks.

**(h)** Bars which are cracked from the top of the bar through to, for example, only a percentage of the total depth of the bar.

**Conclusions:** Due to the diverse variation in cage defects, it is impossible to predict the exact severity of cage defects/number of broken bars or a broken end ring from the magnitudes in dB of the $\pm 2sf$ sidebands with respect to $f$. It is only possible to give an estimated "equivalent broken bar severity indicator" for the condition of the cage winding, due to a broken rotor bar or bars. As already indicated, the only information available is the nameplate data and hopefully the number of rotor bars, which may or may not be provided by an OEM, hence advanced theoretical predictions are non-starters.

## 4.4 SPECTRUM ANALYSIS OF CURRENT

The supply current is modulated at twice slip frequency due to cage winding breaks and an illustration of a current waveform at a frequency of 60 Hz is shown in Figure 4.4 with the modulating frequency due to $2sf$ at 2.4 Hz (slip of 0.02 on a supply of 60 Hz).

In practice, the voltage signal would be taken from a non-inductive low resistive shunt, across the output of a current transformer. To obtain absolute values for

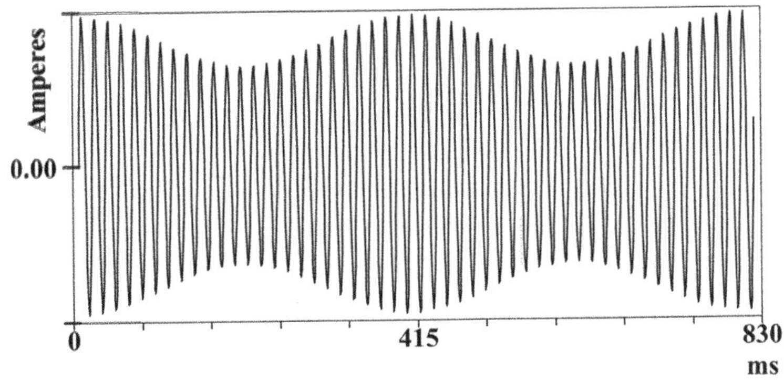


Figure 4.4 An "idealized illustration" of a time domain current waveform, due to amplitude modulation caused by $\pm 2sf$ sidebands around the supply frequency produced by broken rotor bars.

the magnitude (in dB) and frequency content of the time domain signal in Figure 4.4 requires a spectrum analysis to produce a Fourier spectrum. A Fourier spectrum is produced via digital signal processors that use fast Fourier transform (FFT) algorithms to compute the spectrum. Suffice to state that a spectrum, produced by a modern digital signal processor, can have a very narrow frequency resolution (analogous to the bandwidth of an analogue filter) with a large dynamic amplitude range. The terminology normally used is as follows:

**(i)** The frequency range ($x$-axis), over which the spectrum is computed from the input analogue current waveform, is called the baseband frequency span, for example, 0–100 Hz.

**(ii)** The dynamic amplitude range ($y$-axis) for MCSA is given in dB, so that a large dynamic range is displayed, for example, spanning 80 dB (factor of 10,000 in linear units).

**(iii)** The frequency resolution is defined as the baseband frequency span divided by the number of lines selected for the analysis of the input signal to display the spectrum between 0 and 100 Hz. It is given in Hz/line.

The typical number of lines of frequency resolution in modern digital signal processors is selectable and is normally 400, 800, 1600, 3200, 6400, 12,800 lines, and higher. The greater the number of lines of resolution selected the narrower the effective bandwidth of each spectral line but the longer the analysis time ($t = 1/(\text{Hz/line})$). For example, a baseband mode of 0–100 Hz with 400 lines gives a line resolution frequency of 0.25 Hz/line. Whereas with 6400 lines the line resolution frequency would be 0.0156 Hertz/line and for the "avoidance of doubt" that accuracy is achieved by modern digital signal processing. The $2sf$ frequency due to cage winding defects is much smaller in frequency and amplitude than the supply frequency, for example, for a motor with a full-load slip of 1% on a 60 Hz supply, $2sf$ is 1.2 Hz, and for a motor with a full-load slip of 0.33% on a 50 Hz supply, $2sf$ is only 0.33 Hz. The magnitude

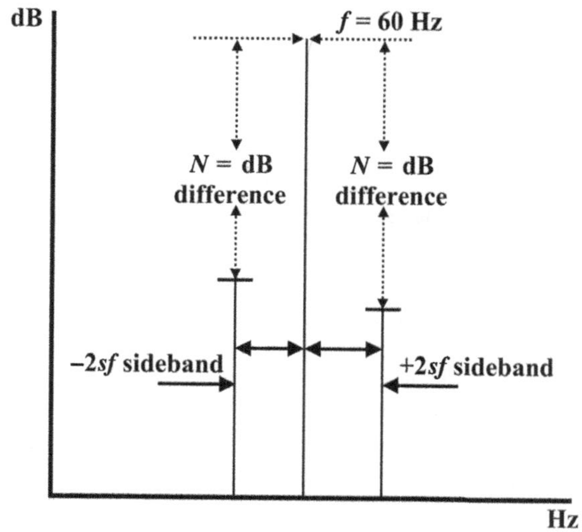


Figure 4.5 Illustration of dB difference $N$ between sidebands and supply component.

of the current components at $f$ and $\pm 2sf$ must be simultaneously displayed by the Fourier spectrum as shown in Figure 4.5.

The typical dB difference ($N$) between the supply component and the sidebands can vary between, for example, 26–60 dB, as shown in the case histories, and this dB difference is a function of the amount of cage winding asymmetry caused by either faulty bar to end ring joints, broken rotor bars, or a broken end ring. This means that the supply frequency component's magnitude can be 20–1000 times greater than the magnitude of the sidebands, so a logarithmic rather than a linear scale, is required for the $y$-axis of the current spectrum.

### 4.4.1 Measurement of Current and MCSA in Industry

A schematic illustration of the instrumentation set-up used to measure the current from LV and HV SCIMs is shown in Figure 4.6. In commercially available MCSA instruments the signal conditioner is an integral part of the instrument.

Only one current transformer (CT) is required to measure the current in any one of the three phases, since the rotating flux waves at $f(1 \pm 2s)$, due to cage winding defects, induce equal emfs and corresponding currents at these frequencies in each of the three stator windings. The MCSA clip-on CT should have an in-built, non-inductive, low resistance shunt ($R_S$) and these CTs are commercially available. Thus the input volts to the spectrum analyzer or MCSA instrument is a voltage $V_{(t)} = R_S\, i_{(t)}$, therefore the time domain waveform to be analyzed is the current.

**High Voltage SCIM:** A typical switchgear room is shown in Figure 4.7. A small clip-on CT (CT2 in Figure 4.6) can be temporarily installed around the secondary side of the *in situ* CT, which as shown in Figure 4.8, supplies current to the ammeter, and is normally mounted on the front panel of the LV instrumentation cubicle in the HV motor switchgear room.

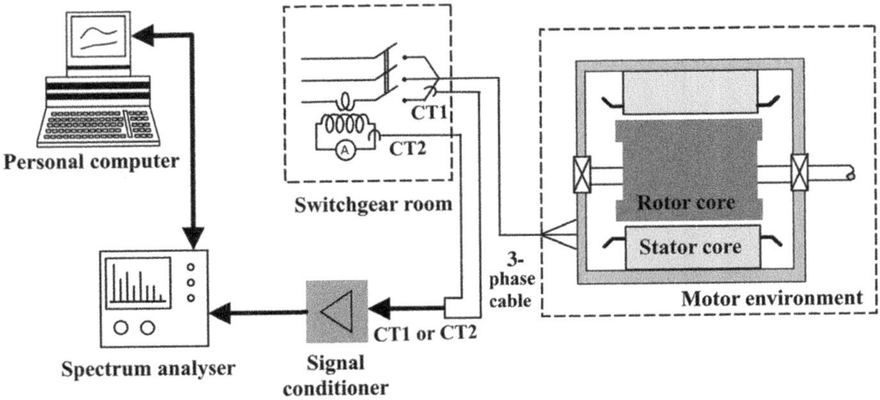


Figure 4.6 Illustration of measurement of current and MCSA instrumentation. *Source:* Thomson et al. [4.11]. Reproduced with permission of IEEE.

In some cases, the phase cables feeding the HV circuit breaker are available and a clip-on CT (CT1) can be installed as shown in Figure 4.9, a photo taken in a power station.

**Low Voltage SCIM:** It can be very difficult and often impossible without proper pre-planning to get on-line access, to install a clip-on CT when the motor is running, since in many cases a composite 3-phase cable supplies the starter for an LV motor and access to individual phase cables is inside the motor's starter cubicle. The cubicle cannot normally be opened while the motor is running, since it is good and normal safety practice to prevent on-line access by having interlocking relays, which are only deactivated if the starter is switched off and the supply isolated. If the motor can be switched off to install a temporary clip-on CT for the test the starter door cannot then be closed, since the MCSA CT cable has to exit the starter.

Thus careful pre-planning is required before going on-site to carry out MCSA on an LV SCIM. The best method is to get a permanent clip-on CT installed in the

Figure 4.7 Typical layout of HV circuit breakers and instrumentation cubicles for HV motors.

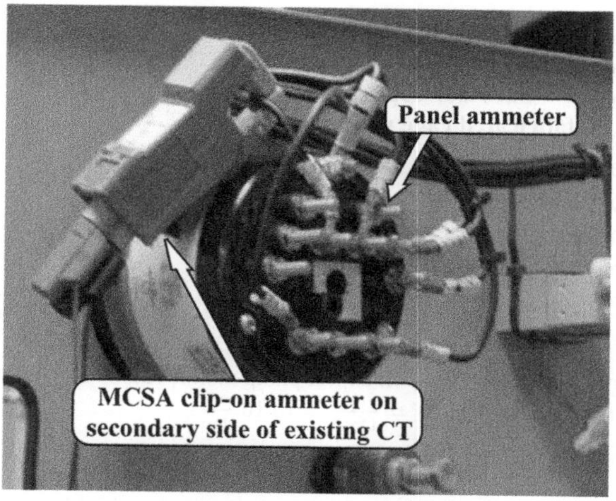


Figure 4.8 MCSA clip-on CT around the secondary side of the *in situ* CT, which feeds the panel ammeter displaying the motor's current.

starter during an outage and the cable from it can be terminated on the front panel of the starter. An example of an actual current spectrum (dB vs. frequency) from a motor driving an oil export pump is shown in Figure 4.10.

### 4.4.2 Introductory Sample of Current Spectra from Industrial SCIMs

The motor is a 3-phase SCIM, 6.6 kV, 38 A, 370 kW/496 HP, 60 Hz, 3575 r/min, with a nominal full-load slip of 0.69%. The twice slip frequency sidebands at

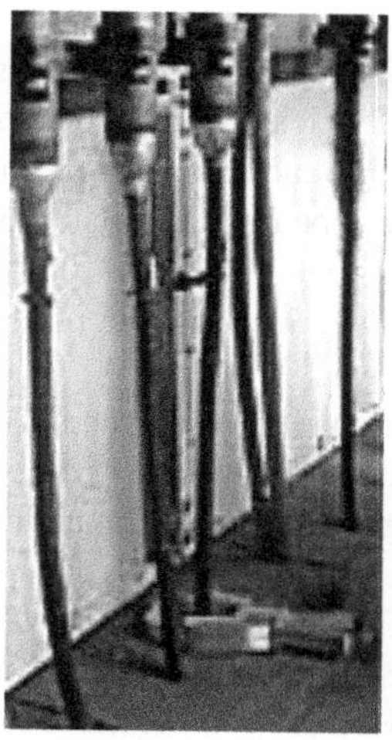

Figure 4.9 MCSA clip-on CT around one of the 3-phases external to the motor's circuit breaker in a power station.

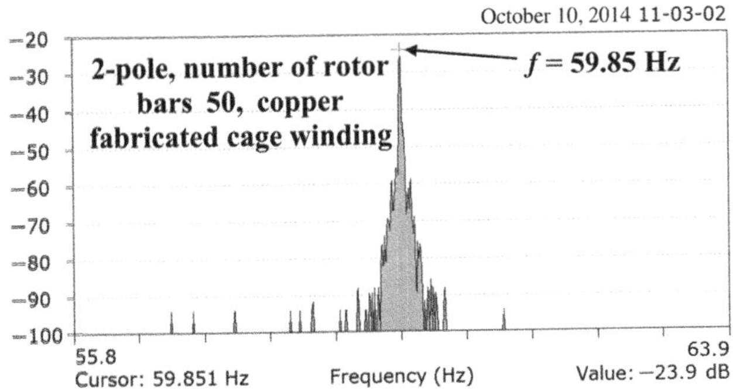


Figure 4.10 Current zoom spectrum oil export pump motor, 12,800 line spectrum, 0–130 Hz, frequency resolution = $\Delta f = \frac{130}{12{,}800} = 10.16\ \frac{\text{mHz}}{\text{line}}$.

full-load should be at ±0.83 Hz around the supply frequency, in this case the operating current was 28 A (±1%) compared to the full-load current of 38 A, thus the frequency of the $2sf$ sidebands will be less than the full-load value of 0.83 Hz. The selected baseband was 0–130 Hz to observe the complete spectrum content in that range and using 12,800 lines gives a line resolution of 0.01 Hz/line, which was selected for subsequent zoom analysis to identify any $\pm 2sf$ sidebands around $f$. An 80 dB dynamic range was used to accurately identify the magnitude in dB and the exact supply frequency, $f$, and likewise the magnitude and frequency of $\pm 2sf$ sidebands as much as 60 dB down on $f$.

1. For the "*avoidance of doubt*" this is the frequency resolution ($10.16\ \frac{\text{mHz}}{\text{line}}$), to which the frequencies are computed (up to three decimal places).
2. The computed frequency on the current spectra could be rounded up to one or two decimal places.
3. However, it is "*emphasized that MCSA requires a very accurate measurement of frequency*" to truly identify $\pm 2sf$ sidebands around $f$ and to separate them from components that can come from rotor dynamics downstream of the motor. This should not be underestimated (see case histories in Chapters 7–9).

There was no evidence of any $\pm 2sf$ sidebands around the supply component, as shown in Figure 4.10, which was 59.85 Hz in comparison to the nameplate rated value of 60 Hz and thus the cage winding is normal and this is therefore a base-line spectrum.

The current spectrum in Figure 4.11 is from an oil booster pump motor which has $\pm 2sf$ sidebands around the supply component, the motor's details are as follows: 3-phase, 6.6 kV, 445 kW/596 HP, 48 A, 60 Hz, 1790 r/min, star connected SCIM, and the nominal full-load slip should be 0.56%. The number of rotor bars is 68. The twice slip frequency sidebands at full-load should be at ±0.67 Hz at a nameplate

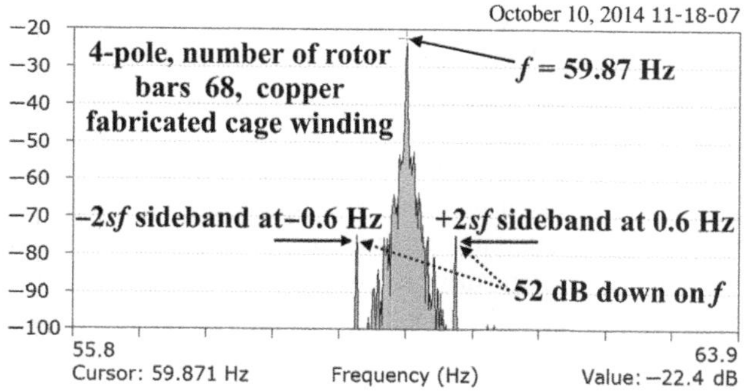


Figure 4.11 Current zoom spectrum oil booster pump motor, 12,800 line spectrum, 0–130 Hz, frequency resolution = $\Delta f = \frac{130}{12{,}800} = 10.16\ \frac{\text{mHz}}{\text{line}}$.

frequency of 60 Hz and the operating current was 45 A (±1%), as displayed on the panel ammeter, compared to the full-load current of 48 A, thus the frequencies of any $\pm 2sf$ sidebands will have less separation from $f$ than the nominal full-load values of ±0.67 Hz, that is, $2sf$ at full-load is greater than $2sf$ at any lesser load.

Figure 4.11 is a zoom spectrum around the supply component from the baseband spectrum of 0–130 Hz and a very small frequency line resolution accurately identifies the sidebands and gives the reader a very clear visual display. The $\pm 2sf$ sidebands are both 52 dB (i.e., $10^{-(52/20)} = 0.25\%$ of the supply component) down on the supply component, which is in fact at 59.87 Hz and the sidebands are at ±0.6 Hz around $f$, giving an operational slip of 0.5% (i.e., 89% of the nominal full-load slip and output) compared to a full-load slip of 0.56%, which is to be expected since the motor was on reduced load.

Recall that as the load on the motor drops, the slip reduces and the current in the rotor bars drops. On no-load, the current in the bars is very small indeed compared to the rated full-load current and even if there is a substantial number of broken bars, it is still not possible to detect any $2sf$ sidebands since their magnitude is negligible (even using an 80 dB dynamic range for the spectrum analysis). The frequency separation at $2sf$ from the fundamental frequency $f$ is very small indeed, since the slip on no-load is very close to zero. For the spectrum (Figure 4.11) being analyzed, suffice to state at this stage that since the sidebands are 52.0 dB down on $f$ (i.e., a factor of 400 times smaller than $f$, or 0.25% of the supply current) this equates to an acceptable and healthy cage winding condition for this particular motor with 4-poles and 68 rotor bars.

### 4.4.3 The dB Scale for MCSA and Specification for Spectrum Analysis of Current

It is often the case that electrical and mechanical engineers working in the electrical power industry find the use of a dB scale very alien since they are used to volts, amperes, watts, mm/sec, inches/sec, etc., but it is essential to use a dB scale for MCSA. For completeness, basic manipulations are now presented using the dB scale followed by conversion back to the actual input current to the motor. This may seem somewhat trivial but it is important for industrial engineers to gain familiarity with dB manipulations for MCSA work.

$$\text{Recall that } dB = 20\log_{10}(V/V_{\text{ref}})$$

Where $V$ rms is the actual measured voltage from the output of the MCSA CT and $V_{\text{ref}}$ is the reference voltage used by the spectrum analyzer or MCSA instrument. Modern spectrum analyzers often use, as the reference in the instrument, the zero dB line, which equates to 1.0 V. A positive dB is above 1 V and a negative dB is below 1 V. The clip-on MCSA CT for the spectrum shown in Figure 4.11 was around the secondary side of the *in situ* CT (ratio of 300:5), which gives an output of 0.1 V/A. Thus the actual input current to the motor is given by

$$I_{\text{rms}} = (\text{Volts rms from the MCSA CT} \times \textit{in situ}\ \text{CT ratio})/0.1\ \text{V/A}$$

The magnitude of the supply frequency component in Figure 4.11 is −22.4 dB with the zero dB line being the reference $V_{\text{ref}}$ at 1.0 V. Thus the volts output from the MCSA CT is

$$-22.4 = 20\ \log_{10} V/1;\ \log_{10} V = -22.4/20;\ V = 10^{-1.12} = 0.076\ \text{V}$$

The *in situ* CT was 300:5 or 60:1, thus the current at the supply frequency to the motor is

$$I_{\text{rms}} = (0.076 \times 300/5)/0.1 = 45.6\ \text{A}$$

This is compared to the panel ammeter of 45 A.

Now consider the $\pm 2sf$ sidebands which are both at −52 dB down on the supply component, therefore the magnitude of the sidebands as a proportion of the supply component's magnitude in absolute units is given by

$$\begin{aligned} M_{\text{sb}} &= 10^{(-N/20)} \\ M_{\text{sb}} &= 10^{(-52/20)} = 0.0025 \end{aligned} \tag{4.3}$$

Thus the sidebands are (1/0.0025) = 400 times smaller than the supply component, which means the sidebands are 0.25% of the magnitude of the mains frequency current to the motor. A guide, given in Table 4.1 can be useful when applying MCSA.

To conclude, an MCSA instrument, designed to accurately display the magnitudes in dBs and frequencies of each required spectral component, should have a specification comprising at least the following features.

- **(i)** A minimum of an 80 dB dynamic range (i.e., a range of 10,000 in absolute units).
- **(ii)** Selectable baseband frequency ranges.
- **(iii)** Selectable line resolutions up to at least 12,800 lines for the baseband frequency span.
- **(iv)** A selectable zoom spectrum analysis feature down to as low as 1% of the baseband spectrum display, which gives a span of 1.3 Hz in a baseband span of 130 Hz. If the zoom was centered on the 60 Hz component, the spectrum would be from 59.35 to 60.65 Hz. The zoom percentage in Figure 4.10 was 6.23% of the baseband span of 0–130 Hz.

**TABLE 4.1 dB to Percentage Equivalences**

| Difference $N$ in dB between supply component and sidebands | Equates to $2sf$ sidebands as a percentage of the mains frequency component |
|---|---|
| 6 | 50.0% |
| 10 | ≈32% |
| 20 | 10.0% |
| 30 | ≈3.2% |
| 40 | 1.0% |
| 50 | ≈0.32% |
| 60 | 0.1% |
| 70 | ≈0.032% |
| 80 | 0.01 |

Therefore, when suppliers of portable MCSA instruments or on-line MCSA instrumentation systems state that their instruments can detect broken bars, the industrial engineer should ask questions about and make comparisons with the specification given above. Suffice to state that the authors of this book use such a specification and the numerous industrial case histories still to be presented justify that selection.

## 4.5 SEVERITY INDICATORS FOR ASSESSING CONDITION OF CAGE WINDINGS AT FULL-LOAD

In 1982, Hargis et al. [4.13] proposed a formula (equation 4.4) using only the magnitude in amperes of the lower sideband relative to the supply current component, to estimate the number of broken rotor bars on the assumption that the motor's speed was constant. It recognized that design details of industrial motors are not available and therefore used the number of rotor bars together with the measured magnitude of the lower sideband with respect to the supply component. This formula assumed that the number of broken rotor bars ($n$) was much less than the total number of rotor bars. The prediction underestimates the severity of the number of broken rotor bars. However, no mention was made of the operating load and slip in the prediction from the formula and comparisons were not given between a current spectrum presented in this paper [4.13], from a motor that had three broken rotor bars and a prediction using the proposed formula. However, in the very early 1980s this was the only available formula, which used the minimum of information to give an estimated prediction of the number of broken rotor bars in a cage winding.

From Reference [4.13]:

The ratio of the magnitude of the lower sideband $I_{\mathrm{LSB}}$ to that of the supply component $I_{\mathrm{p}}$ is given *approximately* by

$$\frac{I_{\mathrm{LSB}}}{I_{\mathrm{p}}} \cong \sin\alpha/2p(2\pi - \alpha) \tag{4.4}$$

where $\cong$ means approximate equivalence

$I_{LSB}$ = lower sideband magnitude in amperes

$I_p$ = supply frequency component magnitude in amperes

$\alpha = 2\pi np/R$ = electrical angle spanned by a group of adjacent broken bars, when $n \lll R$

$p$ = pole-pairs

$R$ = number of rotor bars

$n \cong$ estimated number of broken rotor bars

Thomson et al. [4.14] proposed, in 1987, a formula based on equation (4.4) but converted into a dB format for direct use with a dB versus frequency spectrum to estimate the condition of a cage winding.

For the avoidance of doubt, the following derivation starting from equation (4.4) is now presented.

$$\frac{I_{LSB}}{I_p} \cong \frac{\sin\alpha}{2p(2\pi - \alpha)} \text{ or } \frac{I_p}{I_{LSB}} \cong \frac{2p(2\pi - \alpha)}{\sin\alpha}$$

Apply 20 $\log_{10}$ to both sides of this equation to give

$$20\log_{10}\frac{I_p}{I_{LSB}} = 20\log_{10}I_p - 20\log_{10}L_{SB} \cong 20\log_{10}[2p(2\pi - \alpha)/\sin\alpha]$$

$$\text{let } N = 20\log_{10}I_p - 20\log_{10}L_{SB}$$

where:

$N$ = dB difference between the fundamental and the lower $-2sf$ sideband.

$$N \cong 20\log_{10}[2p(2\pi - \alpha)/\sin\alpha]$$

It is well known that when $\alpha$ is small then $\sin\alpha \cong \alpha$

For $\alpha = 2\pi np/R$

With $\sin\alpha \cong \alpha$, and $N \cong 20\log_{10}[2p(2\pi - \alpha)/\alpha] \cong 20\log_{10}((4p\pi - 2p\alpha)/\alpha)$

$$\alpha = 4p\pi/(10^{N/20} + 2p) \text{ but } \alpha = 2\pi np/R$$

The estimated number of broken rotor bars is given by

$$n \cong 2R/(10^{(N/20)} + 2p)$$

However, based on the application of MCSA to thousands of induction motors, Thomson prefers to use an equivalent broken rotor bar factor $BB_f$ in place of $n$ since a "*definitive number of broken bars cannot be predicted.*"

$$BB_f = 2R/(10^{(N/20)} + 2p) \tag{4.5}$$

The $\cong$ has been replaced by = for ease of presentation of equation (4.5), which refers to full-load output and slip. However, it *should always be remembered* it is an approximation that underestimates the severity of cage winding breaks.

$R$ = number of rotor bars

$p$ = pole-pairs

$N$ = dB difference between the magnitude of lower $-2sf$ sideband and the magnitude of the supply frequency component, $f$

Recall also that the magnitudes of the $\pm 2sf$ sidebands are a function of numerous factors, such as

**(a)** Type and severity of cage winding defects.

**(b)** Design of the rotor cage winding.

**(c)** Number of rotor bars.

**(d)** Low or high pole numbers.

**(e)** The rotor's inertia and speed oscillation of the rotor at $2sf$ due to broken rotor bars, which affects the magnitude of the upper $+2sf$ sideband and can reduce the magnitude of the lower $-2sf$ sideband.

**(f)** The magnitude of the upper sideband at $+2sf$ can be enhanced by modulation of the third time-harmonic flux.

Due cognizance should be taken of a number of these factors when selecting the value of $N$ from the actual levels of the $\pm 2sf$ sidebands with respect to $f$. This will be fully discussed and justified via the industrial case histories to be presented in Chapters 5 and 6. The $BB_{\mathrm{f}}$ underestimates the severity of cage winding defects and breaks but it can be effectively used as an equivalent broken rotor bar severity indicator. As already stated, the design details of induction motors in industry are commonly not available hence detailed theoretical predictions such as presented by Williamson and Smith [4.4] are just not feasible.

A series of papers by Bellini et al. [4.15–4.18], included in-depth theoretical analyses, followed by valid assumptions, which led to a simplified equation for a broken rotor bar index equation ($n$)

$$I_{\mathrm{n}}/I_{\mathrm{p}} \cong n/R \tag{4.6}$$

$I_{\mathrm{n}}$ = the sum of the magnitudes of the $\pm 2sf$ sidebands = $I_{\mathrm{LSB}} + I_{\mathrm{USB}}$, amperes

$I_{\mathrm{p}}$ = supply frequency current to the motor, amperes

$n \cong$ estimated number of broken rotor bars

$R$ = number of rotor bars

Equation (4.6) does not require detailed design information for the prediction, since the authors recognized that such design details are not available when MCSA is applied in industry. Equation (4.6) also underestimates the severity of cage winding defects but in practice it was proposed by Bellini et al. that a broken rotor bar threshold ($n$) can be set at 0.5 from equation (4.6), above which a cage winding break normally exists. However, this was for the motor operating at full-load current and slip and no account was taken of motors operating on reduced loads and corresponding slips.

The case histories by the authors of this book also confirm that Bellini et al.'s papers proposed threshold of 0.5 is indeed a sensible value when also applied to equation (4.5) but always referred to full-load operation and slip. Note that the case histories by the authors go back to 1982 and the threshold value of 0.5 was retrospectively applied to these early case histories and was indeed found to be a valid threshold above which a cage winding defect exists.

Finally, the magnitudes of the $\pm 2sf$ sidebands change with different types of cage defects (see Section 4.3) and therefore it has to be accepted that MCSA cannot distinguish between the actual types of defect but it can give an overall assessment of the condition of the cage winding, which in reality is what the end user requires.

### 4.5.1 Severity Indicator for Assessing Condition of Cage Windings on Reduced Load

The estimated severity indicators used to assess the condition of a cage winding given in Section 4.5 were all based on a motor operating at full-load output power and slip but in industry it is certainly not the case that motors being MCSA tested are always on full-load. When the MCSA test is carried out at a reduced current and slip, these severity indicators have to be corrected to a reference operating condition, which has to be the full-load current and corresponding slip calculated from the nameplate data.

In industry, the MCSA test has to be carried out at the operational load and speed at the time of testing and an end user will not normally disturb the production process to change the load to full-load for an MCSA measurement. This operational norm will be substantiated via the industrial case histories presented in Chapters 5–9. Consider an MCSA test carried out on a SCIM with a broken rotor bar, when the measurement is recorded at reduced load at a slip of 50% of full-load slip. The magnitudes in dB of the twice slip frequency sidebands will be lower with respect to the load current than if it were tested at full-load, consequently the estimated condition of the cage winding will be better than it actually is in practice. Williamson and Smith [4.4] showed, via a theoretical prediction, that there was a linear relationship between the magnitudes of the lower $2sf$ sideband versus slip up to a slip of 5% for the particular motor they modeled with one broken rotor bar, shown in Figure 4.1.

### 4.5.2 Experimental Tests at the Robert Gordon University, Scotland

To thoroughly investigate MCSA for the diagnosis of broken rotor bars at full-load and reduced load, abnormal levels of airgap eccentricity, and the influence of fluctuating mechanical loads, Thomson initiated in 1980, an extensive research program, via controlled experimental tests using five different test rigs.

It is the authors' opinion "*that experiments are as equally important as theoretical modeling of SCIMs*" and this is particularly true when analyzing the current spectra due to broken rotor bars, abnormal levels of airgap eccentricity, and the effects of mechanical load dynamics downstream of the motor. Since theoretical models normally always have to include certain simplifying assumptions, it could be argued that actual results from experimental tests, for the interpretation of current spectra as a function of different faults and load conditions, can give undisputed results.

(i) The first test rig shown in Figures 4.12a and 4.12b was designed and constructed in a modular style, so that subsequent test rigs were all of a standard design and all the component parts, which were purchased from an OEM, were used for the manufacture of a standard SCIM. The bedplate was made of sufficient length to allow for the installation of rotors with long shafts, as shown in Figure 4.12a.

The nameplate data for this SCIM was as follows: 3 phase, delta connected, 415 V, 11 kW (14.7 HP), 20.5 A, 50 Hz, 1430 r/min (with a 51 slot aluminum die-cast

Figure 4.12a Test rig for testing a SCIM with rotors having broken rotor bars. Constructed by Donald Sutherland, formerly of the Robert Gordon University, Aberdeen, Scotland.

rotor). The OEM supplied additional rotors with cage windings having 51 copper fabricated bars, an aluminum die-cast cage rotor with 51 bars, and a deep bar aluminum die-cast rotor having 28 slots. The presentation of some of the details (not previously published) of the test rigs may well be informative and useful to young researchers, who may still wish to investigate problems in SCIMs. The test rigs had the facilities to carry out controlled experiments such as

**(a)** The measurement and analysis of current to a SCIM with a range of aluminum die-cast and copper fabricated squirrel cage rotors with no broken rotor bars was

Figure 4.12b Test rig for testing a SCIM to clearly show all coil ends were available for connecting the stator with different pole numbers. Constructed by Donald Sutherland, formerly of the Robert Gordon University, Aberdeen, Scotland.

Figure 4.13 Test rig with a scaled down frame of a typical design used in a large high power SCIM. Used for airgap eccentricity experiments. Constructed by Donald Sutherland, formerly of the Robert Gordon University, Aberdeen, Scotland.

undertaken. A special copper fabricated rotor with up to three high resistance joints could be introduced and reinstated as reported by Thomson et al. [4.19] in 1983, and rotors with different numbers of broken rotor bars all operating at different loads and with different stator windings such as 4-pole and 6-pole were investigated.

**(b)** The measurement and analysis of current from the SCIM with different amounts of static airgap eccentricity, was achieved by moving the stator frame structure independent of the rotor since the latter was mounted on separate bearing pedestals as shown in Figures 4.12 (a) and (b), 4.13 and 4.14.

**(c)** In addition, a rotor with a fixed amount of dynamic airgap eccentricity (i.e., a total indicated rotor run out TIR of 0.001 inches/1.0 mil/≈25 μm) was used as a reference for the airgap eccentricity experiments with varying levels of static eccentricity.

**(d)** Also different rotors were used with various levels of dynamic eccentricity and fixed amounts of static airgap eccentricity and combinations of static and dynamic airgap eccentricity. The test rigs for the airgap eccentricity experiments are shown in Figures 4.13 and 4.14.

To enable versatility in carrying out a range of experiments, the following component parts were purchased from the OEM of the motor, for the design and construction of the various test rigs.

**(a)** Separate stator frames with unwound cores inserted so that they could be wound such that all the coil ends were available for changing the stator winding connections from a 4-pole to a 6-pole configuration.

**(b)** Stator cores only for the construction of a scaled down frame structure to represent the frame of a large HV SCIM that could also be wound, see Figure 4.13.

**(c)** Aluminum die-cast and copper fabricated rotors with standard shafts inserted.

Figure 4.14 Test rig for airgap eccentricity experiments using a 51 slot rotor with a long flexible rotor assembly.

**(d)** A 51 slot aluminum die-cast cage rotor with no shaft into which was inserted a long flexible shaft as shown in Figure 4.14. This facilitated investigating the effects of varying the amount of airgap eccentricity (static and dynamic independently and with combinations thereof) on the current spectra with a short stiff rotor and a long flexible rotor, see Figures 4.13 and 4.14. Results from these experiments are presented in an MPhil Thesis by Murray [4.20].

### 4.5.3 MCSA Tests at Full-Load—Normal Cage Winding, 1, 2, and 10 Broken Bars

The nameplate data for this new SCIM was as follows: 3 phase, delta connected, 415 V, 11 kW (14.7 HP), 20.5 A, 50 Hz, 1430 r/min (with a 51 slot aluminum, die-cast rotor having round bars). The following MCSA results were personally taken by Thomson (author) in 1982 under controlled experiments in a laboratory and are presented for completeness.

**A. No Cage Winding Defects in Laboratory Based SCIM:** Figure 4.15 shows the current spectrum for the normal cage winding with the motor operating at full-load slip (4.67%) and rated full-load current and the $-2sf$ and $+2sf$ sidebands are 60 and 66 dB down (1000 and 2000 times smaller), respectively, with respect to the supply component, at 50 Hz. They are of a negligible value and it is obvious the cage winding is perfectly normal as it should be for a new cage rotor.

**B. One Broken Rotor Bar in Laboratory Based SCIM**: A nominally identical and brand new cage rotor had one bar completely broken by drilling out the bar next to one of the end rings and Figure 4.16 gives the current spectrum at the nominal full-load current of 20.5 A and the rated nameplate speed measured as 1430 r/min giving a full-load slip of 4.67%.

The current spectrum in Figure 4.16 for one broken bar clearly shows the lower sideband ($-2sf$) has increased by 16 dB (an increase in absolute units by a factor of 6.3 times) compared to the spectrum in Figure 4.15.

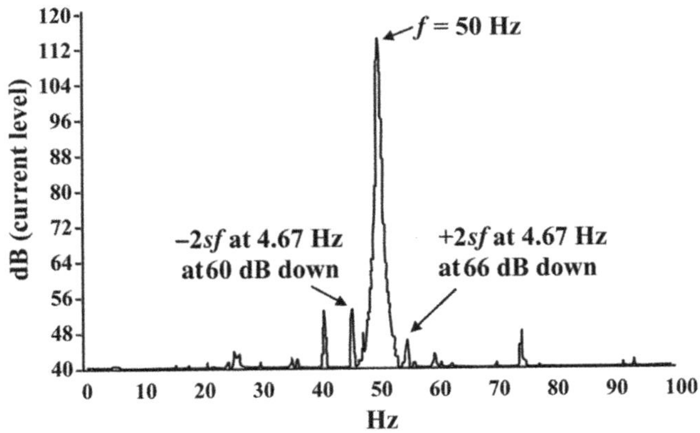


Figure 4.15 Current spectrum, no broken bars, full-load slip of 4.67%, at a full-load current of 20.5 A.

The dB difference of $N = 44$ dB *between the lower* $(-2sf)$ *sideband and the supply component* $f$ is used to calculate the equivalent broken bar factor $BB_f$ by equation (4.5).

$$BB_f = 2R/(10^{(N/20)} + 2p) = 102/(10^{(44/20)} + 4) = 0.63$$

As expected this is less than the actual one broken rotor bar since $BB_f$ is an estimate which underestimates the severity of a cage winding break/s. However, it is above the threshold index of 0.5 at which a cage defect such as a broken rotor bar will normally exist as originally presented by Bellini et al. [4.16–4.18] and also verified in the case histories in this book.

For completeness, now apply Bellini et al.'s equation (4.6) to the measured spectrum at full-load with one broken rotor bar.

$$I_n/I_p \cong n/R$$

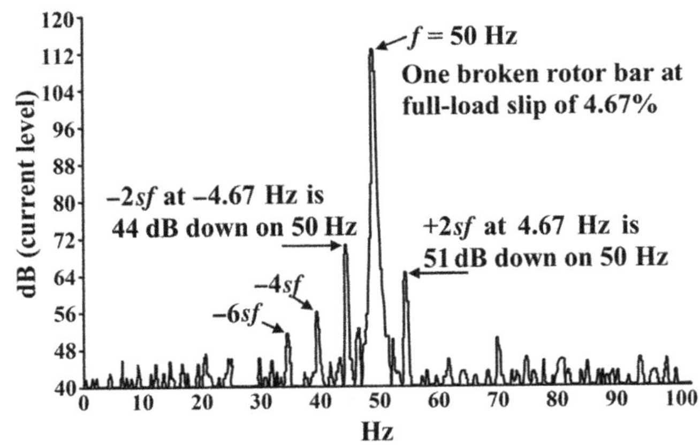


Figure 4.16 Current spectrum, one broken bar, and full-load current at 20.5 A and a full-load slip of 4.67%.

**EXTRACT ONE—TABLE 4.2 Broken Rotor Bar Severity "*ESTIMATOR*" Based on the Average dB Difference ($N_{av}$) Between the $\pm 2sf$ Sidebands and the Supply Component**

| $N_{av}$ = dB Difference | Estimated condition of cage winding—motor should ideally be operating at full-load output ($s_{FL}$) but this is very often not the case |
|---|---|
| 40–45 dB | There is a high probability of a cage winding fault. |
| *$N_{av}$ is replaced by* $N = 44$ dB | When $N_{av} = 40$ dB in low pole numbers (2 and 4) there is a high probability of one or even two broken rotor bars. At 40 dB with higher pole numbers the probability of more broken bars increases. |

This equation requires the use of the magnitudes of the currents in amperes of both sidebands and the supply component, $f$.

$$(I_{LSB} + I_{USB})/I_p \cong n/R$$

$$10^{-\left(\frac{44}{20}\right)} + 10^{-\left(\frac{51}{20}\right)} = 0.0063 + 0.0028 \cong n/51; n \cong 0.91\% \text{ of } 51 \ \cong 0.46$$

The broken bar index ($n$) using equation (4.6) is 0.46 compared to the $BB_f$, which is 0.63 using equation (4.5) when, in this case, the dB difference (44 dB) between the $-2sf$ lower sideband and the supply component was 7 dB higher than the $+2sf$ upper sideband. In this "particular case" this shows that using equation (4.5) and *only* the dB difference between the lower sideband and $f$ gives a closer prediction to the fact that one broken bar exists, since the lower sideband was 7 dB larger than the upper one.

**Assume Number of Rotor Bars is Unknown**: It is all very well using equation (4.5) when the number of rotor bars is known but in practice when MCSA is applied in industry, it is very often the case that the number of rotor bars is unknown and an *estimate of the condition of the cage winding* can only be made by using what is referred to as the dB difference chart and there is no other option but to do so—this is fully discussed in Section 4.6. Note: The dB difference is as illustrated in Figure 4.5 and in subsequent spectra, which is the difference in dB amplitudes between the $\pm 2sf$ sidebands and the supply component, $f$.

It is shown in the spectrum of Figure 4.16 that the dB differences between the lower and upper $\pm 2sf$ sidebands and the supply component were 44 and 51 dB, respectively, therefore $N_{av}$* (see Extract One—Table 4.2) is replaced by $N$ the dB difference between the lower sideband and $f$, since it is 7 dB higher (factor of 2.3 greater in absolute amperes) than the upper sideband.

***FOR THE AVOIDANCE OF DOUBT:**

In Table 4.2, $N_{av}$= average dB difference between the two $\pm 2sf$ sidebands and $f$ but there are results where the $N_{av}$ should not be used and these include

- When the $-2sf$ sideband is >2 dB compared to the $+2sf$ sideband then only the dB difference ($N$) between $-2sf$ and $f$ is used and replaces $N_{av}$ in Table 4.2

- But NOTE when the $+2sf$ is >2 dB compared to the $-2sf$ then $N_{av}$ is used in Table 4.2

It is the arithmetic dB average which is used and this is justified in Section 4.5.3 C.

**At 44 dB Down, the dB Chart Estimates**: There is a high probability of a cage winding fault and actually there is with one completely broken bar.

In practice, the commonsense approach of "Look, Feel, and Listen" *should always be part of information gathering* when an MCSA test is carried out. In the controlled, laboratory MCSA test the analogue ammeter was slowly oscillating about a mean position and the classical low frequency acoustic noise beat could be heard from the SCIM with one broken bar, which was not the case with no broken bars.

However, in industry the end user would be very reluctant indeed to remove a strategic and large HV SCIM based solely on a *Look, Feel, and Listen* approach which is a subjective assessment, for example, each individual has a different response to acoustic noise and the ear is in fact a very non-linear transducer. Also large HV SCIMs can be operating in close proximity in a generally noisy environment and it can be very difficult to determine which one is producing a particular noise. The application of MCSA can provide scientific information for an assessment of the condition of a cage winding for specific SCIMs. However, there is an added complication in SCIM drive trains, since it will be shown via a selection of industrial case histories in Chapters 7–9 that rotor dynamics from the mechanical load, which are reflected back into the cage rotor, can induce current components that can be easily confused with $\pm 2sf$ sidebands.

**C. Two Broken Bars in Laboratory Based SCIM**: The current spectrum at full-load slip with two broken rotor bars in a nominally identical cage rotor is given in Figure 4.17.

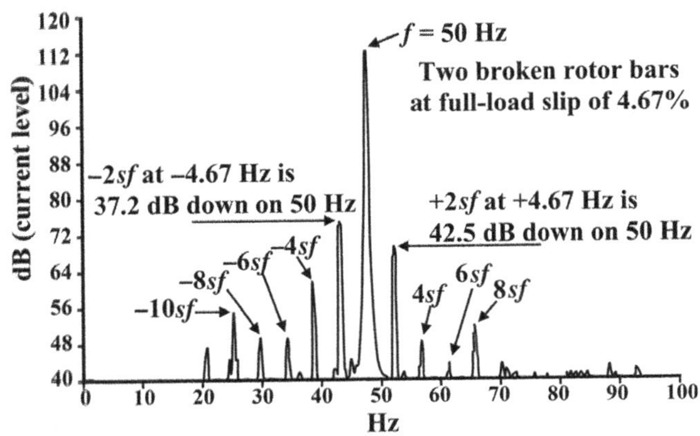


Figure 4.17 Current spectrum, two broken bars, full-load current 20.5 A, full-load slip of 4.67%.

A close interpretation of Figure 4.17 with two broken bars reveals certain spectral features.

**(a)** The lower sideband at $-2sf$ increased by 6.8 dB compared to its value with one broken bar at full-load and the upper sideband $+2sf$ increased by 8.5 dB.

**(b)** There are now more definitive $\pm 4sf, \pm 6sf$, and $\pm\, 8sf$ sidebands around the supply component due to the increased number of broken bars.

At the full-load slip and rated input current with two broken bars, the equivalent broken bar factor from equation (4.5) using only the 37.2 dB difference between the lower sideband and the supply component, gives

$$BB_{\mathrm{f}} = 2R/(10^{(37.2)} + 2p) = 1.3$$

Using equation (4.6), then the broken bar index ($n$) is given by

$$I_{\mathrm{n}} = (I_{\mathrm{LSB}} + I_{\mathrm{USB}})/I_{\mathrm{p}} = 1.38\% + 0.75\% = n/51; n = 2.13\% \text{ of } 51$$

$$n = 1.1$$

The two estimates, $BB_{\mathrm{f}}$ and $n$ at 1.3 and 1.1, respectively, differ by 0.2 but the key decision-making factor is that they are 2.6 and 2.2 times greater than the threshold value of 0.5 and recall equations (4.5) and (4.6) both underestimate the severity of cage winding breaks, in this case two broken bars.

**Assume Number of Rotor Bars is Unknown:** For completeness, again consider the number of rotor bars was unknown and the spectrum in Figure 4.17 was recorded, an extract from the dB difference chart in Table 4.2, Section 4.6 is presented in Extract Two—Table 4.2.

The spectrum in Figures 4.17 showed the dB differences between the lower and upper $2sf$ sidebands and the supply component were 37.2 and 42.5 dB, respectively, therefore in this case, $N_{\mathrm{av}}$ (in Table 4.2) is replaced by $N$, the dB difference between the lower $-2sf$ sideband and $f$ is applied to Table 4.2, since in this case it is 4.3 dB higher (factor of 1.7 greater in absolute amperes) than the upper sideband.

**With $N = 37$ dB the dB Chart Estimates**: A broken rotor bar problem exists and at $N = 40$ dB in a 4-pole motor normally several (e.g. one to three) bars are broken and of course there are two broken bars. If the arithmetic average of the dB difference between $\pm 2sf$ sidebands and $f$ is used, $N_{\mathrm{av}} = 39.9$ (40 dB) and an estimate of the condition of the cage winding using Table 4.2 would be in the same dB band, as when using only the lower $-2sf$ sideband to obtain the dB difference $N$.

**EXTRACT TWO—TABLE 4.2 Broken Rotor Bar Severity *"ESTIMATOR"* based on the Average dB Difference ($N_{\mathrm{av}}$) Between the $\pm 2sf$ Sidebands and the Supply Component**

| $N_{\mathrm{av}}$ = dB Difference | Estimated condition of cage winding—motor should ideally be operating at full-load output ($s_{\mathrm{FL}}$) but this is very often not the case |
|---|---|
| 35–40 dB | A broken rotor bar problem exists |
| $N$ measured = 37 dB | When $N_{\mathrm{av}} = 40$ dB in a 4-pole motor normally several (e.g., one to three) bars are broken |

It is accepted that to be strictly correct the dB difference between each sideband and $f$ should be firstly converted into a relative difference (in linear terms) with respect to the magnitude of $f$. The average is then taken and converted back to the average dB difference, $N_{av}$. For this case (Figure 4.17)

**(a)** Convert to absolute differences and take the average

$$\left\{10^{-\left(\frac{37.2}{20}\right)} + 10^{-\left(\frac{42.5}{20}\right)}\right\}/2 = 10.65 \times 10^{-3}$$

Convert back to obtain the average dB difference

$$N_{av} = 20\{\log_{10}(1/(10.65 \times 10^{-3}))\} = 39.5 \text{ dB}$$

There is a difference of only 0.5 dB between the two averages, even with a 7 dB difference between the sidebands this makes no difference to the dB severity band in Table 4.2 when the arithmetic $N_{av}$ is applied. Even with a 10 dB difference between the $\pm 2sf$ sidebands there is only a dB difference of 1.4 dB (a factor of only 1.2 in absolute terms) between the two averages. Therefore, when the average dB ($N_{av}$) is quoted in this book it refers to the arithmetic average of the dB differences between the $\pm 2sf$ sidebands and $f$ since the difference is negligible and is easier to use in practice, as demonstrated in these examples.

**D. Ten Broken Bars in Laboratory Based SCIM:** A nominally identical cage rotor with 10 broken bars was MCSA tested in the same motor and the rotor is shown in Figure 4.18.

The current spectrum of Figure 4.19 shows that the $-2sf$ sideband is 22.2 dB down and the upper sideband is 28.3 dB down on the supply component. The $\pm 2sf$ sidebands have therefore increased by 21.8 and 22.7 dB relative to the supply component compared to their magnitudes in the case with one broken bar in a 51 slot rotor in the same test motor.

The number of healthy bars is now only 80% of the total number of bars and the electrical resistance of the cage winding will increase since the bars are effectively in

Figure 4.18 The 51 slot aluminum die-cast rotor with 10 broken rotor bars.

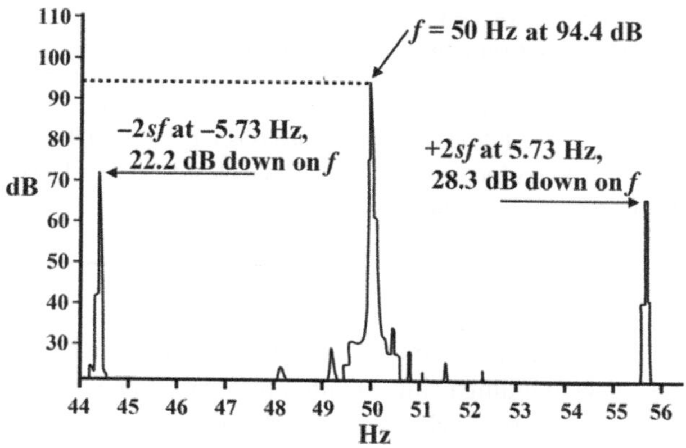


Figure 4.19 Current spectrum, 10 broken bars, and a slip of 5.73%.

parallel via the short circuiting end rings. The cage winding is no longer as per the original design. Therefore, the operational characteristics of input current versus slip and output power and torque versus slip curves will be different for the motor with a rotor having 10 broken rotor bars, compared to a healthy cage winding in the same motor. The rated full-load slip with no broken bars was 4.67%, whereas the slip from the measured $2sf$ sidebands in Figure 4.19 has increased to 5.7% with 10 broken rotor bars at the rated input current of 20.5 A, due to the change in resistance of the cage winding since the rotor cage has now only 41 rotor bars. Applying equation (4.5) to predict the equivalent broken rotor bar factor ($BB_f$) and using only the difference $N$ of 22.2 dB between the lower sideband and $f$, gives the $BB_f$ as

$$BB_f = 2R/(10^{(22/20)} + 2p) = 6.0$$

Applying equation (4.6) gives a broken bar index of $n = 5.9$ and although both predictions (using equations 4.5 and 4.6) underestimate the actual number of broken bars, the values for the $BB_f$ and $n$ from two different formulae are much higher, by a factor of very close to 12 times, for both predictions, compared to the threshold index of 0.5. Interestingly, the $BB_f$ at full-load for 1, 2, and 10 broken bars is 0.63, 1.3, and 6.0, respectively, and in *this particular case,* the increase in $BB_f$, is virtually proportional to the number of broken bars times the $BB_f$ with one broken bar.

**Assume Number of Rotor Bars Unknown:** For completeness, again consider the number of rotor bars is unknown and the current spectrum in Figure 4.19 is recorded. An extract from the dB difference chart of Table 4.2, Section 4.6 is presented in Extract Three—Table 4.2.

The spectrum in Figure 4.19 showed the dB differences between the lower and upper $\pm 2sf$ sidebands and the supply components were 22.2 and 28.3 dB, respectively. In this case, since the magnitude of the lower $-2sf$ sideband is 6 dB greater than the upper $+2sf$ sideband, it is only the lower sideband at 22 dB down on $f$, which is selected for Table 4.2.

**With $N_{av}$ < 30 dB the dB Chart Estimates:** A very serious broken rotor bar problem exists for all pole numbers. There will normally be multiple broken rotor bars, which there is, with 10 broken bars.

**EXTRACT THREE—TABLE 4.2 Broken Rotor Bar Severity *"ESTIMATOR"* Based on the Average dB Difference ($N_{av}$) Between the $\pm 2sf$ Sidebands and the Supply Component**

| $N_{av}$ = dB Difference | Estimated condition of cage winding—motor should ideally be operating at full-load output ($s_{FL}$) but this is very often not the case |
|---|---|
| <30 dB<br>Measured lower sideband<br>$N$ = 22 dB | A very serious broken rotor bar problem exists for all pole numbers. There will normally be multiple broken rotor bars. |

The results presented in Section 4.5.3 show that an *estimate* can be made of the condition of the cage winding using equation (4.5) and when the number of rotor bars is unknown the dB difference chart (Table 4.2) can also be applied. However, all these MCSA tests were carried out at full-load output under controlled experimental conditions with completely broken rotor bars.

This is all very well but SCIMs operating in industry do not necessarily operate on full-load output and this will now be considered.

### 4.5.4 Correction Factor to Estimate Cage Winding Condition During Reduced Load

The majority of SCIMs in industry operate at less than the nominal full-load rated current, output power, torque, speed, and slip, for example, of the 50 industrial case histories in this book, only 20% were operating at the rated full-load output. This is also borne out by the thousands of SCIMs, which the authors have MCSA tested during the past 34 years, for example, typically only 20–30% of the number of SCIMs were actually on full-load in a given industrial plant.

The following laboratory MCSA tests were carried out by Thomson (in 1982) at a reduced load, which verified the pronounced influence that the reduced load has on the $\pm 2sf$ sidebands. A correction factor is proposed, which can be applied to compensate for the estimate of $BB_{fs}$ at reduced load and refer it to the nominal full-load output and slip to obtain $BB_f$.

The current spectrum in Figure 4.20 is for the cage rotor with one broken rotor bar but with the motor now operating on a reduced input current of 71% of the full-load current and a corresponding slip of 2.2%, which is only 47% of the full-load slip ($s_{FL}$). The $\pm 2sf$ sidebands are of the same magnitude at 50 dB down with respect to $f$ and the dB difference ($N$) between the lower sideband and $f$ has increased by 6 dB (a factor of 2) compared to the dB difference (44 dB down on $f$) at the full-load rated current and slip. This verifies that the magnitude of the sidebands is affected by the operating load on the motor and its corresponding slip and the $2sf$ frequency has dropped from 4.67 Hz to 2.2 Hz

First, apply the estimated equivalent broken rotor bar factor, $BB_f$ from equation (4.5), but redefined as $BB_{fs}$ to indicate this refers to the actual reduced operating load and slip, namely at an input current of 14.5 A and a slip of 2.2% (i.e., 47% of $s_{FL}$).

$$BB_{fs} = 2R/(10^{(N/20)} + 2p) = 0.32$$

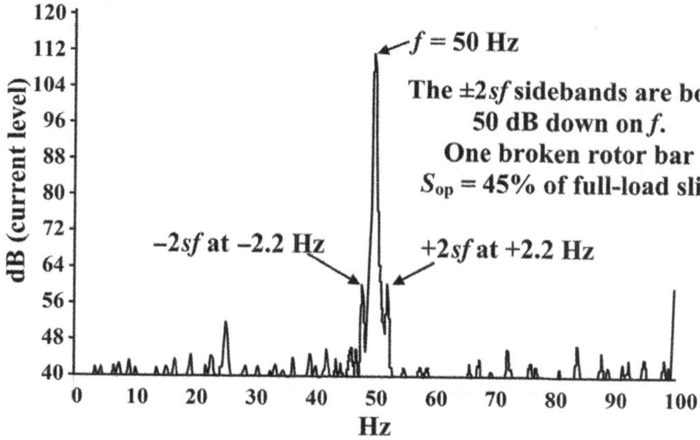


Figure 4.20 Current spectrum, one broken bar, and operating current 14.5A at a slip of 2.2%, that is, 45% of the full-load slip.

where

$BB_{\text{fs}}$ = the equivalent broken bar factor at the reduced operational output and slip

$N = 50$ dB = dB difference between the lower $-2sf$ sideband and $f$

$R = 51$

$p$ = pole-pairs

The $BB_{\text{fs}}$ at 0.32 with one broken rotor bar at a slip of 47% of the full-load slip has now *grossly underestimated* the actual condition of the cage winding. This is expected since the rotor current is smaller and the magnitude of the backward rotating magnetic field at $f(1 - 2s)$ with respect to the stator winding is less, due to the asymmetric rotor current distribution caused by one broken rotor bar. The induced emf and current in the stator winding at $f(1 - 2s)$ therefore drops in magnitude. An industrial SCIM with a *constant* cage winding break, when it is MCSA tested at different operating loads and slips, will give different spectra but the fault severity has not changed. Consequently, a correction factor has to be applied to give an estimate of the equivalent broken bar factor $BB_{\text{f}}$ with reference to full-load output and full-load slip.

Returning to the theoretical plot of the magnitude of the lower sideband in amperes versus speed for a 186.5 kW/248 HP SCIM shown in Figure 4.1, by Williamson and Smith [4.4] there is a linear relationship for the amplitude of the lower sideband in amperes versus the slip between 2% and 5%. Similarly, an actual experimental plot produced by Thomson (author) in 1982 from MCSA measurements with one broken bar as a function of load is shown in Figure 4.21. This shows that the dB difference $N$, between the lower sideband and the supply component versus input current and speed (slip between 1.8% and full-load slip of 4.67%) can be taken as a linear relationship as shown by the best straight line fit.

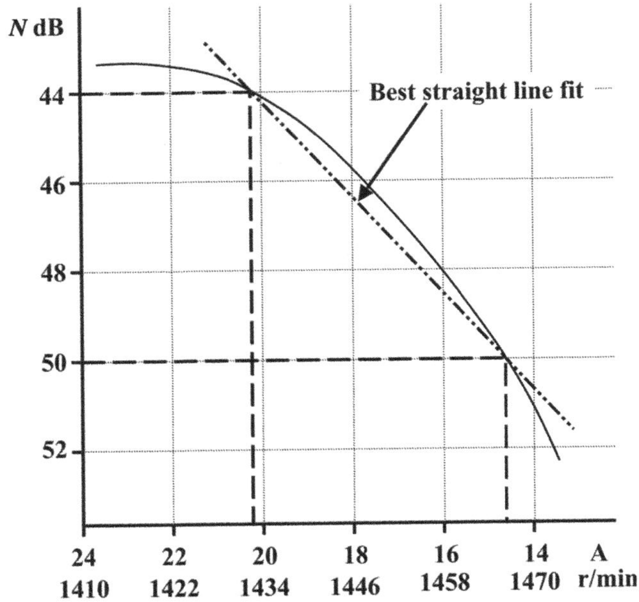


Figure 4.21 Plot of $N$, the dB difference between the lower $2sf$ sideband and the supply component versus input current, and output speed with one broken rotor bar.

The plot in Figure 4.21 is not as linear as the plot in Figure 4.1 since the former was a small motor (415 V, 11 kW) and the no-load current, which is dominated by the magnetizing current was 11 A thus the no-load current as a percentage of the full-load amperes was 54%. A correction factor, $BB_c$ can therefore be applied, based on using a linear relationship (see Figures 4.1 and 4.21), which is valid from typically 35% of full-load slip up to full-load slip with respect to the magnitude of the lower sideband, as a function of slip in that region. The "*Broken Bar Correction Factor*" is therefore given by

$$BB_c = s_{FL}/s_{op} \tag{4.7}$$

where

$s_{FL}$ = full-load slip

$s_{op}$ = full-load slip %

The estimated equivalent broken rotor bar factor $BB_f$ referred to the full-load operating condition is determined as follows:

Calculate the actual current in amperes of the $-2sf$ sideband when the load current was 14.5 A. This component ($I_{LSB}$) was 50 dB down on $f$ and in absolute amperes is

$$I_{LSB} = 10^{(-50/20)} \times 14.5 = 0.046\ \text{A}$$

$$BB_c = \frac{s_{FL}}{s_o} = \frac{4.67}{2.2} = 2.0$$

$$I_{LSBr} = I_{LSB} \times BB_c = 0.046 \times 2.0 = 0.092\ \text{A}$$

where $I_{LSBr}$ is the magnitude in amperes of the $-2sf$ component referred to the full-load slip and it is this value of current, which is then used to get the corrected dB difference value ($N_c$) for recalculating $BB_f$ at the full-load current and slip.

The corrected dB value $N_c = 20\log_{10}(I_{FL}/I_{LSBr}) = 47$ dB

This corrected dB value $N_c$ is then used in equation (4.5) to get the estimated $BB_f$ referred to the full-load slip $BB_f = 2R/(10^{(47/20)} + 2p) = 0.45$

where

$BB_f$ = the equivalent broken bar factor referred to the full-load slip

$N_c = 47$ dB = the corrected dB difference between the lower $-2sf$ and $f$

$R = 51$

$p$ = pole-pairs

Therefore, the estimated equivalent broken bar factor has increased from 0.32 at the reduced load to 0.45 when referred to the full-load slip using, $BB_c = s_{FL}/s_o$. From the MCSA test result with one broken bar this further underestimates the severity of one broken bar but it is only 10% lower than the threshold value of 0.5 for $BB_f$.

Thomson's $BB_c$ correction factor applied to $BB_{fs}$ to give the corrected $BB_f \cong 0.5$ in the experimental test gives a closer estimate to the $BB_f = 0.63$ from the actual full-load current spectrum with one broken bar, compared to *no correction factor* being applied to $BB_{fs} = 0.32$.

### 4.5.5 Application of a Broken Bar Correction Factor in Industry

It is accepted that the theory in Section 4.5.4 demonstrated that the correction factor ($BB_c$) should be applied to the actual $-2sf$ sideband current in amperes at the reduced load. However, from a practical and commonsense stand point Thomson's approach in industry is to apply $BB_c$ directly to the $BB_{fs}$ obtained from the reduced load result since the estimate of $BB_f$ is closer to the condition of the cage winding in practice.

The estimate of the equivalent broken bar factor from equation (4.5) is therefore $BB_f = BB_c \times BB_{fs}$, which is based on experience of applying MCSA for 34 years in industry, since it provides a more realistic compensation factor for the reduced load and slip condition than the theoretical example given in Section 4.5.4.

The MCSA test result at full-load with one broken rotor bar gave $BB_f = 0.63$ and if the $BB_c$ is applied directly to the reduced load and slip spectrum this gives

$$BB_f = BB_c \times BB_{fs} = 0.32 \times 2.0 = 0.64$$

This $BB_f$ corrected for the reduced load result differs by only 0.01 (negligible) compared to the $BB_f = 0.63$ obtained directly from the measured spectrum at full-load. Of particular relevance is that the industrial case histories to be presented in Chapters 5–7 prove that this gives a more representative estimate of the cage winding condition when the motor is on reduced load. It is however, recognized that when *$BB_c$ is applied directly to $BB_{fs}$* to get the $BB_f$ referred to the full-load condition, then $BB_c$ is really an "*empirical correction factor*" which clearly works in practice. It is also shown in the industrial case histories that the following rules have successfully diagnosed cage winding breaks.

- When the upper and lower $\pm 2sf$ sidebands differ by less than 2 dB the average dB difference ($N_{av}$) can be used in equation (4.5) to estimate the equivalent broken rotor bar factor ($BB_f$). Of course equation (4.6) can also be used.
- When the magnitude in dB of the lower sideband ($-2sf$) is greater than 2 dB compared to the upper sideband, the dB difference $N$ between the lower sideband and the supply component should be used in equation (4.5).

For completeness, the reader may wish to apply Bellini's equation (4.6) which takes account of the magnitude of each sideband in amperes to estimate the broken bar index $n$, however, it was applied [4.16–4.18] when the motors (as reported) were operating at rated, full-load input current, full-load output power, torque, and speed. However, on reduced load and slip, Bellini's equation (4.6), defined in this book as ($n_{fs}$) can be multiplied by $BB_c$, to give the corrected ($n$), which gives a closer estimate to the severity of cage winding breaks and this is also demonstrated in a selection of the industrial case studies.

## 4.6 THE dB BROKEN BAR SEVERITY CHART

It is often very difficult indeed for the end user or condition monitoring company to obtain/track down the number of rotor bars from an OEM and therefore the easiest option is to use the dB difference between the $\pm 2sf$ sidebands and the supply component $f$. However, at a much later stage in the book, in Section 11.11, a case history is presented, which verifies that the number of rotor bars can be predicted via advanced interpretation of the current spectrum and a knowledge of the range of typical numbers of rotor bars for a SCIM of a specific voltage, rating, and pole number. However, this is not easy and it is not 100% reliable unless it is being carried out by an expert in MCSA and induction motors.

A search on the *World Wide Web* for guidance on using only the dB difference between the sidebands and the supply component to estimate the condition of a rotor cage winding will indicate that the conclusions arrived at in their broken bar severity charts are very vague. An example is shown in Table 4.2A, which makes no mention of the influence of number of poles or the number of rotor bars. Thus broken bar severity charts on the web must be treated with considerable caution, for example, in Table 4.2A a statement such as moderate is vague and meaningless. Also rated load is not defined but it should not be interpreted as 70% of the rated full-load current since any quoted percentage should refer to the nominal rated output power, torque, and speed at the shaft.

The authors would *not use the chart in Table 4.2A* and they do not agree with the value of 70% and suggest the motor should be operating above 85% of the rated output and slip. It is impossible to definitively predict an exact number of broken bars (see amplitude 36–42 dB in Table 4.2A, which states two bars may be cracked or high resistance joints likely) and the amplitude difference stated in dB does not specify whether this is determined from an average or just the lower $-2sf$ sideband around $f$ or what?

**TABLE 4.2A Assessing Rotor Condition for the Avoidance of Doubt This Table Is Not Being Proposed By the Authors of This Book—It Is Purely An Example of What Can Be Found on the World Wide Web**

| Amplitude Difference (dB) | Rotor Condition (with at least 70% of rated load) |
|---|---|
| >60 | Excellent |
| 54–60 | Good |
| 48–54 | Moderate |
| 42–48 | Bar crack may be developing or high resistance joints |
| 36–42 | Two bars may be cracked or high resistance joints likely |
| 30–36 | Multiple cracked or open bars or end rings probable |
| <30 | Multiple broken bars and/or end rings very likely |

Assessing rotor bar condition by comparing the amplitude difference between line frequency and the first pole pass sideband below line frequency.

An estimate of the condition of a cage winding proposed by the authors using the dB difference concept is presented in Table 4.2, which is based on experience gained (and justified) from industrial case histories presented in this book.

In Table 4.2, $N_{av}$ = average dB difference between the two $\pm 2sf$ sidebands and $f$ but there are results where the $N_{av}$ should not be used and these include

- When the $-2sf$ sideband is >2 dB compared to the $+2sf$ sideband then only the dB difference ($N$) between $-2sf$ and $f$ is used and replaces $N_{av}$ in Table 4.2
- When the $+2sf$ is >2 dB compared to the $-2sf$ then $N_{av}$ in Table 4.2 is used.
- Also, for the avoidance of doubt the dB average $N_{av}$ in Table 4.2 is taken to be the average of the arithmetic sum of the dB differences between each of the $\pm 2sf$ sidebands and the supply component, $f$.
- To be strictly correct the dB difference between each sideband and $f$ should first be converted into a relative difference (in linear terms) with respect to the magnitude of $f$. The average is then taken and converted back to the average dB difference $N_{av}$ as was demonstrated in Section 4.5.3 C. Calculations show that the difference between using the arithmetic average and the strictly correct method lies between 0 and 1.4 dB (a factor of only 1.2) for a difference of 0 to 10 dB (factor of 3.16) between the sidebands. It is therefore justifiable to use the arithmetic dB average.
- In addition, the dB difference on reduced load can also be compensated via the $BB_c$ but this is better demonstrated via the case histories and the procedure for doing so is given in the first case history in Chapter 5.

## 4.7 INFLUENCE OF NUMBER OF ROTOR BARS AND POLE NUMBER ON THE EQUIVALENT BROKEN BAR FACTOR WITH MEASURED dB DIFFERENCE VALUES

Consider MCSA tests carried out at full-load output and rated slip, on two HV SCIMs operating in the same plant, for example, a 2-pole, 11 kV, 7150 kW/9584 HP, 60 Hz,

**TABLE 4.2 Broken Rotor Bar Severity "*ESTIMATOR*" Based on the Average dB Difference ($N_{av}$) Between the $\pm 2sf$ Sidebands and the Supply Component**

| $N_{av}$ = dB Difference | Estimated condition of cage winding—motor should ideally be operating at full-load output ($s_{FL}$) but this is very often not the case. |
|---|---|
| >60 dB | A normal and healthy cage winding. |
| 55–60 dB | A normal rotor cage winding. |
| 50–55 dB | Normal cage winding asymmetry since the $\pm 2sf$ sidebands are normally due to differences in bar to end ring joints in, e.g.,, 2-, 4-, 6-pole motors. |
| 45–50 dB | This is the difficult boundary range for using the dB difference.<br>When $N_{av}$ = 45 dB in lower pole number motors (2 and 4) there is a possibility of either faulty joints or perhaps a broken rotor bar and with higher pole numbers of 6, 8, 10, or 12 and above there is a higher probability of a broken rotor bar. |
| 40–45 dB | There is a high probability of a cage winding fault.<br>When $N_{av}$ = 40 dB in low pole numbers (2 and 4) there is a high probability of one or even two broken rotor bars. At 40 dB with higher pole numbers the probability of more broken bars increases. |
| 35–40 dB | A broken rotor bar problem exists.<br>When $N_{av}$ = 35 dB in a 2-pole motor having typically 38, 46, or 50 bars (actual designs of HV, high power SCIMs) then the probability is high that several bars are broken.<br>When $N_{av}$ = 40 dB in a 4-pole motor normally several (e.g., one to three) bars are broken.<br>When $N_{av}$ = 35 dB in, e.g., a 36-pole (240 slot) motor a serious broken bar problem with multiple broken bars, would exist. |
| 30–35 dB | A severe broken bar problem exists.<br>When $N_{av}$ = 30 dB in a 2-pole motor this normally means that a serious broken rotor bar problem exists and for 4-pole motors and above normally multiple broken bars exist. |
| <30 dB | A very serious broken rotor bar problem exists for all pole numbers. There will normally be multiple broken rotor bars. |

3564 r/min, which has 50 rotor bars and a 4-pole, 11 kV, 580 kW/7507 HP, 60 Hz, 1750 r/min SCIM, which has 86 rotor bars.

Assume that both MCSA tests give a dB difference ($N_{av}$) of 40 dB between the sidebands and the supply component but since the number of rotor bars is quite different the potential cage winding problem/broken rotor bars will be different although the dB difference $N_{av}$ is the same. When the number of bars is known, this helps with an assessment of the cage winding's condition, since the equivalent broken rotor bar factor at full-load slip is $BB_f = 0.98$ with $N_{av} = 40$ dB for the 2-pole motor with 50 rotor bars but with the same measured $N_{av}$ of 40 dB for the 4-pole motor with 86 bars at full-load slip, $BB_f = 1.65$ (a severity factor increase of 59%) but the crucial point is that the dB difference $N_{av}$ value was the same but the number of, for example, broken bars, is very likely to be quite different.

The typical range of stator and rotor slot combinations used by OEMs for SCIMs is presented in Table 4.3 for 2- to 12-pole motors. There is a wide variation in the number of rotor bars for each pole number and this type of data

**TABLE 4.3 Typical Combinations of Number of Stator, S, and Rotor, R, Slots for Different Pole Numbers**

| 2-pole | | 4-pole | | 6-pole | | 8-pole | | 10-pole | | 12-pole | |
|---|---|---|---|---|---|---|---|---|---|---|---|
| S | R | S | R | S | R | S | R | S | R | S | R |
| | | | | | | | | 84 | 68 | | |
| 36 | 28 | 48 | 34 | 54 | 40 | 60 | 46 | 84 | 70 | 72 | 86 |
| | | 48 | 38 | | | | | | | 72 | 52 |
| 42 | 34 | 48 | 40 | 63 | 50 | 66 | 46 | 90 | 70 | | |
| | | 60 | 44 | 63 | 48 | | | 90 | 68 | | |
| 48 | 34 | 60 | 46 | 63 | 46 | 72 | 56 | 90 | 110 | | |
| 48 | 38 | 60 | 50 | | | 72 | 58 | 96 | 118 | 99 | 120 |
| 54 | 34 | 66 | 46 | 72 | 58 | 72 | 86 | 102* | 82 | 99 | 124 |
| 54 | 38 | 66 | 52 | 72 | 86 | 72 | 88 | 102* | 122 | | |
| 54 | 44 | | | | | | | | | 108 | 82 |
| 54 | 46 | 72 | 56 | 81 | 62* | 78 | 62 | 105 | 86 | 108 | 132 |
| | | 72 | 58 | | | 78 | 94 | 108 | 86 | 108 | 134 |
| 60 | 44 | 72 | 86 | 90 | 70 | | | 108 | 128 | 120 | 94 |
| 60 | 46 | 120 | 84 | 90 | 74 | 84 | 62 | | | | |
| 60 | 50 | 78 | 62 | 90 | 68 | 84 | 66 | 120 | 94 | | |
| | | 78 | 58 | | | 84 | 68 | | | 126 | 98 |
| 66 | 50 | 78 | 56 | 99 | 82 | | | | | 126 | 102 |
| | | 96 | 80 | | | 90 | 68 | | | 126 | 94 |
| | | 84 | 62 | | | 90 | 70 | | | | |
| | | 84 | 66 | | | | | | | | |
| | | 84 | 68 | | | 96 | 74 | | | | |
| | | | | | | 96 | 112 | | | | |
| | | 90 | 74 | | | | | | | | |
| | | 96 | 74 | | | | | | | | |

*These combinations are normally only used with skewed rotor bars.

is very rarely, if ever, available in technical papers or text books. The authors do not have records for the nameplate data for all the slot combinations given in Table 4.3 but motors have been manufactured with these stator and rotor slot combinations. In Table 4.4 there are samples of the nameplate ratings and the actual number of rotor bars in HV (3.3 up to13. 8 kV) SCIMs of ratings between 580 kW/778 HP up to 12,800 kW/17,024 HP, 2- and 4-pole since these pole numbers are the most widely used in industry.

Figures 4.22, 4.23, and 4.24 give plots of the equivalent broken rotor bar factor versus the number of rotor bars with various average dB difference ($N_{av}$) values for a given pole number of motor. This clearly demonstrates that, when the number of rotor bars is unknown for a given pole number of SCIM, the broken rotor bar severity chart (Table 4.2) based solely on the average ($N_{av}$) dB difference can easily give inaccurate predictions.

The results in this chapter have shown that a threshold of 0.5 for the $BB_f$ (equation 4.5) or $n$ (equation 4.6) is a sensible level, above which a cage fault will normally

**TABLE 4.4 Ratings and Typical Rotor Slots for 2- and 4-Pole, 3-Phase SCIMs**

(i) All motors are single cage, DOL start.

(ii) For the avoidance of doubt the OEMs are anonymous and no motor serial numbers are given.

**2-pole SCIMs**

| Volts kV | Power kW/HP | Amperes | Hz | r/min | No. of poles | **Rotor slots Single Cage R** | OEM A, B, C, D, etc. |
|---|---|---|---|---|---|---|---|
| 13.8 | 6800/9115 | 326 | 60 | 3570 | 2 | **50** | A |
| 13.8 | 6800/9115 | 333 | 60 | 3567 | 2 | **58** | B |
| 13.8 | 5650/7573 | 292 | 60 | 3577 | 2 | **50** | B |
| 13.8 | 4960/6649 | 240 | 60 | 3570 | 2 | **50** | A |
| 13.8 | 3010/4035 | 157 | 60 | 1769 | 2 | **62** | B |
| 11 | 7150/9585 | 447 | 60 | 3564 | 2 | **50** | B |
| 11 | 4100/5496 | 249 | 60 | 3550 | 2 | **50** | C |
| 11 | 4100/5496 | 249 | 60 | 3550 | 2 | **50** | C |
| 11 | 3550/4758 | 233 | 50 | 2940 | 2 | **58** | B |
| 11 | 3200/4290 | 196 | 50 | 2970 | 2 | **46** | C |
| 11 | 1950/2614 | 120 | 50 | 1485 | 2 | **60** | D |
| 11 | 3180/4263 | 195 | 50 | 2974 | 2 | **34** | E |
| 11 | 2600/3485 | 163 | 60 | 3563 | 2 | **40** | D |
| 6.6 | 3150/4223 | 200 | 60 | 3575 | 2 | **50** | E |
| 6.6 | 3600/4826 | 376 | 60 | 3580 | 2 | **46** | A |
| 6.6 | 1222/1630 | 127 | 60 | 3577 | 2 | **46** | C |
| 6.6 | 760/1019 | 77 | 60 | 3577 | 2 | **50** | C |
| 6.6 | 580/778 | 59 | 60 | 3580 | 2 | **50** | B |
| **4-pole SCIMs** | | | | | | | |
| 13.8 | 12,800/17024 | 644 | 60 | 1775 | 4 | **86** | A |
| 11 | 10,160/13,619 | 574 | 60 | 1771 | 4 | **62** | E |
| 11 | 355/476 | 23 | 60 | 1792 | 4 | **54** | B |
| 6.6 | 1200/1608 | 145 | 60 | 1784 | 4 | **46** | D |
| 6.6 | 6300/8445 | 633 | 50 | 1493 | 4 | **82** | F |
| 6.6 | 5980/8016 | 615 | 60 | 1775 | 4 | **70** | D |
| 6.6 | 6714/9000 | 638 | 60 | 1780 | 4 | **84** | B |
| 4.16 | 2984/4000 | 470 | 60 | 1792 | 4 | **155** | G |
| 3.3 | 1230/1649 | 240 | 50 | 1485 | 4 | **58** | F |
| 3.3 | 4500/6032 | 914 | 50 | 1486 | 4 | **94** | H |

exist. The $BB_f$ threshold line is included in these curves and its validity is further substantiated in the numerous case histories in Chapters 5 and 6.

There are three equivalent broken bar factors ($BB_f$) values $\geq$ the threshold of 0.5 with rotors having 46, 50, and 52 bars, respectively, but with rotors having 28, 34, 38, and 44 bars the $BB_f$ values fall below the threshold of 0.5. This is the problem when using only the dB difference between the $\pm 2sf$ sidebands and the supply component

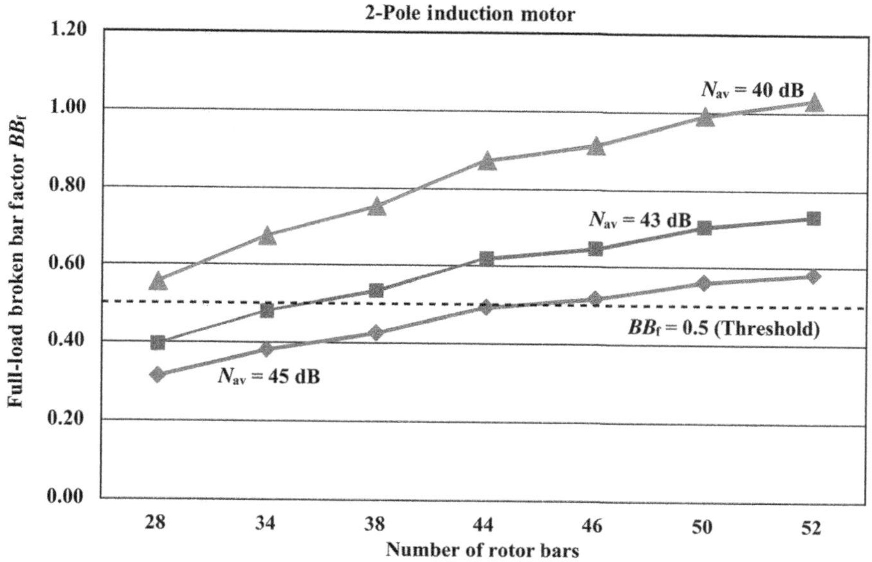


Figure 4.22 $BB_f$ versus number of rotor bars for 2-pole motors with various $N_{av}$ dB differences.

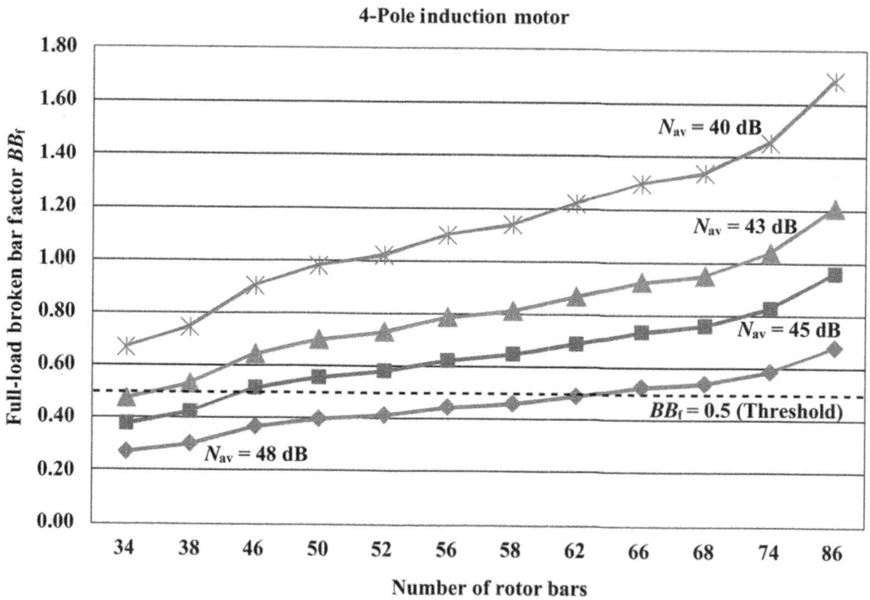


Figure 4.23 $BB_f$ versus number of rotor bars for 4-pole motors with various $N_{av}$ dB differences.

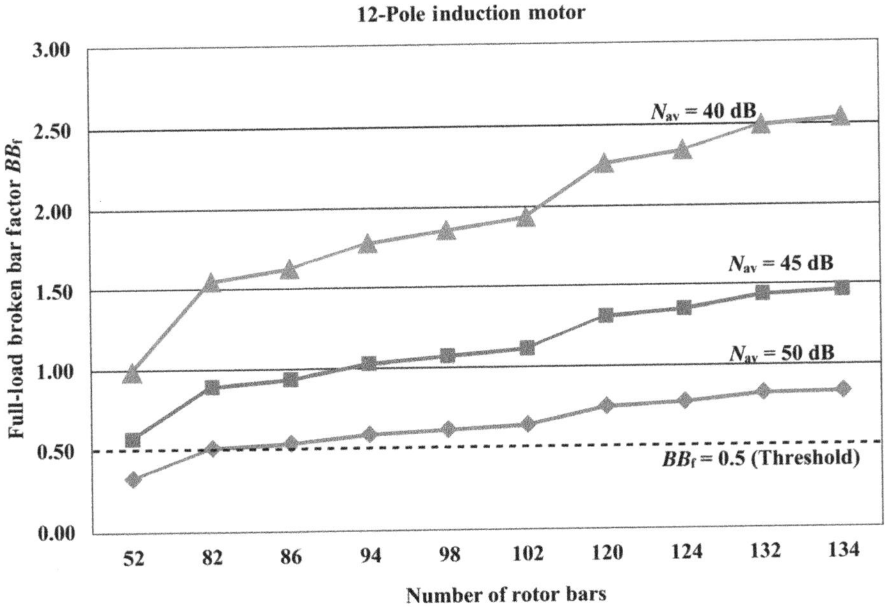


Figure 4.24 $BB_f$ versus number of rotor bars for 12-pole motors with various $N_{av}$ dB differences.

to estimate the condition of the cage winding, since it cannot be generally applied to all rotors with the same number of poles, because the number of rotor bars is a variable. With $N_{av} = 40$ dB for a 2-pole rotor all the values for $BB_f$ are above the threshold of 0.5.

Figure 4.23 shows that for a 4-pole motor with $N_{av} = 48$ dB there are 5 $BB_f$ *values* $\geq$ to the threshold of 0.5 and at $N_{av}$ equals 45 dB there are 10 $BB_f$ values $\geq$ to the threshold of 0.5. Of particular note is at $N_{av} = 45$ dB for the 2-pole motor there are three with $BB_f$ values $\geq 0.5$, whereas with $N_{av} = 45$ dB for the 4-pole motor there are 10 $BB_f$ values $\geq$ the threshold of 0.5.

Figure 4.24 shows that for a 12-pole motor with $N = 50$ dB there are nine $BB_f$ values $\geq$ to the threshold of 0.5 and at $N_{av}$ equal to 45 dB there are 10 $BB_f$ values $\geq$ to the threshold of 0.5. This demonstrates the limitations of using the dB difference chart particularly with higher pole numbers which have higher numbers of rotor bars.

## 4.8 QUESTIONS

**4.8.1** The magnitude in dB of the lower $-2sf$ sideband current component below the supply component of current at $f$ does not change as the operational slip and load on the motor decreases when there are broken rotor bars. Is this statement correct or incorrect? Explain your answer.

**4.8.2** Why is there an upper sideband component of current at $+2sf$ above the supply component at $f$ when there are broken rotor bars? What parameters influence the magnitude of the $+2sf$ component of current?

**4.8.3** Experiments in research laboratories used rotors with completely broken rotor bars to measure and analyze the current spectrum from small SCIMs. Explain, with respect to much larger SCIMs operating in industry, why cage winding faults very often do not have such clear cut broken bars. What is meant by inter-bar currents, when broken bars occur and explain why this can reduce the dB magnitudes of the $\pm 2sf$ sidebands and give a false diagnosis that no broken bars exist?

**4.8.4** Why is it not possible via MCSA to distinguish between numerous faulty joints (not broken), broken rotor bars, or a broken end ring and also discuss why it is impossible to determine which actual bars are broken?

**4.8.5** Explain why the Fourier spectrum from MCSA used to identify the $\pm 2sf$ sidebands around the supply component, $f$ is in dB versus frequency and not in absolute amperes versus frequency?

**4.8.6** If the baseband frequency span for the measured spectra was set to 0–65 Hz, what is the frequency resolution if a 400-line spectrum is used? What frequency resolution should be used if the operating slip is 0.5% and the supply frequency is 60 Hz? Is this frequency resolution suitable for detecting the $\pm 2sf$ sidebands around the supply component from a SCIM fed at 60 Hz operating at a slip of 0.5%?

**4.8.7** Define what is meant by the dB difference $N_{av}$ in a current spectrum displaying $\pm 2sf$ sidebands around the supply component.

**4.8.8** Why is it often difficult to carry out on-line, temporary measurements using a clip-on CT when an LV SCIM is supplied by a motor starter?

**4.8.9** The average dB difference ($N_{av}$) between the $\pm 2sf$ sidebands and the supply component of current $f$ is 46 dB. What is that as a percentage of the supply component?

**4.8.10** A 3-phase, 13,800 V, 5000 HP/3730 kW, 60 Hz, 3564 r/min, p.f. = 0.92, efficiency = 96.7% SCIM drives one of three strategic centrifugal compressors on an offshore oil production platform and it is a major task to remove the rotor for inspection. There is a spare rotor in a storage facility onshore.

**(i)** Calculate the full-load input line current.

**(ii)** Calculate the full-load slip.

**(iii)** An MCSA measurement was carried out on this motor when it was operating at its full-load current and full-load rated speed and the $\pm 2sf$ sidebands were both 44 dB down on the supply component. At this stage of the analysis the number of rotor bars was unknown. What would be your recommendation to the owner of the motor?

**(iv)** It was subsequently established via the OEM that this motor had 50 rotor bars with no ERRs, calculate the equivalent broken rotor bar factor ($BB_f$) at full-load and then give your recommendation to the owner of the plant.

**(v)** An identical motor doing the same duty was subsequently tested at the next annual MCSA survey and the motor was found to be operating on reduced load and $\pm 2sf$ sidebands were present at 0.6 Hz around $f$ and at 50 dB down on the supply component. What would be your recommendation to the owner of the plant?

## REFERENCES


[4.1] G. Kron, *Equivalent Circuits of Electric Machinery*, Dover Publications, Inc., New York, 1967.

[4.2] C. V. Jones, *Unified Theory of Electrical Machines*, Butterworth, 1967.

[4.3] H. Vickers, *The Induction Motor*, Sir Isaac Pitman and Sons Ltd., London, (1st edition, 1924; 2nd edition, 1953).

[4.4] S. Williamson and S. C. Smith, "Steady-state analysis of 3-phase cage motors with rotor-bar and end-ring faults," *IEE Proceedings B Electric Power Applications*, vol. 129, no. 3, 1982.

[4.5] M. Liwschitz-Garik and C. C. Whipple, *Electric Machinery Vol. II, A-C Machines*, Van Nostrand Company, (1st published, September, 1946).

[4.6] G. B. Kliman and J. Stein, "Methods of motor current signature analysis," in *Electric Machines and Power Systems*, Hemisphere Publishing, New York, Volume 20, 1982.

[4.7] G. B. Kliman, R. A. Koegl, J. Stein, R. D. Endicott, and M. W. Madden, "Non-invasive detection of broken rotor bars in operating induction motors," *IEEE Transactions on Energy Conversion*, vol. 3, no. 4, December, 1988, pp. 873–879.

[4.8] F. Filippetti, G. Franceschini, C. Tassoni, and P. Vas, "AI techniques in induction machines diagnosis including the speed ripple effect," *IEEE Transactions on Industry Applications*, vol. 34, no. 1, January/February, 1998, pp. 98–108.

[4.9] N. C. Paterson, "The Analysis and Detection of Faults in Three Phase Induction Machines using Finite Element Techniques," Ph.D. Thesis, The Robert Gordon University, Aberdeen Scotland, 1993

[4.10] J. F. Watson and N. C. Paterson, "The Development of an Accurate Finite Element Model to Investigate the Factors which Influence the Fault-Indicating Components of Current in 3-Phase Induction Motors," IEE 9th International Conference on Electrical Machines and Drives, Pub. No. 468, 1999, pp. 247–252.

[4.11] W. T. Thomson and M. Fenger, "Current signature analysis to detect induction motor faults," *IEEE Industry Applications Magazine*, vol. 7, no. 4, 2001, pp. 26–34.

[4.12] C. F. Landy and P. J. McCully, "Evaluation of Current and Vibration Signals for Squirrel Cage Induction Motor Condition Monitoring," IEE 8th International Conference on Electrical Machines and Drives, Pub. No. 444, 1997, pp. 331–335.

[4.13] C. Hargis, B. Gaydon, and k. Kamish, "The Detection of Rotor Defects in Induction Motors," Proceedings of 1st IEE International, Conference on Electrical Machines, Design and Application. London, UK, pp. 216–220.

[4.14] W. T. Thomson and D. Rankin, "Case Histories of Rotor Winding Fault Diagnosis in Induction Motors," Proceedings of 2nd International Conference on Condition Monitoring, University College of Swansea, March 1987.

[4.15] A. Bellini, F. Filippetti, G. Franceschini, C. Tassoni, and G. B. Kliman, "Quantitative evaluation of induction motor broken bars by means of electrical signature analysis," *IEEE Transactions on Industry Applications*, vol. 37, no. 5, September/October, 2001.

[4.16] A. Bellini, F. Filippetti, G. Franceschini, C. Tassoni, R. Passaglia, M. Saottini, G. Tontini, M. Giovannini, A. Rossi,, "On-field experience with On-line diagnosis of large induction motors cage failures using MCSA," *IEEE Transactions on Industry Applications*, vol. 38, no. 4, July/August, 2002.

[4.17] A. Bellini, G. Franceschini, C. Tassoni, and A. Toscani, "Assessment of Induction Machines rotor fault Severity by Different Approaches," 31st IEEE Conference Proceedings of Industrial Electronics Society, ICON 2005, pp. 1461–1466.

[4.18] A. Bellini, C. Concari, G. Franceschini, E. Lorenzani, C. Tassoni, and A. Toscani, "Thorough Understanding and Experimental Validation of Current Sideband Components in Induction Machines Rotor Monitoring," 32nd IEEE Conference Proceedings of Industrial Electronics Society, ICON 2006, pp. 4957–4962.

[4.19] W. T. Thomson, N. D. Deans, R. A. Leonard, and A. J. Milne, "Monitoring Strategy for Discriminating Between Different Types of Rotor Cage Faults," Proceedings 18th Universities Power Engineering Conference Proceedings, University of Surrey, April, 1983.

[4.20] K. Murray, "The Effects of Rotor Dynamics on Vibration and Current Signals for Detecting Airgap Eccentricity in Induction Motor," MPhil Thesis, 1988, The Robert Gordon University, Garthdee, Aberdeen, Scotland.

CHAPTER 5

# MCSA INDUSTRIAL CASE HISTORIES—DIAGNOSIS OF CAGE WINDING DEFECTS IN SCIMs DRIVING STEADY LOADS

*William T. Thomson*

## 5.0 INTRODUCTION AND SUMMARY OF CASE HISTORIES

The focus of this chapter is on the presentation of case histories, in which the SCIMs are driving steady loads and where the identification of the $\pm 2sf$ sidebands around the supply component is not being influenced by mechanical load dynamics downstream of the motor. The case histories in this chapter are self-standing in that the reader does not have to continually scroll through the previous chapters to find information. A list of the MCSA case histories subsequently presented in detail is as follows. This should assist the reader to select the ones of personal interest. Note that further nameplate data for each case history, which is relevant for MCSA testing, is presented later. NB No case histories are presented in which MCSA detected partial cracks (i.e. in a proportion of the depths of the bars) in rotor bars inboard of the joints (not broken bars) and the bars are still making full contact via healthy bar to end ring joints. The opinion and experience of the authors is that MCSA cannot detect cracked bars unlike its successful detection of broken bars.

**Section 5.1**: A large 6.6 kV, 10,000 kW/13,400 HP, 4-pole SCIM driving a centrifugal compressor. Normal cage winding asymmetry (no broken bars), caused by normal differences in resistances between a number of bar to end ring joints. This is an introductory case history.

**Section 5.2**: Four broken rotor bars in a 415 V, 125 kW/168 HP, 2-speed, 4- and 8-pole SCIM with a copper fabricated cage winding driving a food mixer.

**Section 5.3:** Faulty aluminum die-cast cage rotor due to die-casting porosity and severe arcing in a 440 V, 134 kW/180 HP, 4-pole SCIM, driving a centrifugal compressor.

*Current Signature Analysis for Condition Monitoring of Cage Induction Motors: Industrial Application and Case Histories*, First Edition. William T. Thomson and Ian Culbert.

**Section 5.4**: One broken rotor bar in a copper fabricated cage rotor in an 11 kV, 3100 kW/4155 HP, 4-pole SCIM, driving a seawater injection pump on a production platform. A broken bar crashed into the stator winding but the stator winding did not fail but a rewind was required.

**Section 5.5:** Twenty broken rotor bars in a copper fabricated cage rotor in an 11 kV, 3100 kW/4155 HP, 4-pole SCIM, driving a seawater injection pump but fortunately there was no damage to the stator winding.

**Section 5.6:** Numerous faulty bar to end ring joints in a copper fabricated cage rotor in a 3.3 kV, 500 kW/670 HP, 6-pole SCIM, driving a seawater lift pump.

**Section 5.7:** Seven broken bars in a fabricated aluminum cage rotor in a 6.6 kV, 2238 kW/3000 HP, 4-pole SCIM driving a condensate pump. The end user could not simply stop this strategic motor after the MCSA diagnosis but, unfortunately, before it could be removed a piece of broken bar impacted the HV stator winding causing it to fail.

**Section 5.8:** Four broken bars in a copper fabricated cage rotor in a 4.16 kV, 2984 kW/4000 HP, 4-pole SCIM, driving a blower.

For each case history, when the number of rotor bars is unknown, an estimate of the condition of the cage winding is given using the average dB difference ($N_{av}$) between the $\pm 2sf$ sidebands and the supply component ($f$) of current using Table 4.2. When the number of rotor bars is known, the equivalent broken rotor bar factor or index ($BB_f$ or $n$, respectively) is predicted using equations (4.5) and (4.6) and these predictions are compared with the actual condition of the cage windings in rotors that were removed and made available for inspection.

## 5.1 CASE HISTORY (2000–2014)—SUMMARY AND KEY FEATURES

This SCIM was driving a strategic centrifugal compressor on an offshore oil and gas production platform operating in the North Sea off the coast of Scotland. The analysis that follows was the outcome of an annual MCSA survey of the HV SCIMs and is presented in considerable detail as an introductory case history.

The following data are from the nameplate of the motor: 3-phase SCIM, 11 kV, 10,000 kW, (13,400 HP), 574 A, 60 Hz, 1782 r/min, star connected with a nominal full-load slip of 1%. The rotor has 62 rectangular copper bars placed into annular grooves in the end rings. These connections are often referred to as butt joints, which are puddle brazed or joined by induction heating to the end rings. Interpretation of the measured current spectrum indicates that the $\pm 2sf$ sidebands were due to a normal level of cage winding asymmetry indicative of expected minor and normal differences due to, for example, different resistances across the bar to end ring joints.

### 5.1.1 MCSA Diagnosis

The motor was operating at 580 A compared to a full-load current of 574 A, thus the motor was on full-load and the operating slip from the measured $2sf$ sidebands should therefore be close to the full-load slip of 1%.

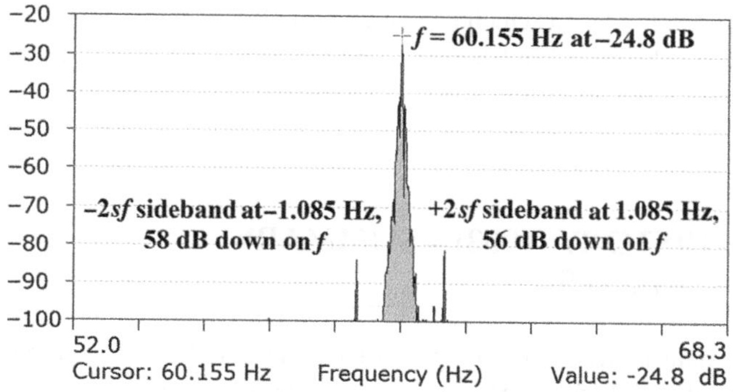


Figure 5.1 Current zoom spectrum, full-load amperes, 12,800 lines, baseband 0–130 Hz, 10.16 mHz/line.

For the avoidance of doubt, the digital signal processor in the MCSA instrument produces the frequency components to three decimal places. It is quite incorrect to simply round up or down any of the frequency values to a whole number. Neither the measured supply frequency nor the $\pm 2sf$ sidebands should be presented to less than two decimal places, since they must be accurately measured, as already explained at the start of Section 4.4.2.

The current spectrum in Figure 5.1 shows there are sidebands at ±1.09 Hz around the supply component $f$, at 60.16 Hz. The measured supply frequency is then used to calculate, from the measured sidebands, the operating slip which is 0.9% and this is close to the nominal full-load slip of 1% at a nominal frequency of 60 Hz. Note that the synchronous speed of the rotating magnetic field from the stator at an operating frequency of 60.16 Hz is 3610 r/min, compared to 3600 r/min at 60 Hz. The average dB difference ($N_{av}$) from Figure 5.1 is 57 dB and Table 5.1 is an extract from Table 4.2, which states that the cage winding is normal. The spectrum shown in Figure 5.1 has remained virtually constant from 2000 to 2016.

With an average dB difference ($N_{av}$) of 57 dB, the sideband currents as a proportion of the supply component is: $10^{-(57/20)} = 0.0014$, that is, the supply component is 714 times greater than the sidebands. Since the number of rotor bars is known, the equivalent broken rotor bar factor using equation (4.5) gives

$$BB_f = 2R/(10^{(N_{av}/20)} + 2p) = 2 \times 62/(10^{(57/20)} + 4) = 0.17$$

As a comparison, using Bellini's equation (4.6) from Chapter 4

$$n = (10^{(-58/20)} + 10^{(-56/20)}) \times R = 0.17$$

**TABLE 5.1 Extract from Table 4.2. Broken Rotor Bar Severity Estimator Based on Only the Average dB Difference ($N_{av}$) Between the $\pm 2sf$ Sidebands and the Supply Component**

| $N_{av}$ = dB Difference | Estimated condition of cage winding when the motor is operating at full-load output and speed at the shaft |
|---|---|
| 55–60 dB<br>"*Measured $N_{av}$ was* 57 dB *at full-load slip*" | A normal rotor cage winding for this 4-pole motor |

The threshold index at which a cage winding break normally exists for $BB_f$ or $n$ was proposed as being 0.5 in Chapter 4 and the $BB_f$ equal to 0.17 in this case history is only 30% of the proposed threshold index of 0.5, demonstrating cage winding normality.

## 5.2 CASE HISTORY (1983)—SUMMARY AND KEY FEATURES

The following data are from the nameplate of a 3-phase SCIM, 415 V, 125 kW/168 HP and 70 kW/94 HP, 200/150 A, 50 Hz, 2-speed, 1480/740 r/min, star connected. The cage winding had 88 bars, and was copper fabricated with rectangular bars brazed on top of the outer periphery of the end rings. The motor was directly coupled to a food mixer, used in the food processing industry to produce tinned rice. This was a very straightforward and classic MCSA case history that successfully detected broken rotor bars. The motor had an arduous duty cycle, which required it to start, between 15 and 18 times per hour and the plant owner reported that the cage windings were failing about every 2 years, almost certainly caused by too many direct-on-line (DOL) starts.

MCSA was initially applied to verify that the $\pm 2sf$ sidebands could be reliably detected and if so, that they were due to broken rotor bars. This knowledge would be used to decide on planned action, thus avoiding random failures and unplanned downtime of the process. As shown in Figure 5.2, these sidebands were both 38 dB down on $f$ which is indicative of a serious broken rotor bar problem. The rotor was removed at the next process shutdown and the faulty cage winding is shown in Figure 5.3. This confirmed the diagnosis and as expected, excessive heat had been generated

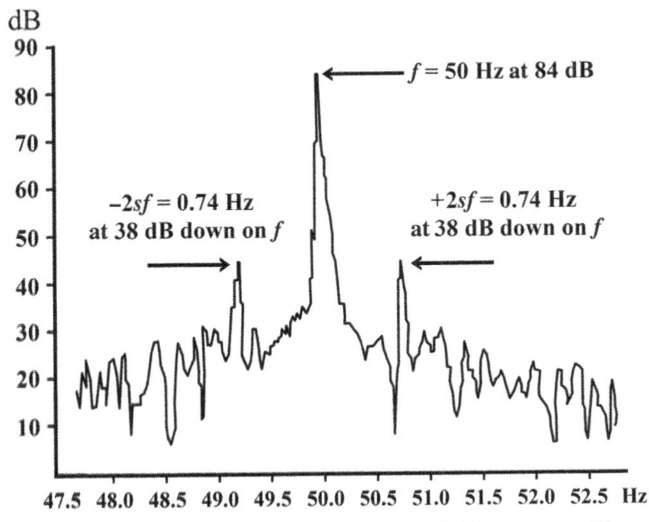


Figure 5.2 Sidebands due to broken rotor bars, 10.16 mHz/line. *Source*: Thomson and Fenger [5.1]. Reproduced with permission of IEEE.

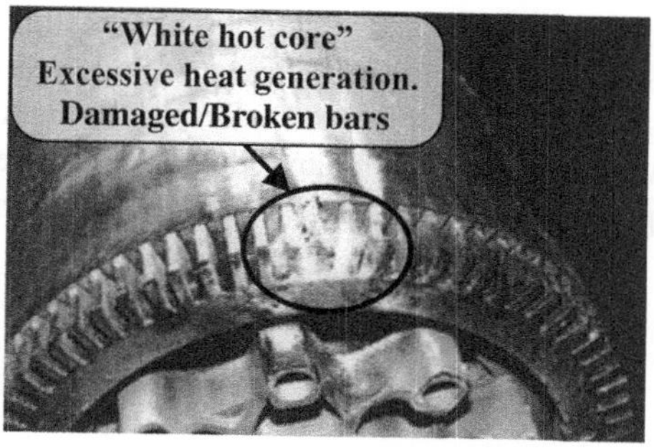


Figure 5.3 Severely damaged cage winding, overheated rotor core. *Source*: Thomson and Fenger [5.1]. Reproduced with permission of IEEE.

because of too many DOL starts. The owner was advised to invite OEMs, to provide a motor with a new rotor design that could meet the stringent starting requirements. These arrangements were left entirely to the plant owner. The current spectrum and its interpretation, together with a photograph of the rotor showing the broken bars are presented in Section 5.2.1

### 5.2.1 MCSA Diagnosis and Inspection

The analysis is deliberately presented in a step-by-step style, since it is the first industrial case history in the book which highlights a faulty cage winding and is useful as an initial learning guide and reference for an industrial engineer. The nominal full-load slip at the higher speed range was 1.33% and Figure 5.2 shows the current spectrum with $\pm 2sf$ sidebands ($\pm$0.74 Hz) around $f$ at 38 dB down on the supply component ($f$) which gives an operational slip of 0.74%.

The motor was operating at 130 A compared to the full-load current of 200 A, on the higher speed setting. Both sidebands are at the same magnitude and Table 5.2 is an extract from Table 4.2, with $N_{av}$ = 38 dB, the estimate is that a broken rotor bar problem exists.

**TABLE 5.2 Extract from Table 4.2. Broken Rotor Bar Severity Estimator Based on Only the Average dB Difference ($N_{av}$) Between the $\pm 2sf$ Sidebands and the Supply Component**

| $N_{av}$ = dB Difference | Estimated condition of cage winding when the motor is operating at full-load output and speed at the shaft. |
|---|---|
| 35–40 dB | A broken rotor bar problem exists. |
| Measured = 38 dB | $N_{av}$ = 40 dB in a 4-pole motor normally means that 1 or 2 bars are broken. |
| "*But note the measured value for* $N_{av}$ *was* 38 dB *but at* 56% *of the full-load slip, therefore the measured* dB *difference had to be corrected to* $N_{cav}$." | |
| The correct dB difference value is $N_{cav}$ = 36.0 dB | The dB difference is now closer to the lower limit of this band (35 dB value)—there is definitely a broken rotor bar problem that would merit stopping the motor for an inspection. |

Since the motor was on 56% of its rated output and slip, the broken bar correction factor ($BB_c$) is applied to get the dB difference referred to the full-load slip and rated full-load current. Now calculate the actual current in amperes of the $-2sf$ sideband (both sidebands were of equal magnitude) when the measured load current was 130 A. This component ($I_{LSB}$) was 38 dB down on $f$ and in absolute amperes

$$I_{LSB} = 10^{(-38/20)} \times 130 = 1.64 \text{ A}$$

$$BB_c = \frac{s_{FL}}{s_o} = \frac{1.33}{0.74} = 1.8$$

$$I_{LSBr} = I_{LSB} \times BB_c = 1.64 \times 1.8 = 3.0 \text{ A}$$

where $I_{LSBr}$ is the referred value of the $-2sf$ component with respect to the full-load slip and rated current, therefore a corrected dB difference value ($N_{cav}$) can replace $N_{av}$ in the dB difference chart of Table 4.2 as shown in Table 5.2. The full-load current = 200 A.

The corrected dB value is $N_{cav} = 20\log_{10}(I_{FL}/I_{LSBr})$

$$N_{cav} = 20\log_{10}(200/3.0) = 36 \text{ dB}$$

The corrected dB difference ($N_{cav}$) is now closer to the lower limit of this band (35–40 dB value in Table 4.2). There is definitely a broken rotor bar problem that merits stopping the motor for an inspection.

For completeness and since the number of rotor bars is known, the equivalent broken rotor bar factor ($BB_{fs}$) at the reduced load and operating slip of 0.74% is

$$BB_{fs} = 2R/(10^{(N_{av}/20)} + 2p) = 2.2$$

with $N_{av} = 38$ dB, $R = 88$, and $p = 2$.

It was verified in Section 4.5.4 that a broken bar correction factor should be applied to $BB_{fs}$ to account for the reduced slip operation, and is given by

$$BB_c = s_{fl}/s_{op} = (1.33/0.74) = 1.8$$

Thus the equivalent broken rotor bar factor corrected for the reduced slip operation and referred to the full-load slip is

$$BB_f = BB_c \times BB_{fs} = 4$$

Equation (4.6) can also be used to give an equivalent broken rotor bar index for the reduced load condition, and is termed ($n_{fs}$)

$$n_{fs} = \left[\frac{(I_{LSB} + I_{USB})}{I_p} \times R\right] = (10^{(-38/20)} + 10^{(-38/20)}) \times 88 = 2.2$$

Corrected to the reference slip of $s_{FL}$, using the broken bar correction factor ($BB_c$) gives $n = 4$.

The $BB_f$ and $n$ values are eight times higher respectively than the threshold index of 0.5. This confirms the rotor cage winding has a serious broken rotor bar problem. The recommendation was to stop the motor as soon as possible for an inspection.

The photo in Figure 5.3 confirms the following faults in the rotor cage and core:

**(i)** Evidence of excessive heat on the bars and the rotor core.

**(ii)** Bars melted and broken.

The rotor was beyond economical repair (BER, an acronym commonly used by the oil industry). The importance of this case is due to its simplicity. The history of regular breakdowns, an onerous duty cycle with excessive DOL starts, together with an MCSA test showing $\pm 2sf$ sidebands at 38 dB down on $f$, giving a $BB_f$ of 4, all meant that this case history was relatively easy to analyze.

Whether the measured dB difference or corrected dB difference (it was on reduced load) is applied to the broken bar or estimates $BB_f$ and $n$ via equations (4.5) and (4.6), respectively, the diagnosis and recommendation in each case are the same. If every case history was as straightforward as this one, there would perhaps be no need for this book, but of course that is not the case, as is subsequently shown in case histories in Chapters 5–9.

## 5.3 CASE HISTORY (1982)—SUMMARY AND KEY FEATURES

The motor's nameplate data are as follows: 3-phase SCIM, 134 kW/180 HP, 440 V, 210 A, 1478 r/min, 50 Hz, 0.88 p.f., delta, aluminum die-cast, single cage, deep bar design.

The number of rotor bars was unknown.

Two of these motors were used to drive centrifugal compressors, via step up gearboxes, on a small offshore gas production platform. One of the motors (A) had $\pm 2sf$ sidebands around $f$ at an average ($N_{av}$) of 34 dB down on $f$ when operating on reduced load (49% $s_{FL}$) and this equates to cage winding defects which merit the motor being shut down for inspection. An initial inspection of the rotor showed that there was no external evidence of any abnormalities, such as overheating of the rotor core close to the end rings, or any damage to the end rings such as cracks.

The rotor was then dissected and at the DE there were numerous defects such as porosity (blow holes) in the die-casting of the end ring, broken bars, and severe spark erosion, where the bars should have been an integral part of the cage winding. The motor was only 9 months old and a brand new rotor was supplied by the OEM. The current spectrum and its interpretation and a photograph of the faulty cage winding are presented in Section 5.3.1.

### 5.3.1 MCSA Diagnosis and Inspection

Motor A was previously removed and returned onshore because the operator suspected that a beating noise, coming from this motor, was due to a bearing fault. New bearings were fitted by an electric motor repair shop but the motor was only tested during an uncoupled, no-load run. Of course, any symptoms of a faulty cage rotor would not be exhibited on no-load, since there is negligible current in the cage winding.

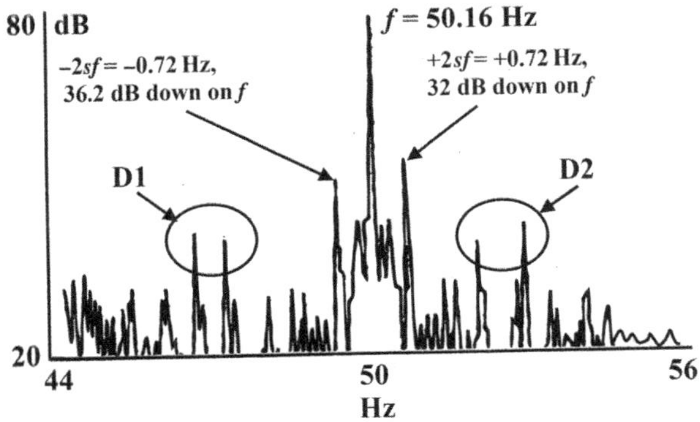


Figure 5.4 Current spectrum of motor A, 10.16 mHz/line.

The authors of this book very strongly recommend that repaired motors should be full-load tested via a heat run and the opportunity also taken to carry out an MCSA test to confirm the integrity of the cage winding before the motor is returned to the end user. The motor was returned offshore and the same low frequency beating noise existed, when on load, but disappeared when run uncoupled.

MCSA was applied to both motors to assess the condition of the cage windings and the current spectra are shown in Figures 5.4 and 5.5, using a frequency resolution of 10.16 mHz/line and a dynamic range of 80 dB.

Interpretation of both spectra indicates that there is a clear distinction between them in the vicinity, where $\pm 2sf$ sidebands from broken rotor bars, could exist. However, there are also similarities between both spectra indicated by regions D1 and D2 in each spectrum. These D-components are due to normal mechanical drive dynamics, downstream of the motors, being reflected back into the cage rotors. These reflections disturb the rotor and its magnetic field and hence affect the input current spectrum, but in this case, not in the vicinity of the $\pm 2sf$ sidebands.

The motors were driving centrifugal compressors via gearboxes but there was no information from the end user on either the construction and design of the gearboxes, such as the number of stages, number of teeth on pinions and gear wheels, or

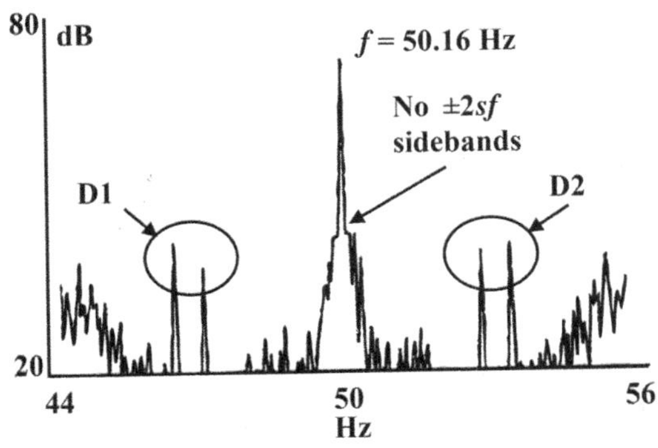


Figure 5.5 Current spectrum of motor B, 10.16 mHz/line.

**TABLE 5.3 Extract from Table 4.2. Broken Rotor Bar Severity Estimator Based on Only the Average dB Difference ($N_{av}$) Between the $\pm 2sf$ Sidebands and the Supply Component**

| $N_{av}$ = dB Difference | Estimated condition of cage winding when the motor is operating at full-load output and speed at the shaft |
|---|---|
| 30–35 dB<br>"*Measured* $N_{av}$ = 34 dB *but at 72% of the full-load slip. Compensated* $N_{cav}$ = 32. 8 dB" | A severe broken rotor bar problem exists |

the number of stages, impellors, and diffusers in the compressors, and therefore it is impossible to identify the definitive source of the components circled as D1 and D2.

There are sidebands at ±0.72 Hz around the supply component from the problem motor (A) and the nominal full-load slip is 1.47%, so at full-load the $\pm 2sf$ sidebands should nominally be at ±1.47 Hz around $f$. The motor, however, was operating on a considerably reduced load, at a current of 120 A compared to a full-load current of 210 A and hence the operational slip was only 0.72%. From the spectra, a serious fault in the cage rotor of motor A was predicted since the average dB difference, $N_{av} = 34$ dB. Table 5.3 clearly shows that this falls into the category of a severe broken rotor bar problem.

The average dB value at reduced load ($N_{av}$) can be corrected to ($N_{cav}$) which refers the measured $N_{av}$ on reduced slip to full-load slip to compensate for the reduced load.

$$I_{LSB} = 10^{(-34/20)} \times 120 = 2.4 \text{ A}$$

$$BB_c = \frac{s_{FL}}{s_o} = \frac{1.47}{0.72} = 2.0$$

Magnitude of lower $-2sf$ sideband referred to full-load slip and the current is therefore

$$I_{LSBr} = I_{LSB} \times BB_c = 2.4 \times 2.0 = 4.8 \text{ A}$$

The full-load current = 200 A

The corrected $N_{cav} = 20\log_{10}(I_{FL}/I_{LSBr})$

$$N_{cav} = 20\log_{10}(210/4.8) = 32.8 \text{ dB}$$

The compensated dB difference $N_{cav}$ is now very clearly in the dB difference band (30–35 dB), which indicates that severe cage winding breaks exist.

The recommendation was to stop the motor and inspect the rotor, a photograph of which is shown in Figure 5.6. There were no visual signs of overheating on the rotor core, or on the laminations close to the end rings or any external abnormalities on the end rings. The rotor core was then machined, through the full diameter of the rotor, so that an axial view of the bars to end ring region was visible at the drive end (DE) and non-drive end (NDE).

Figure 5.7 shows that after dissection it was very clear that this rotor cage winding had very serious defects such as numerous broken bars, porosity caused by faulty

Figure 5.6 Cage rotor, no signs of overheating or external damage. *Source*: Thomson and Fenger [5.2]. Reproduced with permission of IEEE.

die-casting, spark erosion, and also partially broken bars and blow holes thus proving the diagnosis via MCSA to be valid.

## 5.4 CASE HISTORY (2002)—SUMMARY AND KEY FEATURES

The nameplate provided the following data: 3-phase 11 kV, 3100 kW/4155 HP, 195 A, 50 Hz, 2974 r/min (2-pole), star connected SCIM. The rotor which is shown in Figure 5.8 has 38 bars with rectangular copper bars placed into annular grooves in the end rings. These are often referred to as butt joints and puddle brazing or induction heating is used to manufacture them.

Four of these motors were driving seawater injection pumps on an offshore oil production platform, with three of the four drives required to meet the highest

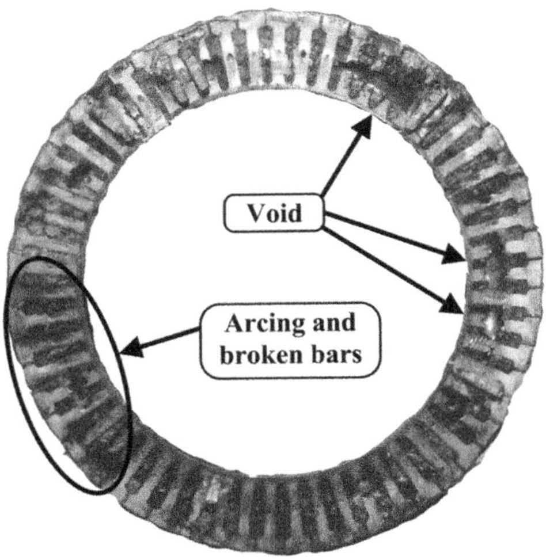


Figure 5.7 Shows the faulty cage winding after the dissection was completed.

Figure 5.8 Photo of rotor.

demand. In 2000, for the first time, all four motors were MCSA tested and one of the four was found to have twice slip frequency sidebands ($\pm 2sf$) at an average of 46.5 dB down on the supply component. With no previous MCSA data, this MCSA result suggests that it is very much a borderline case. However, the analogue ammeter was very slowly and steadily swinging about a mean position and a beating noise (but not particularly pronounced) could be heard from the motor.

In this case, the interpretation of the spectrum is considerably more subtle than simply using a dB difference to assess the condition of the cage winding. Certain key features in the current spectrum are identified in the analysis that follows in Section 5.4.1 that suggests that there was a broken rotor bar problem and it was not a borderline case.

When the number of rotor bars is unknown, the use of the dB difference between the $\pm 2sf$ sidebands and the supply component to categorize the condition of cage winding, as proposed in Table 4.2, is the only available option. In borderline cases, this dB difference is a first stage indicator and other characteristics in the spectrum need to be considered. If there was a broken rotor bar, there was an inherent risk of a broken bar lifting and damaging the stator, due to lack of end ring retaining rings covering the bar to end ring joints, so the operator was advised to replace the rotor as soon as possible. An inspection revealed that one completely broken bar had lifted out of its slot and crashed into the stator winding and core, that damage necessitated a complete stator rewind. The owner decided to purchase a new rotor. The current spectrum and its interpretation with photographs of the faulty rotor and damaged stator winding are presented in Section 5.4.1.

### 5.4.1 MCSA Diagnosis and Inspection

The motor was operating on a reduced load of 170 A compared to the full-load current of 195 A, thus the operational slip was less than the full-load slip (0.87%) and likewise any $\pm 2sf$ sidebands due to broken rotor bars were less in frequency than the prediction at full-load, which is 0.87 Hz. That prediction assumes that the operational supply frequency is 50 Hz and that the full-load nameplate speed is correct. The current spectrum is shown in Figure 5.9 and there are sidebands at $\pm 0.69$ Hz around the supply component $f$ at 50.03 Hz.

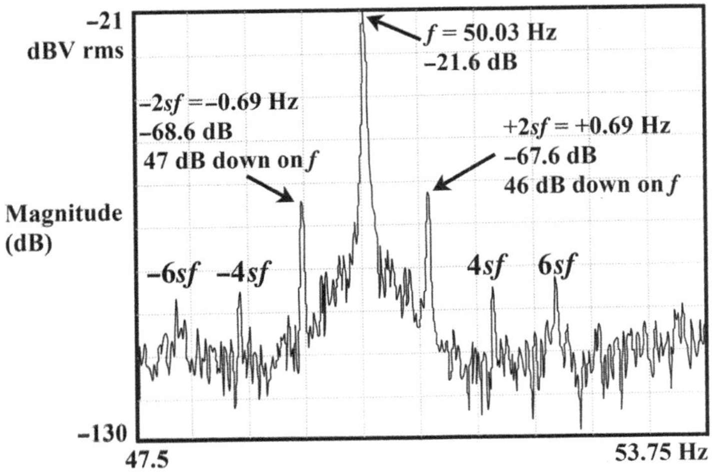


Figure 5.9 Current zoom spectrum, line resolution 10.16 mHz/line.

If these sidebands are considered to be $\pm 2sf$ sidebands, this gives an operational slip of 0.69%. Using only the average dB difference ($N_{av}$) between the sidebands and $f$, this initially suggests (see Table 5.4, extract from Table 4.2) an acceptable operational level of rotor cage asymmetry for a 2-pole motor but the motor was operating at a reduced load of 80% of the full-load slip, this should still be considered. Initially, at the time of the MCSA measurement the number of rotor bars was unknown, thus the only option was to use the dB difference guidance chart for the MCSA analysis.

There was too much uncertainty to make a diagnosis via only the dB difference approach when the $N_{av}$ value was 46.5 dB (compensated to $N_{cav}$ = 45.8 dB) and secondly this was a critical, high speed 2-pole motor that did not have any end ring retaining rings, which covered the bar to end ring joints. The OEM subsequently provided the number of rotor bars (38) to the owner of the motor. Using equation (4.5), the equivalent broken rotor bar factor $BB_{fs}$ at the operational slip is given by

$$BB_{fs} = 2R/[10^{(N_{av}/20)} + 2p] = 2 \times 38/[10^{(46.5/20)} + 2] = 0.36$$

**TABLE 5.4 Extract from Table 4.2. Broken Rotor Bar Severity Estimator Based on Only the Average dB Difference ($N_{av}$) Between the $\pm 2sf$ Sidebands and the Supply Component**

| $N_{av}$ = dB Difference | Estimated condition of cage winding when the motor is operating at full-load output and speed at the shaft. |
|---|---|
| 45–50 dB<br>"*Measured* $N_{av}$ *was* 46.5 dB *but at an operational slip of* 80% *of the full-load slip*"<br>$N_{cav}$ = 45.8 dB, *very little change from* $N_{av}$ *since the motor was on* 80% $s_{FL}$ | This is the difficult boundary range for using the dB difference. At 45 dB in lower pole number motors (2 and 4) there is a possibility of either faulty joints or perhaps a broken rotor bar. |

Taking into account that the spectrum was recorded during reduced load and at a slip of 80% $s_{\text{FL}}$ the $BB_{\text{c}}$ correction factor is applied to $BB_{\text{fs}}$.

$$BB_{\text{f}} = (BB_{\text{c}}) \times (BB_{\text{fs}}) = (0.87/0.69) \times 0.36 = 0.45$$

As a comparison, use equation (4.6) which takes the magnitude in amperes of both sidebands into account relative to $f$ and the broken bar index at the reduced load $n_{\text{fs}}$ is given by

$$n_{\text{fs}} = \left[\frac{(I_{\text{LSB}} + I_{\text{USB}})}{I_p} \times R\right] = (10^{(-46/20)} + 10^{(-47/20)}) \times 38 = 0.36$$

Applying the $BB_{\text{c}}$ correction factor gives $n = 0.45$.

In this case, the two predictions are the same since the difference in magnitude between the $\pm 2sf$ sidebands relative to $f$ is negligible. The equivalent broken bar factor ($BB_{\text{f}}$) or broken bar index ($n$) is very close to the threshold value of 0.5 above which a cage winding break is normally present but it is very much a real borderline case, with dubiety as to whether there is a cracked or broken rotor bar or a number of high resistance, faulty joints.

The normal recommendation, since this was the first time the motor had been MCSA tested, would be to repeat the MCSA test in 3 months time to start trending the current signature to identify if the condition is stable. However, a more detailed interpretation of the current spectrum in Figure 5.9 indicates the following:

There are second and third order harmonic sidebands of $\pm 2sf$ around $f$ at $\pm 4sf$ and $\pm 6sf$. Although the sidebands at $\pm 4sf$ and $\pm 6sf$ are $\geq 64$ dB down on $f$ and $\geq 18$ dB down on the $\pm 2sf$ sidebands, nevertheless they are present and in this case history, where there is no evidence of disturbances from the mechanical load (see Chapter 7) this is a second order effect of an oscillating rotor at $2sf$ due to a cage winding break.

Their presence provides supporting evidence that a rotor cage problem exists, probably a broken bar.

**Conclusions and Recommendations:** Based on all these additional facts, including the ammeter swing and acoustic noise with a low frequency beating characteristic and the quantifiable MCSA result of an equivalent broken bar factor at 0.45 ($\cong 0.5$, the threshold value), there was sufficient evidence to indicate there was a very high probability that a broken rotor bar existed.

It was recommended that the motor should be stopped and inspected as soon as practically possible, since a broken rotor bar or part of a bar could lift out of its slot and because end ring retaining rings were not fitted over the joints between the bars and the end ring, a broken rotor bar could crash into the stator.

The risks were high that consequential damage to the 11 kV stator winding and possibly the stator core could occur, incurring both downtime and an expensive repair. Figure 5.10 confirmed there was one completely broken rotor bar but unfortunately the broken part had already lifted out of the rotor slot and damaged the 11 kV stator winding, as shown in Figures 5.11 and 5.12.

The stator winding had not actually failed while in operation, unfortunately there was mechanical damage to the main wall insulation of the coils in the end

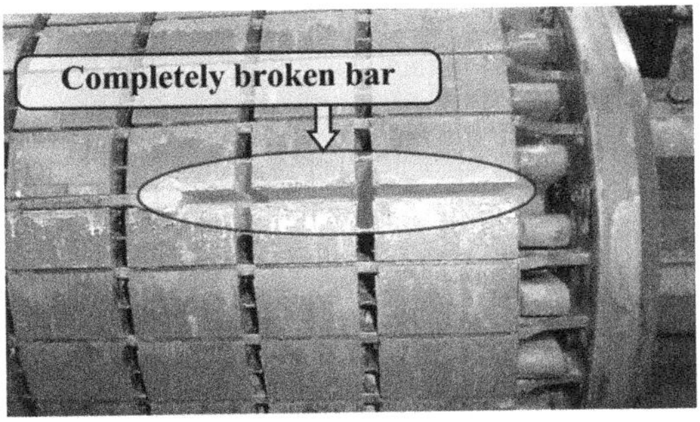


Figure 5.10 Close up of the cage rotor with one broken rotor bar.

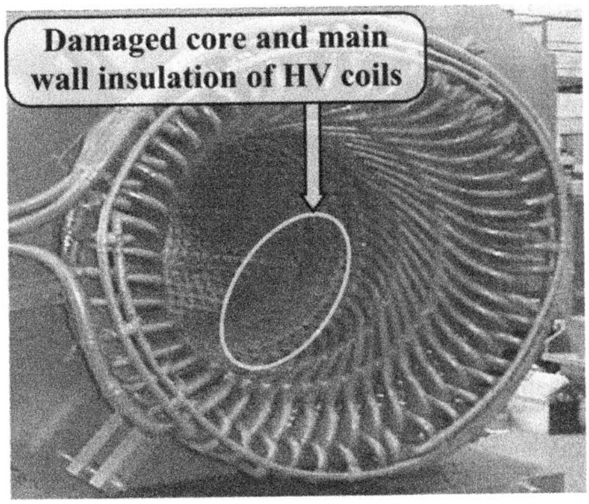


Figure 5.11 Overview of damaged stator winding and core.

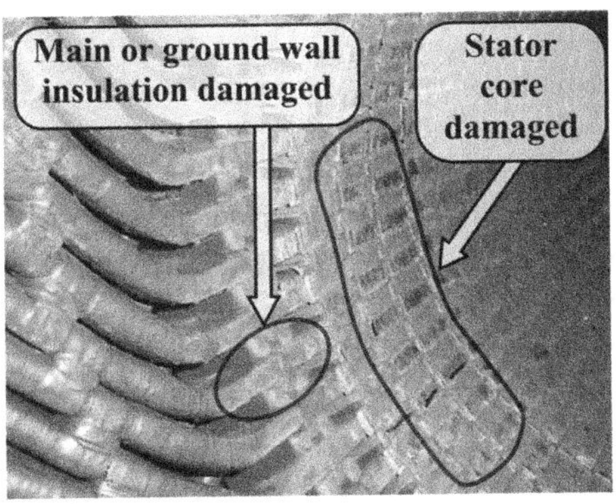


Figure 5.12 Close up of the damaged stator winding and core.

winding region but only minimal surface damage to the stator core laminations. This surface damage was repaired and a new core was not required but the stator was rewound. The diagnosis was in January 2000 but the motor was not stopped until an outage in April 2000 and of course the bar could have lifted in the interim period between January and April 2000.

With reference to the reason for the broken rotor bar, there was no evidence of heating on the rotor core or on adjacent bars caused by too many DOL starts in a short period of time, or starts not in accordance with the OEM's starting specifications. It can only be speculated that the bar failed due to mechanical stresses and fatigue, possibly enhanced by a weak bar to end ring joint, and the bar lifted out of the slot due to centrifugal forces (high speed 2-pole SCIM) since it was not tight enough.

## 5.5 CASE HISTORY (1985–1987) – SUMMARY AND KEY FEATURES

The nameplate provided the following data: 3-phase SCIM, 6.6 kV, 60 Hz, 3600 kW/4825 HP, 376 A, 3580 r/min, star connected. The rotor was copper fabricated with a single 46 bar cage using rectangular bars and butt type joints via annular grooves in the end rings. No end ring retaining rings were fitted.

Three motors were used to drive seawater injection pumps on an offshore oil and gas production platform. Annual MCSA measurements commenced in 1985 and one of the motors consistently gave $\pm 2sf$ sidebands at 48 dB down on the supply component. This was considered to be due to inherent differences in the resistances of a number of the bar to end ring joints, producing some cage asymmetry, but at an acceptable level. There were no $\pm 4sf$ sidebands. The other two motors had no sidebands.

In 1987, the owner reported that the overall vibration displacement of the shaft, measured via eddy current type displacement probes in the bearing housings, had increased from 50 μm (or 2 mils or thou) pk to pk to 100 μm (or 4 mils/thou) pk to pk, nearing the alarm level of 110 μm pk to pk for this motor. This increase occurred during a 3-month period after a stall in the drive train, followed by DOL starts during that time. Neither the reasons for the stall nor the number of starts were given. The owner then requested an MCSA test to be carried out and it was found that the $\pm 2sf$ sidebands were now both 26 dB down on the supply component and that $\pm 4sf$ sidebands were also present.

There was therefore a very serious broken rotor bar problem, so, for health and safety reasons, this motor was immediately stopped. Inspection of the rotor revealed that there were 20 broken bars out of a total of 46. The current spectra were recorded by a Polaroid camera, from the screen of a spectrum analyzer from 1985 to 1987, and their interpretation, accompanied by a photograph of the rotor showing the broken bars, are presented in Section 5.6.1.

### 5.5.1 MCSA Diagnosis and Inspection

Figure 5.13 gives the current spectra from three annual routine surveys recorded from 1985 to 1987. The motor was operating at its full-load current of 375 A, hence the nominal full-load slip from the nameplate data was 0.55% and the predicted $\pm 2sf$ sidebands were at ±0.66 Hz. In Figure 5.13, there are clearly sidebands at ±0.66 Hz but they are 48 dB down on the supply component and had been for 2 years.

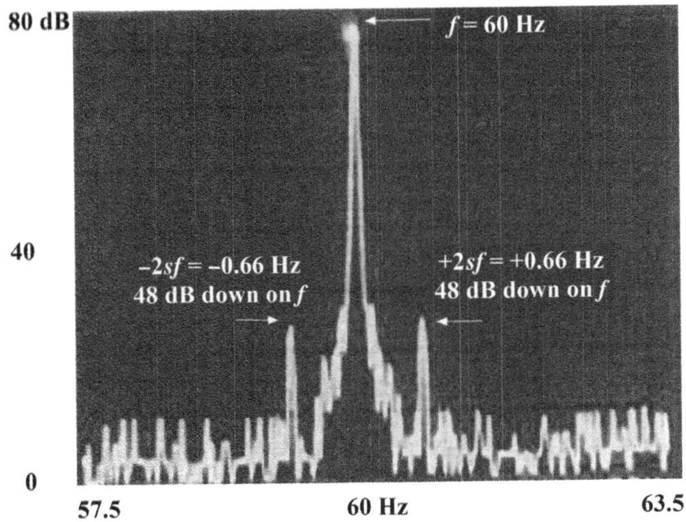

Figure 5.13 Current spectrum, 1985.

The spectrum in Figure 5.13 gives an equivalent broken bar factor using equation (4.5), which is,

$$BB_{\mathrm{f}} = 2R/(10^{(N_{\mathrm{av}}/20)} + 2p) = 0.36$$

with $N_{\mathrm{av}} = 48$ dB, $R = 46$, and $p = 1$

with a $BB_{\mathrm{f}}$ of 0.36, this is considered to be an operationally acceptable rotor cage asymmetry and it had been stable for 2 years.

Five months after the annual survey in 1987, the vibration had doubled as explained in Section 5.5 and a repeat MCSA survey was carried out and the spectrum at full-load current is shown in Figure 5.14. The $\pm 2sf$ sidebands are now 26 dB down on $f$ and there are $\pm 4sf$ sidebands, which were 49 dB down on $f$ and 24 dB down on the $\pm 2sf$ sidebands. It should also be observed that the $\pm 2sf$ sidebands have changed from $\pm 0.66$ Hz ($s = 0.55\%$) to $\pm 0.9$ Hz (0.75%) around the supply component, $f$. Suffice to state (with hindsight) that after the rotor was inspected, it was known there were 20 broken bars which therefore caused an increase in the rotor's resistance and reactance, since there were 26 nominally identical bars in parallel in the cage winding compared to 46.

The rotor was still required to deliver the full-load rated output being demanded by the pump, thus the rotor current had to increase to try to produce the required output power and torque for the pump, therefore the SCIM was no longer performing as per its design value of rotor impedance and consequently the speed drops, the slip and frequency of $2sf$ increases.

An extract from Table 4.2 in Table 5.5 indicates there was a very serious broken rotor bar problem since the $\pm 2sf$ sidebands were both 26 dB down on $f$.

In this case history, there was no requirement to know the number of rotor bars or do any estimates via the equivalent broken rotor bar factor. The motor was immediately stopped.

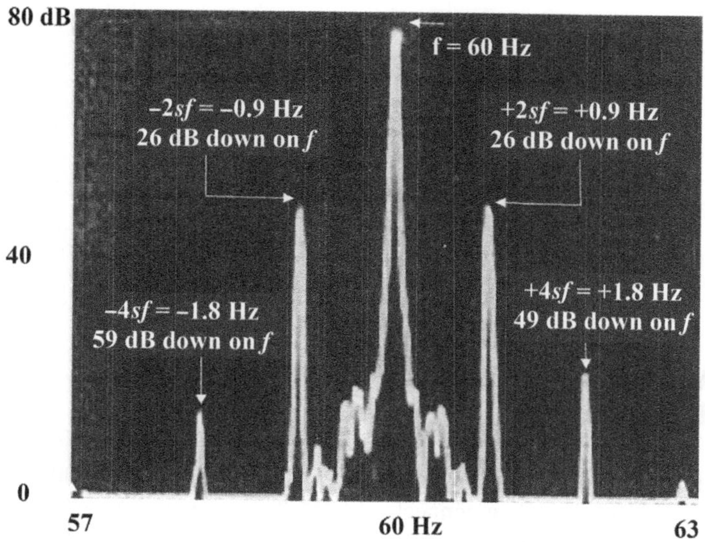

Figure 5.14 Current spectrum, multiple broken rotor bars, 1987.

The rotor was inspected and a photograph is shown in Figure 5.15. On inspection, 20 of the 46 bars had broken, but none of them had lifted out of the slots, which indicates they were tight when the motor was manufactured, as they should be. The repair company reported that there was no obvious stator winding or stator core damage, which was very fortunate indeed.

(i) The added value from presenting this case history is to show that a large, HV SCIM can still operate with 20 broken bars and it was still able to produce sufficient torque and power to keep the motor turning the pump.

(ii) Although the rotor shaft displacement was close to the alarm level, this indicates that even with this serious broken rotor bar problem the shaft displacement was still at an operationally acceptable level. Thus the operator was really unaware of the seriousness of the cage winding breaks by monitoring only vibration but MCSA clearly demonstrated that it had to be shut down immediately.

(iii) The increase in the overall shaft displacement, as reported by the owner, at the 1 × r/min frequency component was clearly due to an increase in mechanical unbalance of the rotor due to the large number of broken bars.

**TABLE 5.5 Extract from Table 4.2. Broken Rotor Bar Severity *ESTIMATOR* Based on the Average dB Difference ($N_{av}$) Between the $\pm 2sf$ Sidebands and the Supply Component**

| $N_{av}$ = dB Difference | Estimated condition of cage winding—motor should ideally be operating at full-load output ($s_{FL}$) but this is very often not the case. |
|---|---|
| <30 dB | A very serious broken rotor bar problem exists for all pole numbers. There will normally be multiple broken rotor bars. |

Figure 5.15 Rotor showing broken bars.

## 5.6 CASE HISTORY (2006)—SUMMARY AND KEY FEATURES

The nameplate provided the following data: 3-phase, 3300 V, 109 A, 500 kW/670 HP, 60 Hz, 6-pole, 1190 r/min, star connected SCIM. The rotor cage winding was copper fabricated with 60 rotor bars, with bar to end ring joints as shown in Figure 5.16; end ring retaining rings were not fitted.

Two nominally identical induction motors were driving seawater lift pumps (*note that the motors are not accessible*) on an offshore oil and gas production platform. The motors A and B were MCSA tested and motor A had $\pm 2sf$ sidebands both at 46 dB down on $f$ and there were no sidebands whatsoever in motor B. It was recommended that the motor continue to run and MCSA should be carried out at 3 monthly intervals and the rotor should be inspected when convenient for the end user—the detailed analysis and reasons for this recommendation are given in Section 5.6.1.

### 5.6.1 MCSA Diagnosis and Inspection

For these motors, when operating at full-load, the $\pm 2sf$ sidebands should nominally occur at $\pm 1.0$ Hz, but motors A and B were operating at 85 A and 75 A, respectively, compared to the full-load current of 109 A. Figures 5.17 and 5.18 give the current spectra for motors A and B. Motor A has $\pm 2sf$ around $f$ and since motor A was operating on a reduced load at 85 A, the operational slip will be less than the full-load slip of 0.83% and likewise any $\pm 2sf$ sidebands, due to broken rotor bars, will be less in frequency than the prediction at full-load.

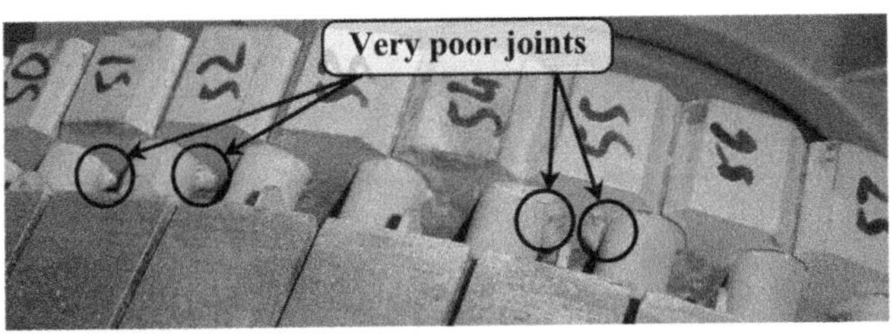


Figure 5.16 Visual inspection after cleaning and prior to NDT dye penetrant test.

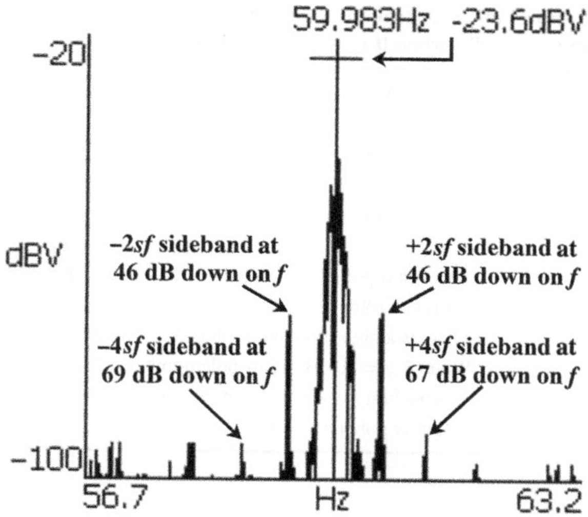


Figure 5.17 Motor A) current spectrum, 10.16 mHz/line.

The first step was to compensate the measured $N_{av}$, since the motor was only on 60% $s_{FL}$.

$$I_{LSB} = 10^{(-46/20)} \times 85 = 0.43 \text{ A}$$

$$BB_c = \frac{s_{FL}}{s_o} = \frac{0.83}{0.5} = 1.67$$

$$I_{LSBr} = I_{LSB} \times BB_c = 0.43 \times 1.66 = 0.7 \text{ A}$$

The full-load current = 109 A

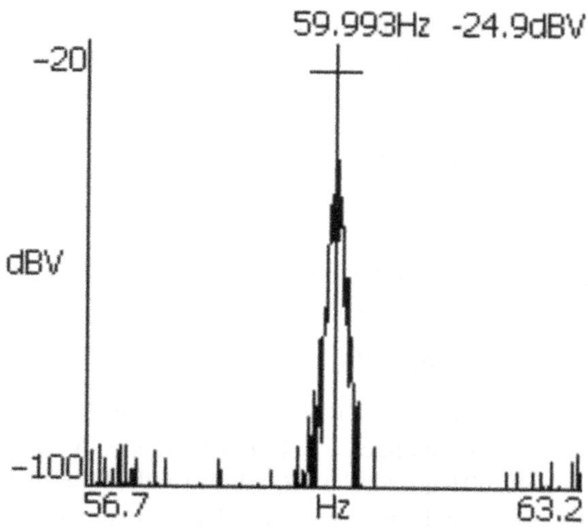


Figure 5.18 Motor B current spectrum, 10.16 mHz/line.

**TABLE 5.6 Extract from Table 4.2. Broken Rotor Bar Severity Estimator Based on Only the Average dB Difference ($N_{av}$) Between the 2*sf* Sidebands and the Supply Component**

| $N_{av}$ = dB Difference | Estimated condition of cage winding when the motor is operating at full-load output and speed at the shaft. |
|---|---|
| 45–50 dB | This is the difficult boundary range for using the dB difference. |
| "*Measured* $N_{av}$ = 46 dB *but note at only 60% of the full-load slip.* | When $N_{av}$ = 45 dB in lower pole number motors (2 and 4) there is a possibility of either faulty joints, or perhaps a broken rotor bar. |
| *Both sidebands were equal in magnitude.*" | This is a 6-pole SCIM—there is therefore a higher probability of a broken rotor bar. |
| 40–45 dB | There is a high probability of a cage winding fault. |
| $N_{cav}$ = 44 dB. | When $N_{av}$ = 40 dB in low pole numbers (2 and 4) there is a high probability of one or even two broken rotor bars. At 40 dB with higher pole numbers the probability of more broken bars increases. |

The corrected $N_{cav} = 20\log_{10}(I_{FL}/I_{LSBr})$

$$N_{cav} = 20\log_{10}(109/0.7) = 43.8\,\text{dB}$$

This now brings the dB difference $N_{cav}$ into the next severity band as shown in Table 5.6 and this suggests there is a high probability of a cage winding fault.

Recall the case history in Section 5.4 with 38 bars in a 2-pole rotor where the dB difference between the $\pm 2sf$ sidebands and $f$ was an average of 46.5 dB but there were clearly $\pm 4sf$ and $\pm 6sf$ sidebands. This information was also used to interpret the spectrum and finally predict there was a broken rotor bar problem in that motor which interpretation was found to be correct.

Examination of the spectrum for this case history shows that there are harmonics of $\pm 2sf$ at $\pm 4sf$ but none at $\pm 6sf$ in the spectrum shown in Figure 5.17 and the $\pm 4sf$ sidebands are very low at 69 dB down (i.e., 2800 times smaller) on $f$ and are considered to be negligible.

For completeness and since the number of rotor bars is known, the equivalent broken rotor bar factor ($BB_{fs}$) at the reduced load and operating slip of 0.5% is

$$BB_{fs} = 2R/(10^{(N_{av}/20)} + 2p) = 0.58$$

with $N_{av}$ = 46 dB, $R$ = 60, and $p$ =3

$$BB_c = s_{FL}/s_{op} = (0.83/0.5) = 1.67$$

Thus the equivalent broken rotor bar factor corrected for the reduced slip operation and referred to the full-load slip, is

$$BB_f = BB_c \times BB_{fs} = 0.97$$

Thus the effect of the number of rotor bars on the estimated broken bar factors is significant, since with 38 bars in case 5.4 the $BB_f$ was 0.45 with $N_{av}$ = 45.6 dB. Whereas with this 6-pole, 60 bar rotor the $BB_f$ is 0.97 (with $N_{av}$ = 46 dB) which seems to suggest this case history will have more cage defects than the one in Section 5.4. However, that is not necessarily the case since as explained previously the features in this case history do not provide enough supportive evidence to make that necessarily

Figure 5.19 (a) and (b) A sample of 1 out of 10 faulty joints is shown by the leakage of the dye penetrant—that is the dark parts shown.

true. It was recommended that the motor should be MCSA surveyed at 3 monthly intervals to trend the spectrum to ascertain if this was a stable condition.

However, before the next MCSA survey, due to a pump failure, this motor was scheduled for removal for a complete overhaul and refurbishment. As part of the end user's planned maintenance (PM) strategy on an offshore oil production platform, it was decided to carry out a thorough inspection of all bar to end ring joints to determine the condition of the cage winding.

There was no visual evidence of any cracked or broken bars prior to the dye penetrant test but there was clear evidence of numerous sub-standard joints, as shown in Figure 5.16. The paint over the bar to end ring joints was removed and the cage winding was cleaned for an NDT dye penetrant test.

The dye penetrant test confirmed that there were numerous very poor quality bar to end ring joints and Figures 5.19a and 5.19b show a close-up sample of 1 of the 10 sub-standard brazed joints to the DE end ring. The rotor was re-barred and new end rings were also fitted.

- **(i)** This case history verifies that MCSA cannot distinguish between a high percentage of faulty bar to end ring joints and a rotor with actual broken bars since both can estimate rotor cage asymmetry of the same order and this is a disadvantage of MCSA.
- **(ii)** It could be argued that this motor could have continued to operate but oil companies operating offshore oil production platforms prefer to have their motors in a reliable condition.

## 5.7 MCSA CASE HISTORY (2004)—SUMMARY AND KEY FEATURES

The nameplate on the motor for this case study provided the following data: 3-phase, 6600 V, 2238 kW/3000 HP, 60 Hz, 1775 r/min, 0.9 p.f., 96.5% 228 A. The number of rotor bars was unknown.

This motor was driving a condensate pump in a coal fired power station and it was suspected that it had a broken rotor bar problem because, as "reported" by the end user, it emitted an uncharacteristic low frequency modulated humming noise during motor operation. It had abnormal vibration levels during start-up and while it was running (but note the end user gave no quantitative vibration levels) and the coupling between the motor and pump was found to be oscillating, when viewed with a stroboscope, again as reported by the end user but the end user could not determine the problem via vibration measurements and analysis.

The end user reported that the plant had a vibration data logger that reputedly had a current signature analysis capability, but it could not zoom into the 60 Hz region of the spectrum to establish the presence or absence of $\pm 2sf$ sidebands around the supply component ($f$) in the current spectrum. If the sidebands were at a fault level, this would confirm the presence of cage winding breaks. The firm's data logger had been incorrectly specified and with no 60 Hz zoom capacity was useless for MCSA testing.

An MCSA analysis with an instrument that did have the correct specification to display the $\pm 2sf$ sidebands around $f$ was recommended to scientifically establish if the uncharacteristic acoustic noise and vibration symptoms, reported by the end user, were in fact due to cage winding breaks.

### 5.7.1 MCSA Diagnosis and Inspection

An MCSA test was performed in January 2004. The full-load slip from the nameplate speed is 1.4% but the motor was on a reduced load at a current of 137 A compared to the full-load current of 228 A. Therefore, the operating slip will be much less than the full-load slip.

There are symmetrical sidebands around $f$ at 59.34 Hz and 60.66 Hz and these were considered to be the $\pm 2sf$ sidebands ($\pm 0.66$ Hz) giving an operational slip of 0.55% (39% of the full-load slip) and a corresponding speed of 1490 r/min.

The dB difference ($N_{av}$) between the fundamental 60.0 Hz power supply frequency and the sidebands is 29 dB. Figure 5.20 also establishes the presence of $\pm 4sf$ second harmonics of the $\pm 2sf$ sidebands. This is a characteristic feature of the current spectra from motors with cage winding breaks as already shown in previous case histories in this chapter, which provides further confirmation, if any is needed, of rotor bar failure.

There is no requirement to correct the $N_{av}$ from the lightly loaded value to a corrected $N_{cav}$ at full-load and an extract of the broken bar severity chart is given in Table 5.7.

The diagnosis is obvious: The motor has multiple broken rotor bars and should be stopped. However, before this could be done, a piece of rotor bar broke free and impacted the HV stator winding causing it to fail. This was a clear cut case, which shows that reports by the end user of abnormal vibration and an uncharacteristic acoustic noise were insufficient for the *end user to make a decision to stop the motor but it should have been.*

If MCSA had been correctly applied much earlier, when for example, only one broken bar existed this would have been detected and the consequential expensive

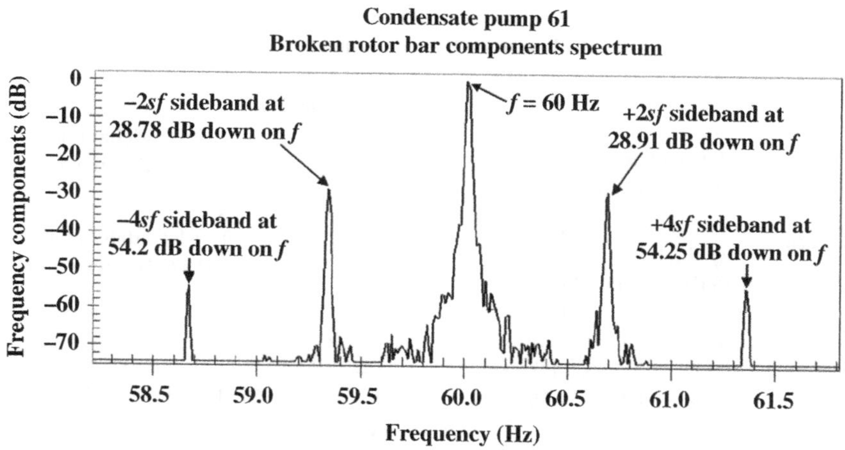


Figure 5.20 Current spectrum indicating $\pm 2sf$ and $\pm 4sf$ sidebands, 10.16 mHz/line.

**TABLE 5.7 Extract from Table 4.2. Broken Rotor Bar Severity Estimator Based on Only the Average dB Difference ($N_{av}$) Between the $\pm 2sf$ Sidebands and the Supply Component**

| $N_{av}$ = dB Difference | Estimated condition of cage winding when the motor is operating at full-load output and speed at the shaft |
|---|---|
| <30 dB<br>*"In this case* $N_{av}$ = 29 dB *and at only* 39% *of the full-load slip"* | A very serious broken rotor bar problem exists for all pole numbers, normally multiple broken rotor bars |

stator winding failure could have been prevented. The motor was taken out of service and sent to an electric motor shop for inspection in March 2004. When the motor was disassembled 7 bars were found to be broken in the mid-section of the rotor cage winding.

The original rotor cage was of fabricated aluminum type, which is more likely to crack under the thermal stresses induced during motor starting (see Section 3.10.6). On this basis, the cage winding was replaced by a new one made from copper alloy and the motor was put back in service in May 2004 and has operated reliably since then.

## 5.8 MCSA CASE HISTORY (2004)—SUMMARY AND KEY FEATURES

The nameplate on the motor, for this case study, provided the following data: 3-phase, 4.16 kV, 4000 HP/2984 kW, 60 Hz, 1792 r/min (4-pole) star connected with a full-load current of 470 A and the cage winding has 155 rotor bars. This motor was driving a blower in a gas separation plant and the plant maintenance engineers requested an MCSA test by an external contractor, with significant MCSA experience, without giving any background information as to whether the motor exhibited any external symptoms of developing a broken rotor bar problem.

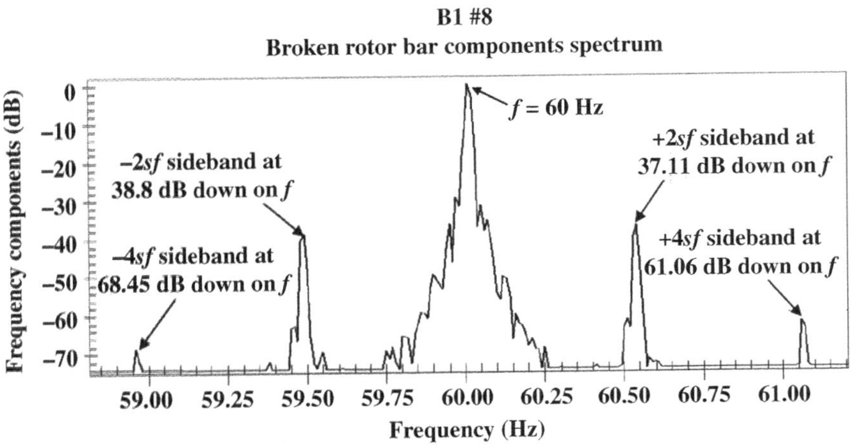


Figure 5.21 Current spectrum, note the zoom spectrum analysis gives a frequency resolution of 0.063 Hz/division to cope with a full-load slip of 0.44% and the line resolution is 0.10 Hz/line.

## 5.8.1 MCSA Diagnosis and Inspection

The motor was operating at its full-load rated current and the full-load nameplate speed is 1792 r/min which equates to a slip of 0.44%. The current spectrum in Figure 5.21 shows there are $\pm 2sf$ sidebands at $\pm 0.52$ Hz around the supply component of current ($f$) at 60 Hz which gives a slip of 0.43% which ties in with the full-load slip of 0.44%.

With an average dB difference of $N_{av} = 38$ dB, the extract from Table 4.2 given in Table 5.8 indicates that there is a broken rotor bar problem.

The end user requested more definitive information on the predicted number of broken rotor bars and since the number of bars was known at 155—this seemed to be on the high side for a 4.16 kV, 4000 HP/2984 kW, 4-pole SCIM. The estimated equivalent broken bar factor is

$$BB_{f} = (2 \times 155)/(10^{38/20} + 4) = 3.7\ (\cong 4.0)$$

The recommendation was to shut down the motor and carry out an inspection of the cage winding. The plant engineer indicated that they had already inspected the

**TABLE 5.8 Extract from Table 4.2. Broken Rotor Bar Severity Estimator Based on Only the Average dB Difference ($N_{av}$) Between the $\pm 2sf$ Sidebands and the Supply Component**

| $N_{av}$ = dB Difference | Estimated condition of cage winding when the motor is operating at full-load output and speed at the shaft. |
|---|---|
| 35–40 dB | A broken rotor bar problem exists. |
| Measured $N_{av} = 38$ dB | $N_{av} = 40$ dB in a 4-pole motor normally means that 1 or 2 bars are broken. However, in this case at $N_{av} = 38$ dB it is highly likely that at least 2 or 3 bars are broken. |

cage winding immediately after the on-site MCSA test and it had 4 broken bars but unfortunately no photos were provided by the end user of the damaged cage winding.

For this motor the calculated $BB_f$ gave a good estimate of the number of broken bars (3.7) compared to the actual 4 broken bars. The reason for the relatively good correlation, in this case history, is because in Section 4.5 it was stated that the value estimated by the $BB_f$ will be closest to the actual provided the *number of broken bars is* $\ll R$, [5.2 to 5.4], and in this case this is true since $R = 155$ and the number of broken bars was 4.

## 5.9 QUESTIONS

**5.9.1** What information should be obtained about an HV SCIM and the driven mechanical load before even attempting to carry out an MCSA measurement and analysis to estimate the operational condition of its cage winding? Justify your answer.

**5.9.2** Why should the end user contact the motor manufacturer to *try* to obtain the type of cage winding (copper fabricated or aluminum die-cast), the number of rotor bars, the type of bar to end ring joints, if there are end ring retaining rings and if there are, do they cover the bar to end ring joints and the bars between the end rings up to the rotor core ends? Comment on the significance of the different types of end ring retaining rings.

**5.9.3** Why is the operational slip of a SCIM such an influential parameter when carrying out MCSA to assess the operational condition of a cage winding in a SCIM?

**5.9.4** If the end user informed you that the SCIM to be tested would be operating at a much reduced load, typically at 20% of its rated output power, when you were due to be on-site to take MCSA measurements to estimate the condition of the cage winding. What would be your response and justify it?

**5.9.5** A 3-phase SCIM has a full-load power output of 100 HP/74.6 kW and takes a full-load current of 100 A at a slip of 1% and its rated output speed is 1782 r/min. When you are arranging to visit the end user's process plant to carry out an MCSA measurement to assess the condition of the cage winding, the end user's on-site engineer states that the motor is operating at 50% load.

When you ask the engineer what he/she means by that, the reply is "because the ammeter displaying the input current shows 50 A and the full-load current is 100 A."

Has the on-site engineer made a correct or incorrect statement and hence justify your answer?

**5.9.6** Explain to an on-site mechanical maintenance engineer, responsible for the maintenance of all rotating equipment in the production plant, why MCSA cannot provide any information on the condition of the cage winding in a SCIM, when the motor is operating on an uncoupled no-load run.

**5.9.7** A 3-phase, 4160 V, 500 HP/373 kW, 58.6 A, 60 Hz, 3582 r/min, 0.92 p.f., efficiency, of 96%, star connected, SCIM drives a centrifugal compressor. The load demand is less than its full-load rating and the input current is around 45 A as reported by the on-site electrical engineer.

The current spectrum indicated $\pm 2sf$ sidebands at ±0.4 Hz around the supply component and both sidebands were 47 dB down on the supply component.

Initially the number of rotor bars was unknown and the load on the motor could not be changed, what would be your interpretation of the result and recommendation to the owner of the motor?

It was subsequently established that the cage winding was copper fabricated with 46 rotor bars and there were no ERRs, thus part of a broken rotor bar could crash into the HV stator winding. What would now be your interpretation and recommendation to the owner?

**5.9.8** A 3-phase, 460 V, 300 HP/224 kW, 329 A, 60 Hz, 1782 r/min, 0.9 p.f., efficiency 95%, star connected, SCIM drives a centrifugal compressor. The number of rotor bars is 52. The cage winding is copper fabricated and no ERRs are fitted that cover the bar to end ring joints.

The motor was operating at its full-load slip and the current spectrum indicated the upper $+2sf$ sideband above $f$ was 49 dB down on $f$ and the lower sideband at $-2sf$ was 46 dB down on $f$. There were also $\pm 4sf$ sidebands at 68 dB down on $f$ and the compressor was running at a steady load.

Interpret the current spectrum and estimate the condition of the cage winding and give a definitive recommendation to the end user.

**5.9.9** A 3-phase, 13,800 V, 4000 HP/2984 kW, 2980 r/min, 50 Hz, 144 A, 0.9 p.f., efficiency 96%, star connected, SCIM drives a centrifugal pump on an offshore oil production platform. The copper fabricated cage winding has 38 rotor bars and there are no ERRs.

The current spectrum indicated $\pm 2sf$ sidebands at $\pm 0.5$ Hz around the supply component $f$ at 50.4 Hz, which were both 46 dB down on $f$ and there were also sidebands at $\pm 4sf$ and $\pm 6sf$ at 65 and 72 dB down on $f$, respectively.

Analyze the current spectrum and predict the equivalent broken bar factor referred to the full-load operating slip and hence recommend the action to be taken by the end user.

**5.9.10** A 3-phase, 11 kV, 50 Hz, 1150 HP/858 kW or 1620 HP/1209 kW, 62.5 A/78.5 A, 50 Hz, 425/495 r/min, PAM (2-speed pole amplitude modulated) SCIM, drives an ID fan in a coal fired power station.

The cage winding is copper fabricated, has 112 rotor slots, and there are no ERRs. The motor operates in the high speed mode of connection during normal operation but is started on its low speed mode of connection. During an MCSA test the panel ammeter reading was oscillating at a very low frequency from 58 to 60 A rms. For the 12-pole mode connection the nominal full-load current is 78.5 A (nameplate value). There are $\pm 2sf$ sidebands $\pm 0.7$ Hz around the supply component at 50 Hz and the lower $-2sf$ sideband is 48 dB down on the supply component but the upper $+2sf$ sideband is 53 dB down on $f$.

**(i)** What can cause the lower $-2sf$ sideband to be higher than the upper one in this 12-pole SCIM with a cage winding fault?

**(ii)** Interpret the current spectrum and estimate the operational condition of the cage winding.

## REFERENCES

[5.1] W. T. Thomson and M. Fenger, "Industrial Application of Current Signature Analysis to Diagnose Fault in 3-phase Squirrel Cage Induction Motors," IEEE Pulp and Paper Industry Technical Conference, Atlanta, GA, June 19–23, 2000, pp 205–211.

[5.2] W. T. Thomson and M. Fenger, "Current signature analysis to detect induction motor faults," *IEEE Industry Applications Magazine*, vol. 7, no. 4, 2001, pp. 26–34.

[5.3] C. Hargis, B. Gaydon, and K. Kamish, "The Detection of Rotor Defects in Induction Motors," Proceedings of 1st IEE International Conference on Electrical Machines, Design and Application, London, UK, pp. 216–220.

[5.4] W. T. Thomson and D. Rankin, "Case Histories of Rotor Winding Fault Diagnosis in Induction Motors," Proceedings of 2nd International Conference on Condition Monitoring, University College of Swansea, Wales, UK, March 1987.

CHAPTER 6

# MCSA CASE HISTORIES—DIAGNOSIS OF CAGE WINDING DEFECTS IN SCIMs FITTED WITH END RING RETAINING RINGS

*William T. Thomson*

## 6.0 INTRODUCTION AND SUMMARY OF CASE HISTORIES

Two case histories are presented to illustrate the difficulty of detecting broken rotor bars in cage rotors with end ring retaining rings (ERRs), which were used in large, HV, 2- and 4-pole, 60 Hz, squirrel cage induction motors (SCIMs), and these are:

**Section 6.2—One case history:** Ten broken rotor bars in a copper fabricated cage winding (ERRs fitted) of a 6800 kW/9115 HP, 2-pole SCIM driving a seawater injection pump. A root cause failure analysis (RCFA) is also presented in this case history.

**Section 6.3—Three case histories:** One broken rotor bar and 46% faulty bar to end ring joints in a copper fabricated cage winding (ERRs fitted) in a 4270 kW/5724 HP, 2-pole SCIM driving a seawater injection pump.

Cage rotors with ERRs that cover the bar to end ring joints can prevent broken bars lifting and crashing into the stator winding. This is particularly the case (see Section 6.3) when the axial length of the ERRs also cover the bars from the inside of the end rings up to the end of the rotor core. With this design the cage winding is fault tolerant to broken rotor bars.

*Current Signature Analysis for Condition Monitoring of Cage Induction Motors: Industrial Application and Case Histories*, First Edition. William T. Thomson and Ian Culbert.

## 6.1 CASE HISTORY (2006)—SUMMARY

Four SCIMs were used to drive water injection pumps on an offshore oil production platform: 3-phase, 13.8 kV, 6800 kW/9115 HP, 326 A, 3570 r/min, 60 Hz, class F, p.f. = 0.89, efficiency = 97%, star connected. The rotor cage windings were copper fabricated and ERRs were fitted. This was a very challenging case history indeed, first, due to the fact that the full-load rated speed on the nameplate was 9 r/min lower than the actual speed when the motor was taking its rated full-load current as stated on the nameplate. This certainly caused considerable confusion since the $\pm 2sf$ sidebands around the supply component did not match the $2sf$ value using the rated full-load slip. This is discussed in Section 6.1.1, including the reasons given by the OEM for the mismatch in speed.

It was subsequently confirmed that there were $\pm 2sf$ components around and at 41 dB down on the supply component ($f$), which indicated that there was a cage winding fault. It is virtually impossible to give a reliable estimate of the severity of cage winding breaks when ERRs are fitted.

In these motors the ERRs covered the end rings and the bar to end ring joints but they did not cover the total bar length between the end rings and the rotor core ends. If a bar breaks and lifts, it can touch the conducting end ring and form a parallel conducting path, allowing current flow, where none would otherwise exist through a completely broken bar with no ERRs. MCSA can grossly underestimate the severity of the problem.

Broken rotor bars may well still be making partial face-to-face contact with the end rings, thus allowing current to flow but through high resistance paths, producing an MCSA result which would imply that there were less broken bars than, in fact, is the case.

The OEM will normally provide the number of rotor bars to the end user provided the motor's serial number is given, but it is not simply a phone call to the OEM for the following reasons:

1. The serial number on the motor's nameplate may not be visible due to rust and ageing.
2. The end user may not have a record of the serial number.
3. The end user may not have records on old motors.
4. The OEM may no longer exist or has been bought over by a much larger OEM.
5. The person who can locate and release the number of rotor bars for a motor has to be tracked down and this can be much more difficult than it would seem.
6. The OEM can simply refuse to provide the information.

When this rotor was initially inspected, there was clear evidence of numerous cracked and damaged joints where the bars entered the end ring, and high temperature effects were very obvious. The rotor was subsequently dissected by removing the end rings and it was found that there were 10 broken rotor bars, clearly a serious broken rotor bar problem. The current spectrum and its interpretation with photos of the rotor showing the broken bars are presented in Section 6.1.1.

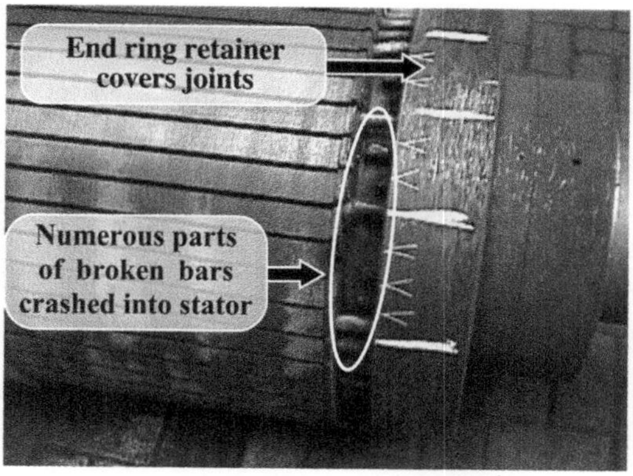


Figure 6.1 Multiple broken rotor bars.

### 6.1.1 MCSA Diagnosis

In 2005, prior to MCSA being routinely applied on an annual basis to estimate the operational integrity of the cage windings in these four motors, there was a catastrophic stator winding failure caused by broken rotor bars crashing into the HV stator winding. The faulty rotor with numerous broken rotor bars is presented in Figure 6.1 and the failed stator winding is shown in Figure 6.2.

The ERRs covered the bar to end ring joints but did not cover the bars up to the end of the rotor core, if it had, the broken bars would have been trapped within an extended ERR preventing them from crashing into the HV stator. Such a rotor would have a high fault tolerance against damage to the stator windings caused by broken bars and, therefore, a catastrophic failure would not have occurred. However, unless specifically requested by the end user, such a specification would not normally be requested by the pump manufacturer when the OEMs were invited to tender for the new motors but no such request had been made. After this failure in 2005, MCSA was applied, on a yearly basis, to assess the condition of the cage rotors of all HV motors

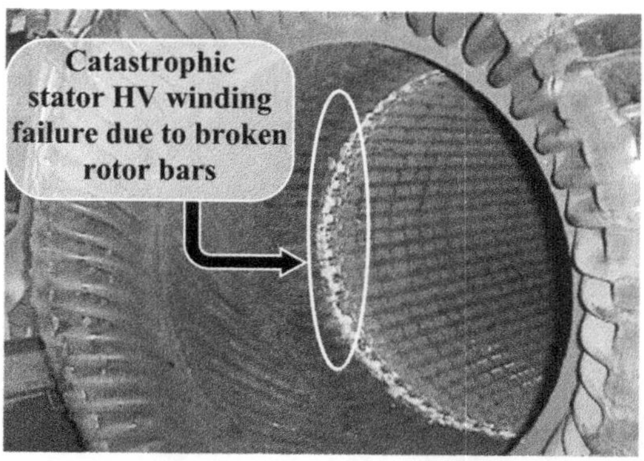


Figure 6.2 Stator winding (13.8 kV) failure.

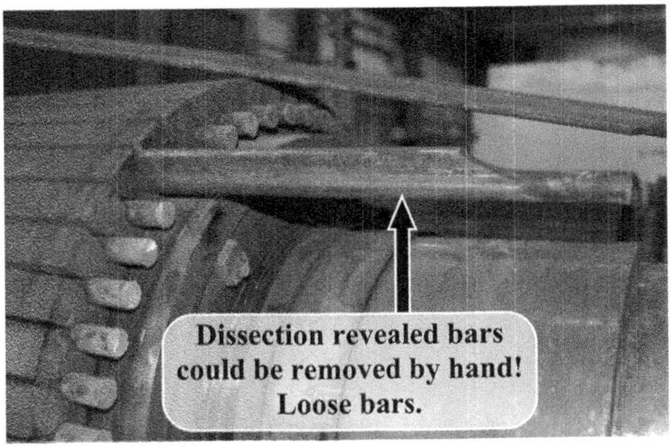


Figure 6.3 Pear or sash bar shape rounded at the end.

on this offshore oil production platform. The following information was provided by the OEM:

**(i)** Copper fabricated cage rotor.
**(ii)** Single cage with 50 rotor bars.
**(iii)** Bars are pear shaped in the slots, as shown in Figure 6.3, but they are made round as they leave the slots, so that they can be inserted into round holes in the end rings for brazing.
**(iv)** ERRs cover the bar to end ring joints as shown in Figure 6.4.

The starting point for MCSA measurements is the nameplate data which may not be completely accurate. For example, in this case history it was subsequently established after the MCSA measurements were taken, that the OEM did the Factory Acceptance Load Test at 50 Hz and extrapolated the results to simulate a 60 Hz full-load heat run. Therefore, the nameplate data is unlikely to be completely accurate, particularly the full-load speed and corresponding current.

The first step in applying MCSA is to record the ammeter reading, whether it is steady, or fluctuating due to random load changes, or whether it has a very slow and

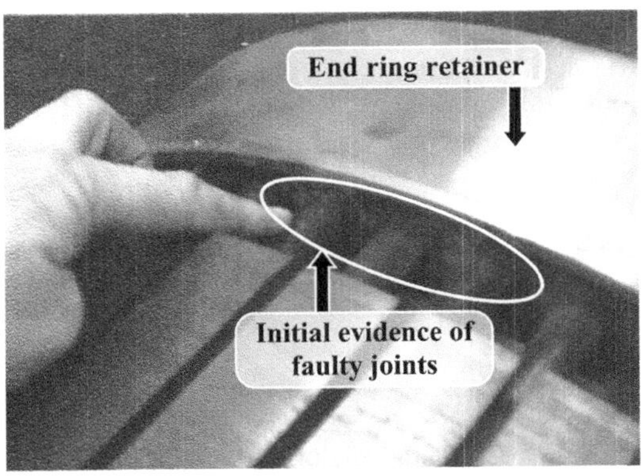


Figure 6.4 Showing ERR, initial evidence of faulty joints.

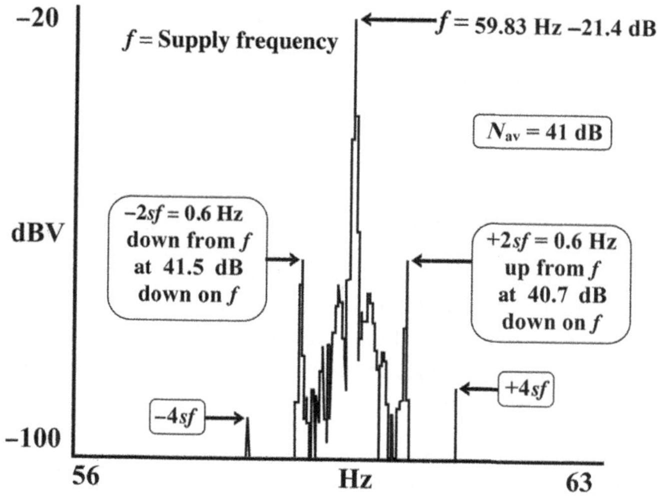


Figure 6.5 Current zoom spectrum, baseband span 0–130 Hz, 12,800 lines, 10.16 mHz/line, 80 dB dynamic range.

regular swing about a central position (only evident on an analogue display). In this case, there was a very slow and steady swing on the analogue ammeter, the frequency of which was estimated to be close to 1 Hz. This is not scientific evidence that the swing was actually caused by twice slip frequency modulation of the current due to broken rotor bars, since it can be caused by rotor disturbances from the mechanical load (see Chapters 7–9). The motor was operating in very close proximity to the other three identical motors and the overall noise level was high; consequently, listening to the acoustic noise was very inconclusive.

An oil company would not remove a large (6800 kW/≅9000 HP) and strategic HV induction motor, from an offshore oil production platform, based solely on an ammeter swing. It is an observation that must be recorded but not one that could, as a single piece of evidence, be used to stop a vital motor.

The current spectrum from the motor is shown in Figure 6.5, with sidebands at ±0.6 Hz and ±1.2 Hz around $f$. Consider that the sidebands at ±0.6 Hz, which are 41 dB down on $f$, are in fact $\pm 2sf$ sidebands due to broken rotor bars. This implies that there is an operational slip of 0.5% at the operational current of 320 A, which is only 1.9% lower than the full-load nameplate current of 326 A. However, a slip of 0.5% is 40% lower than the nominal full-load slip ($s_{FL}$) of 0.83%.

Based on the measured current spectrum and the above analysis, there was a high degree of uncertainty as to whether these sidebands at ±0.6 Hz around $f$ were actually twice slip frequency sidebands or could they be due to an induced oscillation in the pump being reflected back into the motor? The OEM was contacted to comment on the nameplate data and the Factory Acceptance Test (FAT) heat run at full-load and the following questions were submitted:

**(i)** The nominal full-load nameplate speed is 3570 r/min at 60 Hz with a full-load current of 326 A, giving a full-load slip of 0.83%. What was the full-load speed when this motor was full-load FAT tested at the rated volts and frequency?

**(ii)** The MCSA measurements and analysis indicate an operational slip of 0.5%, at a current of 320 A (only 1.9% less than the nameplate amperes). Would you kindly consider commenting on this value of slip at that current?

**Response from OEM:**

Due to power conversion limitations at 60 Hz, the motor could only be full-load tested using a 50 Hz supply. At the end of this heat run, the slip was 0.69% and was then extrapolated to give a slip of 0.57% at 60 Hz on full-load.

**(i)** The slip is affected by the rotor winding resistance, so the extrapolated slip of 0.57% is likely to be higher than that experienced on-site and the motor would run cooler on a 60 Hz supply, due to the higher speed of cooling fans. The full-load slip of 0.5% from your current spectrum is in line with what the OEM would expect on the oil rig.

**Authors' comment:**

The OEM's extrapolated value for the full-load slip being at 0.57% would give an extrapolated full-load speed of 3579 r/min at 60 Hz but the nameplate states a full-load speed of 3570 r/min, however, no explanation was given as to why the wrong speed was on the nameplate. When this occurs, MCSA clearly has an external constraint outwith its control which is therefore a weakness. Consequently, an expert on SCIMs and MCSA interpretation is really required to resolve these difficult situations.

**Measurement of operating slip via MCSA:**

The well-known rotor slot passing flux components have been known to exist for many years, for example, since the first half of the twentieth century, Morrill [6.1], Alger [6.2,6.3], and are given by equation (6.1). These rotating flux waves can induce emfs in the stator winding and thus drive corresponding currents through the stator. These current components can be detected via current spectrum analysis of the current supplying the motor, Cameron et al. [6.4], and the rotor slot passing flux components are given by equation (6.1), see References 6.1–6.3.

$$f_{\text{rs}} = f\left\{\left(\frac{R}{p}\right)(1-s) \pm n_{\omega s}\right\} \tag{6.1}$$

where

$R$ = number of rotor bars

$f$ = supply frequency, Hertz

$p$ = pole-pairs

$s$ = operational slip

$n_{\omega s}$ = 1, 3, 5,... mmf time harmonics

For this motor

$R = 50$

$f$ via MCSA = 59.83 Hz

$p$ = pole-pairs = 1

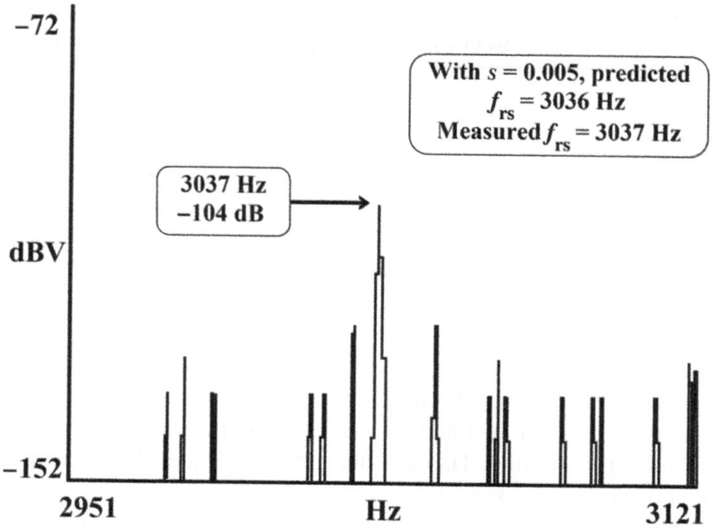


Figure 6.6 Zoomed current zoom spectrum to detect $f_{rs}$, baseband span 70–3400 Hz, 12,8000 line, 0.26 Hz/line.

$s$ = operational slip from the measured sidebands, $s = 0.005$ in equation (6.1) (0.5% cannot be used in equation 6.1)

$n_{\omega s} = +1$, equation (6.1) is referred to as the principal rotor slot passing frequency, $f_{rs(+1)}$

If the slip from the measured $2sf$ sidebands is used in equation (6.1) and if it is the correct operational slip, then the measured rotor slot passing frequency should be the same or very close to the predicted value. With $n_{\omega s} = +1$, equation (6.1) gives a rotating flux wave at $f_{rs(+1)} = 3036$ Hz and a corresponding current component at that frequency should flow in the stator winding.

The measured current spectrum shown in Figure 6.6 confirms that the predicted (3036 Hz) and measured (3037 Hz) $f_{rs(+1)}$ values are virtually identical using the operational slip of 0.005 from the sidebands at $\pm 2sf$ equal to $\pm 0.6$ Hz. This verifies that the operational slip was obtained independently (via MCSA) of the nameplate data but the number of rotor bars was required.

For the avoidance of doubt, and particularly for the benefit of condition monitoring companies, there is in fact a series of these flux and current components spaced at twice the supply frequency ($2f$) apart since ($n_{\omega s}$) in equation (6.1) equals 1, 3, 5, etc. These rotor slotting flux components are inherent in an induction motor [6.1 to 6.4] and are not due to a problem.

It is accepted that there must be pole-pair compatibility between the pole-pairs of each of these flux waves given by equation (6.1) and corresponding pole-pair harmonics from the winding distribution, for an emf and corresponding current at these frequencies to be induced in the stator winding. This is discussed in Chapter 10 and via the industrial case histories presented in Chapter 11 for the diagnosis of airgap eccentricity problems.

**TABLE 6.1 Extract from Table 4.2. Broken Rotor Bar Severity Estimator Based on Only the Average dB Difference ($N_{av}$) Between the $\pm 2sf$ Sidebands and the Supply Component**

| $N_{av}$ = dB Difference | Estimated condition of cage winding when the motor is operating at full-load output and speed at the shaft |
|---|---|
| 40–45 dB | There is a high probability of a cage winding fault. |
| Measured ($N_{av}$) of 41 dB down on $f$ | For example, at 40 dB in low pole numbers (2 and 4) then there is a high probability of one or even two broken rotor bars. At 40 dB with higher pole numbers the probability of more broken bars increases. |

Figure 6.5 shows that the $\pm 2sf$ sidebands are an average ($N_{av}$) of 41 dB down on $f$, and assuming the number of rotor bars *had not been provided* by the OEM, an extract from Table 4.2 given in Table 6.1 is the only option available to estimate the condition of the cage winding. This guidance chart indicates that there was a high probability of one or two broken rotor bars.

For completeness and since the number of bars was 50, the equivalent broken bar factor can be calculated.

$$BB_f = 2R/\left(10^{(N_{av}/20)} + 2p\right) = 0.88$$

$$N_{av} = 41 \text{ dB (see Figure 6.5)}, R = 50, \text{ and } p = 1$$

As a comparison, using Bellini's equation (4.6)), the broken bar index $n$ is

$$n = \left(10^{(-41.6/20)} + 10^{(-40.8/20)}\right) \times R = 0.87$$

The difference is negligible between $BB_f$ and $n$ and since the $BB_f$ is 76% higher than the threshold value of 0.5 (see Chapter 4), there was clearly a faulty cage winding. As already stated, the $BB_f$ underestimates the severity of the fault. With ERRs fitted, broken bars can lift and touch the conducting ERR which constitutes a parallel conducting path, and/or broken bars can still be making face–to-face contact with an end ring at the joints. These two conditions can further mask the severity of the problem. Taking account of all the evidence and particularly the fact that an identical motor had a catastrophic failure in 2005, as shown in Figures 6.1 and 6.2, the recommendation to the oil company was as follows:

**(i)** This motor should be stopped and immediately removed from service to avoid the possibility of broken rotor bars crashing into the HV stator winding causing a catastrophic failure.

The oil company immediately accepted this recommendation and the motor was removed for an onshore inspection.

### 6.1.2 Inspection

The ERRs were removed from each end of the rotor. Figures 6.7 and 6.8 show damaged and broken bars at the drive end (DE), but it should be noted that the broken bars were still making partial electrical contact with the short circuiting end ring.

Figure 6.7 End ring retaining ring removed, faulty joints.

The bars were cut at the DE as shown in Figure 6.9, to establish how many were actually broken at the joints and Figure 6.10 shows that there were 10 broken rotor bars.

If the motor had not been stopped, there was a very high risk that a catastrophic failure would have occurred, which was the case in 2005 in an identical motor (see Figures 6.1 and 6.2), therefore, MCSA plus induction motor expertise prevented a repeat incident. In the short term the rotor was repaired and a new cage winding was fitted, it then became the spare rotor. The oil company required an RCFA to determine the possible cause or causes of broken rotor bars in these motors and this was carried out by Thomson (author).

### 6.1.3 RCFA of Broken Rotor Bars and Solution

The OEM designed the original motor in accordance with the specification from the pump manufacturer, which included a "*start up against a closed valve*" and not an

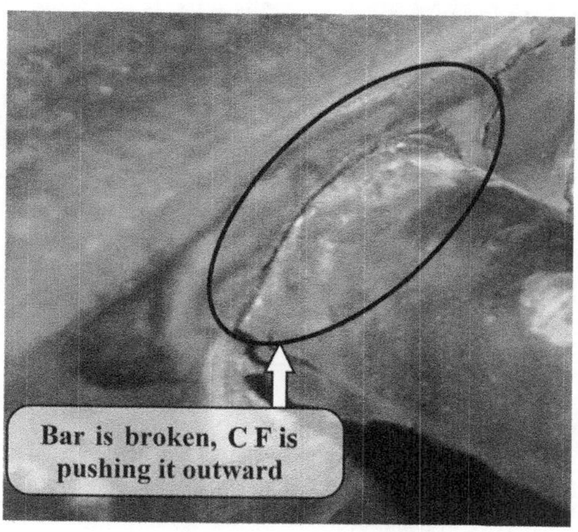


Figure 6.8 Broken rotor bar still making contact with end ring.

Figure 6.9 End ring removed.

"*open valve setting*" in the drive train's control system. If during start-up the voltage dipped to 80% of the rated voltage, there had to be sufficient accelerating torque to reach steady-state operation with a closed valve setting and the OEM's original motor design catered for that condition. Consequently, "*no criticisms whatsoever can be levelled at the OEM*" of the motor. Figure 6.11 gives the torque versus speed curves for starting a centrifugal pump against closed and open valve settings.

Figure 6.12 gives the torque–speed curves for the motor at various applied voltages and also the curves for the closed and open valve setting of the centrifugal pump.

Figure 6.10 Ten broken rotor bars at the DE.

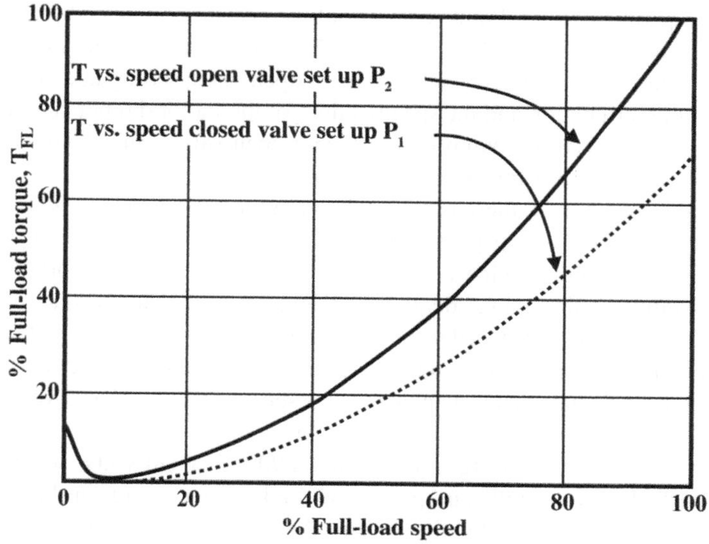


Figure 6.11 Torque–speed curves for a centrifugal pump: open and closed valve settings.

Clearly, if the motor has to start up against an open valve setting, there is less accelerating torque.

The M curves in Figure 6.12 show the torque production capabilities of the motor, and the P curves give the torque required by the centrifugal pump. The difference between M and P curves for M greater than P is the accelerating torque and the greater the area between the two curves, the shorter is the run-up time to the

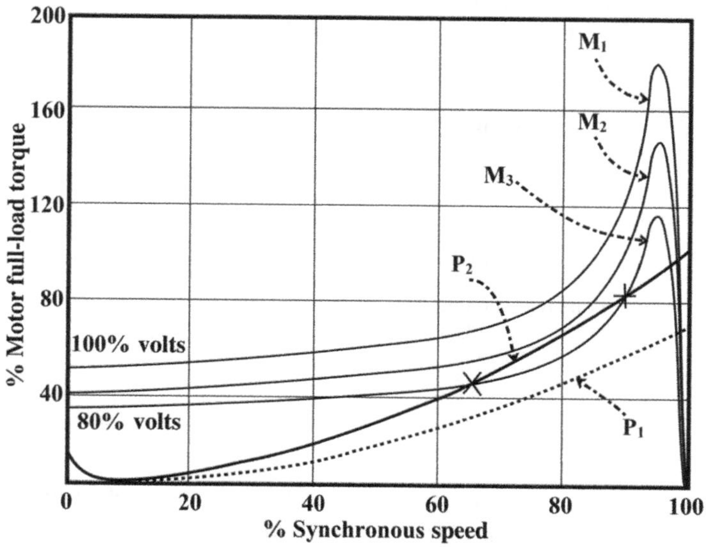


Figure 6.12 Torque–speed curves for motor ($M_1$: 100%, $M_2$: 90%, and $M_3$: 80%) volts and centrifugal pump curves, $P_1$: closed valve setting, $P_2$: open valve setting.

steady-state operating speed. Point X at 67% of the synchronous speed (2400 r/min) is where the motor torque at 80% voltage ($M_3$ curve) intersects the pump torque (open valve) curve ($P_2$). However, recall that an open valve setting at switch on is a control system malfunction and the specification for the motor did not cater for such a malfunction. In fact, if the pump is set up with open valves, the motor should be automatically locked out from a start-up.

Therefore, at point X, there is no acceleration torque, the lack of which causes the motor to stall. It will then trip on over current due to the high slip. The $P_1$ curve is for a closed valve setting, which was the design requirement and is comfortably below the M curves thus easily giving adequate acceleration to the steady-state operating speed. Note also the much reduced torque difference between $M_2$ (90% volts) and $P_2$ (open valve malfunction) giving a much reduced acceleration and long run-up times at 90% voltage. DOL starts against an open valve setting, at this reduced voltage, will therefore lead to overheating of the cage winding.

As an example, if the voltage at start up is 90% of rated volts (see curve $M_2$ in Figure 6.12) the motor will run up but it will take longer to reach the operating speed, due to the reduced accelerating torque. As a result, higher currents will flow for a longer period, and the top section of the rotor bars will heat up to a temperature that may be well above the original design value.

As part of the RCFA investigation, the voltage transient was measured during start-up of an identical seawater injection pump motor and it was found that the supply volts dropped to 84% of its rated value as shown in Figure 6.13a, thus reducing the available accelerating torque to an unacceptably low level, if started against an open valve setting.

The combination of DOL starts and starting against an open or partially open valve system ultimately led to weakening of the rotor cage, due to the abnormally high

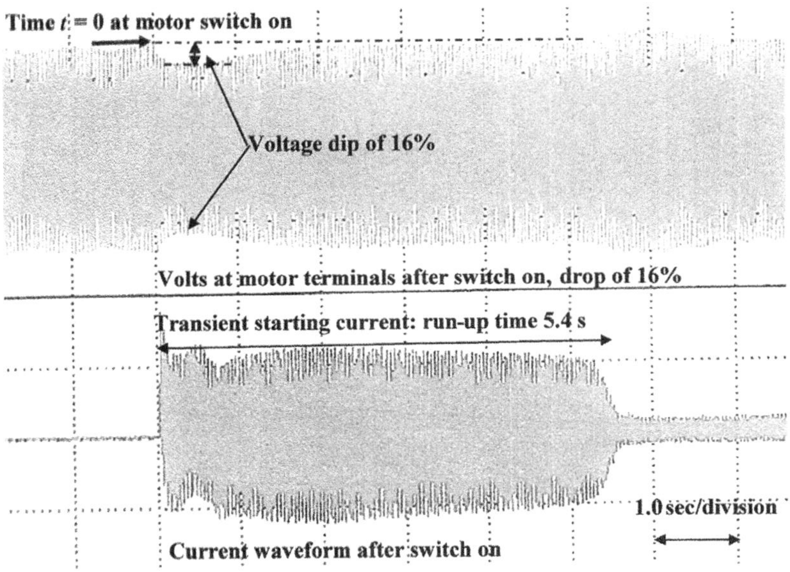


Figure 6.13a Voltage and current waveforms immediately after switch on.

thermal and mechanical stresses, which resulted from that combination. The specification in the ITT for the original SCIM did not include starting against an open valve torque–speed curve for the pump thus the motor was not designed to cope with these stresses. It was confirmed that there had been problems with the valve system and it was accepted by the oil company that the above causes were the fundamental reasons for the broken rotor bars. In fact, MCSA diagnosed a further two cases of broken rotor bars in identical seawater injection pump motors on this platform before any catastrophic failures occurred.

An invitation to tender (ITT) with a new upgraded operating specification was issued to various OEMs of large HV motors. The successful OEM proposed a design, which could be started against an open valve system at 100% and 80% of the rated voltage with corresponding run-up times of 3.5 and 7 seconds. Both motors had starting currents of 5.5 times the full-load current. The new motor has a higher performance specification than the original motors as witnessed by the following data.

At rated voltage, the starting torque in the redesigned motor is 60% of full-load torque as compared to 52% in the original motor, furthermore, their breakdown or pull-out torques are respectively 2.45 and 1.8 times the full-load torque for the new and original motor. The torque versus speed curves for the new motor and the torque–speed curves for the pump with a closed and open valve setting are shown in Figure 6.13b, which shows that the risks of broken rotor bar failures due to low accelerating torque are greatly reduced.

The new motor could still start up against an open valve setting (considered a malfunction) at 80% of the rated voltage, but the volts actually dropped to 84% of the rated voltage at switch on as was proven via measurement of the transient volts

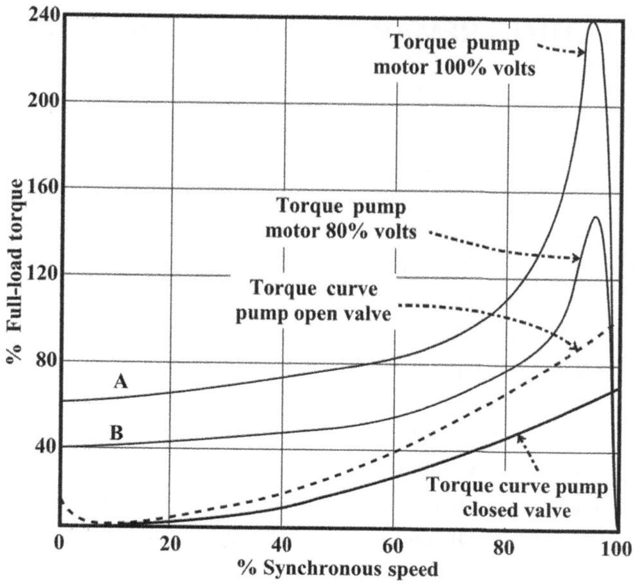


Figure 6.13b Torque versus speed curves for new motor at 100% and 80% rated volts and for the centrifugal pump with closed and open valve settings.

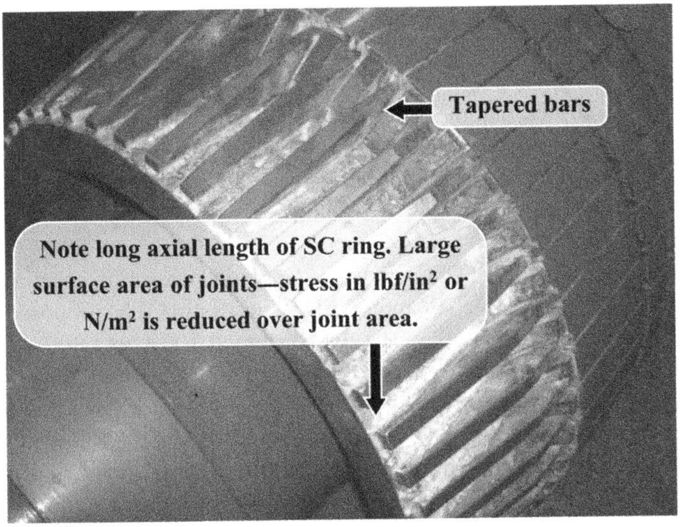


Figure 6.14 Rotor design in brand new motor. Reproduced with permission of Parsons Peebles, Scotland.

at the motor, there is therefore more accelerating torque available than is shown in Figure 6.13b when started against a non-allowed open valve setting. The new motors have been in operation since 2008, with no indications whatsoever of any $\pm 2sf$ sidebands in the current spectra due to rotor cage problems. Figure 6.14 shows the design of the cage rotor in the brand new replacement motors.

## 6.2 CONCLUDING REMARKS ON THIS CHALLENGING CASE HISTORY

**(i)** The rated speed on the nameplate was incorrect and initially caused considerable uncertainty with respect to the interpretation of the current spectrum to verify that $\pm 2sf$ sidebands around the supply component of current were truly present. This is a disadvantage of MCSA if its user relies solely on the nameplate data being perfectly correct, and in this case history it was essential to get the number of rotor bars.

**(ii)** The slip obtained from the measured $\pm 2sf$ sidebands was used to predict and detect the principal rotor slot passing frequency component in the current spectrum. The actual operating slip was then confirmed from the measured principal rotor slot passing frequency and compared with the slip from the sidebands around the supply component and they were the same.

**(iii)** Although there were 10 broken bars, they were still making partial face-to-face contact with the DE end ring and in practice, there was a path for rotor currents to flow. As such, the magnitude of the $\pm 2sf$ sidebands with respect to the supply component grossly underestimated the actual number of broken rotor bars but this was inevitable and cannot be prevented, as already mentioned in Chapter 4 broken rotor bars very often do not occur as clean cut breaks in industrial motors. Although MCSA grossly underestimated the actual number of broken rotor bars, the test and analysis still predicted that there was a cage winding fault which merited the removal of the rotor for inspection.

**(iv)** The fundamental reasons for the broken rotor bars were a combination of two factors, the first was DOL starts which produced a 16% drop in supply voltage as shown in Figure 13a and the second was start-up against an open valve setting. This caused longer run-up times and a stall if the actual voltage dropped to 80% of its rated value thus the cage winding experienced very high thermal, mechanical, and electromagnetic forces on the rotor bars, which ultimately led to broken rotor bars.

## 6.3 CASE HISTORY (1990)—SUMMARY AND KEY FEATURES

The nameplate data on the motor was as follows: 3-phase, 11 kV, 4270 kW/5724 HP, 258 A, 3570 r/min, 60 Hz, p.f. = 0.9, efficiency = 96.4%, star connected SCIM. The number of rotor bars was not available at the time of testing and analysis of the MCSA data.

Three 11 kV induction motors were used to drive seawater injection pumps on an offshore oil production platform. The design of the cage winding had large ERRs that covered the bar to end ring joints and very nearly the full length of the bars between the end rings and the rotor core ends as shown in Figure 6.15. These motors are designed to prevent a broken rotor bar from crashing into the HV stator winding with consequential winding failure.

The MCSA spectrum showed that the $\pm 2sf$ sidebands were an average of 48 dB down on the supply component in one of the motors and this indicates a very borderline case indeed as to whether there is, for example, a cracked or broken rotor bar. However, with this design, a bar can break and lift and touch the conducting ERR thus providing a parallel conducting path, which masks the severity of the cage winding damage due to the reduction in dB magnitude of the $\pm 2sf$ sidebands. The

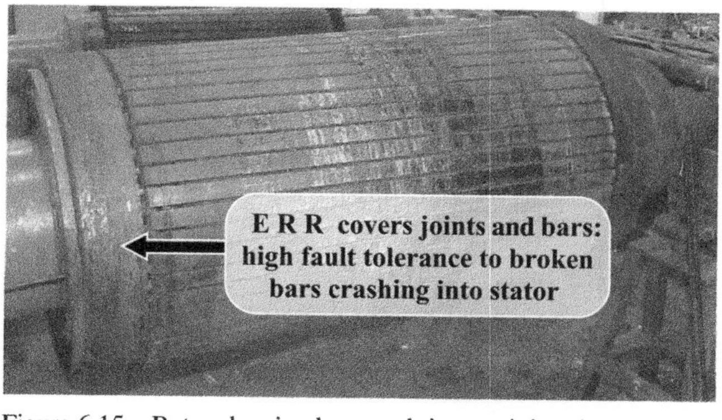


Figure 6.15 Rotor showing large end ring retaining rings (ERRs).

owner of the oil and gas offshore production platform took the opportunity to remove the rotor during an annual outage for major maintenance works throughout the platform.

An inspection and removal of the ERRs revealed one completely broken rotor bar but before the dissection that bar was making partial contact with the end ring and thus providing a high resistance path for current flow. Therefore, the $\pm 2sf$ sidebands at an average ($N_{av}$) of 48 dB down did not indicate an actual broken rotor bar. A dye penetrant NDT inspection was carried out on the end rings and it was found that 46% of the bar to end ring joints at the DE were sub-standard.

Recall that MCSA gives an estimate of the condition of the cage winding via the equivalent broken bar factor ($BB_f$) from equation (4.5) and does not give a definitive prediction of the actual number of broken bars. This is a weakness, particularly when ERRs are fitted and simply using an MCSA instrument to give a diagnosis without knowledge of the actual design features of the rotor is a distinct limitation. The current spectrum and its interpretation with photos of the rotor with the broken bar and the dye penetrant checks on the bar to end ring joints are presented in Section 6.3.1.

### 6.3.1 MCSA Diagnosis

MCSA was applied to three motors as part of an annual survey and the current spectra for motors A, B, and C are shown in Figures 6.16–6.18, respectively. A frequency resolution of 0.01 Hz/line and a dynamic range of 80 dB were used to obtain the current spectra.

The supply frequency ($f$) changed during the measurements from the three motors and the operating currents were 246, 252, and 252 A for motors A, B, and C, respectively, compared to the nominal full-load current of 258 A. Their loads were therefore approximately equal and close to full-load. Since the supply frequency differs for each measurement, the $\pm 2sf$ sidebands will be at slightly different frequencies for each motor.

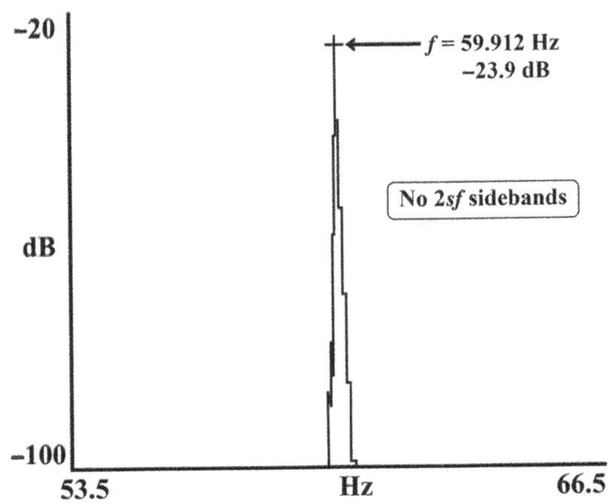


Figure 6.16 Normal current spectrum, motor A.

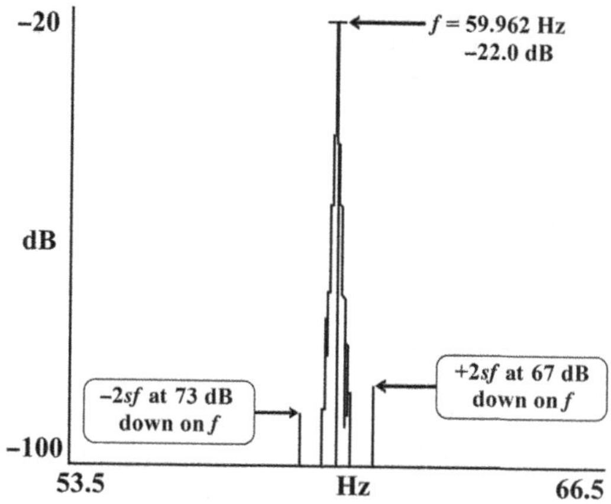


Figure 6.17 Current spectrum with very low level $\pm 2sf$ sidebands, motor B.

**Motor A** had no evidence of any $\pm 2sf$ sidebands as shown in Figure 6.16 and can be deemed to have a perfect rotor cage winding.

**Motor B** had $\pm 2sf$ sidebands as shown in Figure 6.17 but they were an average of 70 dB down on $f$ which equates to 3160 times smaller than $f$ in absolute units of amperes, hence the rotor is normal. The fact that sidebands exist in this particular motor will normally be due to small but normal differences in the resistances of a number of bar to end ring joints. The value of $2sf$ was 0.88 Hz thus giving a slip for this motor of 0.74% with $f$ at 59.96 Hz.

**Motor C** was distinctly different as shown in Figure 6.18 in that the $\pm 2sf$ sidebands were an average of 48 dB down on $f$ but note the lower and upper sidebands were 52 and 44 dB down on $f$, respectively, which is a substantial difference at 8 dB (factor of 2.5 times in amperes). The value of $2sf$ was 0.95 Hz thus giving a slip for this motor of 0.79% with $f$ at 59.98 Hz compared to the full-load slip of 0.83%

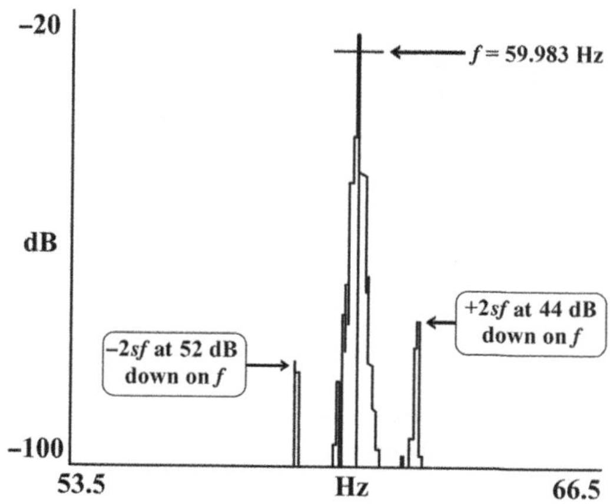


Figure 6.18 Current spectrum with $\pm 2sf$ sidebands from motor C, which are up to 23 dB (factor of 14 in absolute amperes) higher than in motor B.

**TABLE 6.2 Broken Rotor Bar Severity Estimator Based on Only the Average dB Difference ($N_{av}$) Between the $\pm 2sf$ Sidebands and the Supply Component**

| $N_{av}$ = dB Difference | Estimated condition of cage winding when the motor is operating at full-load output and speed at the shaft |
|---|---|
| 45–50 dB<br>Measured ($N_{av}$) of 48 dB down on $f$ | This is the very difficult boundary range for using the dB difference.<br>At 45 dB in lower pole number motors (2 and 4), there is a possibility of either faulty joints or perhaps a broken rotor bar. |

based on the nameplate speed. The motor was on full-load current during the current measurement. An extract from Table 4.2 from Chapter 4 is given in Table 6.2.

Based solely on using the guidance chart in Table 6.2, an average dB difference, $N_{av}$ of 48 dB means it is very clearly in the borderline category for a 4-pole motor and there was considerable doubt as to whether this 2-pole cage winding actually had a cage winding problem such as a cracked or broken rotor bar. It was subsequently established from a previous repair that this cage winding design has a high fault tolerance to prevent broken bars causing secondary damage to the stator winding.

The recommendation was to continue to run the motor and carry out repeated MCSA surveys every 3 months to trend any increase in magnitude of the $\pm 2sf$ sidebands. The end user did not do so but 6 months on during a 4-week planned outage of the platform, the duty operator of the oil rig decided to remove the rotor for inspection and install a healthy spare rotor.

## 6.3.2 Inspection

Figure 6.15 shows the rotor as received by the electric motor repair shop which shows that the bar to end ring joints cannot be viewed and the cage winding was subsequently dissected. At the DE, there was one completely broken rotor bar as shown in Figure 6.19, but as expected the broken part of the bar was contained within the ERR

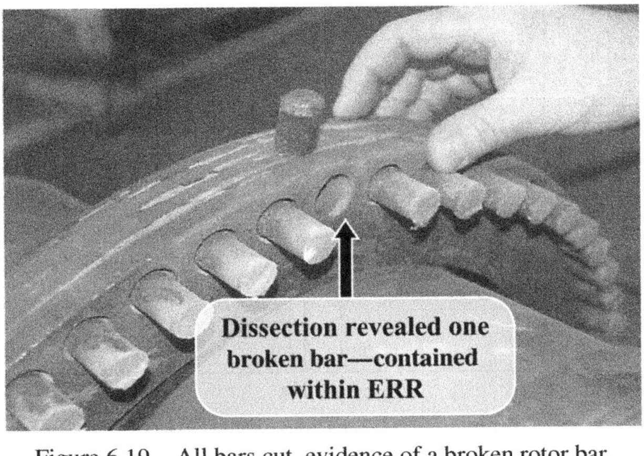


Figure 6.19 All bars cut, evidence of a broken rotor bar.

Figure 6.20 Dye penetrant (dark areas) applied to the end ring.

and no consequential damage was caused to the HV stator winding. It was agreed to carry out a dye penetrant inspection to identify any cracks and faulty joints between the bars and end rings.

The end ring was thoroughly cleaned and dye penetrant was applied as shown in Figure 6.20 to reveal cracks and faulty joints that cannot be seen by the naked eye. The excess dye has to be removed and the results are shown in Figures 6.21 and 6.22.

Figure 6.21 Developer applied, dark areas show 46% of 52 joints were inferior.

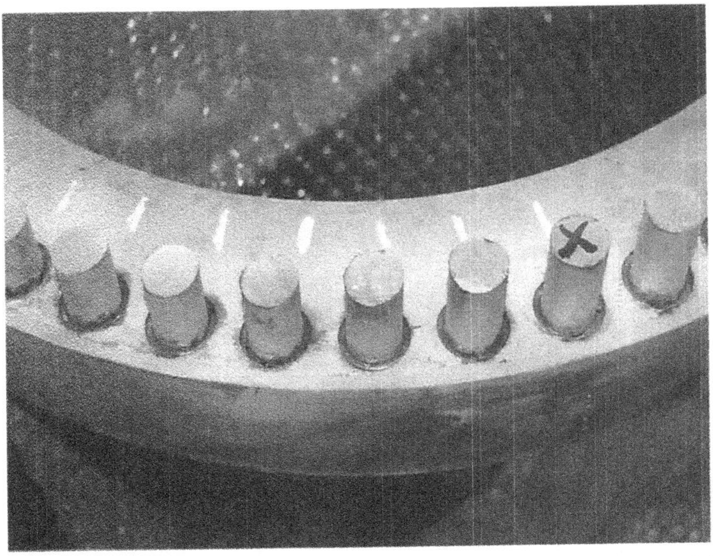

Figure 6.22 Sample of faulty joints (dark portions at the joints) on inside face of the DE ring.

Due to the number of faulty joints identified via the dye penetrant test the cause of the actual broken bar was probably due to a faulty joint and too many sequential DOL starts.

## 6.4 SUMMARY AND LESSONS LEARNED FROM INDUSTRIAL CASE HISTORIES IN CHAPTERS 5 AND 6

In offshore oil and gas production platforms, onshore oil refineries, power stations, etc., the end users have to be convinced that there is a genuine problem in a large, HV SCIM before they take it out of service. The experienced engineer therefore needs to gather all the available back-up information to augment the measurement and interpretation of the current spectrum to truly identify the $\pm 2sf$ sidebands and their relative magnitudes with respect to the supply component, to ensure a reliable diagnosis of the condition of a cage winding. A false prediction that broken bars exist (known as a false positive), when the rotor cage is healthy, is just not acceptable to an end user.

There are very good reasons for this, for example, the removal of a large, HV SCIM (e.g., an 11 kV, 7460 kW/10,000 HP) from an offshore oil production platform involves a major pre-planning exercise, logistical offshore surveys, an assessment of the availability of suitable and safe lifting gear/cranes, additional specialist personnel, and all the consequential costs, in fact, the total cost can be of the order of US$225,000 (GBP£150,000) in the North Sea (between Scotland and Norway), off the coast of Scotland. The photos in Figures 6.23–6.29 indicate what is involved in the removal of a rotor from a large, HV SCIM on an offshore oil production platform. This should provide a real appreciation of what is subsequently involved if an MCSA test concludes that there is a broken rotor bar fault and the rotor is removed but the inspection reveals a perfectly normal cage winding. The CM company's contract with the end user would be terminated with the distinct possibility of a financial claim being made against the service provider for the losses incurred by the end user. Figure 6.30 shows the rotor after it was removed for inspection and the caption shows the cost of a new replacement rotor (approximately GBP£140k/US$150k to £155k/US$170k).

**OEM's Specification**

3-phase, 6800 V, 6316 kW/8467 HP, 604 A, 60 Hz, 1788 r/min, star connected SCIM
Totally enclosed, closed air circuit water-cooled, IP56, max., continuous, insulation: Class F
Weights: stator 6400 kg; rotor: 3600 kg; heat exchanger complete and dry: 1300 kg
Figures 6.27 and 6.28: stator and rotor lift of 10,000 kg (imperial 22,000 lb/9.8 tons)

**Figure 6.23** View of DE of the motor.
Reproduced with permission of Quartzelec, Aberdeen, Scotland.

**Figure 6.24** Survey of module and space for the lift.
Reproduced with permission of Quartzelec, Aberdeen, Scotland.

**Figure 6.25** The cooler has been removed.
Reproduced with permission of Quartzelec, Aberdeen, Scotland.

**Figure 6.26** Photo of the crane set up.
Reproduced with permission of Quartzelec, Aberdeen, Scotland.

**Figure 6.27** The lift in progress. Reproduced with permission of Quartzelec, Aberdeen Scotland.

**Figure 6.28** Motor on upper module deck. Reproduced with permission of Quartzelec, Aberdeen Scotland.

**Figure 6.29** The rotor being removed. Reproduced with permission of Quartzelec, Aberdeen, Scotland.

**Figure 6.30** The rotor onshore for inspection. Reproduced with permission of Quartzelec, Aberdeen, Scotland.

Samples of the costs of new SCIMs are given in Table 6.3.

### 6.4.1 Conclusions

It is recommended, whenever possible, that the following factors should be reviewed, before a final diagnosis and recommendation is made to the owner of the motor.

**(i)** Is there any history of previous broken rotor bars in the motor under test or in an identical one?

**(ii)** If MCSA testing has been carried out on a regular basis, what is the historical evidence of any change in the results?

**(iii)** Has the operational regime for the motor being MCSA tested changed?

**TABLE 6.3 Costs for Foot Mounted SCIMs**

Note: It can be difficult to obtain costs from OEMs/suppliers due to commercial sensitivity.

| Power Rating | Poles/Hz | Voltage | Enclosure | List price |
|---|---|---|---|---|
| **Country: USA** | | | | |
| 150 HP | 2/60 Hz | 460/2300/4000 V | TEFC | US$12,827/25,203/26,883 |
| 1000 HP | 2/60 Hz | 460/2300/4000/6600 V | TEFC | US$108,062/108,062/111,127/187,202 |
| 1000 HP | 12/60 Hz | 460/2300/4000 V | TEFC | US$210,606/210,606/219,030 |
| 4000 HP | 6/60 Hz | 4000 V | TEFC | US$314,550 |
| **Country: UK** | | | | |
| 15 kW | 2/50 Hz | 400/460/690 V | TEFC | £1640 |
| 75 kW | 2/50 Hz | 400/460/690 V | TEFC | £7874 |
| 150 kW | 2/50 Hz | 400/460/690 V | TEFC | £14,825 |
| 200 kW | 4/50 Hz | 400/460/690 V | TEFC | £20,095 |

**(iv)** Have there been any malfunctions in the driven load during start-ups?

**(v)** What are the records on the number of starts?

**(vi)** Was there any evidence of a regular low frequency beat on an *in situ* analogue ammeter reading of the input current to the motor during the MCSA test?

**(vii)** Was the acoustic noise from the motor being modulated at a low frequency?

**(viii)** What are the characteristics of the drive dynamics of the driven mechanical load? This is fully covered in Chapters 7–9 via further industrial case histories.

**(ix)** Are the sidebands being measured around the supply component truly at $\pm 2sf$?

**(x)** What are the relative amplitudes of the $\pm 2sf$ sidebands with respect to the supply component and relative to each other?

**(xi)** Are there any integer and higher order sidebands of $\pm 2sf$ at $\pm 4sf$ and/or at $\pm 6sf$ around $f$?

**(xii)** For comparison, is there the possibility of testing an identical motor doing the same duty and driving an identical mechanical load?

**(xiii)** If the number of rotor bars is known, predict the equivalent broken rotor bar factor $BB_f$ or the broken bar index ($n$) and if it is above the threshold value of 0.5 then there is a high probability that there is truly a rotor cage winding problem. Note that the higher the $BB_f$ or $n$ the greater the probability of broken rotor bars.

**(xiv)** Is there an end ring retainer fitted which covers the bar to end ring joints and if there is, does it also cover the bars between the inner face of the end rings up to the ends of the rotor core, as shown in Figure 6.15? If the latter type is present, the probability is very low that a broken rotor bar could cause any consequential damage to an HV stator winding and the motor can continue to run and be regularly monitored, for example, every 3 months. If more bars break, a possible by-product is that the rotor's mechanical imbalance will increase, thus the vibration in mm/sec (inches/sec) on the bearing housings (or displacement measured by shaft displacement probes) can increase. This will be due to an increase in the 1× r/min frequency component and for that reason the motor may well have to be stopped.

**(xv)** A major factor in the decision making process is the possibility/probability of a catastrophic failure, such as broken bars lifting and crashing into the HV stator winding. This is certain to require an expensive repair but may also cause considerable loss of production and loss (or delay as the case may be) of revenue income. But of course, the implications for safety, particularly in the petrochemical industry, are of the utmost importance and are certainly the most important considerations, which can certainly persuade an oil company to take action and remove a motor from service.

**(xvi)** MCSA can significantly underestimate the severity of cage winding breaks if ERRs are fitted and broken bars are making contact with an ERR.

## 6.5 QUESTIONS

**6.5.1** Why should you check the reading on an *in situ* ammeter before carrying out an MCSA test? If it is an analogue ammeter, what features on the display are worth noting?

**6.5.2** Why is it important to try to obtain definitive records on the number of DOL starts the SCIM has experienced since it was installed or since the last repair?

**(i)** What are the difficulties of obtaining that information?

**(ii)** Why is it essential to check the serial number of a SCIM each time it is MCSA tested?

**6.5.3** Why is it important to establish if there have been any abnormal incidents in the drive train that could affect the integrity of the cage winding in a SCIM?

**6.5.4** Why is it very important to determine the actual DOL starting regime by the end user for a large, HV SCIM with a copper fabricated cage winding?

**6.5.5** Why is it necessary to establish the characteristics of the driven mechanical load downstream of the motor?

**6.5.6** What are the advantages of fitting ERRs to a copper fabricated cage winding that will also cover the bar to end ring joints in a 3-phase, 13,800 V, 10,000 HP/7460 kW, 60 Hz, 3580 r/min SCIM driving a centrifugal pump on an offshore oil and gas production platform?

**6.5.7** ERRs were fitted to a copper fabricated cage rotor in a 3-phase, 13,800 V, 10,000 HP/7460 kW, 60 Hz, 3564 r/min SCIM and the motor had been regularly started against an open valve set up although it was originally designed to start against a closed valve set up, when driving a centrifugal pump. The ERRs did not cover the full length of the 46 rotor bars between the joints up to the ends of the rotor core but the ERRs did cover the joints between bars and end rings. The cage rotor had 50 bars.

(i) If for example, broken rotor bars have occurred at the joints between the bars and an end ring and they lift and touch the conducting ERR, what influence does this have on the magnitude of the $\pm 2sf$ sidebands around the supply component? Of course, at this stage of an investigation you do not know that this has happened.

(ii) An MCSA test is carried out and the $\pm 2sf$ sidebands and $\pm 4sf$ sidebands are 45 and 60 dB down, respectively, on the supply component of current. The motor was operating at its full-load speed and current. Analyze this result and detail the recommendations you would make to the end user.

**6.5.8** A 3-phase, 6600 V, 1000 HP/746 kW, 115 A, 50 Hz, 2970 r/min, 0.9 p.f., 96.8% efficient, star connected SCIM drives a centrifugal compressor. The copper fabricated cage winding has 52 rotor bars and there are ERRs that cover the bar to end ring joints but not the full length of the bars between the core ends and the joints. An MCSA test indicated that there were $\pm 2sf$ sidebands at $\pm 0.75$ Hz around the supply component ($f$) at 50 Hz, and the lower and upper sidebands were 36 and 35 dB down on $f$, respectively, when the motor was operating at 112 A compared to a full-load current of 115 A.

(i) What is the predicted value of $2sf$ at the rated nameplate full-load speed and frequency?

(ii) The measured $\pm 2sf$ sidebands around $f$ do not match the predicted $2sf$ sidebands at full-load, why might this be the case?

(iii) What additional visual check would you make if there were an *in situ* analogue ammeter?

(iv) What questions would you ask the OEM of the motor with respect to the nameplate data and their original FAT results and conditions for the full-load heat run?

(v) If it is established that the sidebands are truly the $\pm 2sf$ sidebands due to cage winding breaks, predict the $BB_f$ and make your recommendations to the end user.

**6.5.9** For the motor in question 6.5.8, the end user decided to remove the rotor for inspection, based on the MCSA result and the recommendations in your report. You are commissioned to oversee the inspection to determine the actual condition of the cage winding which is to be carried out at a motor repair workshop. The end user has written to the motor repair company (at your request) that you have full authority to act on their behalf and make decisions without consulting the end user.

(i) Would you allow the motor repair company to carry out any form of dissection whatsoever without you being present? Give reasons for your answer.

(ii) Propose a sequence of inspection checks of the cage winding that you require the motor repair company to carry out in your presence.

**6.5.10** A 3-phase, 11,000 V, 5000 HP/3730 kW, 221 A, 50 Hz, 2970 r/min, 0.92 p.f., 96% efficiency, star connected SCIM drives a centrifugal pump. The motor is started DOL and was designed to have a torque–speed curve to provide sufficient accelerating torque at 100% and 90% of the rated voltage without abnormal overheating of the rotor bars during the run-up transient to steady-state speed. This is provided that the motor is started against a closed valve setting of the discharge valves and the pump is operating under the corresponding torque–speed curve provided by the pump manufacturer to the manufacturer of the motor. The end user has not been applying MCSA as a condition monitoring technique but they have now experienced two motor failures out of three motors due to broken bars crashing into the stator windings. There are no ERRs fitted.

The end user does not understand why displacement probes that measure shaft movement in a plain (sleeve) bearing are not giving an early warning of a broken rotor bar problem.

**(i)** Explain from fundamental principles, why measuring shaft movement in the motor's bearings is not suitable for diagnosing broken rotor bars and thus convince the plant owner that MCSA is the way forward to detect a broken rotor bar problem and greatly reduce the risks of a broken bar crashing into an HV stator winding.

**(ii)** The owner asks you to comment on the possible reasons for the broken bars, propose possible causes and sources of information that the owner has to provide to assist you with such a review.

**(iii)** Propose an MCSA test program on the three SCIMs driving the centrifugal pumps and the information you require from the on-site engineers and the need for certain data from the OEM of the motor.

## REFERENCES

[6.1] W. J. Morrill, "Harmonic theory of noise in induction motors," *Transactions of American Institute of Electrical Engineers*, vol. 59, no. 8, 1940, pp. 474–480.

[6.2] P. L. Alger, *Induction Machines-Their Behavior and Uses*, Gordon and Breach Science Publications Inc., 2nd Edition, published by OPA Amsterdam, 3rd printing with additions, 1995.

[6.3] P.L. Alger, "The magnetic noise of polyphase induction motors," *Transactions of the American Institute of Electrical Engineers, Power Apparatus and Systems, Part III*, vol. 73, no. 1, Part IIA, 1954, pp. 118–125.

[6.4] J. R. Cameron, W. T. Thomson, and A. B. Dow, "Vibration and current monitoring for detecting airgap eccentricity in large induction motors," *IEE Proceedings*, vol. 133, Part B, no. 3, May 1986, pp. 173–179.

CHAPTER 7

# MCSA CASE HISTORIES—CYCLIC LOADS CAN CAUSE FALSE POSITIVES OF CAGE WINDING BREAKS

*William T. Thomson*

## 7.1 INTRODUCTION AND SUMMARY OF CASE HISTORIES

This chapter focuses on case histories where cyclic loads in SCIM drive trains can cause false positives of cage winding breaks and the following list should assist the readers in selecting the case histories of personal interest.

**Section 7.2—One case history:** Effect of gas recycling in a centrifugal compressor on the spectral content of the current spectrum for detecting broken rotor bars; no broken bars but numerous sidebands in the current spectrum in a 2450 kW/3280 HP, 4-pole SCIM.

**Section 7.3—One case history:** False positive of broken rotor bars due to recycling of gas in a centrifugal compressor; current spectrum indicated sidebands at twice slip frequency but this was a false positive in a 5980 kW/8016 HP, 2-pole SCIM.

**Section 7.4—Two case histories:** Broken rotor bars in a copper fabricated cage winding in the same 1950 kW/2614 HP, 4-pole SCIM, without and with gas recycling in a centrifugal compressor.

**7.4.1** One broken rotor bar in a copper fabricated cage winding and a damaged rotor tooth; no gas recycling.

**7.4.2** Three broken bars in a copper fabricated cage winding, bars lifted out of the slots, slight damage to the stator core; with gas recycling.

*Current Signature Analysis for Condition Monitoring of Cage Induction Motors: Industrial Application and Case Histories*, First Edition. William T. Thomson and Ian Culbert.

**Section 7.5—One case history:** A false positive of cage winding breaks caused by fluid coupling dynamics in the drive train; sidebands coincided with twice slip frequency sidebands in a 1650 kW/2200 HP, 4-pole SCIM.

An emf will be induced in the stator winding of a SCIM by any mechanical disturbance to its squirrel cage rotor which results in perturbations of the magnetic field, which links the rotor and stator windings. This rate of change of flux linkage will induce an emf in the stator winding, whose coils are analogous to a set of search coils. These induced emfs drive currents through the stator and they can be detected in the current spectrum. In the oil and gas industry, particularly on offshore oil platforms, centrifugal gas compressors often operate in a mode in which the available gas has to be recycled due to a temporary depletion of gas. This can result in consequential temperature and pressure changes of the gas, and also gas slugging may occur, Boyce [7.1]. This effect is then imposed on the compressor blades and thus reflected back into the cage rotor and can cause very small fluctuations of the rotor speed which result in modulation of the current. Therefore, sidebands in the current spectrum can be misinterpreted as $\pm 2sf$ sidebands caused by cage winding breaks and hence the possibility of a false positive exists.

In certain drives, when it is necessary to prevent shock loads on the motor at start-up, a transmission device called a fluid coupling is used between the motor and the mechanical load to provide a controlled start up. The motor is started on no-load, but subsequently a controlled amount of load is applied, and in this way a fluid coupling acts like a clutch. As its name implies, the fluid coupling transmits power via hydrodynamic action, and hence there is inherent slippage and a very small percentage of power is lost due to natural friction and turbulence. Perturbations from the fluid coupling can be transmitted to the rotor in a SCIM and can therefore induce emfs in the stator and consequential currents will flow. These current variations can appear in the spectrum as sidebands around the supply component of current and can be misinterpreted as coming from broken rotor bars, giving rise to a possible false positive.

The industrial case histories will substantiate the above statements and demonstrate the complexity of applying MCSA to SCIM drive trains, from which the mechanical dynamics are reflected back into SCIMs and cause sidebands to appear in the current spectrum. This can result in a false positive of broken rotor bars where none exist. This dubiety about the interpretation of the spectrum is clearly a weakness of MCSA but can, in certain cases, be overcome by expertise and experience. An MCSA instrument alone certainly does not have these human attributes. The following case histories should convince the reader that extreme caution should be exercised, when interpreting spectra from SCIM drives with gas compressors in a gas recycling mode, or with a fluid coupling or any mechanical load variations that are reflected back into the SCIM.

### 7.1.1 Effect on Current Spectra due to Oscillations from Mechanical Loads

Recall from Section 1.2 that when the SCIM is operating at a constant load demand and speed, the operational slip is

$$s = (N_s - N_r)/N_s$$

However, if the mechanical load causes a cyclic variation around a mean average speed then the slip is a time varying function and is given by

$$s_{(t)} = (N_s - N_{r(t)})/N_s$$

Assume that the cyclic variation is sinusoidal, then

$$N_{r(t)} = N_a + k\sin(2\pi f_c t)$$
$$s_{(t)} = [N_s - (N_a + k\sin(2\pi f_c t))]/N_s$$

where

$s_{(t)}$ = slip as a function of time
$N_{r(t)}$ = rotor speed time varying function, r/min
$N_a$ = average rotor speed, r/min
$f_c$ = oscillating frequency due to cyclic load change, Hertz
$k$ = peak magnitude of speed oscillation which is a function of load changes, inertia of the rotor drive train, and torque characteristic of the mechanical load.

It has been verified by Thomson [7.2], Schoen and Habetler [7.3–7.5] that current components in the stator winding can be induced at the following frequencies:

$$f \pm f_c$$

A derivation is presented in Appendix 7A.1

It is most unlikely that the disturbance to the rotor is purely sinusoidal so it should be of no surprise to find sidebands, due to harmonics of $f_c$, at $2f_c$, $3f_c$, and $4f_c$ and Figure 7.2 verifies that this does occur in practice. These sidebands, due to the fundamental ($f_c$) and its harmonics, are purely due to load dynamics being reflected back into the rotor and are independent of a cage winding break. If the frequency of $f_c$ or one of its harmonics is equal or very close to the $2sf$ frequency, the sidebands around the supply component could be due to $\pm f_c$ or one of its harmonics or $\pm 2sf$ or both and a false positive of a broken rotor bar can readily occur. Furthermore, one of the current signature features of broken bars (case histories in Sections 5.4 and 5.5) is the presence of sidebands at $\pm 4sf$ and $\pm 6sf$, and that current signature may now be compromised in these circumstances.

Thomson carried out controlled experimental tests on the effects of cyclic loads in combination with and without broken rotor bars in the late 1980s. To investigate cyclic loads, the field current to a dc load dynamometer, driven by a SCIM, was varied in a low frequency mode thus the output volts and power to the resistive load varied and therefore the speed and slip of the SCIM was varied, thus varying its input current. The combination of a broken rotor bar with a cyclic load was investigated under controlled experimental conditions. This work was necessary because MCSA was being evaluated in the offshore oil and gas industry from 1982 onward (BP exploration sponsored research) and this problem of fluctuating loads causing sidebands that could be misinterpreted as being at $\pm 2sf$, due to broken bars, was observed during on-site trials.

In 1992, Thomson [7.2] reported, via experimental results, the effects of fluctuating loads on the current spectrum and in particular, how this fluctuation may lead to a false diagnosis of $\pm 2sf$ sidebands and cage winding breaks. In 1995 and 1997, papers by Schoen et al. [7.3, 7.4] investigated the effects of time varying loads on rotor fault detection in induction machines and proposed a strategy on how to eliminate arbitrary load effects on current monitoring. These papers did not include any industrial case histories but did provide valuable theoretical knowledge supported by experimental tests on small laboratory based SCIMs. Reference 7.4 reported that three currents and three voltages were required. Since this book is dedicated to industrial application of classical MCSA using only one CT, a method supported by numerous industrial case histories, the following points need to be recognized:

(i) The end users (owners) of large HV (4.16, 6.6, 11, and 13.8 kV) SCIMs are not receptive to providing access to measure three voltages and three currents, while the motors are running, since this would normally require access to VTs and CTs used in their protection circuitry. They would not normally permit such an invasive action via the low voltage instrumentation panel, which does contain the protection circuits including protection CTs and VTs, while a large HV SCIM is running. A shutdown would be required for access to install temporary MCSA CTs and to measure three voltages. Even when using only one clip-on CT around the secondary side of one of the cables feeding an *in situ* display ammeter (not part of the protection circuitry) in the LV instrumentation panel, the end user has to be convinced that there will be no risks of tripping the motor. A professional method and risk assessment for classical MCSA, while the motor is running, has to be vetted and approved by the end user.

### 7.1.2 MCSA Laboratory Experiments with a Cyclic Load and a Broken Rotor Bar

For completeness, a sample of the results recorded by Thomson in 1989 using classical MCSA measurements is now provided before presenting the industrial case histories on MCSA applied to SCIMs with fluctuating loads.

The nameplate data for the laboratory based SCIM was as follows: 3-phase, 415 V, 11 kW (14.7 HP), 20.5 A, 50 Hz, 1430 r/min, delta connected, with a 51 slot aluminum die-cast rotor.

Figure 7.1 shows the current spectrum with one broken rotor bar at an operational speed of 1440 r/min and a slip of 0.04 (4%) but with no superimposed oscillation from the load on the motor and therefore only $\pm 2sf$ sidebands are present at $\pm 4$ Hz around the supply component ($f$) and at $N_{av}$ equal to 46 dB down on $f$. The equivalent $BB_f$ referred to the full-load slip of 0.047 (4.7%) is 0.6. As expected, since there is only one broken bar which is much less than the total number of 51 bars, the $BB_f$ underestimates the actual number of broken bars but it is still above 0.5, which was proven in Chapters 4–6 to be the threshold above which a cage winding break normally exists.

With one broken rotor bar, the magnitude of the field current to the dc dynamometer was automatically oscillated and hence the generated volts and power

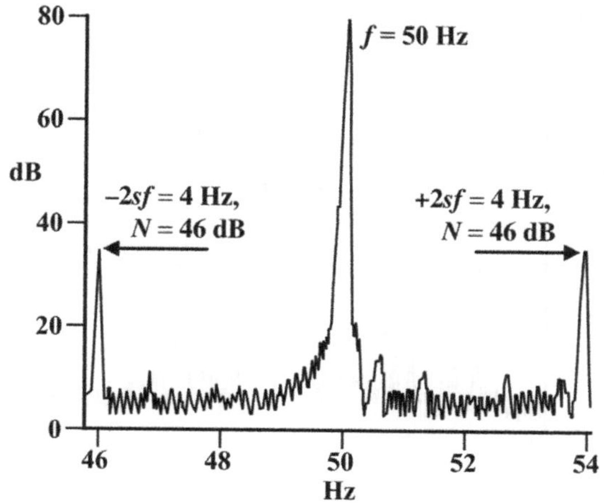


Figure 7.1 Current spectrum, one broken bar (slip 4%), no external load oscillations, line resolution 10.16 mHz/line.

demand by the dynamometer was cyclically varied at a frequency of $f_c$ equal to 0.73 Hz. The current spectrum shown in Figure 7.2 is now considerably more complex due to the combination of the $\pm 2sf$ sidebands from the broken rotor bar and components from the load oscillations but they are from two distinctly different physical phenomena. The $\pm 2sf$ sidebands are still at ±4 Hz around $f$ but the components at $\pm f_c$ around $f$ are dominant and there are also harmonics at $\pm 2f_c$, $\pm 3f_c$, and $\pm 5f_c$, these harmonics are expected since the oscillation at $f_c$ is not likely to be a pure sine wave in a practical drive train. In this case, where the $\pm 2sf$ sidebands and $f_c$ are spaced well apart (at 3.27 Hz), there is no problem in identifying the components due to a broken rotor bar.

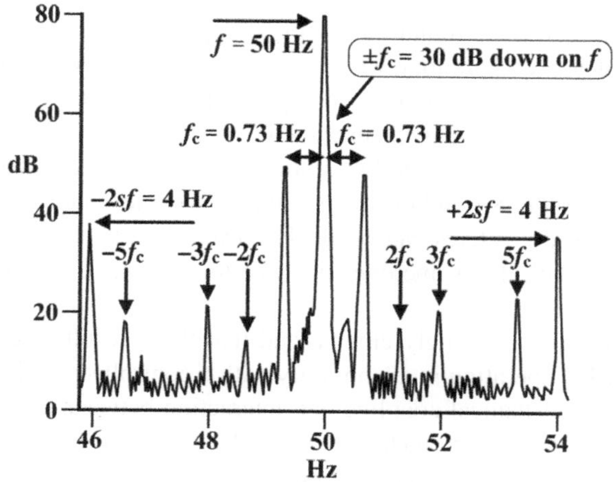


Figure 7.2 Current spectrum, one broken bar (mean slip 4%) and external load oscillations, line resolution 10.16 mHz/line.

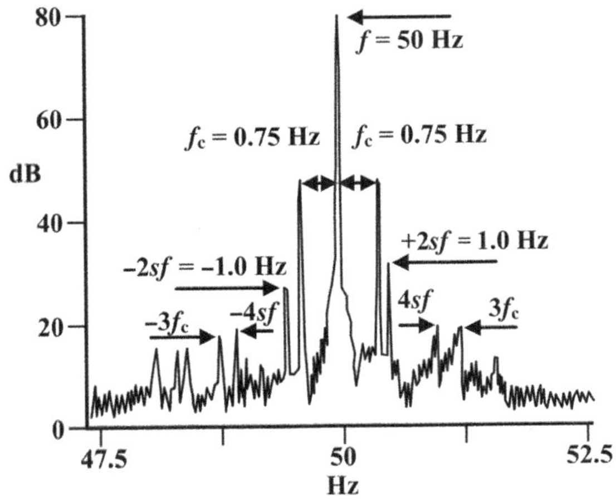


Figure 7.3 Current spectrum, one broken bar (mean slip 1%) and external load oscillations, line resolution 10.16 mHz/line.

In industrial drives, the full-load slips in large, HV SCIMs are considerably lower (typically from 2% down to as low as 0.35%) than the 4.7% slip of this 11 kW motor. Therefore, a repeat test was carried out with one broken bar but at an operational slip of 0.01 (1%) and also with an oscillating external load. The frequency of the $\pm 2sf$ sidebands will reduce to $\pm 1.0$ Hz around $f$ and their dB magnitudes will be considerably lower. Figure 7.3 shows the resulting spectrum and now the sidebands at $\pm 2sf$ and $\pm f_c$ are much closer at only 0.25 Hz apart and in terms of a change in $2sf$ with $f$ constant that equates to a slip of only 0.0025 (0.25%).

In industry, consider a SCIM driving an oscillating load, whose oscillating frequency $f_c$ in Hertz is unknown. There are no broken rotor bars in the SCIM but the load is oscillating and the $\pm f_c$ components are very close to or coincident with the predicted $\pm 2sf$ components around the supply component. In these circumstances, a false positive of broken bars may well be diagnosed. If the end user is willing to change the operational load such that the slip changes, the two phenomena can be separated, but as already stated in Chapters 4–6, end users are very reluctant indeed to alter a production process to merely accommodate an MCSA measurement. This is a real challenge for MCSA and the industrial case histories that follow will clearly demonstrate that false positives can occur. However, by very careful analysis and great attention to detail, it is sometimes possible to avoid false positives of broken rotor bars.

However, it is very strongly emphasized that these results are from a small 11 kW, 415 V SCIM operating in a laboratory with a controlled oscillating load and there is very close mechanical coupling and interaction between the motor and the oscillating load via the dynamometer. It cannot be expected that the same current signature pattern will be produced due to mechanical load dynamics downstream of a large HV SCIM drive train.

## 7.2 CASE HISTORY (2006)—EFFECT OF GAS RECYCLING IN A CENTRIFUGAL GAS COMPRESSOR AND THE DETECTION OF BROKEN ROTOR BARS

### 7.2.1 MCSA Diagnosis

Four motors, driving centrifugal gas compressors at an LNG onshore processing plant, were MCSA tested on an annual basis, commencing in 2004. The nameplate data for each motor was as follows: 11 kV, 2450 kW/3280 HP, 152A, 50 Hz, 2965 r/min, 0.88 p.f., efficiency = 96%, star connected SCIM with a cage rotor having 52 rotor bars. In 2010, one of the motors was MCSA tested when the gas was being recycled.

The motor was operating at a current of 126 A compared to the full-load current of 158 A. The current spectrum in Figure 7.4 shows that there are sidebands spaced at $\pm f_c$ ($\pm 0.15$ Hz) around the supply component and multiple harmonics of $\pm f_c$ at $\pm 2f_c$ ($\pm 0.3$ Hz), $\pm 3f_c$ ($\pm 0.45$ Hz), and $\pm 4f_c$ ($\pm 0.6$ Hz). This substantiates the theory presented in Section 7.1.1 and in this case the fundamental oscillating frequency due to gas recycling in the compressor is $f_c = 0.15$ Hz.

The nominal full-load slip was 1.17% and the calculated slip from the 0.15 Hz sidebands was 0.15%. Clearly, this slip is much too low for the operational current and these sidebands were being induced due to gas recycling.

The MCSA measurement was repeated the following day, when the recycling of the gas in the compressor was removed, and at an operational current of 120 A, the sidebands had disappeared, as shown in Figure 7.5. This proved, that the source of the sidebands in Figure 7.4 was compressor dynamics from gas recycling being reflected back into the rotor of the induction motor, and hence emfs in the stator are induced and corresponding currents can be detected in the supply current.

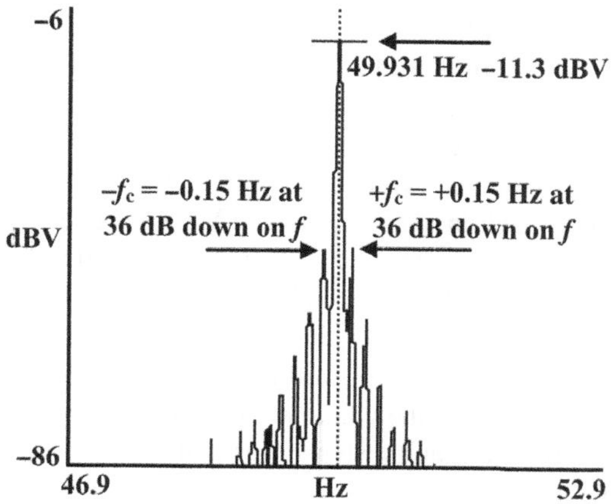


Figure 7.4 Current spectrum, recycling of gas in compressor, $f_c = 0.15$ Hz is the frequency of the cyclic disturbance from gas recycling, line resolution 10.16 mHz/line.

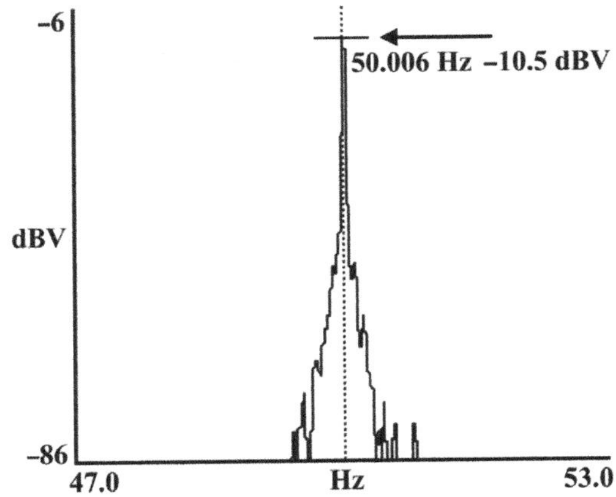


Figure 7.5 Case one, current spectrum, no recycling of gas, line resolution 10.16 mHz/line.

## 7.3 CASE HISTORY: FALSE POSITIVE OF BROKEN ROTOR BARS DUE TO RECYCLING OF GAS IN A CENTRIFUGAL COMPRESSOR

### 7.3.1 Summary and Key Features

In 2006, an HV SCIM was driving a centrifugal compressor which was not recycling the gas and there were $\pm 2sf$ sidebands around the supply component at an average of 63 dB down on $f$. In 2007, the $\pm 2sf$ sidebands increased to an average of 53 dB down on $f$, but the compressor was on recycling. Clearly there had been a change in the magnitude of the sidebands and the client was informed that MCSA measurements had to be taken at a substantially different load current, with the same recycling percentages, or at the same load with recycling removed. Either of these strategies would identify whether the increase in the $\pm 2sf$ sidebands was due to gas recycling, or truly coming from a rotor cage fault.

Unfortunately, neither of these load changes was feasible because this was the main gas compressor, in fact the only one, on the platform and the offshore production and operations division would not change their operating regime for MCSA tests. Their viewpoint was as follows: "Here is the motor for MCSA measurements, we cannot interfere with production to accommodate MCSA." The oil wells on this platform rely on gas lift from this single train gas compressor, and approximately 80% of the total production relies on it being available. The platform has a two-yearly shutdown frequency and the oil company decided to replace the rotor in 2008, based on the business risk of the gas compressor failing in service, in which case the time it would take to replace the rotor or the whole motor would have resulted in significant lost production. There were no faults whatsoever in the cage winding and it was in a perfectly normal condition.

This was a false positive due to gas recycling but the decision to remove the rotor was made solely by the oil company. An MCSA test was carried out in 2009, with a new (spare) replacement rotor installed, and with the compressor on recycling.

As expected, the spectrum was virtually identical to the one previously observed when the original rotor was in place, thus confirming that recycling was producing sidebands, which coincided with $\pm 2sf$ sidebands.

### 7.3.2 MCSA Diagnosis

The motors' nameplate data provided the following relevant information: 3-phase, 6.6 kV, 5980 kW/8016 HP, 615 A, 1775 r/min, star connected SCIM. The OEM provided the following data on the rotor; it is a single cage copper fabricated winding and the number of rotor bars is 70, with butt type bar to end ring joints and no ERRs are fitted.

The motor was MCSA tested on an annual basis, commencing in 2006 and the current spectrum is shown in Figure 7.6, at an operational current of 564 A, compared to a full-load current of 615 A. There were sidebands at ±1.046 Hz, around $f$, at an average dB difference ($N_{av}$) of 63 dB down on $f$ with the upper and lower sidebands at 61 and 65 dB down, respectively. There was no recycling of gas in the compressor train.

If the sidebands at ±1.046 Hz around $f$ are truly $2sf$ sidebands, then this gives an operational slip of 0.87% (at $f = 60.2$ Hz). The nameplate speed and corresponding full-load slip were 1775 r/min and 1.4%. The motor was operating at 92% of its full-load amperes and the slip from the 1.046 Hz sidebands was 0.87%, which seemed on the low side (full-load slip 1.4%) for an input current of 92% of full-load current. The OEM subsequently confirmed that the full-load heat run, at the original Factory Acceptance Tests (FAT), was carried out on a supply of 50 Hz and the nameplate speed was obtained by extrapolation, to estimate the speed at full-load on a 60 Hz supply. Therefore, the nameplate speed at 60 Hz may be incorrect, as was the case with the nameplate data on a motor from the same OEM as reported in case history 6.1.

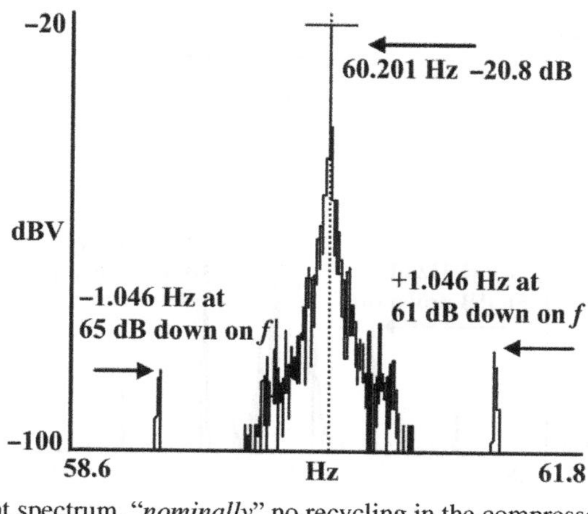


Figure 7.6 Current spectrum, "*nominally*" no recycling in the compressor, 2006, line resolution 10.16 mHz/line.

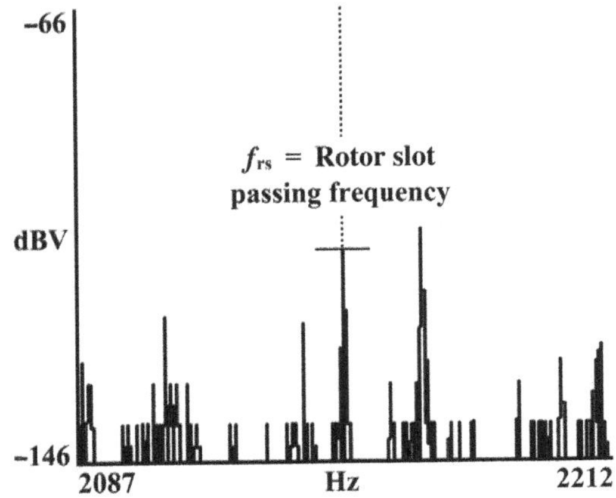


Figure 7.7 Current spectrum showing the rotor slot passing frequency, baseband span 70–2500 Hz, 12,800 lines, 0.19 Hz/line.

For confirmation of the operational slip, the same analysis approach was used as that reported in the first case history of Section 6.1 and for this motor the measured rotor slot passing frequency ($f_{rs}$) was found to be 2149 Hz, as shown in Figure 7.7. Using the measured supply frequency (60.2 Hz) and the other parameters in equation (6.1), the true operational slip, as calculated from the measured $f_{rs}$, was 0.86%, thus confirming that the slip from the $2sf$ sidebands was at 0.87%.

In 2007, the MCSA test was repeated but the compressor was recycling the gas and the analogue ammeter indicated that the current was fluctuating between 565 and 570 A. The sidebands at $2sf$ were now an average of 53 dB down on $f$, which averaged an increase (compared to the 2006 levels) of 10 dB (a factor of 3.16), as shown in Figure 7.8 and the upper and lower sidebands were 50 and 56 dB down on $f$ compared to 61 and 65 dB down without gas recycling.

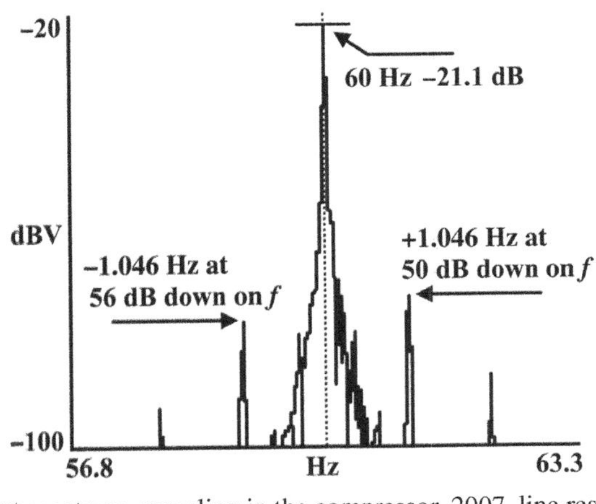


Figure 7.8 Current spectrum, recycling in the compressor, 2007, line resolution 10.16 mHz/line.

Figure 7.9 Photograph of rotor.

This was a substantial increase, but, was the recycling of the gas in the compressor causing sidebands that coincided with the $\pm 2sf$ sidebands and thus gave a false impression that the increase was due to a change in the condition of the cage winding? The end user was requested to reduce the load on the motor by, for example 100 A, to obtain a substantial change in the slip but keep the recycling of gas constant, or to keep the same load on the motor and remove the recycling to separate the two different phenomena. The client could not accommodate either of these requests in April 2008 and the same result was obtained as was the case in 2007, but in August 2008, the oil company decided to remove the rotor, for an inspection, during their major two-yearly outage and the rotor was found to be in a perfectly healthy condition as shown in Figures 7.9 and 7.10. The initial recommendation, based on the current spectrum shown in Figure 7.8 was to carry out an MCSA resurvey in 3 months and arrange for the load to be changed to a different operating current or to significantly reduce the percentage of recycling.

However, this was not arranged by the end user who decided to remove the rotor due to the strategic importance of the drive train and a brand new rotor was installed in 2008. The MCSA test was repeated with the compressor still on recycling and Figure 7.11 confirmed that sidebands, which coincided with $\pm 2sf$ sidebands were still present at $\pm 1.04$ Hz around $f$ at an average of 52 dB down on $f$ and the operating current was swinging between 524 and 530 A.

There are sidebands $\pm 0.26$ Hz, which are in fact the highest of all the sidebands at 47 dB down on $f$ in Figure 7.11 and there are also sidebands at $\pm 0.52$ Hz (60 dB down). Since the compressor was on gas recycling with this new rotor it is therefore concluded that the sidebands are in fact $f \pm f_c$, where $f_c$ (0.26 Hz) is the oscillating

Figure 7.10 No evidence of broken, cracked bars or faulty joints in the cage winding.

frequency caused by the recycling in the compressor and this is further borne out by the fact that there are $f \pm 2f_c$ and $f \pm 4f_c$ components and $4f_c$ is 1.04 Hz These sidebands are not due to cage winding asymmetry. However, an engineer who was required to apply MCSA to this motor for the first time and was unaware of the effects on the current spectrum of recycling in a compressor, he/she could easily give a verdict of a false positive of a cage winding problem with this brand new cage rotor.

For example, the $BB_f$ from equation (4.5) corrected and referred to the full-load slip is 0.5 if the ±1.04 Hz sidebands are taken to be $\pm 2sf$ around $f$. This equals the threshold value, accepted and proven in the case histories in Chapters 5 and 6, as being from a cage winding break, which should not be the case with a brand new rotor. A false positive could certainly occur with this brand new rotor. Suffice to state an OEM would not accept a cage winding break or inferior bar to end ring joints in a brand new rotor. The OEM had guaranteed (in writing) that the cage winding

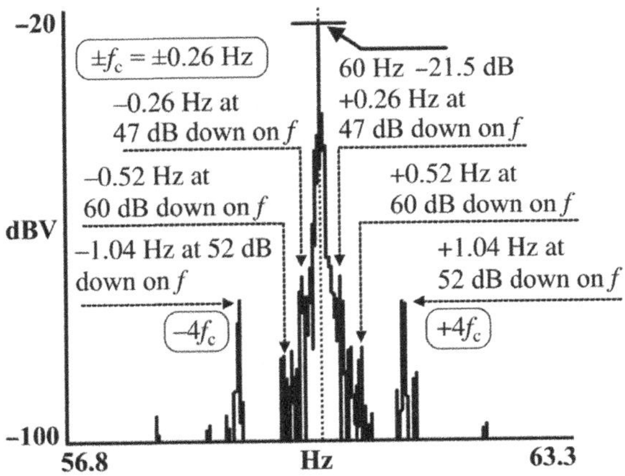


Figure 7.11 Case two, current spectrum, recycling of gas, 2008, brand new rotor, line resolution 10.16 mHz/line.

was very carefully inspected after its manufacture and it was in a perfectly healthy condition before it was shipped to the end user and installed offshore.

The key fact to remember is that when MCSA is being applied to a SCIM driving a centrifugal compressor during gas recycling, it is highly recommended to take MCSA measurements at two significantly different load currents to substantially change the slip, or to remove the recycling of gas to identify, with certainty, the true source of any sidebands that are close to or coincide with $\pm 2sf$ sidebands around $f$.

## 7.4 TWO CASE HISTORIES (2002 AND 2013)—BROKEN ROTOR BARS IN THE SAME SCIM WITHOUT AND WITH GAS RECYCLING IN A GAS COMPRESSOR

### 7.4.1 Case One 2002: Summary and Key Features

The motor nameplate provided the following data: 3-phase, 11 kV, 1950 kW/2614 HP, 120 A, 50 Hz, 1485 r/min, star connected SCIM, having a cage rotor with 60 rotor bars. This motor was used to drive a low pressure (LP), two-stage, gas compressor on an offshore oil and gas production platform. The rotor has rectangular bars in skewed slots and is a copper fabricated cage winding with the bars placed into slots in the outer periphery of the end rings and end ring retaining rings (ERRs) are not fitted.

In 2002, MCSA indicated that twice slip frequency sidebands ($\pm 2sf$) were an average ($N_{av}$) of 43.5 dB down on the supply component ($f$) and there were also $\pm 4sf$ sidebands around $f$ but at 20 dB lower than the main $\pm 2sf$ sidebands. The predicted broken bar factor ($BB_f$) which was corrected for reduced load operation at a slip of 0.64% was 1.2 and this suggests there was at least one broken rotor bar. The rotor was inspected and it was found that there was indeed one completely broken rotor bar and also that an adjacent bar was clearly overheated, where the bar entered the end ring. A brand new rotor was installed in 2002 with exactly the same rotor design. The current spectrum and its interpretation with a photo of the rotor with one broken bar are presented in Section 7.4.3.

### 7.4.2 Case Two 2012–2013: Summary and Key Features

Annual MCSA surveys were carried out after the new rotor was installed in 2002 and there was no evidence whatsoever of any $\pm 2sf$ sidebands until June 2012 when the compressor was in a recycling mode compared to no recycling in previous years. This caused numerous sidebands around the supply component ($f$) and also sidebands which coincided with $\pm 2sf$ sidebands at 60 dB down on $f$. In June 2013, it was proved by taking MCSA measurements at different recycling percentages in the compressor that there were indeed $\pm 2sf$ sidebands at 49 dB down on $f$ and it was recommended that the motor should be MCSA tested every 3 months but the operator and duty holder of the platform did not do so.

Ten months later in March 2014, the operator reported that the vibration was excessive (no values provided) on the motor's bearings, also next to its fixing down

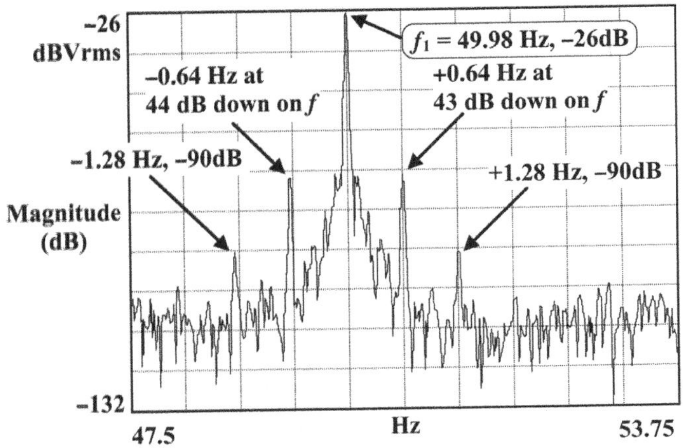


Figure 7.12 Current spectrum in 2002 with $\pm 2sf$ and $\pm 4sf$ sidebands, line resolution 10.16 mHz/line.

bolts and on connected pipework on the compressor and at all positions the vibration spectrum was dominated by the 1× r/min frequency component from the motor. For safety reasons the motor was shut down. The rotor was removed and there were three broken rotor bars but suffice to state that this serious condition had developed between the measurements in June 2013 and March 2014, the most probable cause being too many sequential DOL starts during that 9-month period. Interpretation of current spectra and photos of the rotor with the broken rotor bars are presented in Section 7.4.4

### 7.4.3 Case One 2002: MCSA Diagnosis and Inspection

There is only one drive train on the platform for this production process, hence an assessment of its operational integrity is a key element of the maintenance policy. In December 2002, the compressor was not on recycling and the current spectrum shown in Figure 7.12 indicated that there were $\pm 2sf$ sidebands ($\pm$0.64 Hz) around the supply component at an average of 43.5 dB down from $f$. The motor was operating at 104 A compared to its full-load current of 120 A. No ERRs were fitted and therefore if a broken bar lifts it will not obtain a parallel conducting path to mask the severity of the problem, but it could catastrophically damage the stator.

It was noted that there was a small and regular oscillation (around 2 A) about a mean position on the analogue ammeter. This is an observation and cannot be used to reliably detect broken rotor bars or to quantify the severity of a fault. Similar effects can occur due to normal oscillating loads or recycling effects in compressors. The whole objective is to gather all the evidence when crucial decisions have to be made about a motor's operational condition. Note that routine bearing vibration measurements on the motor's bearing housings by an offshore contractor indicated that no changes had occurred.

The end user had a major concern about possible secondary damage to the stator winding and the crucial questions to be answered were:

**(i)** How severe was the problem?

**(ii)** Could the motor be kept running?

**(iii)** What are the chances of a broken bar hitting the stator winding?

**(iv)** With respect to item (i), an estimate of the severity of the cage defect is given as:

$$BB_{fs} = 2R/(10^{(Nav/20)} + 2p) = 0.78$$

with $N_{av} = 43.5$ dB, $R = 60$, and $p = 2$

The equivalent broken bar factor at full-load is

$$BB_f = BB_{fs} \times BB_c = 0.78 \times (s_{FL}/s_o) = 0.78 \times (1.0/0.64) = 1.2$$

It was therefore predicted that there was at least one broken rotor bar. The recommendation was to replace the faulty rotor with a spare one as soon as possible and the operator had no option but to keep the motor running until the replacement was installed.

Inspection of the faulty rotor, as shown in Figure 7.13 proved that the diagnosis was valid since there was one completely broken rotor bar. In addition, there was approximately only 1 mm (or 40 thou or mils) of laminated steel left at the top of the two adjacent teeth in the slot with the broken rotor bar and there was damage to the rotor core. Also there is evidence of overheating on an adjacent bar next to the completely broken one. There was indeed a very high risk of the broken bar lifting out of the slot with consequential mechanical damage to the HV stator winding. If the broken rotor bar fault had gone undetected and resulted in an HV stator winding failure this would have required a rewind on an offshore oil and gas production platform, because during the 35 years since the platform was commissioned major changes to the modular structure meant that the stator could not be removed for an onshore rewind.

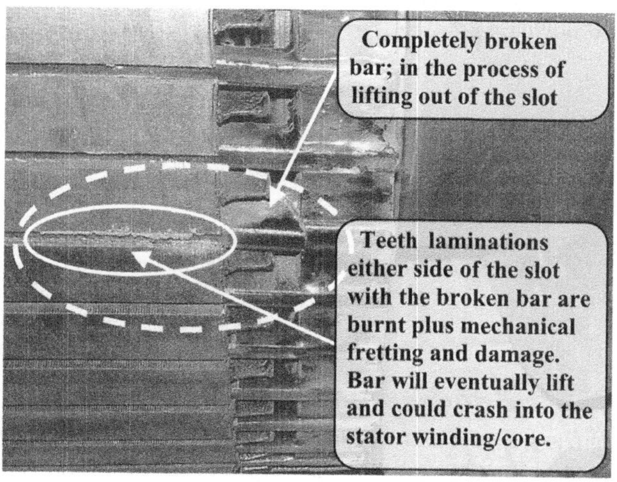


Figure 7.13 One broken rotor bar and an overheated adjacent rotor bar.

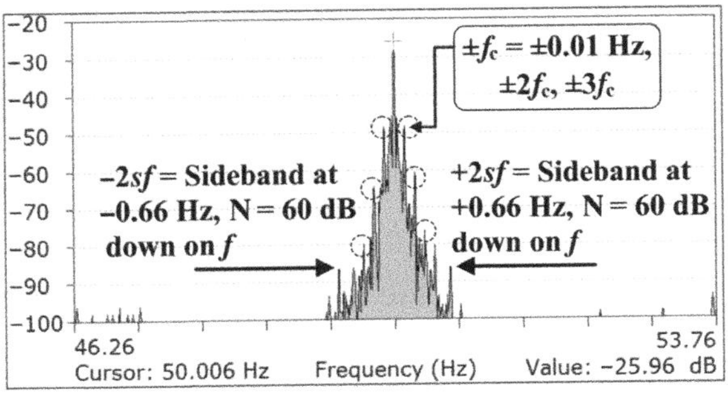


Figure 7.14 Current spectrum in 2012 with numerous sidebands, compressor on recycling, line resolution 10.16 mHz/line.

## 7.4.4 Case Two: MCSA Diagnosis and Inspection, 2012–2013

MCSA surveys of this motor with its new rotor continued on an annual basis and in June 2012, a survey was carried out when the two-stage compressor was recycling the gas with stage one at 70% and stage two at 81% (both high values). The motor was operating on reduced load at a current of 97 A compared to the full-load current of 120 A. Figure 7.14 is the current spectrum, which shows that there are numerous sidebands around the supply component, $f$ at $\pm f_c = \pm 0.1$ Hz and harmonic sidebands at $\pm 2f_c$, $\pm 3f_c$, $\pm 4f_c$ around $f$. These are due to the cyclic disturbances from gas recycling in the compressor, which are transferred into the cage rotor. There are also sidebands at $\pm 2sf$ (not harmonics of $\pm f_c$) around $f$ but at 60 dB down on the supply component the broken rotor bar factor, $BB_f$, corrected for reduced load operation at a slip of 0.66% is 0.18. This is considered to be a normal rotor cage at such a low value of $BB_f$.

An MCSA survey was carried out a year later, in June 2013, with the compressor also in a gas recycling mode with stage one at 61% and stage two at 70% (in June 2012 it was 70% and stage two at 81%). The current spectrum shown in Figure 7.15a now indicates that there are clear sidebands at ±0.67 Hz around $f$ and they are an average of 49 dB down on $f$.

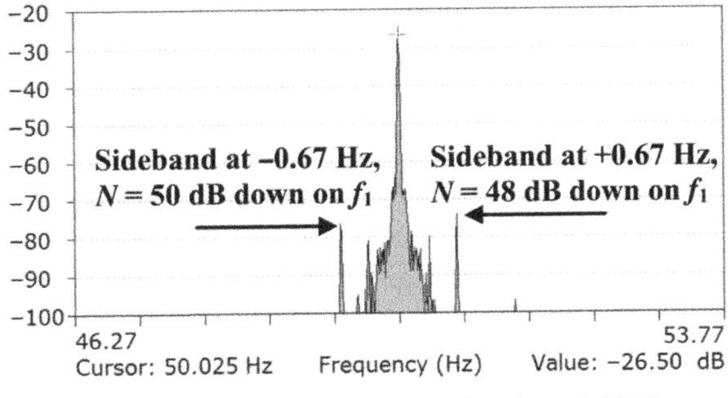


Figure 7.15a Current spectrum in 2013 showing sidebands at 0.67 Hz around $f$, compressor was on gas recycling, line resolution 10.16 mHz/line.

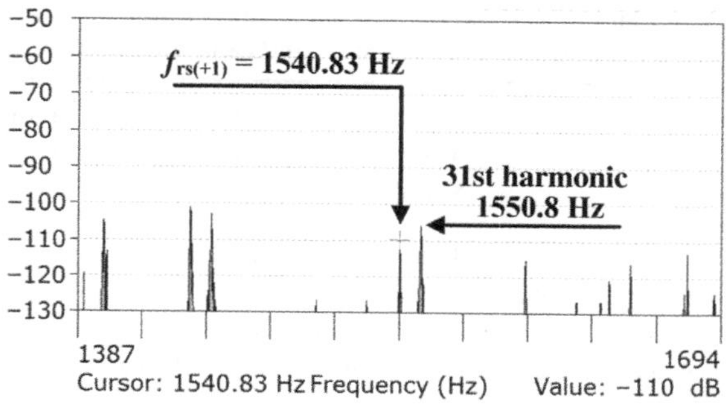


Figure 7.15b Measurement of $f_{rs(+1)}$, baseband span 70–1800 Hz, 12,800 lines, line resolution 135 mHz/line.

It has therefore to be established if they are truly $\pm 2sf$ sidebands. Fortunately, the number of rotor bars was known ($R = 60$) and therefore the principal rotor slot passing frequency can be predicted using the value of slip obtained from the sidebands at 0.67 Hz, ($s = 0.67/(2 \times 50.025) = 0.0067$ (0.67%). The current spectrum can then be searched to find the principal rotor slot passing frequency using $s = 0.0067$ and if present that confirms the sidebands at $\pm 0.67$ Hz are truly at $\pm 2sf$.

The principal rotor slot passing frequency is given by (as per case history 6.1):

$$f_{rs} = f\left\{\left(\frac{R}{p}\right)(1-s) \pm n_{\omega s}\right\} \tag{6.1}$$

where

$R$ = number of rotor bars = 60

$f$ = supply frequency = 50.025 Hz (the measured frequency)

$p$ = pole-pairs = 2

$s$ = operational slip = 0.0067 from the sidebands at 0.67 Hz

$n_{\omega s}$ = 1, 3, 5… mmf time harmonics

The predicted $f_{rs(+1)}$ (with $n_{\omega s} = +1$) component is therefore:

$$f_{rs} = 50.025\left\{\left(\frac{60}{2}\right)(1-0.0067)+1.0\right\} = 1540.8\ \text{Hz}$$

A search of the spectrum is carried out to find $f_{rs(+1)}$ and Figure 7.15b shows it is at 1540.83 Hz (predicted 1540.8 Hz) and there is no dubiety whatsoever that the operational slip is 0.0067 and the sidebands at 0.67 Hz are truly $2sf$ sidebands at an average of 49 dB down on $f$. The $\pm 2sf$ sidebands are an average of 49 dB down on the supply component as shown in Figure 7.15a and an extract from Table 4.2 is given in Table 7.1. This is the difficult band to give an assessment via the broken bar severity chart particularly when the motor is on a reduced load of 67% of full-load slip, so it is not possible to make a decision on the condition of the cage winding from the dB difference chart.

**TABLE 7.1 Extract From Table 4.2. Broken Rotor Bar Severity Estimator Based on Only the Average dB Difference ($N_{av}$) Between the $\pm 2sf$ Sidebands and the Supply Component**

| $N_{av}$ = dB Difference | Estimated Condition of Cage Winding When the Motor is Operating at Full-Load Output and Speed at the Shaft |
|---|---|
| 45–50 dB<br>Measured ($N_{av}$) of 49 dB down on $f$, the motor was operating on reduced load at 67% of the nominal full-load slip. | This is the very difficult boundary range for using the dB difference.<br>At 45 dB in lower pole number motors (2 and 4), there is a possibility of either faulty joints or perhaps a broken rotor bar. |

However, the number of rotor bars is known and the equivalent broken rotor bar factor $BB_{fs}$ can be calculated at the operating slip, which is then referred to the full-load slip:

$$BB_{fs} = 2R/(10^{(Nav/20)} + 2p) = 0.42$$

with $N_{av} = 49$ dB, $R = 60$, and $p = 2$

Referred to the full-load slip and rated output using the $BB_c$ gives the estimated equivalent broken rotor bar factor at full-load as

$$BB_f = BB_{fs} \times BB_c = 0.42 \times (s_{FL}/s_o) = 0.42 \times (1.0/0.67) = 0.63$$

This is an increase from June 2012, by a factor of 3.5 times larger.

For completeness, a further set of current spectra were recorded at different percentages of recycling in the compressor on the June 3–4, 2013, while the current remained the same, these have caused the "*distinct differences in the sidebands inboard*" of the $\pm 2sf$ sidebands as shown in Figure 7.16. The $\pm 2sf$ sideband magnitudes in dB, relative to the dB magnitude of $f$, have remained at $49 \pm 0.12$ dB. It should also be observed that the supply frequency $f$ on the June 3–4, 2013, was very slightly different (by 0.08 Hz); hence there is a difference in frequency in the $\pm 2sf$ sidebands as shown in the circles in Figure 7.16 thus confirming that these $\pm 2sf$

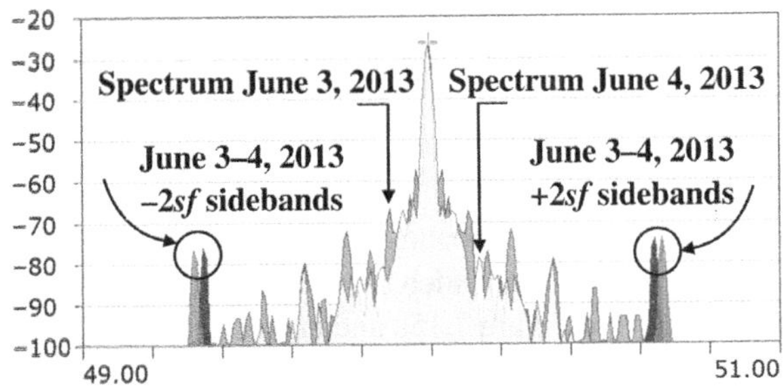


Figure 7.16 Current spectrum (same load at 98 A) on June 3–4, 2013, showing sidebands at $\pm 2sf$ with the compressor on recycling; baseband span of 0–55 Hz, 12,800 lines, frequency resolution of 4.3 mHz/line.

sidebands are purely an electrical phenomenon. The two phenomena, namely: (a) sidebands at $\pm 2sf$ due to a rotor cage break and (b) sidebands due to purely mechanical disturbances from compressor gas recycling, have been separated.

It can therefore be concluded that the $\pm 2sf$ sidebands were truly coming from a rotor cage winding break and the corrected $BB_f$ was 0.63 referred to the full-load slip. A comparison between the $BB_f$ in 2002 of 1.2, corresponding to one completely broken bar, and the $BB_f$ of 0.63 in 2013 would suggest that at this stage, there may not in this case be a completely broken rotor bar, but the result could certainly be due to a broken bar that was still making partial contact with one of the end rings but had not lifted out of that end ring. Due to the rotor design, such a broken bar could lift and crash into the stator winding, causing a stator winding failure and a risk of stator core damage. The following recommendations were given to the owner of the offshore platform:

**(i)** At present, the motor can continue to run, but the number of starts should be kept to a minimum as far as is reasonably practical.

**(ii)** Carry out repeated MCSA measurements every 3 months as from June 4, 2013.

**(iii)** If the $BB_f$ was to subsequently increase to $\geq 1.0$ then there is a high probability that one or two broken rotor bars exist.

Unfortunately, the owner and operator of the offshore platform did not implement the recommendations since MCSA resurveys were not carried out at 3 monthly intervals. In December 2013, it was reported that external vibration on pipework connected to the LP compressor became excessive, as quoted by the operator (actual values were not given) and was also being transmitted to other connected gas pipework on other compressors, which was thus a safety issue.

The compressor was decoupled from the motor and the velocity level at the motor's holding down bolts on the fabricated skid for the motor was up to 5.5 mm/sec (0.22 inches/sec) rms, which was being transmitted throughout the fabricated floor structure of the platform and suggested that the source of high vibration, up to 15 mm/sec (0.6 inches/sec) rms on the pipework was within the motor. This 1× r/min frequency of vibration had to be due to an increase in mechanical unbalance in the rotor and the only possible cause was that one or more broken rotor bars had lifted out of the slots causing an increase in mechanical unbalance.

The operator had to switch the motor off for health and safety reasons in March 2014—10 months after the MCSA tests in June 2013. The rotor was removed and inspected on the platform and photographs in Figures 7.17–7.19 show that two completely broken rotor bars had lifted out of the slots and a total of three bars were broken.

There was also damage to rotor teeth and evidence of heating of the rotor core. Fortunately, there was no damage to the stator winding and only very minor damage to the stator core which was successfully repaired (as reported by the operator and motor repair company). Suffice to state that this situation was very unlikely to have existed 10 months previously and had the motor been MCSA tested every 3 months, the degradation of the rotor cage would have been detected much earlier and thus the high vibration on the inter-connected gas pipework, and the safety hazard could have

Figure 7.17 Rotor construction showing skewed rotor bars and broken rotor bars.

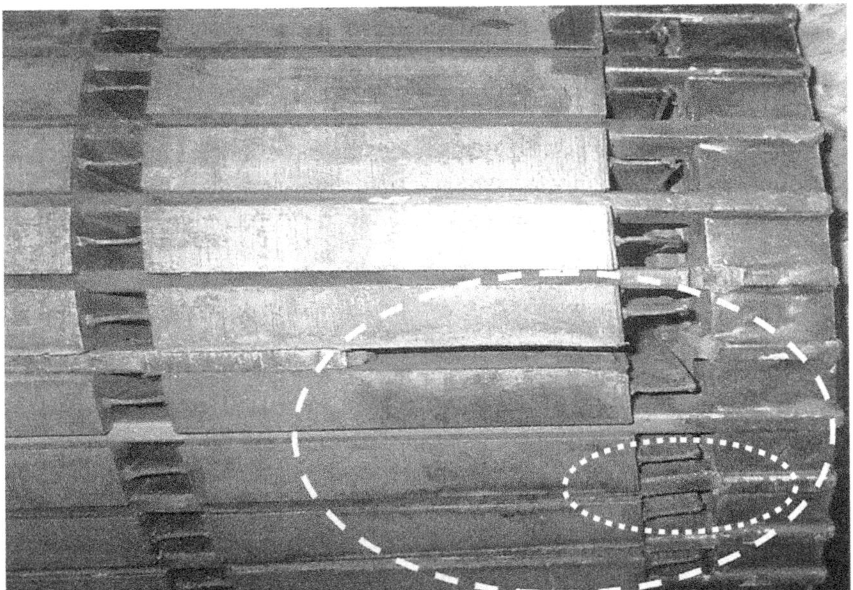

Figure 7.18 NDE: one completely broken bar, and two adjacent broken bars that were still in partial contact with the end ring, core damage.

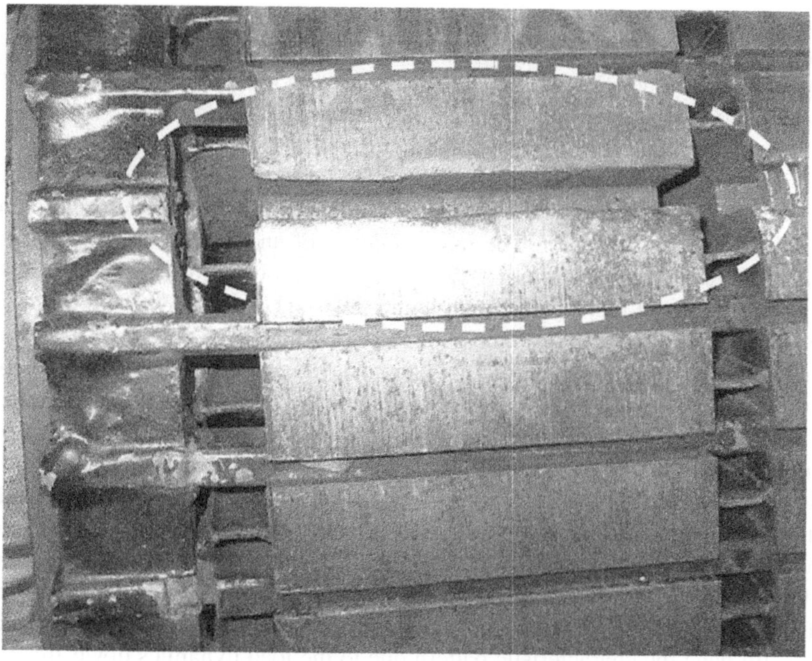

Figure 7.19 DE: one completely broken bar.

been prevented. It was accepted by the operator that too many DOL starts during a 10-month period from June 2013 to March 2014 had accelerated the development of broken rotor bars compared to its condition indicated by MCSA in June 2013 and that they should have tested the motor every 3 months.

## 7.5 CASE HISTORY 1986–FLUID COUPLING DYNAMICS CAUSED A FALSE POSITIVE OF A CAGE WINDING BREAK

### 7.5.1 Summary and Key Features

The nameplate data is as follows: 3-phase, 11 kV, 1650 kW/2211 HP, 110 A, 50 Hz, 1487 r/min, star connected SCIM. The OEM provided the following data on the rotor; it is a single cage copper fabricated winding and the number of rotor bars is 56, with butt type bar to end ring joints and no ERRs are fitted. This motor was one of the six driving crude oil pumps via fluid couplings at an onshore oil tanker loading site. In 1984, when MCSA was in its infancy, MCSA measurements were applied to these motors for the first time and due cognizance was not given to the fact that the operation of a fluid coupling and its fluid dynamics could cause perturbations or slight oscillations to be reflected into the motor.

One of the motors had what appeared to be $\pm 2sf$ sidebands at 37 dB down ($N_{av}$) on the supply component and there was no option other than to bring this to the attention of the owners of the plant. The motors were operating on reduced load and this motor's operating current was 60 A compared to its full-load current of 110 A, the operator would not change the load on the motors to observe the change in the current spectrum since that was the operating demand of the process.

The owner decided to remove this motor and inspect the rotor but there was no obvious evidence of broken or cracked rotor bars nor any discoloration of the paint on the rotor bars and end rings due to overheating nor at the rotor core ends where the bars leave the rotor. The only evidence of an abnormality was that five joints had a very rough external finish where the bars entered the end ring at the NDE. This suggested faulty joints with higher resistance paths. With hindsight, the owner should have been told that unless the load was changed to observe the effects of the fluid coupling, a diagnosis could not be made. However, the owner's perspective was as follows: "Here are the motors for testing on the days arranged, the permits to work have been issued for MCSA measurements and the operator is not willing to affect the delivery of the process to provide a test bed environment for the condition monitoring company." The above case history demonstrates a weakness of MCSA due to external constraints which resulted in a "false positive" with respect to the existence of broken rotor bars. Of course, with the experience gained since then, the presence of a fluid coupling would have been a warning to be very careful. All six motors are still running (34 years later) without any actual rotor cage failures and the current spectra still have the characteristic pattern due to the load dynamics downstream of each motor.

### 7.5.2 MCSA Diagnosis and Inspection

The full-load slip calculated from the nameplate data was 0.87% and Figure 7.20 gives the current spectrum with the motor on 60 A (full-load current 110 A) and there are sidebands at ±0.383 Hz (37 dB down on $f$) around $f$ which would give an

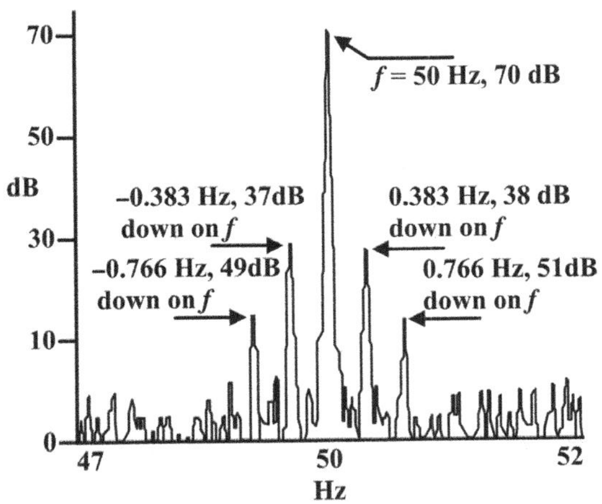


Figure 7.20 Current spectrum in 1984, sidebands incorrectly identified to be at $\pm 2sf$ and $\pm 4sf$ since they were from fluid coupling dynamics.

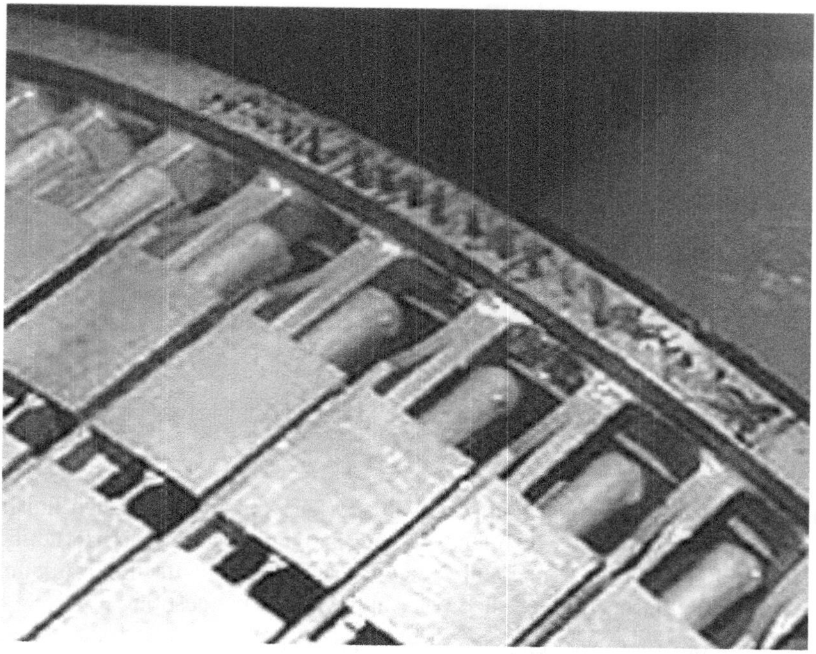

Figure 7.21 Region of rough finished bar to end ring joints at the NDE.

operational slip of 0.38% (i.e., 44% of the full-load slip) if these sidebands were at $2sf$.

The broken bar factor corrected for load $BB_f$ was 3.5 and this certainly suggested a serious broken rotor bar problem, provided of course the sidebands were truly at $\pm 2sf$. The no-load current for this motor would be of the order of 15–20% of the full-load current and using the approximation of a linear relationship for the current versus slip between no-load and full-load, an estimate of the operational band of slip at an operational current of 60 A is between 0.38% and 0.41% which is certainly of the same order as the slip at 0.38% from the sidebands. The owner decided it was wise to inspect the motor (one of the six) with the highest magnitude of sidebands and a photo of the rotor is shown in Figure 7.21. A detailed inspection was carried out and the following was found:

- **(i)** The drive end (DE) surface finish of the bars to end ring joints was normal and the end ring was of uniform thickness. The distance between the rotor core and inner face of the end ring was measured around the circumference and was found to be constant.
- **(ii)** The same measurements were repeated at the non-drive end (NDE) but the results showed that the non-drive, end ring was lying 2 mm (80 thou) offset relative to the rotor core face and the thickness of the end ring varied from top to bottom by 1.5 mm (60 thou). Five joints were suspected in the vicinity, where the distance between the rotor core face and the end ring was largest.

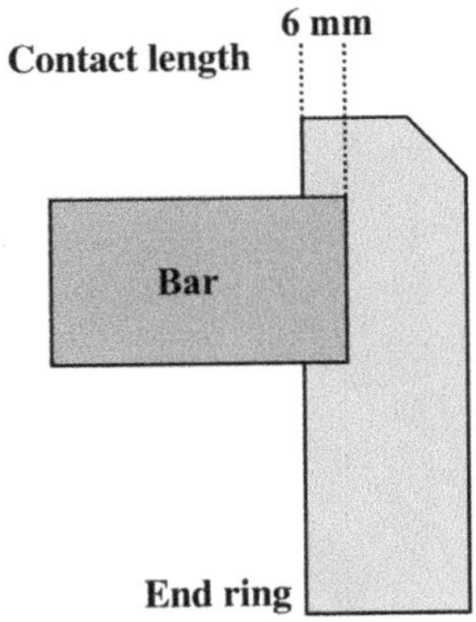


Figure 7.22 Illustration of bar to end ring joint length.

**(iii)** In the bar to end ring region there were surface spikes and a rough finish as shown in Figure 7.21 which tends to suggest the suspect region had not been heated to the correct temperature when the rotor was manufactured. A gamma radiography inspection revealed voids at the region in question.

**(iv)** Illustrations (not exact engineering drawings) are shown in Figures 7.22 and 7.23 and the latter indicates voids were situated 19 mm (0.75 inches) from the front edge of the end ring and these had a width of 1.5–2 mm (60–80 thou). This left a maximum bar length of 3.5 mm (137 thou) since the end ring width is 23.5 mm (0.93 inches) in this region. In the faulty region, the brazing had sunk by 1.5 mm (60 thou), and hence the cumulative effect is that the jointing length may only be 1.5–2 mm (60–80 thou).

No further actions were taken and this motor with the rotor, as it was in 1984, is still running today in 2016. No broken rotor bars have occurred and the current spectrum still displays the sidebands due to the mechanical dynamics from the fluid coupling.

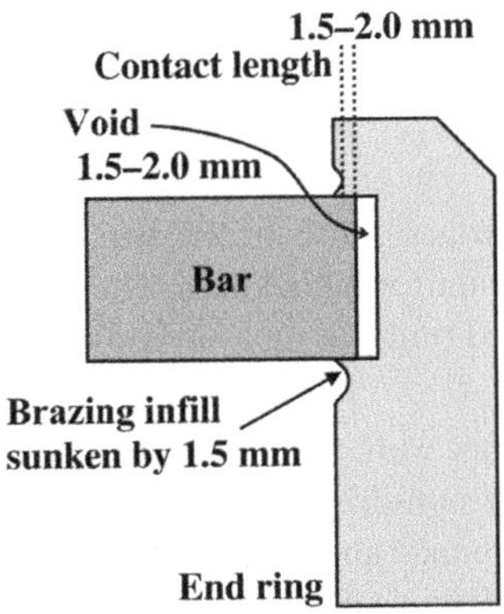


Figure 7.23 Illustration of actual bar to end ring joint length in faulty region.

## APPENDIX 7.A.1 DERIVATION OF SIDEBANDS DUE TO LOW FREQUENCY OSCILLATION FROM THE MECHANICAL LOAD

Following on from Section 7.1.1, assume that the cyclic variation of the rotor's speed is sinusoidal, then

$$N_{r(t)} = N_a + k\sin(2\pi f_c t)$$
$$s_{(t)} = [N_s - (N_a + k\sin(2\pi f_c t))]/N_s$$

where

$s_{(t)}$ = slip as a function of time
$N_{r(t)}$ = rotor speed time varying function, r/min
$N_a$ = average rotor speed, r/min
$f_c$ = oscillating frequency due to cyclic load change, Hertz
$k$ = peak magnitude of speed oscillation which is a function of load changes, inertia of the rotor drive train, and torque characteristic of the mechanical load

The approximate equivalent circuit per phase for a 3-phase induction motor is the starting reference to show that the input current to the motor will be modulated at an oscillating frequency ($f_c$) due to cyclic load changes being reflected into the motor's rotor from the mechanical load dynamics.

The voltage and currents are phasor quantities (rms) in the approximate equivalent circuit. The slip ($s$) is a function of time and it is the referred rotor current to the stator input terminals, which contains that quantity.

For simplicity, assume that the $jX_{eq}$ *and* $R_s$ terms are negligible compared to $(R'_r/s_{(t)})$since the derivation is purely to show the input current will be modulated at the oscillating frequency $f_c$. It is then possible to present the rotor current in the time domain as

$$i'_{2(t)} = \sqrt{2}V\sin(2\pi ft)/(R'_r/s_{(t)})$$

with $s_{(t)} = [N_s - (N_a + k\sin(2\pi f_c t))]/N_s$ it can be easily shown that

$$i'_{2(t)} \cong \frac{\sqrt{2}V\sin(2\pi ft)(N_s - N_a) - \sqrt{2}V\sin(2\pi ft)k\sin(2\pi f_c t)}{(R'_r N_s)} \quad (7.1)$$

Using trigonometric identities, the term $\sin(2\pi ft)k\sin(2\pi f_c t)$ becomes

$$k/2[\cos 2\pi t(f - f_c) - \cos(f + f_c)]$$

The input current to the motor is now modulated with sidebands around the supply component at

$$f \pm f_c$$

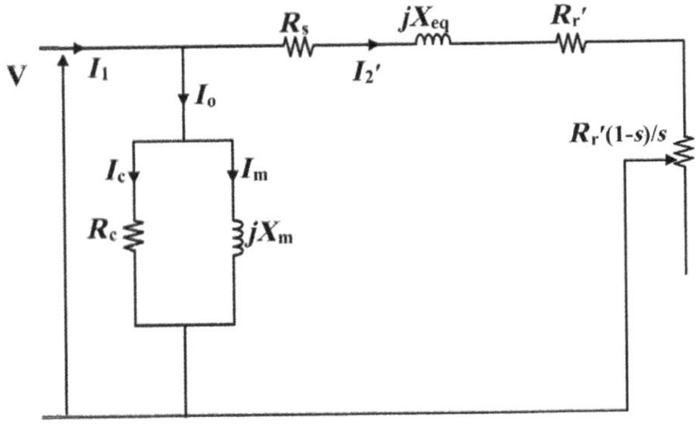


Figure 7.A1 Approximate equivalent circuit per phase referred to the stator input terminal for a 3-phase induction motor.

For completeness and with respect to the approximate equivalent circuit in Figure 7.A1 . It is not necessary to use an exact equivalent circuit, (as in IEEE standard 112) since no performance calculations are presented in the derivations in Appendix 7.A.1.

$R_c$ = fictitious resistor to model core losses in W/phase, Ω

$R_s$ = stator resistance/phase, Ω

$x_m$ = magnetizing reactance/phase, Ω

$x_s$ = stator leakage reactance/phase, Ω

$x_r'$ = referred rotor leakage reactance/phase to stator, Ω

$x_{eq} = x_s + x_r'$ = total equivalent leakage reactance referred to the stator, Ω

$R_r'$ = referred rotor resistance/phase, Ω

$s$ = p.u. slip

$R_r'/(1-s)/s$ = a fictitious resistor used to model the power out in watts at the shaft, function of the slip thus the load on the motor's shaft, Ω

Power output at the shaft is given by $P_{out} = 3 \times I_2'^2(R_r/(1-s)/s)$ watts

## 7.6 QUESTIONS

**7.6.1** Give practical examples of mechanical dynamics downstream of a SCIM, which can produce sidebands in the current spectrum that can result in false positives of cage winding breaks.

**7.6.2** Give the equation for the slip of a SCIM in which the speed of the rotor is oscillating about an average rotor speed.

**7.6.3** What is the effect on the current spectrum when the load is causing the rotor speed to oscillate at ±18 r/min, about a mean speed of 1760 r/min in a 4-pole, 60 Hz SCIM?

**7.6.4** Before an MCSA test is to be carried out on a SCIM, driving a centrifugal gas compressor on an offshore oil and gas production platform, what is a very important question to be answered by the end user?

**7.6.5** You are contacted to carry out MCSA on SCIMs driving pumps via fluid couplings.

**(i)** What is the main reason for using a fluid coupling in a SCIM drive train?

**(ii)** Explain to the end user why it may be difficult to analyze the current spectra to detect any cage winding breaks in SCIMs when fluid couplings are used.

**(iii)** What key questions should you ask with respect to the operational loads on SCIMs with fluid couplings in the drive trains?

**7.6.6** A 3-phase, 6.6 kV, 5000 HP/3730 kW, 377 A, 60 Hz, 3582 r/min, 0.9 p.f., efficiency of 96%, star connected SCIM drives a centrifugal gas compressor that is recycling the gas. The end user has reported that this motor has suffered broken rotor bars on two occasions during the past 10 years with consequential stator winding failures and they now want to apply MCSA to detect cage winding breaks as early as possible to prevent stator winding failures which cause downtime and lost production and are very expensive to repair. The motor operates on full-load but at present it recycles a percentage of the gas. What frequency resolution would you require to separate any $\pm 2sf$ sidebands due to a cage winding break and those caused by recycling of gas if the latter was inducing sidebands in the current spectrum at 0.55 Hz around the supply component? Explain the reasons for your answer.

**7.6.7** With reference to question 7.6.6, the MCSA spectrum indicates that there are sidebands at ±0.68 Hz (both 40 dB down on $f$), ±0.55 Hz (upper sideband 37 dB down on $f$, lower sideband 40 dB down on $f$), ±1.36 Hz (both 40 dB down on $f$), and ±1.1 Hz (upper sideband 70 dB down on $f$, lower sideband 72 dB down on $f$). The operating current was 385 A and the rotor had 46 rotor bars.

**(i)** What is your interpretation of the measured current spectrum?

**(ii)** What is your recommendation to the end user?

**(iii)** Would you consider it prudent to carry out any further MCSA measurements and under what operating conditions?

**7.6.8** A 3-phase, 11 kV, 10,000 HP/7460 kW, 377 A, 60 Hz, 3564 r/min, 0.92 p.f., efficiency of 96.8%, star connected SCIM drives a three-stage centrifugal gas compressor that is recycling the gas. The compressor is on recycling with 30%, 44%, and 55% for stages 1, 2, and 3, respectively. The operational current is 60% of the full-load current. The number of rotor bars is 58.

The current spectrum has sidebands around $f$ at ±0.65 Hz (upper sideband 41 dB down on $f$, lower sideband 44 dB down on $f$), ±1.3 Hz (upper sideband 60 dB down on $f$, lower sideband 64 dB down on $f$), ±1.95 Hz (upper sideband 70 dB down on $f$, lower sideband 72 dB down on $f$), and no other sidebands exist around the supply component, $f$. What is your initial diagnosis of the condition of the cage winding? Hint, you need to apply the $BB_f$ and take account of reduced load operation.

What is your recommendation to the client bearing in mind the gas compressor is recycling the gas and the motor was operating on a reduced load?

**7.6.9** As the senior electrical engineer with a condition monitoring company that offers MCSA, you are invited to attend an initial exploratory meeting with the maintenance manager (a mechanical engineer) and two electrical maintenance engineers at a power station to discuss the case for using MCSA to assess the operational condition of cage windings in HV SCIMs driving boiler feed pumps, FD and ID fan motors in a 2000 MW coal fired power station. Propose a technical marketing strategy to convince the power station that your company has the technical knowledge on induction motors and the reliable application of MCSA so that your company is the one that should be used to provide the service. Your plan should include case studies of your company's success in the field.

**7.6.10** Research has shown [7.4] that, by measurement, advanced analysis and signal processing of three voltages and three currents from a 10 HP/7.46 kW, 230 V, 3-phase induction motor in a university laboratory, it is possible to eliminate load induced current components so that any $\pm 2sf$ sidebands (around $f$) due to broken rotor bars can be detected, when load oscillations and broken bars occur simultaneously. Discuss the practical installation and cost implications of actually measuring three voltages and three currents from HV (4160–13,800 V), high power, strategic SCIMs to provide an on-line condition monitoring service to assess the condition of cage windings in, for example, a power station, a petrochemical refinery, or on an offshore oil and gas production platform.

## REFERENCES

[7.1] M. P. Boyce, "Principles of Operation and Performance Estimation of Centrifugal Compressors," Proceedings of the Twenty-Second Turbomachinery Symposium, Texas A&M University Turbomachinery Laboratory, 1993.

[7.2] W. T. Thomson, "On-Line Current Monitoring—The Influence of Mechanical Loads/Unique Rotor Designs on the Detection of Broken Rotor Bars in Squirrel Cage Induction Motors," International Conference on Electrical Machines (ICEM '92), UMIST, September 1992.

[7.3] R. R. Schoen and T. G. Habetler, "Effects of time varying loads on rotor fault detection in induction machines," *IEEE Transactions on Industry Applications*, vol. 31, no. 4, pp. 900–906, July–August 1995.

[7.4] R. R. Schoen and T. G. Habetler, "Evaluation and implementation of a system to eliminate arbitrary load effects in current-based monitoring of induction machines," *IEEE Transactions on Industry Applications*, vol. 33, no. 6, pp. 1571–1577, November–December 1997.

[7.5] L. Wu, T. G. Habetler, and R. G. Harley, "Separating Load Torque Oscillation and Rotor Fault Effects in Stator Current-Based Motor Condition Monitoring," IEEE International Conference on Electrical Machines and Drives, IEMDC, 2005, pp. 1889–1894.

CHAPTER 8

# MCSA CASE HISTORIES—SCIM DRIVES WITH SLOW SPEED GEARBOXES AND FLUCTUATING LOADS CAN GIVE FALSE POSITIVES OF BROKEN ROTOR BARS

## 8.1 INTRODUCTION AND SUMMARY OF CASE HISTORIES

This chapter focuses on the influence of slow speed gearboxes in SCIM drive trains which are driving fluctuating mechanical loads and is followed by the presentation of seven industrial case histories. A list of the case histories which are presented in detail in the remainder of this chapter is given below to assist the reader to make a choice (if so desired) of the ones of personal interest.

**Section 8.2.1—One case history:** A slow speed (nominal full-load speed of 12 r/min) coal conveyor was driven by a gearbox, which was powered by a 4160 V, 336 kW/450 HP 4-pole SCIM. Due to the fluctuating load, a false positive of broken rotor bars was prevented by taking account of the drive dynamics downstream of the motor.

**Section 8.3.1—One case history:** A slow speed (nominal full-load speed of 66 r/min) coal conveyor was driven by a gearbox, which was powered by a 415 V, 127 kW/170 HP, 4-pole SCIM. Numerous current sidebands around the supply component existed with a high probability of a false positive which was also avoided.

**Section 8.4.1—One case history:** A coal crusher which had a very random and highly fluctuating load was driven by a 3300 V, 336 kW/450 HP, 4-pole SCIM via a 3-stage reduction gearbox to give a crusher speed of nominally

*Current Signature Analysis for Condition Monitoring of Cage Induction Motors: Industrial Application and Case Histories*, First Edition. William T. Thomson and Ian Culbert.

41 r/min at full-load. It was *impossible to analyze the spectrum* even after applying a large number of spectral averages.

**Section 8.5.1—Two case histories:** These coal crushers (48 r/min) had a less severe fluctuating load than case history 8.4.1 and were driven by 3300 V, 470 kW/630 HP, 6-pole SCIMs via a 2-stage reduction gearbox. It was possible to analyze the current and the cage windings were healthy.

**Section 8.6.1—Two case histories**: Two 6600 V, 2400 kW/3200 HP, 6-pole SCIMs drive a ship's starboard and port side thrusters at 234 r/min via two separate step down gearboxes. This case history predicts current components due to mechanical drive dynamics. There was coincidence with components from the drive dynamics and $\pm 2sf$ sidebands around $f$ possibly due to broken rotor bars in one of the motors but this was proven to be from damaged propeller blades.

Thomson (author) initiated his research and development work on MCSA in 1981 at The Robert Gordon University, Aberdeen, Scotland. The research was sponsored by the offshore oil and gas industry and a major power generation company and on-site trials were carried out using MCSA to diagnose cage winding breaks in large, HV SCIMs via the detection of the $\pm 2sf$ sidebands around the supply component. During the evaluation trials which commenced in 1983, it was observed that the current spectrum contained components related to the input and output shaft speeds to and from speed reducing gearboxes in the drive train and also shaft speeds within a gearbox with two and more reduction stages. In particular, when the output speed of the gearbox shaft is of the order of 30–150 r/min, as can be the case from coal conveyors and crushers in coal fired power stations, sidebands around the supply component could be misinterpreted as being $\pm 2sf$ sidebands due to cage winding breaks. Further research work via controlled experiments was therefore required to better understand why current components were induced in the stator winding due to mechanical drive dynamics downstream of the motor.

### 8.1.1 Influence of Mechanical Misalignment on the Current Spectrum

A significant cause of forces in a SCIM drive train applicable to all drive trains is misalignment across the coupling between the shaft of the electric motor and the driven load, causing an external mechanical force to be transmitted to the cage rotor in the motor and likewise to the driven load. The higher the misalignment, the larger the preload forces on the shaft and rotor of the electric motor and the driven equipment. Thus great attention is taken at the commissioning of SCIM drive trains to keep misalignment as low as is practically possible, for example, 0.05 mm (2 mils) is considered to be a very good level to achieve in different types of misalignment.

It was therefore appropriate to carry out MCSA tests, as part of the R&D work, to analyze the current spectrum as a function of misalignment. For background information on this topic, a very brief overview of misalignment is now presented. The study of three-dimensional misalignment forces in drive trains is a complex subject and beyond the scope or objectives of this book and the reader is referred to an excellent white paper (62 pages), "A Practical Guide to Shaft Alignment" by Pruftechnik Ltd. [8.1]. These types of misalignment are shown in Figures 8.1, 8.2 and 8.3.

**Parallel Misalignment**

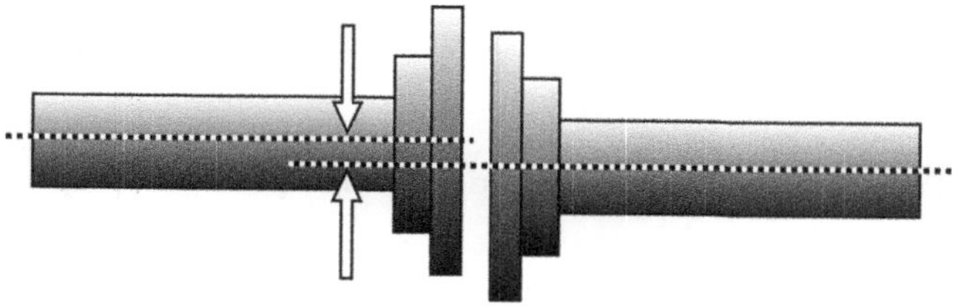

Figure 8.1 Illustration of parallel misalignment.

**Axial Misalignment**

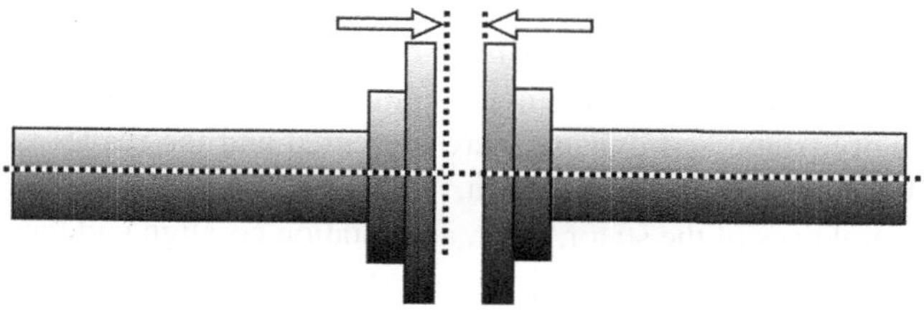

Figure 8.2 Illustration of axial misalignment.

**Angular Misalignment**

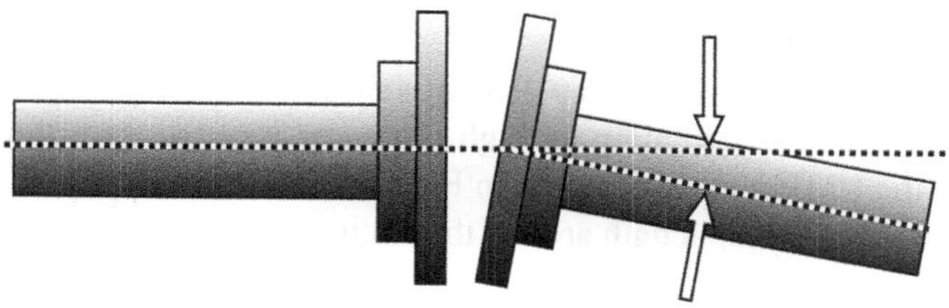

Figure 8.3 Illustration of angular misalignment.

Horizontal parallel misalignment is often the most critical and gives rise to shaft reaction forces in the direction of displacement as shown in Figures 8.4 and 8.5.

One of the major effects of misalignment between rotors and load shafts in a drive train is the production in a specific radial direction of *rotor pre-load*, which produces a *radial force* that can push the rotor to the side. The rotor can in fact (with high levels of misalignment) become displaced from its original position resulting in a higher eccentricity level inside the seals and bearings. In extreme cases, it is also possible for the rotor to become bowed and rotate in a bow configuration. This means that misalignment can change the radial airgap length due to a change in dynamic

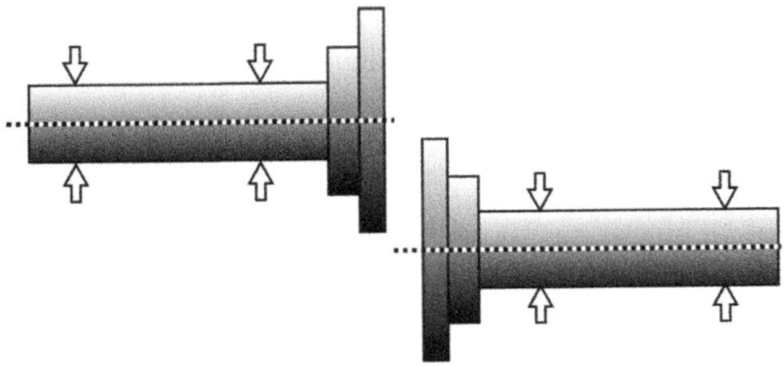

Figure 8.4 Misaligned shafts prior to coupling being made up.

airgap eccentricity. A drive with a flexible coupling running at 1800 r/min, with a coupling offset of −0.046 mm (−1.8 mils) and 0.03 mm (1.2 mils) is excellent since these misalignments fall within the limits of 0.05 mm (2 mils) in the horizontal and vertical directions, respectively, which are typical target values in industry [8.1].

As mentioned previously, misalignment in the drive train can cause changes to airgap eccentricity within a SCIM, therefore it is necessary to define the different types of airgap eccentricity before Chapters 10 and 11 on airgap eccentricity are presented. At this stage, only schematic illustrations of the two types of operational airgap eccentricity (static and dynamic) are presented and the reader is referred to Section 10.1 for more detailed information.

If the central axes of the stator, rotor, and rotation co-align and the rotor core's diameter and stator bore are perfectly round (not possible in a practical SCIM) there will be no unbalanced magnetic pull (UMP) since the air gap is constant over 360 degrees.

**Definition of Static Airgap Eccentricity:** This occurs when the central axes of rotation and of the rotor co-align through (Z) but the stator bore central axis through (X) is displaced from that co-alignment. Figure 8.6 shows that static eccentricity ($e_s$) is in the horizontal direction.

**Definition of Dynamic Airgap Eccentricity:** This is when the central axes of rotation and stator bore co-align through (X) but that of the rotor through (Z) is displaced and the UMP rotates as shown in Figure 8.7.

The total radial airgap length around the circumference can be mathematically expressed as

$$g_{(\theta,t)} = g(1 - e_s \cos(\theta) - e_d \cos(\omega t - \theta)) \quad (8.1)$$

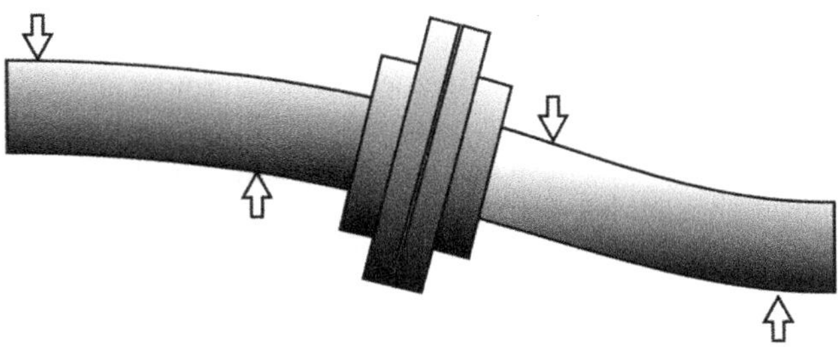

Figure 8.5 Deflected shape after coupling is made up.

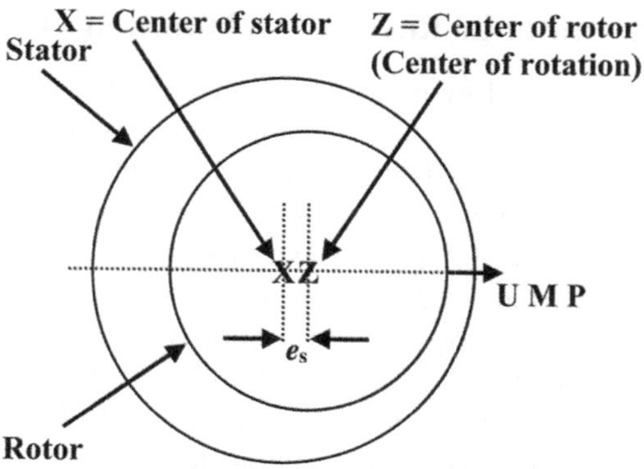


Figure 8.6 Schematic illustration of only static airgap eccentricity.

where

$g_{(\theta,t)}$ = the airgap length as a function of static and dynamic airgap eccentricity, millimeter

$\theta$ = angular position in degrees around the circumference

$t$ = time, seconds

$g$ = nominal design value for the airgap length, millimeter

$e_s$ = level of static eccentricity, millimeter

$e_d$ = level of dynamic eccentricity, millimeter

$\omega$ = angular speed in rad/sec

Equation (8.1) describes the combination of static and dynamic airgap eccentricity. For purely static eccentricity, $e_d = 0$ and equation (8.1) becomes

$$g_{(\theta)} = g(1 - e_s \cos(\theta))$$

UMP = unbalanced magnetic pull in Newtons or lbs force.

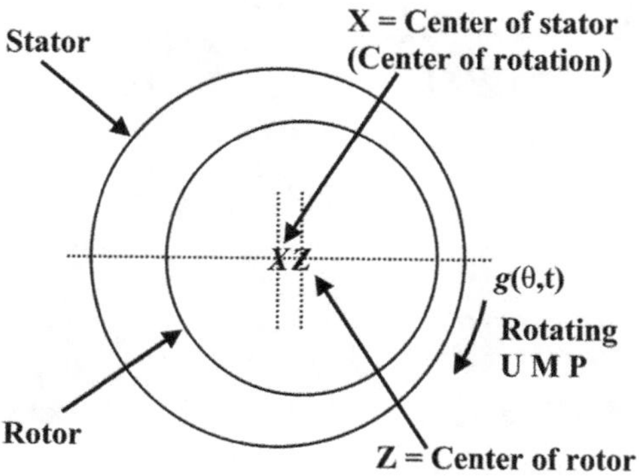


Figure 8.7 Schematic illustration of only dynamic airgap eccentricity.

### 8.1.2 MCSA Experimental Results as a Function of Misalignment

Since the R&D was sponsored by industrial companies, they required results relatively quickly; it was Thomson's strategy to carry out experiments that were directly relevant and applicable to the needs of industry, rather than theoretical modelling, which may often be perceived by industry as more an academic exercise than work of practical benefit. Thomson [8.2] published a short paper in 1994 on the influence of misalignment on the current spectrum but more informative results and explanations are now presented. These experimental results were obtained from tests using a purpose built test rig, the details of which are now described. The rig is basically a SCIM driving a dynamometer with special features designed to allow for variation of misalignment between the motor and dynamometer shafts.

The nameplate data for the SCIM was as follows: 3-phase, delta connected, 415 V, 11 kW/14.74 HP, 20.5 A, 50 Hz, 1470 r/min using a 28-slot, deep bar aluminum die-cast cage rotor with no cage winding defects.

For the benefit of researchers who may be commencing research into CM of induction motors a description of the experimental test rig is now presented.

**(i)** The rotor was mounted on separate bearing pedestals, which could be moved horizontally and likewise the stator frame in the directions shown at positions A and B in Figure 8.8. The static airgap eccentricity could be independently adjusted by moving the stator frame assembly in the direction of A in Figure 8.8 with respect to a fixed reference position on the baseplate. The minimum static airgap eccentricity of the stator bore was ±0.05 mm (2 mils) or 8.33% of the nominal radial airgap of 0.6 mm (24 mils).

**(ii)** The rotor was machined to give a total indicated run out (TIR) of 0.025 mm (1.0 mil) of the nominal radial airgap length giving a dynamic airgap eccentricity of 4.2%.

Figure 8.8 Close up of SCIM under test.

**(iii)** The total airgap eccentricity was ±12.5% (0.076 mm/3 mils)) of the 0.6 mm (24 mils). This was the reference airgap eccentricity in the motor.

**(iv)** The test rig could be set up with a minimum, horizontal parallel misalignment of 0.05 mm (2 mils) between the dynamometer and the SCIM. A flexible coupling was used between the motor and dynamometer shafts and the angular and axial misalignments were negligible in comparison to the horizontal parallel misalignment.

**(v)** In SCIM industrial drive trains there is always a combination of inherent airgap eccentricity (static plus dynamic) within the motor plus some degree of misalignment between the motor and driven load via the coupling.

### 8.1.3 Interpretation of Current Spectrum as a Function of Misalignment

The test rig was set up with the inherent airgap eccentricity 12.5% (combination of static and dynamic) within the motor and the minimum, parallel misalignment was 0.05 mm (2 thou/mils).

The current spectrum in Figure 8.9 shows there are current components at $f \pm f_r$ around the supply component with inherent airgap eccentricity (static and dynamic) and external parallel misalignment across the coupling between the motor and dynamometer. This is in fact the base line current spectrum, where

$f$ = the supply frequency, Hertz

$f_r$ = rotational speed frequency of the rotor $= N_r/60$ Hz or $f(1-s)/p$ Hz, at 1470 r/min

$f_r = 1470/60 = 24.5$ Hz (Note: For the avoidance of doubt this value should not be rounded up or down to a whole number, e.g., 24 Hz = 1440 r/min, a difference of 30 r/min!)

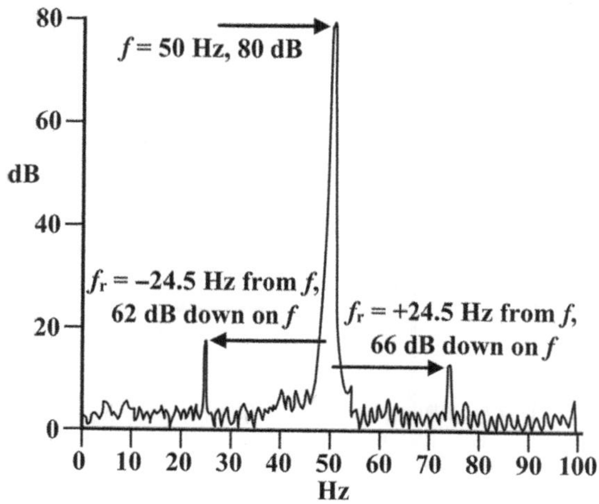


Figure 8.9 Current spectrum—no broken bars, minimum parallel misalignment of 0.05 mm (2 mils), 25 mHz/line.

**(i)** The current spectrum in Figure 8.9 shows that the $-f_r$ and $+f_r$ components are 62 and 66 dB down from the magnitude of the supply component at 50 Hz. The magnitudes are very low indeed at less than 1000$^{th}$ smaller than the supply component.

**(ii)** The horizontal, parallel misalignment between the motor and the dynamometer was increased from 0.05 mm (2 mils) to 0.2 mm (8 mils) by moving the bearing pedestals and the stator frame by 0.15 mm (6 mils) from the minimum value of 0.05 mm (2 thou/2 mils) so that the airgap eccentricity within the motor remained the same. The eccentricity was set when the motor was stationary and the ambient temperature was 20°C (68°F).

The spectrum in Figure 8.10 shows the $-f_r$ component has now increased by 10 dB (a factor of 3.2 times higher) and the $+f_r$ has increased by 12 dB (a factor four times higher). Current spectra were measured with various amounts of parallel misalignment and graphs of the $f \pm f_r$ components in dB versus misalignment are shown in Figure 8.11. The $f \pm f_r$ components increased by 6 dB (a doubling in absolute amperes) between a misalignment of 0.05 mm (2 mils) and 0.15 mm (6 mils), a change of 0.1 mm (4 mils), which indicates the sensitivity of these components to an increase in misalignment. There was a 20 dB increase (a factor of 10) in the components between misalignments of 0.05 mm (2 mils) and a high level of 0.4 mm (16 mils), respectively.

For currents to flow in the stator winding at $f \pm f_r$ Hz, rotating magnetic fields at these frequencies must induce voltages in the stator winding at the same frequencies. However, for many years, static and dynamic eccentricity flux waves were considered to be independent of each other and their interrelationships were not initially considered. As reported by, for example, Freize and Jordan [8.3], Fruchericht et al. [8.4], Kron [8.5], ERA survey report, [8.6], Yang [8.7], De Bortoli, [8.8] that dynamic

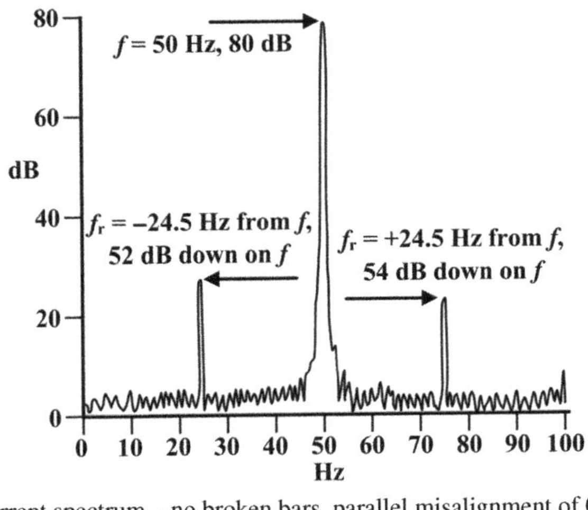


Figure 8.10 Current spectrum—no broken bars, parallel misalignment of 0.2 mm (8 mils), 25 mHz/line.

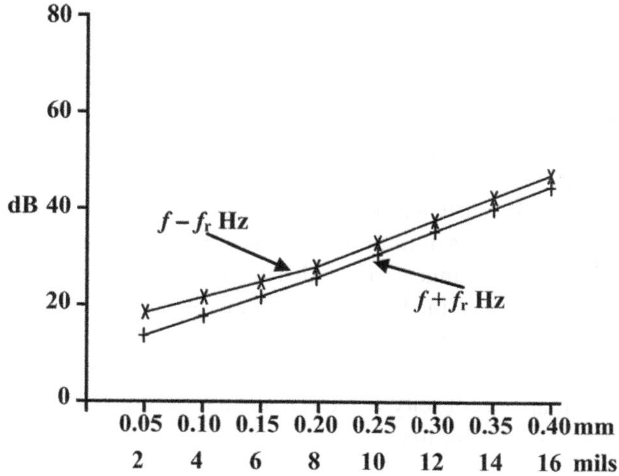


Figure 8.11 Magnitude in dB of $f \pm f_r$ (Hz) components as a function of external misalignment, but with a fixed airgap eccentricity as set internally within the SCIM.

eccentricity will cause rotating flux waves at frequencies given by $f \pm f_r$ but with only pole-pairs equal to $p \pm 1$, where $p$ is the fundamental number of pole-pairs.

Rotating flux waves having $p \pm 1$ pole-pairs cannot induce emfs in the stator winding since there is no compatible number of pole-pairs in the winding. Without these emfs there can be no currents at these frequencies, yet experiments and analysis of current spectra clearly proved that there were current components at frequencies given by $(f \pm f_r)$ Hz. This demonstrates how experimentation prior to theoretical analysis can provide new information, which prevents misconceptions arising, sometimes rendering that theory invalid. Thomson's research guide is to "*carry out experiments and see what happens*" which was clearly justified on this occasion.

A subsequent paper by Dorrell and Thomson et al. [8.9] in 1997 theoretically proved that the interaction between static and dynamic airgap eccentricity flux waves causes flux components given by $f \pm f_r$ with $p$ pole-pairs. It can therefore be concluded that the $f \pm f_r$ components in the current spectrum in Figure 8.10 and the graphs presented in Figure 8.11 are due to an increase in the operational dynamic airgap eccentricity within the SCIM, but the fundamental cause was the forces external to the motor due to an increase in parallel misalignment. From an industrial perspective this means that if the $f \pm f_r$ components increase due to an increase in dynamic eccentricity it could be due to a change within the motor such as bearing wear or, as just demonstrated, it could be due to an external cause in the drive train, downstream of the motor, such as an increase in shaft-coupling misalignment.

If the coupling is split and the SCIM is MCSA tested uncoupled and the $f \pm f_r$ components significantly drop in dB magnitude or disappear then misalignment is the problem, but to reiterate, the end user will be very reluctant indeed to stop a large, HV SCIM, which is driving a strategic mechanical load to carry out an MCSA test during an uncoupled run. In practice, such a test has to be preplanned with skilled personnel available to uncouple and recouple the motor and carry out alignment checks. The end user will simply state: "The motor is running, please carry out your MCSA test under the operational conditions provided."

In Chapters 10 and 11 it will be shown and supported by industrial MCSA case histories, that there is a current signature pattern independent of the $f \pm f_r$ components which can be used to diagnose the operational airgap eccentricity (static plus dynamic). In practice it is not possible to separate static from dynamic airgap eccentricity using classical MCSA when the only information available is the nameplate data and hopefully the number of rotor bars from the OEM. What the end user needs to know is: can the motor continue to run with a given level of total airgap eccentricity or should action be taken to remedy the cause of an unacceptable level? The OEMs of SCIMs do not separate the level of static and dynamic airgap eccentricity, when asked by the client for the airgap eccentricity as a percentage of the nominal radial airgap. The total airgap eccentricity is the reference term used by OEMs, who have been manufacturing SCIMs for many years.

### 8.1.4 MCSA Experimental Results with a Speed Reduction Gearbox in the SCIM Train

Thomson and Campbell [8.10] published a short paper in 1991 on the influence of a gearbox in a SCIM drive train, when analyzing the current to detect $\pm 2sf$ sidebands around $f$ but the results that follow are more informative than in Reference 8.10. With no broken rotor bars in the cage winding, tests were carried out with a 2-stage reduction gearbox inserted between the motor and dynamometer in the test rig used for the misalignment experiments. The gearbox details follow, supported by a schematic diagram in Figure 8.12.

The overall speed reduction was 2.702:1 (as stated on the nameplate of the gearbox). Note that gearboxes never have exact integer numbers as their step down or step up ratios, Taylor [8.11]. Since the dynamometer was rated for a speed of 1500 r/min to give 230 V dc, the cooling was not designed to cope with a speed of 556 r/min (drop of 63%) and likewise, with rated field current, the generated volts dropped by the same percentage, therefore the test with the gearbox in the drive train was carried out during a run of the full drive train but with no load on the dynamometer. Recall that the main objective was to observe the effect on the current spectrum of a gearbox

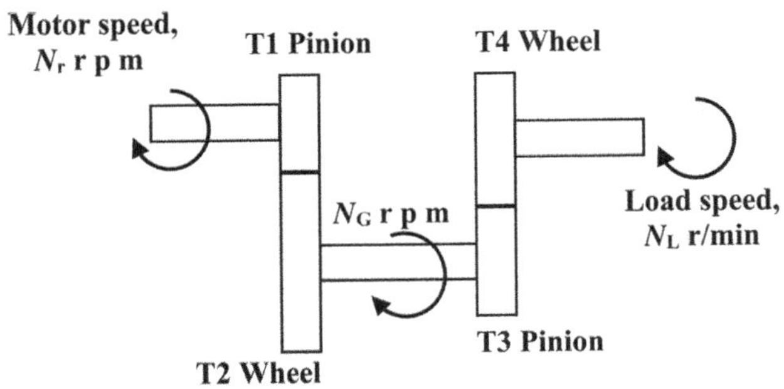


Figure 8.12 Illustration of gearbox information; T1 = 38 teeth, T2 = 56 teeth, T3 = 30 teeth, and T4 = 55 teeth.

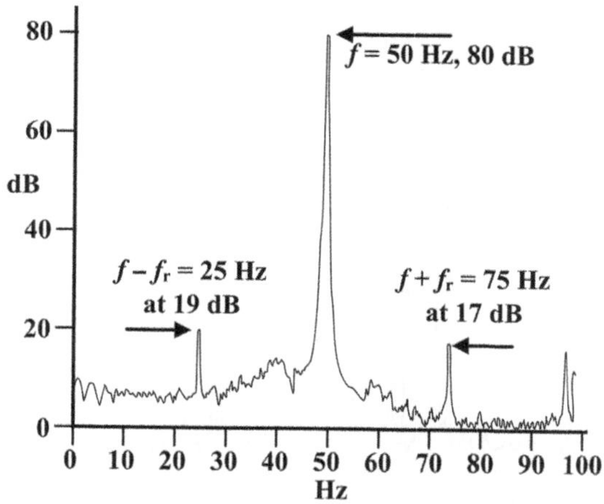


Figure 8.13 Current spectrum, no gearbox, 25 mHz/line.

in the drive train. From Figure 8.12, the rotational speed frequencies of the three shafts are as follows:

$f_r = N_r/60$ (very close to 25 Hz on no-load)

$f_g = (N_r/60)(38/56) \approx 17$ Hz

$f_L = (N_g/60)(30/55) = 9.27$ Hz (Note: For the avoidance of doubt this value should not be rounded down, 9 Hz = 540 r/min, compared to 9.27 Hz = of 556 r/min, which is a difference of 16 r/min.)

The prediction was that current components at the following frequencies should exist in the stator current with the gearbox in the drive train.

$$(f \pm f_r)\ \text{Hz}$$

$$(f \pm f_g)\ \text{Hz}$$

$$(f \pm f_L)\ \text{Hz}$$

A comparison between Figures 8.13 (no gearbox) and 8.14 (with a brand new gearbox) shows that the predictions of the frequency content with the gearbox were verified.

The magnitudes of the components with the gearbox inserted were as follows:

$(f \pm f_r) = 35$ dB (45 dB down on $f$, a factor of 178 times smaller)

$(f \pm f_g) = 32$ dB (48 dB down on $f$, a factor 200 times smaller)

$(f - f_L) = 20$ dB (60 dB down on $f$, a factor of 1000 times smaller)

$(f + f_L) = 16$ dB (64 dB down on $f$, a factor of 1585 times smaller)

This experiment was to confirm, under controlled experimental conditions, the influence on the current spectrum of a gearbox in the drive train since this phenomenon of components appearing in the spectrum at $f$ plus or minus rotor, gear, and load sideband frequencies had been observed during field trials of MCSA to

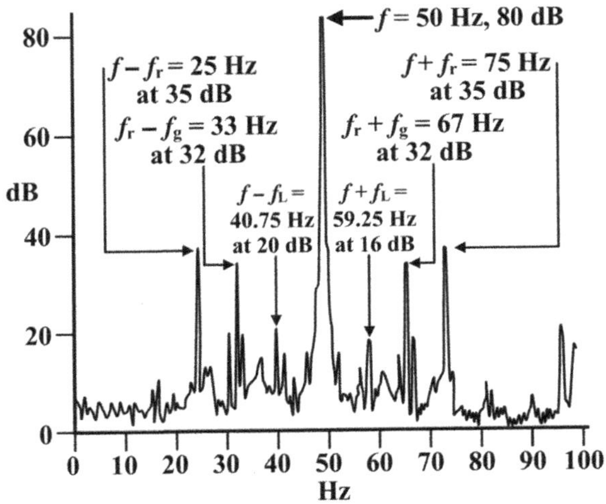


Figure 8.14 Current spectrum, with brand new gearbox, 25 mHz/line.

detect cage winding breaks, the main focus of the on-site trials. The confirmation of the effects of gearing was achieved by laboratory experiments and here again, as in the case of misalignment, mechanical disturbances from the gearbox, which cause small disturbances to the operational level of dynamic airgap eccentricity, were being transmitted to the rotor in the SCIM.

Where it is clearly uneconomical to fit vibration sensors, for example, to numerous induction generator driven trains in a wind farm it was suggested in 2011 by Salon et al. [8.12] that MCSA could be applied to give an indication of the general health of the drive train including bearing and gearbox problems but no MCSA results were presented in that paper. In 2012, Dinggua Lu et al. [8.13] studied current based diagnosis of gear tooth breaks which could be applicable to wind turbine gearboxes, a novel test rig in a university laboratory was used to simulate the dynamics of a wind turbine rotor and the results indicated that the spectral content changed when one tooth was completely removed. In 1991, 21 years before Reference 8.13 was published, Thomson et al. [8.10] presented a paper on MCSA with a gearbox having one worn tooth in a SCIM drive train. This is a much less severe fault than a completely broken gear tooth and the results verified that the current spectrum to the motor was distinctly different with a worn tooth compared to that when the gearbox was brand new. It is clear that there is now a need for extensive on-site trials of MCSA applied to induction generator wind turbines to obtain industrial case histories and to report on them, since there appears to be no actual industrial case histories in the public domain. The operators of large wind turbine farms have to be convinced that MCSA is the way forward to prevent catastrophic failures, reduce downtimes, and prevent lost income before they will invest in the installation costs and apply MCSA as a CM technique for wind turbines.

In conventional SCIM drives, where gearbox vibration can be easily measured and analyzed, MCSA will certainly not replace vibration analysis (VA) for the detection of gearbox problems since VA is a well-proven diagnostic technique and numerous case histories have been published [8.11]. Another important point is that CM of gearboxes comes under the domain of mechanical engineers who are very comfortable and familiar with vibration analysis. The authors of this book are certainly not recommending MCSA for detecting faults in gearboxes in place of vibration analysis in conventional SCIM drive trains. The focus of the results in Sections 8.1.1 and 8.1.2 is to provide knowledge via MCSA to help avoid false positives of broken rotor bars via MCSA.

## 8.2 CASE HISTORY (1989)—SLOW SPEED COAL CONVEYOR, LOAD FLUCTUATIONS, AND GEARBOX IN THE DRIVE TRAIN

### 8.2.1 Summary and Key Features

The motor's nameplate provided the following information: 3-phase, 4160 V, 336 kW/450 HP, 58 A, 60 Hz, 1788 r/min, star connected SCIM. The number of rotor bars was unknown. The motor was driving a very low speed coal conveyor via a 3-stage reduction gearbox to give a nominal full-load output speed of 12.4 r/min. The conveyor delivered payloads of coal at a rate of 42 per minute and the load was therefore changing, since each payload is not exactly the same weight. The combination of a fluctuating load and a gearbox in the drive train resulted in complex mechanical dynamics being reflected back into the rotor of the SCIM with the appearance of multiple sidebands around the supply component of current. An illustration of the drive train layout is shown in Figure 8.15. The analysis that follows will verify the need to take account of the mechanical drive dynamics, otherwise a false positive of broken rotor bars could occur.

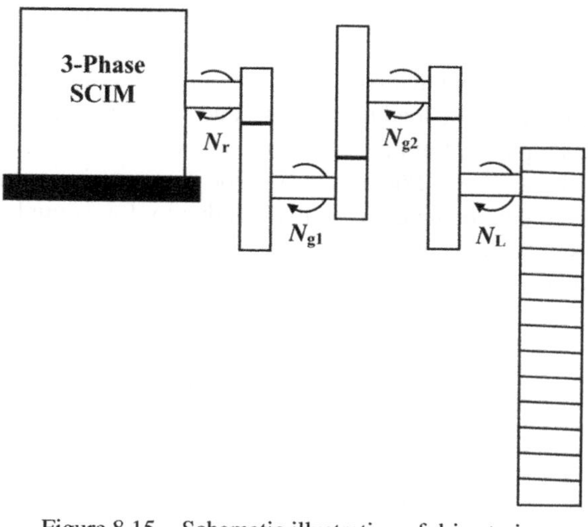


Figure 8.15 Schematic illustration of drive train.

### 8.2.2 MCSA Diagnosis

At full-load and using the information provided by the power station engineers about the gearbox ratios, the rotational speed frequencies of the different shafts are as follows:

$$N_r = 1788\ \text{r/min}$$

$$f_r = 29.8\ \text{Hz}$$

$$N_r/N_{g1} = 4,\ N_{g1} = 447\ \text{r/min}$$

$$f_{g1} = 7.5\ \text{Hz}$$

$$N_{g1}/N_{g2} = 5,\ N_{g2} = 89\ \text{r/min}$$

$$f_{g2} = 1.5\ \text{Hz}$$

$$N_{g2}/N_L = 7.2,\ N_L \approx 12\ \text{r/min}$$

$$f_L = 0.2\ \text{Hz}$$

Note that the step down ratios of 4.0 and 5.0 given by the power station staff will not be exact integers since as already stated, a step down ratio in a gearbox is normally never an exact integer [8.11] but that was the information provided. Based on the results presented in Section 8.1.2 the gearbox dynamics can be transmitted into the rotor of the SCIM and produce current components at the following frequencies at full-load.

$60 \pm 29.8$ Hz $\qquad 60 \pm 7.5$ Hz $\qquad 60 \pm 1.5$ Hz $\qquad 60 \pm 0.2$ Hz

At full-load, the slip is 0.67% and any sidebands at $\pm 2sf$ would nominally be at $\pm 0.8$ Hz around $f$, but the motor was operating at a reduced load of 47 A (full-load 58 A), therefore the slip drops and the output speed of the gearbox increases but the percentage change in rotor speed relative to full-load speed is negligible compared to the percentage change in slip relative to full-load slip.

The nominal number of payloads of coal per minute was quoted as being 42, this means in practice there can be load fluctuations due to the intervals of time between the dropping of each payload of coal onto the conveyor. Hence the load oscillation frequency should be at $f_c = 42/60 = 0.7$ Hz. For the motor's current to be lower means the volume of coal being conveyed is less due to, for example, each payload of coal being of a smaller volume. Examination of the current spectrum in Figure 8.16 indicates the following:

$$f \pm f_{sb} = 60 \pm 0.67\ \text{Hz}$$

$$f \pm 2f_{sb} = 60 \pm 1.34\ \text{Hz}$$

$$f \pm 3f_{sb} = 60 \pm 2.0\ \text{Hz}$$

$$f \pm 4f_{sb} = 60 \pm 2.68\ \text{Hz}$$

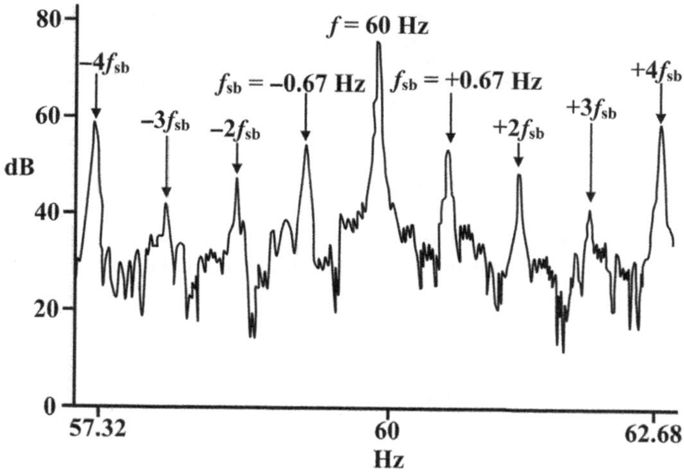


Figure 8.16 Current zoom spectrum, baseband span 0–130 Hz, 12,800 lines, 10.16 mHz/line.

The signature characteristic of the current spectrum in Figure 8.16 is not typical of that from a cage rotor with broken rotor bars. Recall from previous case histories in Chapters 5 and 6 that with broken rotor bars there can be sidebands at $\pm 4sf$ but they are normally much lower in dB magnitude than the $\pm 2sf$ sidebands and if $\pm 6sf$ sidebands also exist they are considerably lower (in dB) than the $\pm 4sf$ sidebands. In the spectrum from this conveyor motor the magnitudes of the sidebands with respect to $f$ are as follows:

(i) The $\pm f_{sb}$ sidebands at $\pm 0.67$ Hz are 30 dB down on $f$

(ii) The $\pm 2f_{sb}$ sidebands are 38 dB down on $f$ and 8 dB down on $f_{sb}$

(iii) The $\pm 3f_{sb}$ sidebands are 45 dB down on $f$ and 15 dB down on $f_{sb}$

(iv) The $\pm 4f_{sb}$ sidebands are 24 dB down on $f$ but this is 6 dB higher than $f_{sb}$

A key feature of the current spectrum shown in Figure 8.15 is that there are components at $\pm 4f_{sb}$ which are 6 dB (a times 2 factor) higher than the $\pm f_{sb}$ sidebands and this would not be the case in a current spectrum from a cage rotor with broken bars. The $\pm 2f_{sb}$ and $\pm 3f_{sb}$ sidebands are also too high compared to $\pm f_{sb}$ at $\pm 0.67$ Hz.

Some of the MCSA test equipment presently in use for the detection of broken rotor bars concentrates solely on finding and establishing the level of $\pm 2sf$ components around $f$, which in cases like this, could indicate a false positive due to the fact that the motor is on reduced load and the $\pm 0.67$ Hz sidebands could be incorrectly identified as being $\pm 2sf$ sidebands. The analysis confirms the multiple sidebands around the supply component are due to load fluctuations from the conveyer and its transmission characteristics.

## 8.3 MCSA CASE HISTORY (1990)—POSSIBLE FALSE POSITIVE OF BROKEN ROTOR BARS IN A SCIM DRIVING A COAL CONVEYOR VIA A SLOW SPEED GEARBOX

### 8.3.1 Summary and Key Features

Twin induction motors were used to drive coal conveyors, via low, output speed gearboxes, at a coal fired power station. The motors' nameplate data provided the following relevant information 3-phase, 415 V, 211 A, 127 kW/170 HP, 50 Hz, 1480 r/min SCIM. The only available information, about the gearboxes, was their overall speed reduction ratio of 3/67 and the number of stages, or teeth on pinions and gear wheels was unknown. The results presented in Section 8.3.1 will confirm that mechanical load dynamics downstream of the motor are the source of the multiple sidebands in the current spectrum.

### 8.3.2 MCSA Diagnosis

The analysis of one of the motors (A) is presented but both exhibited the same characteristic current signature patterns. Motor A was operating on a reduced load of 130 A compared to its full-load current of 211 A, therefore the operating slip was considerably lower than the full-load slip of 1.33%. The current spectrum between 0 and 100 Hz is shown in Figure 8.17 and there are multiple sidebands around the supply component.

The zoom current spectrum of Figure 8.18 shows sidebands at ±1.3 Hz ($N_{av}$) = 48 dB down on *f* and additional sidebands at second and third harmonics of ±1.3 Hz, which are due to drive dynamics, being reflected back into the rotor of the motor. Inboard of the ±1.3 Hz sidebands, there are no sidebands which is where sidebands at less than 1.33 Hz would occur if caused by broken rotor bars. It was indeed very fortuitous that the motor was on reduced load (130 A), since had it been operating at,

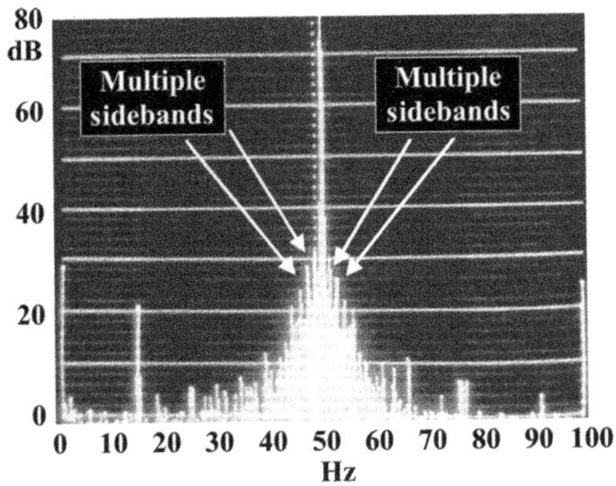


Figure 8.17 Current spectrum, baseband span 0–100 Hz, 6400 lines, 15.6 mHz/line.

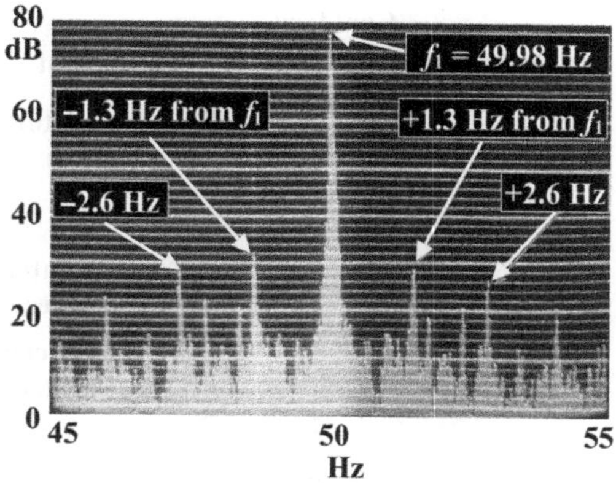


Figure 8.18 Current zoom spectrum 45–55 Hz, 10% of spectrum in Figure 8.17, 15.6 mHz/line.

or close to, full-load (211 A), the $\pm 2sf$ sidebands at full-load would nominally be at $\pm 1.33$ Hz, which would have been very close indeed, or coincident with, those caused by drive dynamics at $\pm 1.3$ Hz as shown in Figure 8.18. This case again emphasizes the need to be very aware of the effects on the current spectrum of mechanical load dynamics downstream of the motor.

## 8.4 CASE HISTORY (1992)—IMPOSSIBLE TO ANALYZE MCSA DATA DUE TO SEVERE RANDOM CURRENT FLUCTUATIONS FROM THE MECHANICAL LOAD DYNAMICS FROM THE COAL CRUSHER

### 8.4.1 Summary and Key Features

Motor details: 3-phase, 3.3 kV, 336 kW/450 HP, 70 A, 50 Hz, 1482 r/min SCIM. The number of rotor bars is unknown.

Due to the coal crushing process the load on the motor was very variable and the current was swinging from 25% (17 A) to 96% (65 A) of full-load current. This is a very challenging signal to analyze and spectrum averaging is essential to produce a characteristic spectrum from an induction motor driving a coal crusher due to that dramatic load variation. A schematic diagram of a coal crusher is shown in Figure 8.19, which confirms it is a highly complex item of mechanical plant that produces numerous mechanical dynamics that can be reflected back into the gearbox and the rotor of the SCIM.

The current spectra presented in Section 8.4.1 confirms that components from the mechanical load dynamics downstream of the motor dominated the spectrum. Although there is an outside possibility of detecting any $\pm 2sf$ sidebands around $f$, this is really only feasible, when there are multiple broken bars in the cage winding. In that situation the $\pm 2sf$ sidebands may be sufficiently high in magnitude, typically 30 dB

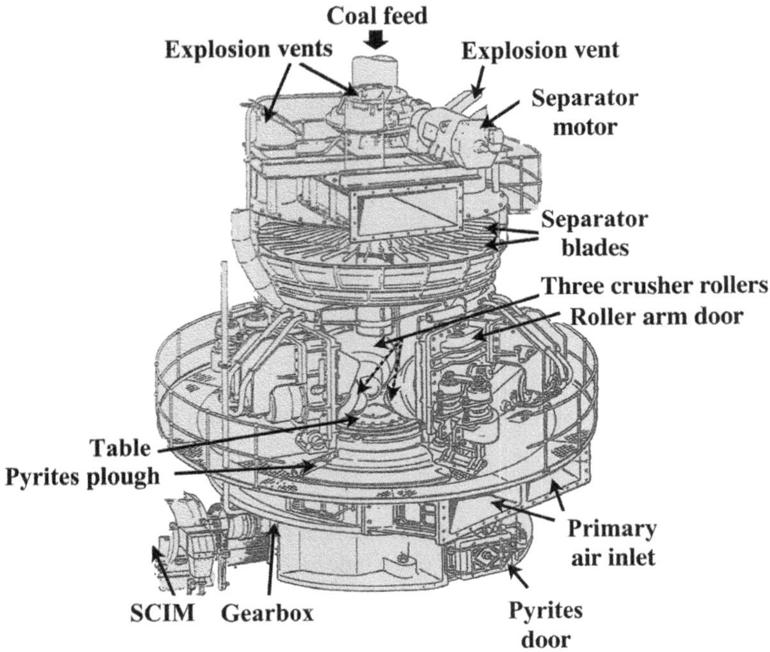


Figure 8.19 Typical coal crusher in a coal fired power station. Reproduced with permission of EM Diagnostics Ltd.

and less down on $f$, so that they are not swamped by the sidebands from the mechanical load dynamics. With SCIMs driving coal crushers in power stations, there is a higher probability of predicting broken rotor bars when none exist—a false positive.

This is a challenge for further research into classical MCSA and the development of signal processing algorithms to detect cage winding breaks in motors, driving slow speed coal crushers with highly fluctuating loads. It is impractical for coal crusher motors to be uncoupled during the generation of electricity and to carry out an analysis of the transient current in an attempt to detect a broken rotor bar. Such a stoppage would affect the production of electricity causing consequential loss of income generation and could only be done during a major outage of the power plant.

### 8.4.2 MCSA Diagnosis

A sample of the time domain current waveform from one of the motors (A) is shown in Figure 8.20 and confirms that the current is continually changing, and therefore spectrum averaging must be applied to produce a spectrum for possible interpretation.

For the avoidance of doubt this gearbox ratio is typical of the accuracy to which a gearbox OEM* will specify on the nameplate. An illustration of the coal crusher drive is given in Figure 8.21, and at the motor's full-load speed and using the gearbox

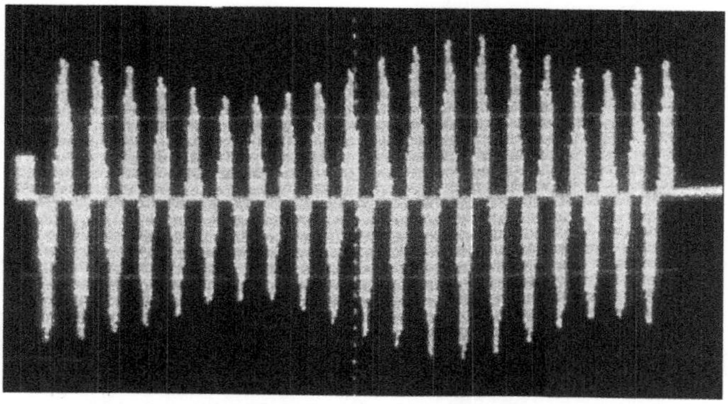

Figure 8.20 Current waveform, 20 cycles of 50 Hz supply.

ratios provided by the power station engineers, the rotational speed frequencies of the different shafts are as follows:

$$N_r = 1482 \text{ r/min}$$

$$f_r = 24.7 \text{ Hz}$$

$$N_r/N_{g1} = 4.4, \; N_{g1} = 337 \text{ r/min}$$

$$f_{g1} = 5.6 \text{ Hz}$$

$$N_{g1}/N_{g2} = 2.78, \; N_{g2} = 121 \text{ r/min}$$

$$f_{g2} = 2.0 \text{ Hz}$$

$$N_{g2}/N_{gL} = 2.95, \; N_L = 41 \text{ r/min}$$

$$f_{gL} = 0.68 \text{ Hz}$$

The motor has a nominal full-load slip of 1.2% and if it was (which it is not) operating at full-load current and nominal full-load slip then the $\pm 2sf$ sidebands would be at $\pm 1.2$ Hz on a 50 Hz supply. The load current is very variable and therefore the slip and also sideband frequency is continually varying. A sample spectrum from one of the coal crusher motors is shown in Figure 8.22 after 124 spectrum averages

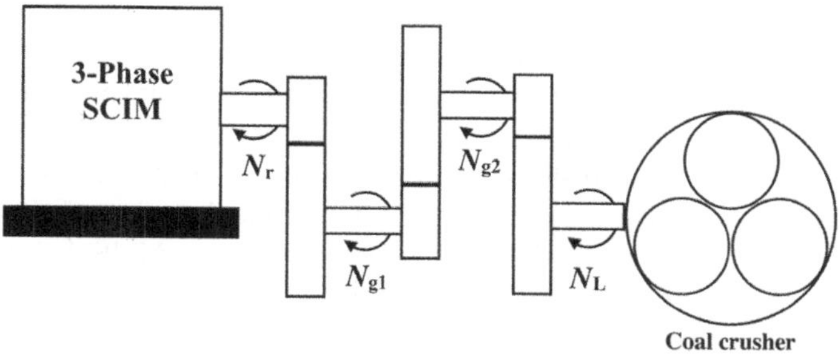


Figure 8.21 Three-stage reduction gearbox, overall step down ratio of 36.146*.

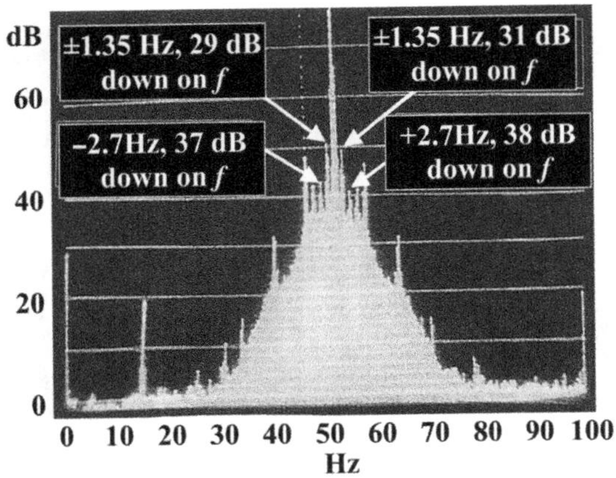


Figure 8.22 Motor A, current spectrum, span 0–100 Hz, line resolution 15.6 mHz/line.

using a base band span of 0–100 Hz and a spectral line resolution of 25 mHz/line. There are multiple sidebands around the supply component and a large skirting effect in the spectrum around $f$ due to the continually changing current during spectrum averaging.

The observations from Figure 8.21 indicate there are sidebands around $f$, at

$$\pm 2f_{gL} = \pm 1.35\text{ Hz(an average of 30 dB down on } f)$$

$$\pm 4f_{gL} = \pm 2.7\text{ Hz(an average of 38 dB down on } f)$$

$$\pm 6f_{gL} = \pm 4.05\text{ Hz(an average of 42 dB down on } f)$$

$$\pm 8f_{gL} = \pm 5.4\text{ Hz(an average of 30 dB down on } f)$$

This confirms that the spectrum is swamped by sidebands from the load dynamics and any $\pm 2sf$ sidebands have to be inboard of the sidebands at $\pm 1.35$ Hz, recall the steady-state full-load slip is 1.2% but the motor is not operating at a constant slip. A zoom spectrum is shown in Figure 8.22 (span 45–55 Hz) in an attempt to see if any sidebands existed between $f$ and the sidebands at $\pm 1.35$ Hz.

There are sidebands at $\pm 0.6$ Hz around $f$, but they are 47 dB down on $f$ and their origin is uncertain, because they could be $\pm 2sf$ sidebands or a function of modulation of the current due to mechanical dynamics being reflected back into the cage rotor. The only option is to trend the spectrum using the same number of spectrum averages over a period of time (say, every 3–6 months) to identify any changes in the spectrum that could be due to broken bars.

In SCIMs driving coal crushers in power stations, classical MCSA cannot really be used as an early warning diagnostic technique for the detection of, for example, one or two broken rotor bars in SCIMs driving coal crushers. Multiple broken rotor bars would need to exist before MCSA could yield any possibility of detecting the problem. Even when using spectrum averaging techniques, as in this particular case

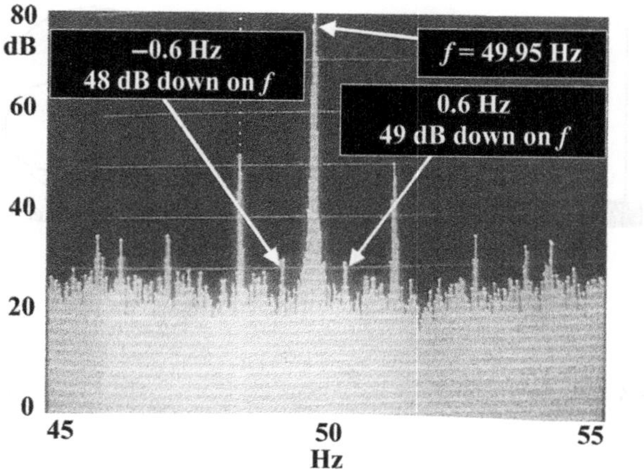


Figure 8.23 Motor A, current zoom spectrum 45–55 Hz, 10% of spectrum in Figure 8.22, 15.6 mHz/line.

history, a current swing of 70% of full-load current virtually precludes the use of classical MCSA in this instance.

## 8.5 CASE HISTORY (1995)—SUCCESSFUL ASSESSMENT OF CAGE WINDINGS WHEN THE LOAD CURRENT FLUCTUATIONS ARE NORMAL FROM A SCIM DRIVING COAL CRUSHER

### 8.5.1 Summary and Key Features

Two induction motors driving coal crushers (sometimes referred to as mill motors) via 2-stage, step down gearboxes were MCSA tested in a 2000 MW coal fired power station. The motor's nameplate data provided the following information: 3-phase, 3.3 kV, 97 A, 470 kW/630 HP, 50 Hz, 985 r/min SCIM. The rotors had double cage windings with 84 rotor bars and as already confirmed classical MCSA can only be applied to assess the condition of the inner cage run winding. The loads on these two coal crusher motors were not varying as much as the previous case history in Section 8.4 and it was possible to determine the current components related to the dynamics of the shafts downstream of the motor. Thus the operating speed could be accurately measured via the current spectrum analysis and since the number of rotor bars was known it was cross checked with the operational slip from the measured rotor slot passing frequency. The frequency of any $2sf$ sidebands could be reliably predicted and the current spectra for the two motors confirmed that none existed and the inner cage windings were perfectly normal. Thus it is possible to apply classical MCSA to assess the operational condition of the inner cage run windings of coal crusher SCIMs when the loads are not dramatically fluctuating.

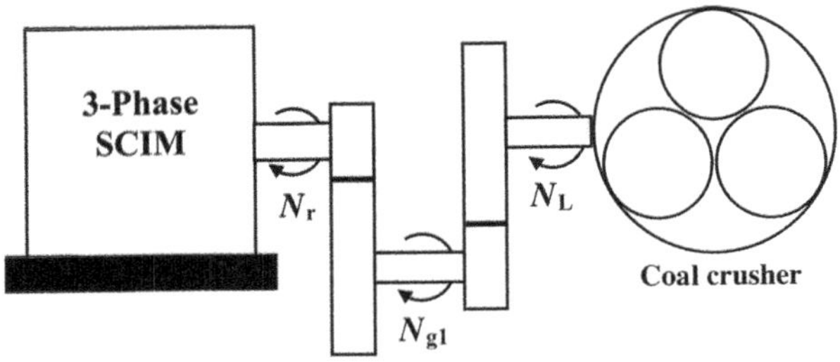


Figure 8.24 Illustration of coal crusher drive train.

### 8.5.2 MCSA Diagnosis

An illustration of the coal crusher drive train is shown in Figure 8.24 showing three coal crushing rollers but the actual coal crusher unit is much more complex as is shown in Figure 8.19.

The predicted current components that could occur as sidebands around $f$ are as follows:

$$N_r = 985\ \text{r/min}$$

$$\pm f_r = \pm 16.42\ \text{Hz}$$

$$N_r/N_{g1} = 3.038,\ N_{g1} = 324\ \text{r/min}$$

$$\pm f_{g1} = \pm 5.4\ \text{Hz}$$

$$N_{g1}/N_L = 6.75,\ N_L = 48\ \text{r/min}$$

$$\pm f_L = \pm 0.8\ \text{Hz}$$

There are three coal crusher rollers downstream of the output shaft from the gearbox which is at a rotational speed frequency, $f_L$ of 0.8 Hz, therefore,

$$\pm 3f_L = \pm 2.4\ \text{Hz}$$

Samples of the time domain current waveforms are presented in Figures 8.25, 8.26, 8.27 and 8.28 for motors A and B, recall that the measured current from the

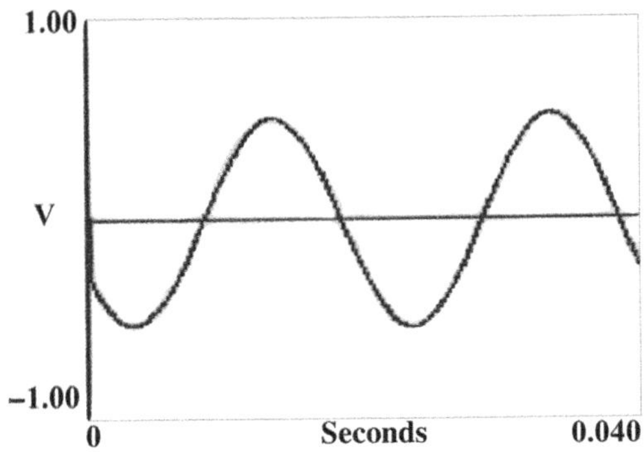


Figure 8.25 Time domain current, two cycles from motor A.

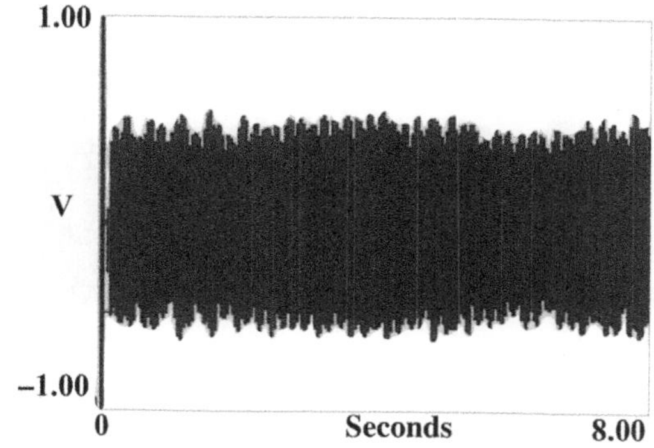


Figure 8.26 Time domain current, 400 cycles from motor A, current swinging between 65 and 90 A on an *in situ* panel ammeter.

MCSA CT is converted to a voltage since spectrum analyzers and MCSA instruments operate with an input voltage signal and not the actual current in amperes.

Figures 8.29 and 8.30 show the current spectra for motors A and B using a baseband span of 0–100 Hz, a line resolution of 15.63 mHz/line, and 20 spectrum averages.

**(i)** A general observation of the spectra in Figures 8.29 and 8.30 is that motor A's components from 5.44 Hz inboard toward *f* are more distinct than is the case with motor B since in the latter the current is swinging between 50% and 100% of full-load current (a 50% swing) whereas motor A's current has only half the swing at between 65 and 90 A (a 26% swing with reference to $I_{FL}$ ) and the same number of spectrum averages was deliberately used to demonstrate via a direct comparison that more averages were really required for motor B due to its greater current swing.

**(ii)** The 1× r/min frequency components ($f_r$) of the shaft speeds (input to the gearboxes) are evident at ±16.55 and ±16.53 Hz around the supply component for motors A and B, respectively, and the magnitudes of these components in both motors are of the same order.

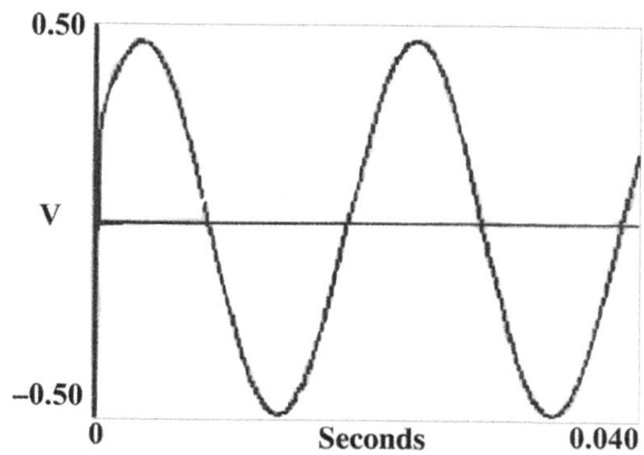


Figure 8.27 Time domain current, two cycles from motor B.

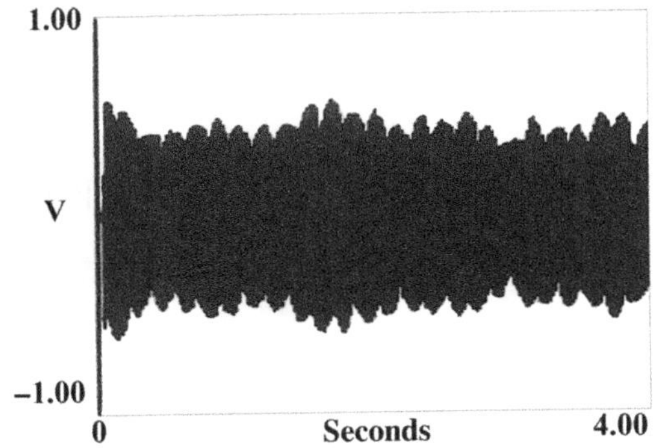


Figure 8.28 Time domain current, 200 cycles from motor B, current swinging between 50 and 100 A on an *in situ* panel ammeter.

**(iii)** There are components at ±2.42 Hz corresponding to $\pm 3f_L$ (i.e., three crushing rollers times output speed frequency of 0.8 Hz of the gearbox shaft) around the supply component which dominate the spectra at an average of 33 dB down on $f$, for both motors. This is understandable since the greatest mechanical disturbance is reflected back into the rotor of the SCIM from the rollers where the actual crushing of the coal occurs.

**(iv)** The components at ±5.44 Hz around $f$ are equal to $f_{gL}$, the internal shaft speed of the gearbox.

**(v)** These results verify the predictions and confirm the relevance and validity of the original experiments by Thomson, as presented in Section 8.1.2.

The sidebands at ±16.55 (i.e., $\pm f_r$) and ±16.53 (i.e., $\pm f_r$) Hz around the supply component in the current spectra (Figures 8.29 and 8.30) are for motors A and

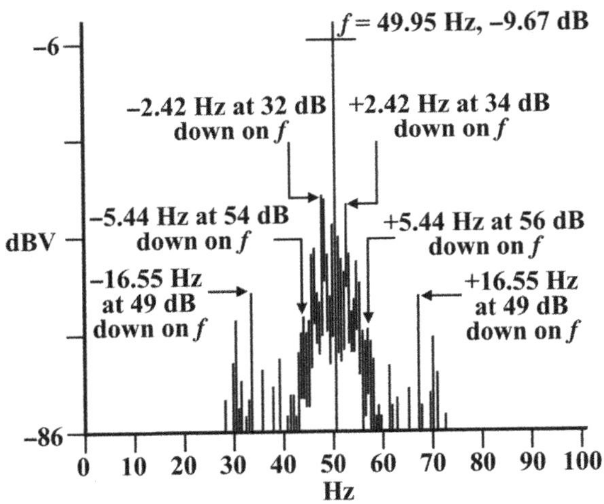


Figure 8.29 Current spectrum, motor A, 6400 lines, baseband span 100 Hz, frequency resolution 15.63 mHz/line, current swing 65–90 A.

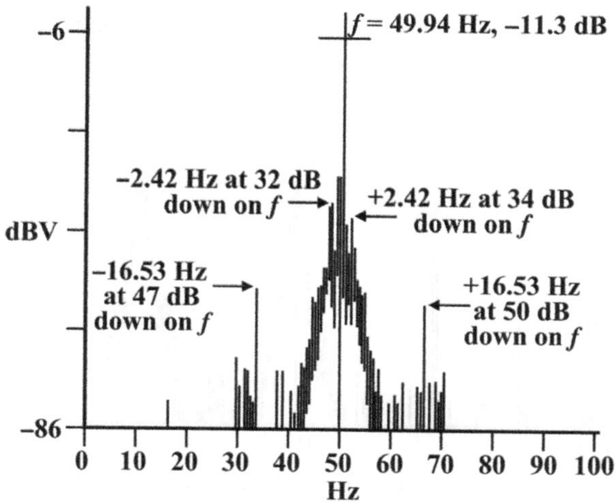


Figure 8.30 Current spectrum, motor B, 6400 lines, baseband span 100 Hz, frequency resolution 15.63 mHz/line, current swing 50–100 A.

B, respectively, which means that the rotational speed and operational slip can be obtained.

Motor A:

$f_r$ = 16.55 Hz, $N_r$ = 993 r/min, the operational slip = 0.007 (0.7%), assume $N_s$ = 1000 r/min

Motor B:

$f_r$ = 16.53 Hz, $N_r$ = 992 r/min, the operational slip = 0.008 (0.8%), assume $N_s$ = 1000 r/min

$N_s$ is actually very slightly less (e.g., 1 or 2 r/min) than 1000 r/min since the supply frequency during each test was not equal to 50 Hz, but the difference is negligible.

The predicted ±2*sf* sidebands for motors A and B are ±0.7 and ±0.8 Hz, respectively. As a further cross check, and since the number of rotor bars was known, the principal rotor slot passing frequency ($f_{rs(+1)}$) given by equation (6.1) and also in Chapter 10 on airgap eccentricity is now applied to measure $f_{rs}$ using the measured slips.

$$f_{rs(+1)} = f\left\{\left(\frac{R}{p}\right)(1-s) \pm n_{\omega s}\right\}$$

Using motor A as a sample, where

$f$ = 49.953 Hz; $R$ = 84 slots; $p$ = pole-pairs = 3; $s$ = 0.007; $n_{\omega s} = +1$

Therefore, predicted $f_{rs(+1)}$ = 1439 Hz with $n_{\omega s} = +1$

And predicted $f_{rs(+3)}$ = 1539 Hz with $n_{\omega s} = +3$

These are the inherent and normal rotor slot passing frequencies spaced 2*f* apart.

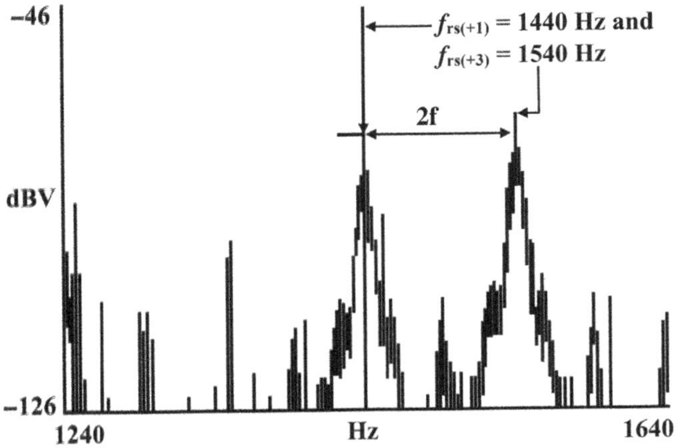


Figure 8.31 Motor A, current zoom spectrum, rotor slot passing frequencies, baseband span 70–1800 Hz, 12,800 lines, 135 mHz/line.

The spectrum shown in Figure 8.31 shows the measured $f_{rs(+1)}$ is 1440 Hz compared to the predicted value of 1439 Hz which is a negligible difference, this confirmed the operational slip for motor A was 0.7% and the $\pm 2sf$ sidebands should be very close to ±0.7 Hz around $f$ and the $\pm 2sf$ sidebands for motor B should be close to ±0.8 Hz. Figures 8.32 and 8.33 show the zoom current spectra for motors A and B and there are no $\pm 2sf$ sidebands at ±0.7 Hz or at ±0.8 Hz (or any sidebands even close to these values) around $f$ for motors A and B, respectively.

The run windings are normal, and thus the goal was achieved in that the operational condition of the cage windings was assessed. The components at ±1.29 Hz around $f$ for both motors must be from reflected mechanical dynamics downstream of the motors, but due to the complex structure of a coal crusher (Figure 8.19) it is

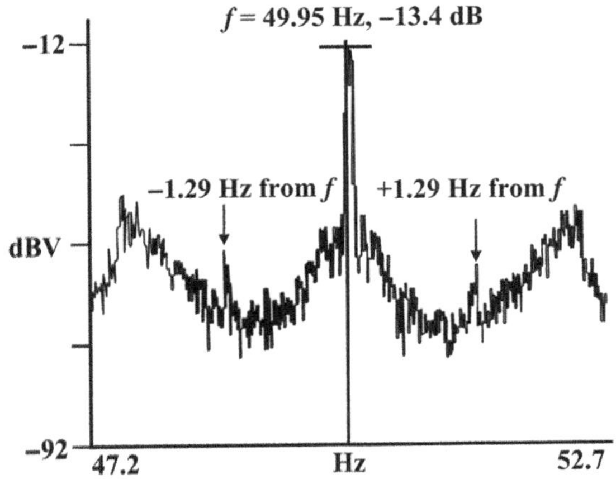


Figure 8.32 Current zoom spectrum, motor A, 15.63 mHz/line.

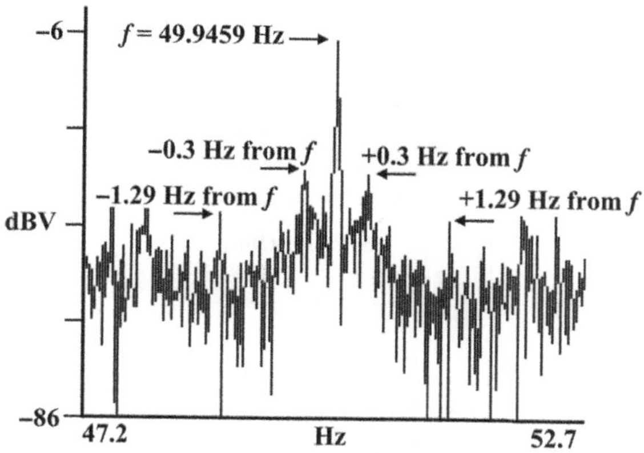

Figure 8.33 Current zoom spectrum, motor B, 15.63 mHz/line.

impossible to identify the source, likewise for the ±0.3 Hz components in the spectrum from motor B as shown in Figure 8.33.

## 8.6 TWO CASE HISTORIES (2015)—FALSE POSITIVE OF BROKEN BARS IN ONE OF THE SCIMS DRIVING THRUSTERS ON AN FPSO IF INFLUENCE OF DRIVE DYNAMICS IS DISCOUNTED

### 8.6.1 Summary and Key Features

Two case histories on MCSA applied to SCIMs driving port and starboard thruster propellers on an offshore Floating Production, Storage and Oil Offloading floating ship (FPSO) used for the production of hydrocarbons. An illustration of the layout of the main electrical drives is shown in Figure 8.34.

The owner of the FPSO had received a report from a vendor which stated that there was asymmetry in the rotor cage winding of the port-side thruster motor (PTM). The client considered that the term "cage asymmetry" was neither defined nor quantified and was a somewhat meaningless statement. Due to the strategic importance of the thruster motors, the owner required an independent investigation by specialists in the design, operation, and condition monitoring of electrical machines, this was carried out by Thomson (author).

The port and starboard thruster drives are the most important drives on an FPSO since they keep it in a stable position to compensate for wind and wave variations, in order to sustain safe production of hydrocarbons. The thrusters are controlled by an automated position control system and the pitch of the propeller blades is varied depending on the sea state. This is a much more complex drive system compared to that of a SCIM, which is directly coupled to a centrifugal pump or an FD fan in a power station, hence the analysis of the current is indeed a challenge.

The motor's nameplate included the following: 3-phase, 6.6 kV, 2400/1245 kW, (i.e., 3210/1670 HP), 2-speed PAM (pole–amplitude modulation) stator winding,

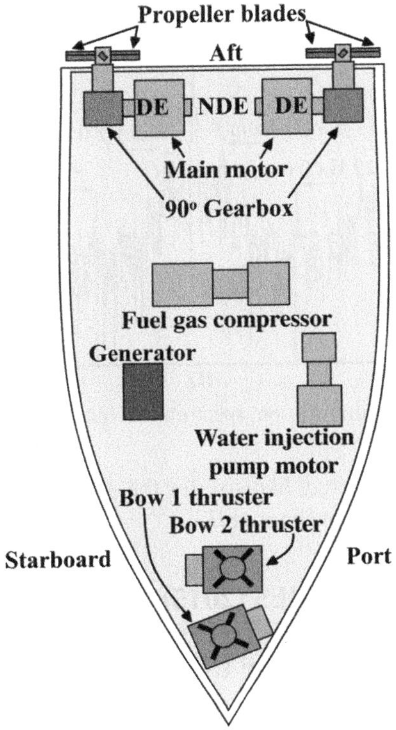


Figure 8.34 Schematic illustration of the main HV SCIM drives on this FPSO.

252/155 A, 60 Hz, 1188/894 r/min, p.f. 0.87/0.74, efficiency 94.6/95% SCIM. The client obtained the number of rotor bars ($R = 58$) from the OEM.

A schematic illustration of the drive train is shown in Figure 8.35 and for each thruster drive, there are two, separate, single stage gearboxes, one top side (photo shown in Figure 8.35) which is a right angled bevel gear unit and one subsea which drives the 4-bladed propeller.

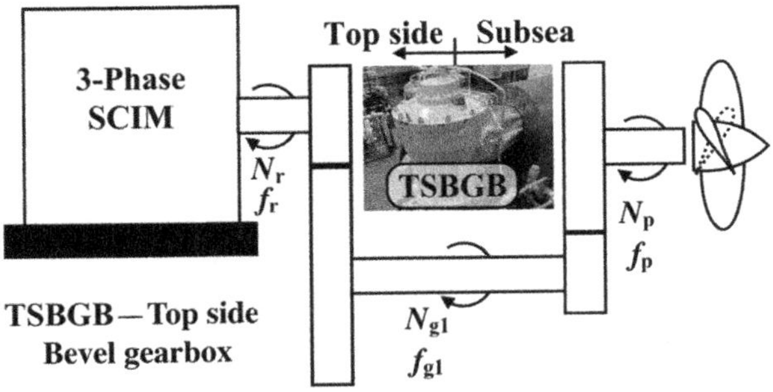


Figure 8.35 Illustration of drive train and photo of the top side gearbox (TSBGB). Reproduced with permission of EM Diagnostics Ltd.

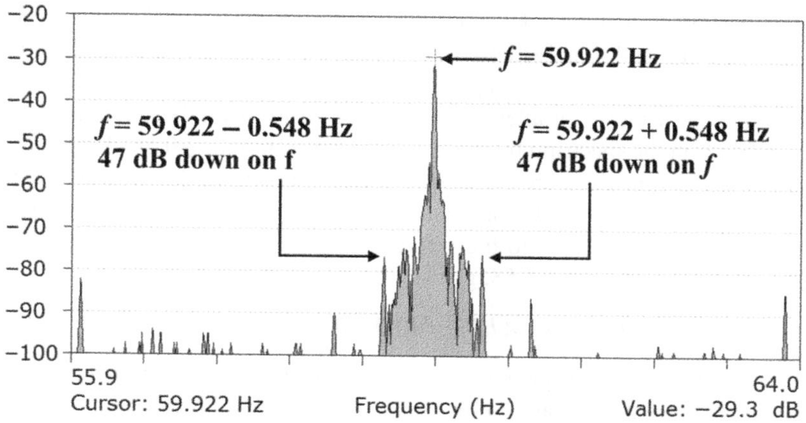


Figure 8.36 Current zoom spectrum for PTM, 10.16 mHz/line.

## 8.6.2 MCSA Diagnosis (2013–2015)

**(a)** First Stage Analysis—Detection of $\pm 2sf$ sidebands due to cage winding breaks

The classical MCSA approach is applied in the first instance before considering the influence of the rotor dynamics on the current spectrum due to two gearboxes and a propeller downstream of the motor. Both motors were operating at 106 A compared to a full-load current of 252 A, with a rated speed of 1188 r/min, a corresponding full-load slip of 0.01 or 1%, and the propeller pitch was 50°. At full-load, the $\pm 2sf$ sidebands should nominally be at ±1.2 Hz around $f$ but the motors were on reduced load and therefore the $2sf$ sidebands will be considerably lower in frequency than 1.2 Hz.

The typical approach, for example, via "*a press the button type*" of MCSA instrument is to display a zoom current spectra around the supply component to search for $\pm 2sf$ sidebands in the expected region around $f$. Figures 8.36 and 8.37 show the spectra for PTM and STM and there are sidebands at ±0.548 Hz

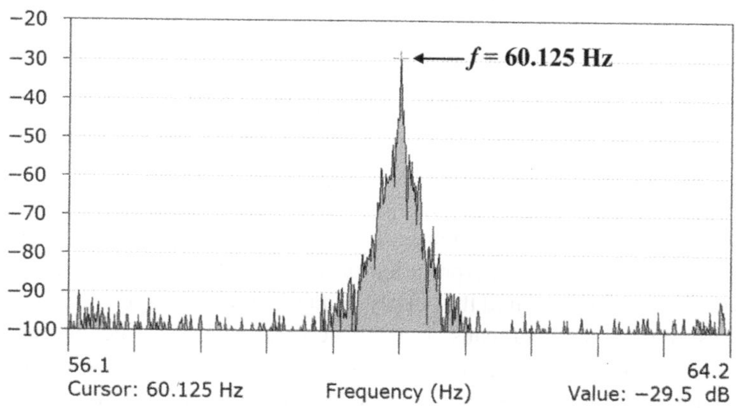


Figure 8.37 Current zoom spectrum for STM, 10.16 mHz/line.

around $f$ and both these sidebands are 47 dB down on $f$ from PTM (Figure 8.36) and there are additional sidebands at $\pm(2 \times 0.548)$ Hz and sidebands inboard of the ones at $\pm 0.548$ Hz. But are these sidebands at $\pm 0.548$ Hz truly due to a rotor cage winding problem such as broken rotor bars? There are no distinct sidebands in the spectrum from STM (Figure 8.37). If the $\pm 0.584$ Hz sidebands are at $\pm 2sf$, this gives an operational slip of 0.00457 at the measured supply frequency of 59.92 Hz. Using equation (4.5) to predict the equivalent broken rotor bar factor referred to the reference slip at full-load, with sidebands at 47 dB down on $f$, and $R = 58$, gives

$$BB_{\mathrm{f}} = BB_{\mathrm{c}} \times BB_{\mathrm{fs}} = (0.01/0.00457) \times 0.5 = 1.1$$

This is 2.2 times greater than the 0.5 threshold (see Section 4.5 and case histories in Chapters 5 and 6) for the $BB_{\mathrm{f}}$. A diagnosis based on only the spectrum shown in Figure 8.36 would be that there is at least one broken rotor bar or several broken bars still making partial contact as has been shown in case history 6.1.

Such a diagnosis would be premature indeed since no consideration has been given to the influence of the drive dynamics downstream of PTM since it is possible that sidebands from drive dynamics can be coincident with the presumed $\pm 2sf$ sidebands and therefore a false positive is possible.

**(b)** Second Stage Analysis—Influence of drive dynamics on current spectrum

The top side gearbox has a ratio of 21/33 and the subsea one a ratio of 13/42. The full-load speed of the motor is 1188 r/min, therefore the predicted current components that could occur around the supply component $f$, are as follows:

$$N_{\mathrm{r}} = 1188 \text{ r/min}$$

$$\pm f_{\mathrm{r}} = \pm 19.8 \text{ Hz}$$

$$N_{\mathrm{g1}} = 1188 \times (21/33) = 756 \text{ r/min}$$

$$\pm f_{\mathrm{g1}} = \pm 12.6 \text{ Hz}$$

$$N_{\mathrm{p}} = 756 \times (13/42) = 234 \text{ r/min}$$

$$\pm f_{\mathrm{p}} = \pm 3.9 \text{ Hz}$$

with four propeller blades, hence possible components at:

$$\pm 4f_{\mathrm{p}} = \pm 15.6 \text{ Hz}$$

Due to the complex drive dynamics, the second stage of the analysis is to search for these components and their harmonics, which are predicted above. A lateral thinking approach to the interpretation of the current spectra is necessary compared to merely zooming around the supply component to search for $\pm 2sf$ sidebands around the supply component, caused by a cage winding break. The current spectra between 0 and 130 Hz from the star board and port side thruster motors (STM and PTM) are shown in Figures 8.38 and 8.39. At this stage of the analysis the frequency

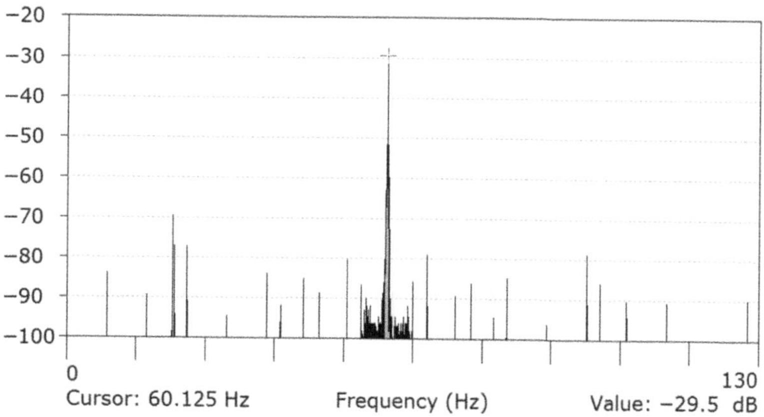


Figure 8.38 Current spectrum from STM, 106 A, 50° pitch, baseband span (0–130 Hz), 12,800 lines, 10.16 mHz/line.

components have deliberately not been identified in these spectra so that any obvious differences can be identified by visual inspection only.

The current spectra are distinctly different even although the MCSA tests were carried out at the same input current and propeller pitch. There are numerous components in the spectrum from STM but at dB levels much lower than the supply component *f*, whereas the spectrum from PTM has less components but there are components (circled in Figure 8.41) at more than 10 times greater (20 dB higher) than the highest ones from STM. This initially suggests there are atypical rotor dynamics downstream of PTM.

The current spectra have now been annotated as shown in Figures 8.40 and 8.41 to identify the measured current components due to drive dynamics downstream of the SCIM.

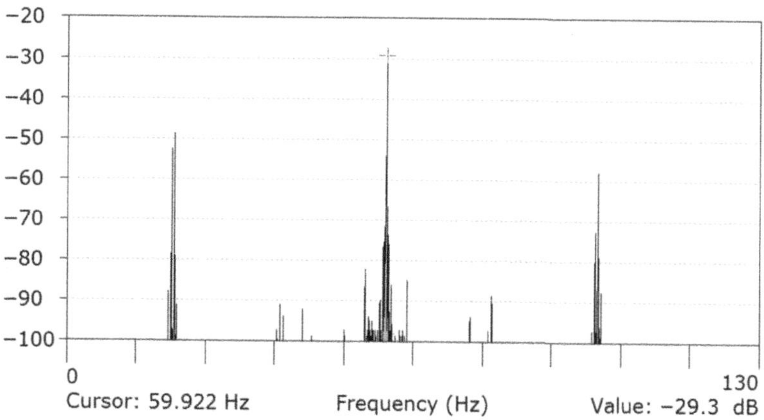


Figure 8.39 Current spectrum from PTM, 106 A, 50° pitch, baseband span (0–130 Hz), 12,800 lines, 10.16 mHz/line.

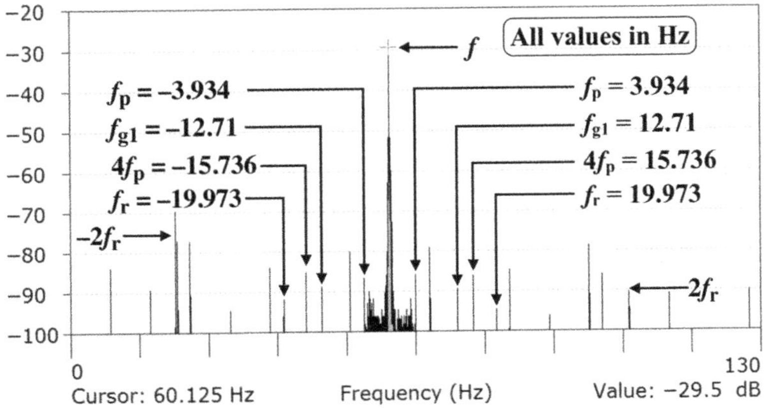


Figure 8.40 Current spectrum from STM, 106 A, 50° pitch, baseband span (0–130 Hz), 12,800 lines, 10.16 mHz/line.

## STM

Using rated speed (1188 r/min), current (252 A), and frequency (60 Hz) for the prediction:

*Predicted* $f \pm f_r = 60 \pm 19.8$ Hz
*Measured* $f \pm f_r = 60.125 \pm 19.97$ Hz

$$N_{g1} = 1188 \times (21/33) = 756 \text{ r/min}$$

*Predicted* $f \pm f_{g1} = 60 \pm 12.6$ Hz
*Measured* $f \pm f_{g1} = 60.125 \pm 12.7$ Hz

$$N_p = 756 \times (13/42) = 234 \text{ r/min}$$

*Predicted* $f \pm f_p = 60 \pm 3.9$ Hz
*Measured* $f \pm f_p = 60.125 \pm 3.93$ Hz

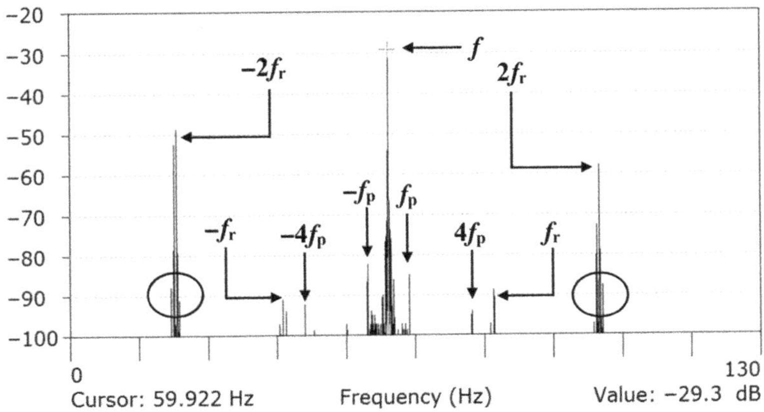


Figure 8.41 Current spectrum from PTM, 106 A, 50° pitch, baseband span (0–130 Hz), 12,800 lines, 10.16 mHz/line.

With four propeller blades,
*Predicted* $f \pm 4f_p = 60 \pm 15.6$ Hz
*Measured* $f \pm 4f_p = 60 \pm 15.74$ Hz
The measured supply frequency = 60.125 Hz, which means that the synchronous speed is

$$N_s = 1202 \text{ r/min}$$

**PTM**

Using rated speed (1188 r/min), current (252 A), and frequency (60 Hz) for the predictions:
*Predicted* $f \pm f_r = 60 \pm 19.8$ Hz
*Measured* $f \pm f_r = 59.922 \pm 19.89$ Hz

$$N_{g1} = 1188 \times (21/33) = 756 \text{ r/min}$$

*Predicted* $f \pm f_{g1} = 60 \pm 12.6$ Hz
*Measured* *No evidence*

$$N_p = 756 \times (13/42) = 234 \text{ r/min}$$

*Predicted* $f \pm f_p = 60 \pm 3.9$ Hz
*Measured* $f \pm f_p = 59.922 \pm 3.89$ Hz
With four propeller blades,
*Predicted components at:* $f \pm 4f_p = 60 \pm 15.6$ Hz
*Measured components at:* $f \pm 4f_p = 59.922 \pm 15.58$ Hz
The measured supply frequency = 59.92 Hz, which means the synchronous speed is

$$N_s = 1198 \text{ r/min}$$

These results confirm the presence of current components produced by drive dynamics downstream of the motor and demonstrate the sensitivity of current spectrum analysis. The next stage is to identify any unique differences in the spectra and these are around the $f \pm 2f_r$ components. A direct comparison between PTM and STM indicates that the $f - 2f_r$ from the former is 20 dB higher (factor of 10) than from the latter and there is evidence of additional components (circled in Figure 8.41) around $f \pm 2f_r$ in PTM. A zoom current spectrum around the $f \pm 2f_r$ components from PTM is shown in Figures 8.42 and 8.43.

This is the most interesting result so far since it shows there are also sidebands at ±0.548 Hz around the $f \pm 2f_r$ components, and this characteristic of having the same sidebands at ±0.548 Hz around the supply component $f$ was the crucial breakthrough in the analysis.

### 8.6.3 Conclusions

It was concluded that the ±0.548 Hz sidebands around $f$ were not due to broken rotor bars and were due to some complex interaction within the drive train dynamics

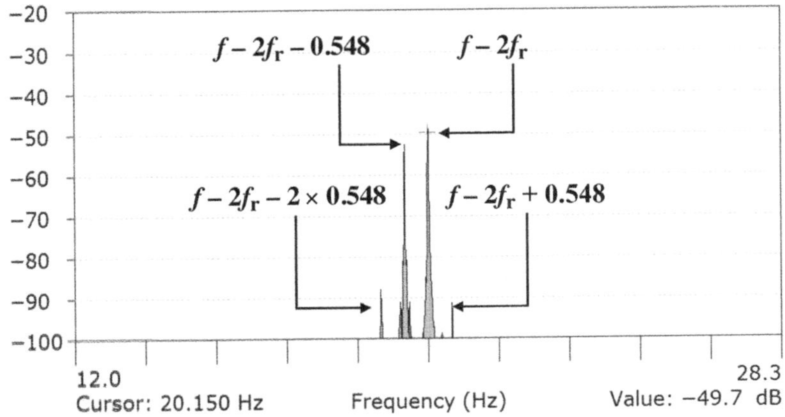


Figure 8.42 Current zoom spectrum around $f - 2f_r$ from PTM, 106 A, 50° pitch, baseband span (0–130 Hz), 12,800 lines, 10.16 mHz/line.

downstream of the motor. It is impossible to identify the source of that interaction because of the highly complex nature of the drive train, since it includes two gearboxes, and a propeller with four blades but it was sufficient to state that there was some form of abnormality.

Photos of the port and starboard thruster propellers are shown in Figures 8.44a and 8.44b while the FPSO was in dry dock.

The end user and owner of the FPSO subsequently confirmed the following:

(i) A very strong rope made of Dextron 12 plus (diameter 40 mm/1.6 inches; weight 1.36 kg/m or 2.73 lb/yard (36 inches); break load 62,200 kg/137,096 lb) had become entangled around the propeller shaft of the portside thruster

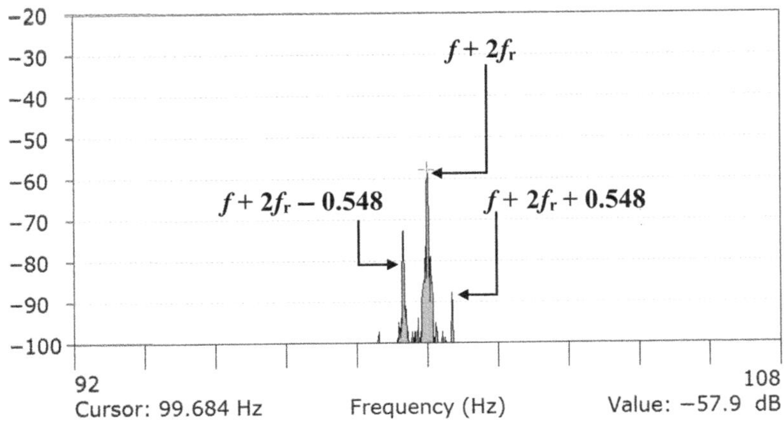


Figure 8.43 Current spectrum around $f + 2f_r$ from PTM, 106 A, 50° pitch, baseband span (0–130 Hz), 12,800 lines, 10.16 mHz/line.

Figure 8.44a Hull of an FPSO in dry dock. Reproduced with permission of BW Offshore (UK) Ltd., Aberdeen, Scotland.

drive train as shown in Figure 8.45. This rope is used for connection between the FPSO and the tanker during oil cargo offload operations as shown in Figure 8.45.

**(ii)** Subsea inspections revealed chips/broken part on the port side thruster propeller as shown in Figures 8.46 and 8.47 which of course should not exist and the owner of the FPSO concluded this was caused by the rope entanglement.

**(iii)** There may also have been an element of distortion on the shaft propeller downstream of the subsea gearbox.

Figure 8.44b Close up, thruster propeller. Reproduced with permission of BW Offshore (UK) Ltd., Aberdeen, Scotland.

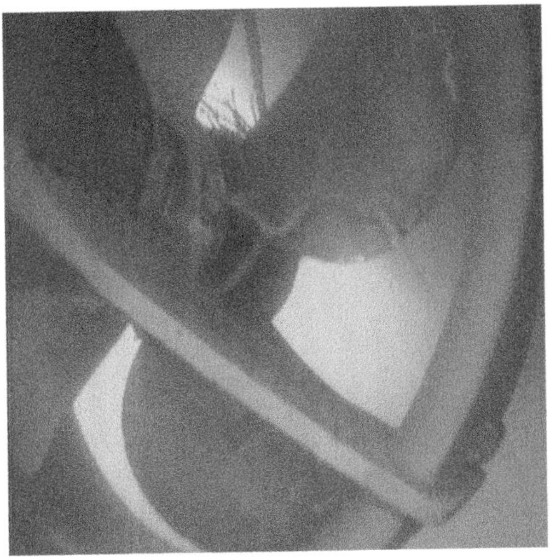

Figure 8.45 Rope wrapped around the propeller. Reproduced with permission of BW Offshore, (UK) Ltd., Aberdeen, Scotland.

**(iv)** It was therefore concluded that the abnormal current spectra shown in Figures 8.39, 8.41 8.42 and 8.43) are due to abnormal drive dynamics downstream of the motor. The features of these current components were caused by damage to the propeller (Figures 8.46 and 8.47) and possibly a distortion between the gearbox output bearing shaft assembly and the propeller.

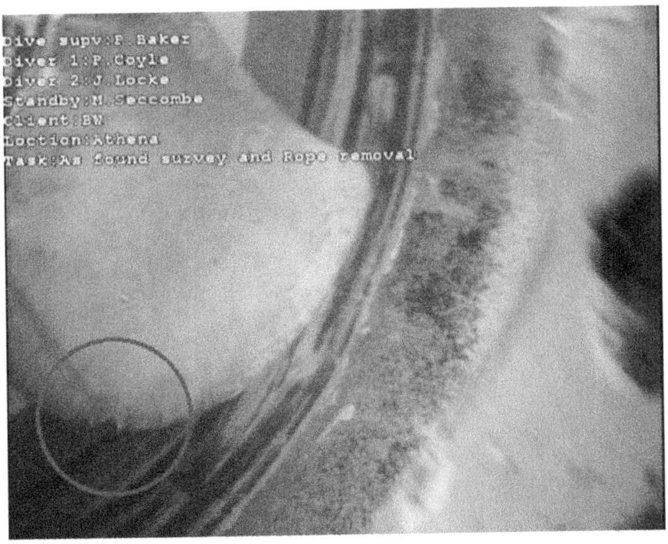


Figure 8.46 Two largish chips on propeller blade. Reproduced with permission of BW Offshore (UK) Ltd., Aberdeen, Scotland.

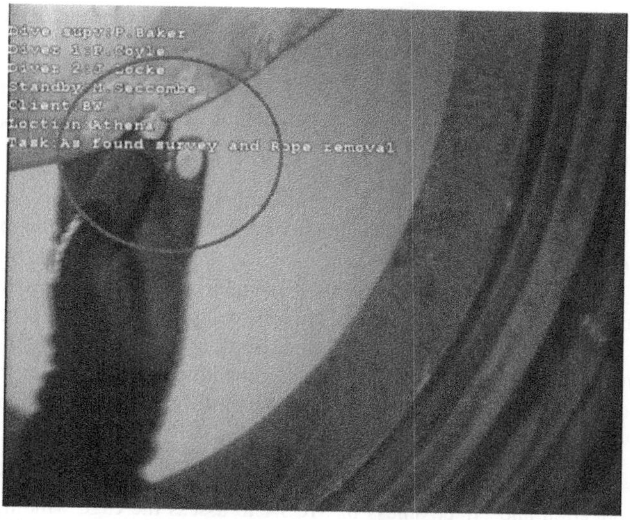


Figure 8.47 Chip on propeller blade. Reproduced with permission of BW Offshore (UK) Ltd., Aberdeen, Scotland.

## 8.7 QUESTIONS

**8.7.1** State the three types of misalignment that can occur across the coupling between a SCIM, which is driving a centrifugal pump.

**8.7.2** Which type of misalignment can cause the rotor in a SCIM to run eccentrically in the bearings or even cause a slight bow on the cage rotor due to pre-load forces?

**8.7.3** Why can mechanical misalignment, external to a SCIM, between it and its mechanical load connected via a coupling result in components in the current spectrum and at what frequencies if the supply frequency is 60 Hz and the motor is operating at 1782 r/min?

**8.7.4** Why can a gearbox in the drive train of a SCIM result in components in the current spectrum?

**8.7.5** An SCIM is driving a conveyor via a 3-stage step down gearbox with ratios of 3.03456, 4.03421, and 3.04345, calculate the frequencies of components in the current spectrum, due to gearbox dynamics, if the motor's speed is 1185 r/min from a 60 Hz supply.

**8.7.6** A 3-phase, 4160 V, 373 kW/500 HP 69 A, 60 Hz, 1788 r/min, p.f. 0.92, efficiency 95.8% SCIM drives a coal conveyor via a 3-stage, step down gearbox. The motor is operating at full-load current and speed and the senior mechanical maintenance engineer at the power station was able to provide the following information about the gearbox because accurate records were kept, when it was previously stripped down and fully inspected.

Stage 1: Pinion 11 teeth; Wheel 47 teeth

Stage 2: Pinion 11 teeth; Wheel 55 teeth

Stage 3: Pinion 13 teeth; Wheel 58 teeth

Predict the components in the current spectrum to the motor, which can be caused by the gearbox dynamics being transmitted to the SCIM.

**8.7.7** In question 8.7.6 there are typically 48 dumps of coal per minute from the conveyor, predict the effect that can have on the spectral content of the current spectrum from the motor.

**8.7.8** Why is it difficult to apply MCSA to a SCIM driving a coal crusher with a 3-stage, low output speed gearbox when the load current is randomly changing between 30% and 100% of the full-load current?

Comment on the signal processing required when trying to produce a current spectrum from a SCIM driving a coal crusher or any other crushing process.

Why is it the case that if the current is swinging by, say 25% of the full-load current compared to 70%, there is a higher probability of analyzing the current signal to produce a more meaningful current spectrum to identify $\pm 2sf$ sidebands from a cage winding break?

**8.7.9** Your company is invited to attend a meeting with power station engineers to discuss the application and effectiveness of MCSA to assess the cage windings in SCIMs driving coal conveyors and coal crushers and to submit a quotation for the work. In preparation for that initial meeting list all the questions you require answers to before you would consider offering your services.

**8.7.10** In case history 8.3 the motor was MCSA tested a year later from the result shown in Figure 8.17 at a new operating current of 180 A and two additional sidebands at $\pm 0.9$ Hz around the supply component $f$ (now 50 Hz) have appeared, and both sidebands are 40 dB down on $f$.

Why are the changes in frequencies of the other components in Figure 8.17 negligible when the operating current has increased by 50 A?

The OEM has now provided the number of rotor bars ($R = 58$) in this single cage design, interpret the new spectrum and recommend the action to be taken by the power station staff.

## REFERENCES


[8.1] Pruftechnik, "A Practical Guide to Shaft Alignment," Distributed in USA by Ludeca Inc., Doral, FL, 2002, pp. 1–62. Available at www.ludeca.com

[8.2] W. T. Thomson, "On-Line Current Monitoring to Diagnose Shaft Misalignment in Three-Phase Induction Motor Drive Systems," ICEM '94, Paris, September 1994.

[8.3] W. Freize and H. Jordan, "Unilateral Magnetic Pulls in 3-Phase Machines," *ETZ-A*, vol. 83, no. 9, 1962. Translation: C. E.Trans. 7836.

[8.4] S. Fruchtenicht, E. Pittius, and H. Seinsch, "A Diagnostic System for Three-Phase Asynchronous Machines," Proceedings of IEEE Fourth International Conference on Electrical Machines and Drives (EMDA' 89), IEE, Savoy Place, London, September 1989, pp. 163–171.

[8.5] G. Kron, *Equivalent Circuits of Electric Machines*, Dover Publications, Inc., New York, 1967.

[8.6] P. von Kaehne, "Unbalanced Magnetic Pull in Rotating Electrical Machines: Survey of Published Work," ERA Report Z/T 142, Electrical Research Association, England, UK, 1963, pp. 1–24.

[8.7] S. J. Yang, 1981,"Low Noise Electric Motors," in *Monographs in Electrical and Electronic Engineering*, IEE Oxford University Press (August 6, 1981) ISBN 10-0198593325, hard cover, 112 pages.

[8.8] M. J. DeBortoli, S. J. Salon, D. W. Burow, and C. J. Slavik, "Effects of rotor eccentricity and parallel windings on induction machine behavior: a study using finite analysis," *IEEE Transactions on Magnetics*, vol. 29, no. 2, March, 1993, pp. 1676–1682.

[8.9] D. G. Dorrell, W. T. Thomson, and S. Roach, "analysis of airgap flux, current and vibration signals as a function of the combination of static and dynamic airgap eccentricity in 3-phase induction motors," *IEEE Transactions on Industry Applications*, vol. 33, no. 1, January/February, 1997, pp. 24–34.

[8.10] W. T. Thomson and C. Campbell, "Current Monitoring for Detecting Broken Rotor Bars—The Influence of Gearboxes," 26th Universities Power Engineering Conference, Brighton Polytechnic, September 1991.

[8.11] James. I. Taylor, *The Gear Analysis Handbook—A Practical Guide for Solving Vibration Problems in Gears*, 1st edition, Vibration Consultants Inc., 1994, ISBN 0-9640517-1-0.

[8.12] S. Salon, S. Salem, and K. Sivasubramaniam, "Monitoring and Diagnostic Solutions for Wind Generators," Power and Energy Society General Meeting, 2011, pp. 1–4.

[8.13] D. Lu, X. Gong, and W. Qiao, "Current-Based Diagnosis for Gear Tooth Breaks in Wind Turbine Gearboxes," IEEE Energy Conversion Congress and Exposition (ECCE), 2012, pp. 3780–3786.

CHAPTER 9

# MISCELLANEOUS MCSA CASE HISTORIES

William T. Thomson

## 9.0 INTRODUCTION AND SUMMARY OF CASE HISTORIES

This chapter presents a diverse range of MCSA industrial case histories obtained during routine MCSA surveys, which were carried out to assess the condition of cage windings in SCIMs. They come under miscellaneous case histories because they do not have a common theme. The following list of case histories in this chapter is presented to assist the reader (if so desired) to make a choice of the ones of interest.

**Section 9.1—Two case histories:** Possible false positives of cage winding breaks in two, 6600 V, 1850 kW/2480 HP, 4-pole SCIMs due to number of poles ($2p$) being equal to number of axial cooling ducts and spider support arms (as shown in Figures 9.3 and 9.4) on the rotor assembly. This design causes $\pm 2sf$ sidebands around the supply component of current even though the cage winding is perfectly healthy.

**Section 9.2—One case history:** A 4160 V, 2230 kW/2990 HP, 8-pole SCIM with number of poles equal to number of kidney shaped axial ducts in the rotor. A false positive of broken rotor bars was prevented by changing the load; if that had not been possible, then a false positive of broken rotor bars would have certainly occurred using classical MCSA.

**Section 9.3—Two case histories:** MCSA detected abnormal pumping dynamics downstream of one of the two, 6600 V, 520 kW/700 HP, 4-pole, submersible SCIMs driving seawater lift pumps (SWLPs). This diagnosis was made possible by comparing the current spectra from SCIMs driving healthy and abnormal pump units. Holes in the caisson of the faulty pump unit were found.

**Section 9.4—Two case histories:** MCSA detected slack and worn belt drives in two, 3300 V, 132 kW/180 HP, 6-pole SCIM cooling fan drives in a cement factory.

---

*Current Signature Analysis for Condition Monitoring of Cage Induction Motors: Industrial Application and Case Histories*, First Edition. William T. Thomson and Ian Culbert.

**Section 9.5.3—One case history:** MCSA measurements and analysis from an inverter-fed LV, 440 V, 450 kW/600 HP, 6-pole SCIM with no cage winding breaks.

**Section 9.5.4—One case history:** MCSA measurements and analysis from a large, inverter-fed, 6600 V, 6300 kW/8450 HP 4-pole SCIM with no cage winding breaks.

**Section 9.6—One case history:** MCSA assessment of the mechanical operational condition of an electrical submersible pump (ESP) used for artificial lift in an offshore oil production platform and driven by 4160 V, (2 × 210 HP/2 × 157 kW), 2-pole SCIMs operating in tandem.

## 9.1 POSSIBLE FALSE POSITIVES OF CAGE WINDING BREAKS IN TWO 1850 kW SCIMs, DUE TO NUMBER OF POLES (2*p*) EQUAL TO NUMBER OF SPIDER SUPPORT ARMS ($S_p$) ON SHAFT (1991)

### 9.1.1 Modulation of Magnetizing Current at Twice Slip Frequency

When the number of poles ($2p$) equals the number of spider support arms ($S_p$) between the rotor shaft and the inside of the rotor core, the magnetizing current is modulated at twice slip frequency. Diagrammatic illustrations of the rotating magnetic field for two instants in time are shown in Figures 9.1a and 9.1b; all leakage fields are neglected.

This phenomenon has been known to manufacturers long before MCSA was initially applied (mid to late 1970s) to diagnose broken rotor bars in cage rotors. This phenomenon can give a false positive of broken bars when none exist. The use of $2p = S_p$ by certain OEMs is normally due to ease of manufacture when the spider support arms are welded onto the rotor shaft, rather than using a higher number of spider support arms for a 4-pole motor, which would be more expensive.

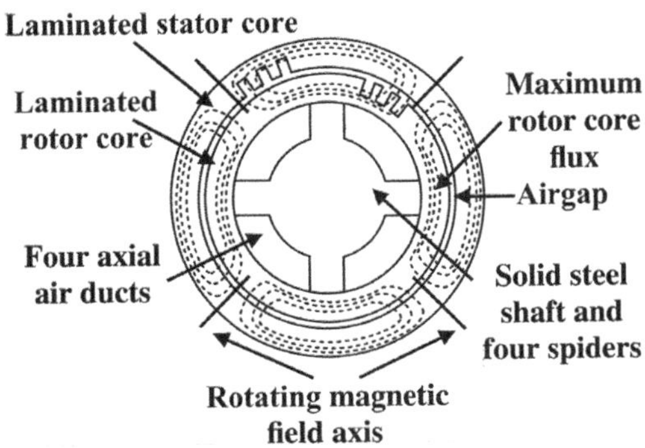


Figure 9.1a 4-pole magnetic field with four spiders at one instant in time.

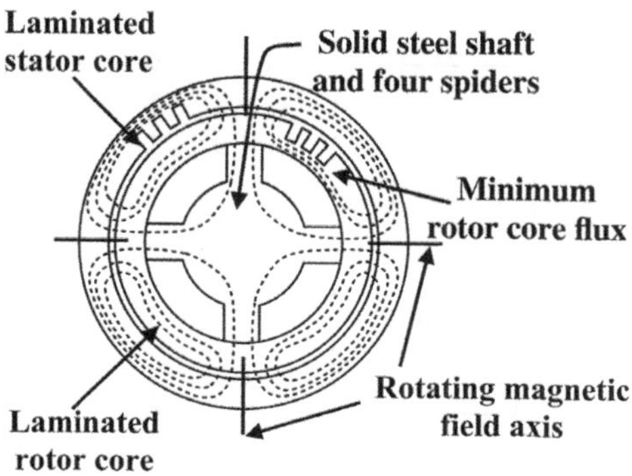


Figure 9.1b 4-pole magnetic field moved forward by 90° electrical.

Total reluctance

$$S_{T1} = S_{stc} + S_{ag} + S_{roc}$$

Total reluctance

$$S_{T2} = S_{stc} + S_{ag} + \frac{(S_{roc} \times S_{spi})}{(S_{spi} + S_{roc})}$$

The crucial starting point to understanding why the magnetizing current is modulated at twice slip frequency ($2sf$) when $2p = S_p$ is that the main flux/pole is constant in a 3-phase SCIM, when the applied voltage and frequency are constant and is given by the emf equation for the stator

$$V = \sqrt{2}\pi f \phi_p T_{ph} k_s k_d \quad (9.1)$$

where

$V$ = applied voltage/phase to the stator winding, Volts

$f$ = supply frequency, Hertz

$\phi_p$ = flux/pole, Webers

$T_{ph}$ = turns/phase in the stator winding

$k_s$ = coil span factor

$k_d$ = winding distribution factor in a group of series and adjacent connected coils

Figures 9.2a and 9.2b are simplified equivalent magnetic circuit models for Figures 9.1a and 9.1b, respectively. The magneto-motive force (mmf) is $F = NI$ and is the driving source of the magnetic flux ($\phi$) and reluctance ($S$) is the opposition to the flow of flux. Recall from basic electromagnetics and circuit theory the following analogy can be used:

$$F = \phi S \leftrightarrow V = IR$$

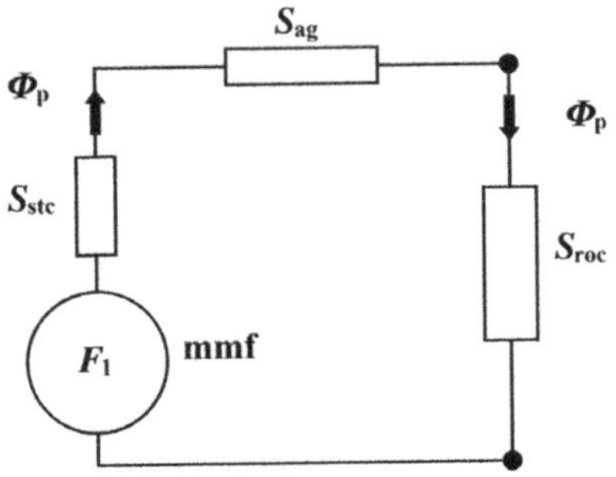


Figure 9.2a Equivalent magnetic circuit for one pole of Figure 9.1a.

$F$, mmf is analogous to volts ($V$)

$\phi$, flux is analogous to current ($I$)

$S$, reluctance is analogous to resistance ($R$)

Consider the simple circuit of a single reluctance ($S$) fed by an mmf ($F$) at constant flux ($\phi$). $F$ must be directly proportional to $S$ ($F = S$ times a constant). For a more complex network of reluctance which is reducible to a single equivalent (total) reluctance $S_T$, $F$ is proportional to $S_T$ but $F = I \times$ (number of turns $N$) and for $N$ constant, $I$ is directly proportional to $S$. Therefore, for the reluctance networks of Figures 9.2a and 9.2b, the magnetizing current is proportional to the reluctance in each case and the following terminology applies to these networks.

The mmfs, $F_1 = T_{ph} I_{1m}$ and $F_2 = T_{ph} I_{2m}$ in ampere-turns, A

$I_{1m}$ and $I_{2m}$ are the magnetizing currents at different instants in time, Amperes

$\phi_p$ = flux per pole ($\phi$/pole) is constant, Webers

$S_{stc}$ = reluctance of stator core magnetic circuit/pole, A/Wb

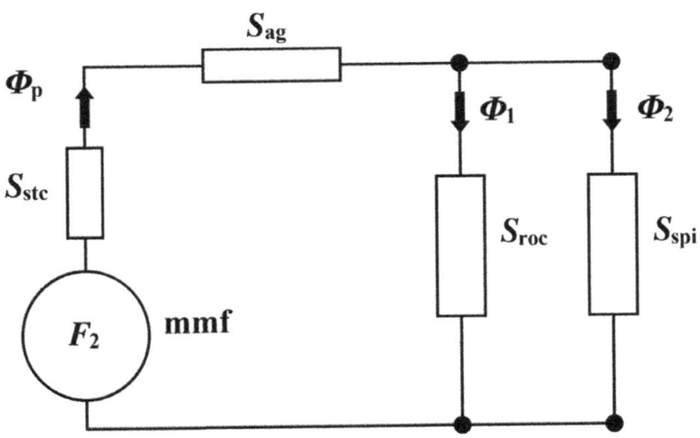


Figure 9.2b Equivalent magnetic circuit for one pole of Figure 9.1b.

$S_{ag}$ = reluctance of the airgap/pole, A/Wb
$S_{roc}$ = reluctance of rotor core magnetic circuit/pole, A/Wb
$S_{spi}$ = reluctance of spider support arms and shaft magnetic circuit/pole, A/Wb
$\phi_1$ = flux in rotor core/pole, Webers
$\phi_2$ = flux in spiders and shaft/pole, Webers

During rotation of a cage rotor having $2p = S_p$, the reluctance of the magnetic circuit of the rotor changes cyclically with respect to time and since the flux/pole is constant the mmf supplied by the stator winding must change since the turns/phase is constant. Therefore, the magnetizing current will vary when there is a change in the rotor's magnetic reluctance path. An inspection of Figures 9.1a and 9.1b shows that the flux lines in both cases are the same in the stator core and across the airgap. In Figure 9.1a, the magnetic flux per pole is carried by the rotor core alone and the magnetic circuit of Figure 9.2a shows that the magnetic reluctance path is a maximum.

When the rotating field advances by 90° electrical relative to the rotor core as shown in Figure 9.1b, the magnetic reluctance is therefore lower as shown in Figure 9.2b compared to Figure 9.2a and the mmf is therefore lower and the magnetizing current drops. With a rotor having $2p = S_p$, the reluctance of the rotor's magnetic circuit is a variable with respect to time and therefore the mmf and the magnetizing current vary.

With reference to Figures 9.2a and 9.2b

$F_1 > F_2$ and $I_{1m} > I_{2m}$ since the turns/phase is constant.

$$S_{T1} > S_{T2}$$

Electrical degrees with respect to flux distribution of north and south poles is given by $\theta_e^\circ = p\theta_m$, where $p$ = pole-pairs and $\theta_m^\circ$ = mechanical degrees. Since 180° electrical relative movement between the rotating field and the rotor results in a complete cycle of magnetizing current swing, the frequency of that current swing will be at twice the slip frequency $2sf$ when $2p = S_p$. Amplitude modulation of the magnetizing current at $2sf$ will cause sidebands at $\pm 2sf$ around the supply component of current.

Mr. John Middlemiss, formerly chief designer of electrical machines at Parsons Peebles, Scotland, is duly acknowledged for his technical advice on this topic during the late 1980s when MCSA was initially applied to a SCIM with $2p = S_p$, Thomson [9.1]. This resulted in a false positive of cage winding breaks in a large, HV SCIM in 1984, which was a salutary lesson for Thomson (author)—more is learned from mistakes than successes.

### 9.1.2 Application of MCSA to SCIMs with $2p = S_p$

With respect to MCSA and current spectra, it was Thomson [9.1] in 1992 who initially reported that a false positive of broken rotor bars can occur when $2p = S_p$. In line with his research strategy, he initially carried out experiments with a rotor having $2p$ equal to $S_p$ with different loads. It was established that the magnitude of the $\pm 2sf$ sidebands as a percentage of the input current at light load (predominantly magnetizing current) is higher than their magnitude as a percentage of the input current at full-load.

Interestingly, 23 years later, this phenomenon was revisited in a paper by Chanseung Yang et al. [9.2] in 2015, who proposed that certain space harmonic, current components are a function of broken rotor bars and are given by the following equation [9.2]:

$$f_{\mathrm{brb,h}} = (k_2(1 - s) \pm s)f \tag{9.2}$$

where

$k_2 = 5, 7, 11, 13, \ldots$

This paper presented results for rotors with $2p = S_{\mathrm{p}}$ by monitoring components at $(5 - 4s)f$ and $(5 - 6s)f$ from equation (9.2) and these are independent of the modulation of the supply component at $2sf$ due to $2p = S_{\mathrm{p}}$. This showed that false positives of broken rotor bars in cage rotors, which have $2p = S_{\mathrm{p}}$, may be prevented and this contribution [9.2] provided new knowledge on the application of MCSA to SCIMs that was previously not available. Results were presented from controlled experiments in a laboratory with rotors, having $2p = S_{\mathrm{p}}$ with and without broken rotor bars which proved that components given by $(5 - 4s)f$ and $(5 - 6s)f$ increased in magnitude at a given load when the number of broken rotor bars was increased, thus avoiding a false positive, which would have occurred if only the $2sf$ sidebands had been monitored.

Reference 9.2 also reported on MCSA tests of two, 6600 V SCIMs rated at 350 kW/470 HP and 280 kW/375 HP within which $2p$ equaled $S_{\mathrm{p}}$ in these motors. With no broken rotor bars they proved that the $\pm 2sf$ sidebands around the supply component, which had been previously monitored (but not by the authors in Reference 9.2) and had given a false positive of broken rotor bars could not be coming from broken rotor bars. An excellent and very thorough paper by Sungho Lee et al. [9.3] in 2013, also presented MCSA results on the industrial application of MCSA with respect to the influence of $2p$ equal to $S_{\mathrm{p}}$ on the diagnosis of broken rotor bars.

Thomson thanks the authors of References 9.2 and 9.3 for referencing his original paper [9.1] of 1992 which did identify and establish, via experimental results, the problem of trying to diagnose broken rotor bars via MCSA in cage rotors in which $2p$ equals $S_{\mathrm{p}}$.

The following points are provided as additional information to that in References 9.2, 9.3:

**(i)** It is Thomson's experience of having applied MCSA to thousands of large, HV SCIMs during the past 34 years that the majority do not have the number of poles equal to the number of spider support arms and axial cooling ducts on the rotor shaft.

**(ii)** A very important point is that 2-pole, high power (say 1000 kW/1340 HP upward) HV SCIMs are very widely used in industry to drive, for example, large centrifugal compressors and high speed centrifugal pumps, and based on the authors' knowledge and experience the rotors of 2-pole cage rotors are not designed with a rotor shaft having $2p = S_{\mathrm{p}}$.

The reason being that a large, 2-pole, HV SCIM with only two spider support arms (and two axial ventilating ducts) would normally be mechanically unfit for the purpose since the mechanical stiffness of the rotor would be much too low and with only two spider support arms, the rotor core could not be adequately supported. Several critical speeds are highly likely during run up when one or more of the lateral natural frequencies of the rotor equals the speed of the rotor, and inevitably there would be serious vibration problems during run up and probably during steady-state operation.

**(iii)** With respect to 4-pole motors and above, many OEMs have discontinued the design of rotors with $2p = S_p$ for large HV SCIMs.

High power (1000 kW/1340 HP and upward), HV, 2-pole and 4-pole SCIMs are more widely used than those with higher pole numbers. For example, one of the largest offshore oil production platforms (commissioned in 2008) operating in the North Sea between Scotland and Norway, produces 200,000 barrels of oil per day. On that installation there are 33 HV (6600 and 11,000 V) SCIMs with ratings between 260 kW/350 HP and 7150 kW/9590 HP. Of these, 34% are 2-pole, 33% are 4-pole, and 33% are 6-pole. None of the 4- or 6-pole motors have $2p = S_p$ and note that these are relatively new motors as of 2016. More industrial case histories are required to increase the level of confidence in the proposed diagnostic strategies put forward in References 9.2 and 9.3 for preventing a false positive of broken rotor bars when $2p = S_p$.

### 9.1.3 Summary and Key Features in Case History 9.1 (In 1991)

Two 3-phase, 6600 V, 1850 kW/2480 HP, 4-pole SCIMs with cage rotors having $2p = S_p$, which equality can give a false positive of a cage winding break due to $\pm 2sf$ sidebands appearing in the spectrum around the supply component. The rotor design of each motor as shown in Figures 9.3 and 9.4 was copper fabricated, with a single

Figure 9.3 Rotor, no broken rotor bars.

Figure 9.4 Rotor shaft to show spider support arms (4 of) and axial cooling ducts (4 of).

cage winding, using 50 rectangular bars and butt type joints, via annular grooves in the end rings. No end ring retaining rings were fitted. These motors were used to drive oil tanker loading pumps on an offshore oil production platform.

This case history verifies that a false positive of broken rotor bars can occur, when $2p = S_p$. At the time of the initial MCSA testing of these motors in 1991, the construction of the rotors was unknown.

### 9.1.4 MCSA Diagnosis and Inspection (1991)

The resulting spectra from the two motors A and B recorded in 1991 when it was initially unknown that $2p = S_p$ are presented in Figures 9.5 and 9.6, respectively. These spectra indicated sidebands which coincided with $\pm 2sf$ (also $\pm 4sf$ and $\pm 6sf$) around the supply component and in each case, the $\pm 2sf$ sidebands were exactly 36 dB down on $f$. If the $2sf$ sidebands from both motors at 0.39 Hz and 0.38 Hz, respectively, were truly due to broken rotor bars, this would mean that the fault severity was effectively identical, the chances of which were virtually zero, particularly when their operating currents differed by only 4 A at 120 A and 124 A, for motors A and B, respectively, compared to the full-load current of 199 A. This corresponds to an operational slip for motors A and B of 0.33% and 0.32% at the measured supply frequency compared to a full-load slip of 0.833%. This is only 38% of the full-load slip, thus the motors are lightly loaded.

It was very unusual for two motors to have exactly the same current spectra with $2sf$ sidebands of the same magnitude, even although these were nominally identical motors of the same age, operating at virtually the same reduced load during the tests. Information gathering also verified that both ammeters were swinging at a steady and very low frequency but there was no low frequency modulation of the acoustic noise from either of the motors, which would certainly be expected, if the $\pm 2sf$ sidebands at 36 dB down on $f$ were truly coming from broken rotor bars in both motors. The equivalent broken rotor bar factor ($BB_f$) from equation (4.5) referred to the full-load

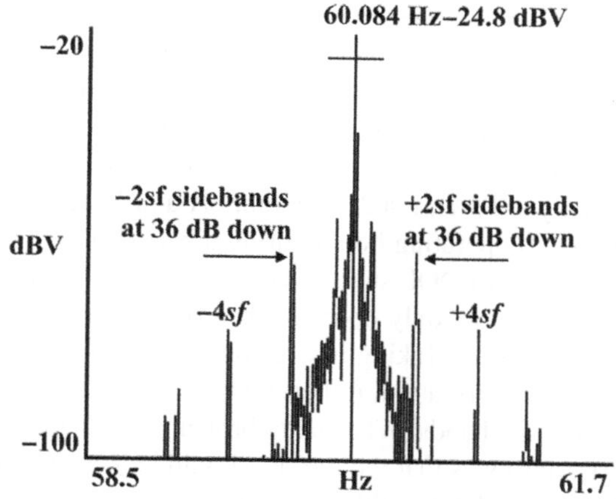


Figure 9.5 Motor A, 120 A, current zoom spectrum, 10.16 mHz/line.

slip, and using $N_{av} = 36$ dB, an operating slip of 0.0032(0.32%), a full-load slip of 0.0083(0.83%) and $R = 50$, gives

$$BB_f = \{2R/(10^{(36/20)} + 4)\} \times (0.83/0.32) = 3.8$$

Recall from Chapter 4, the $BB_c$ correction factor due to the motor being on reduced load is, in this case, $s_{FL}/s_{op} = 0.83/0.32$.

This equates to multiple broken rotor bars, as already verified in Chapter 4 and further proven by the numerous case histories in Chapters 5 and 6. The $BB_f$ is 7.8 times greater than the threshold $BB_f$ of 0.5. Since there was no acoustic noise typical of that coming from SCIMs with multiple broken rotor bars and the exact coincidence of $\pm 2sf$ sidebands in both motors it was necessary to obtain further information on the operational history of these two motors before making a definitive diagnosis.

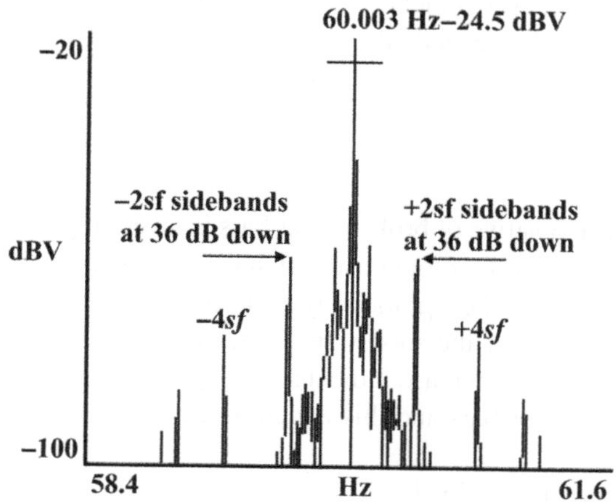


Figure 9.6 Motor B, 124 A, current zoom spectrum, 10.16 mHz/line.

The authors were not convinced that these motors had multiple broken rotor bars but they postulated that the sidebands were due to some other phenomenon; note this was in 1991 and a case history of this type had not previously occurred during the period since the authors started applying MCSA in industry in 1982.

The end user confirmed that the number of DOL starts for each motor, over a 10-year period, was only 1105 and 903 for motors A and B, respectively. They had run for 15,469 and 16,670 operating hours during a period of 87,600 hours. This is certainly not a large number of DOL starts over 10 years and the run hours were very low indeed, because the motors only operate when oil is being delivered to tankers alongside the offshore platform. Taking into account the identical spectra from each motor and this additional information, it therefore seemed to be highly unlikely that the twice slip frequency sidebands were due to broken rotor bars. The conclusion at that time (just after testing) was that the motors should be kept running and a repeat MCSA test carried out if the load could be changed.

The OEM was contacted after the MCSA measurements and kindly confirmed that the $\pm 2sf$ sidebands were due to the magnetic circuit design ($2p = S_p$) and that the value of slip from the sidebands in the spectra was in line with their Factory Acceptance Test results, at the operating currents recorded during the MCSA tests.

If broken bars did subsequently occur in one of these motors, the magnitude in dB of the $\pm 2sf$ sidebands would be a function of two independent phenomena. In this case history, a fortunate situation arose in 1992, 12 months after the first MCSA measurements, when pump A was removed for repairs and the cage rotor was inspected at the same time. The photographs in Figures 9.1 and 9.2 show the cage rotor from motor A and confirmed that $2p = S_p$. The cage winding was perfectly healthy and the construction verified the fundamental source of the twice slip frequency sidebands.

If, in 1991, only one motor had been available for MCSA testing, which can often be the case, and the test could only be carried out at one load condition, as often occurs, there is absolutely no doubt, that the prediction would have been, that a serious broken rotor bar problem existed and thus a false positive would have occurred. It was very fortuitous indeed, that two identical motors were running at the same load and slip, which coincidently gave exactly the same current spectra, which was statistically very unlikely to be caused by broken rotor bars.

### 9.1.5 Conclusions

When $2p = S_p$, a false positive of broken rotor bars is a very high probability, particularly if a single motor, whose rotor construction is unknown, is being MCSA tested. If the load on the motor can be significantly reduced and the magnitude of the $\pm 2sf$ sidebands as a percentage of the supply frequency current component increases as the load tends to zero. This increase indicates that $2p = S_p$. Therefore the sidebands are not due to broken rotor bars and this is demonstrated in the next case history in Section 9.2.

However, changing the load can be very difficult in industry as has already been stated in previous case histories in the book. If there are broken rotor bars in a SCIM with $2p = S_p$ then the opposite occurs to the $\pm 2sf$ sidebands since their magnitude drops relative to the magnitude of the supply frequency component as the load is reduced. However, the reader is strongly encouraged to read the most recent papers [9.2, 9.3] published in 2013 and 2015, respectively, if MCSA is being applied to SCIMs with $2p = S_p$.

## 9.2 CASE HISTORY (2007)—SCIM WITH NUMBER OF POLES EQUAL TO NUMBER OF KIDNEY SHAPED AXIAL DUCTS IN THE ROTOR—FALSE POSITIVE OF BROKEN BARS PREVENTED BY LOAD CHANGES

### 9.2.1 Summary and Key Features

This was an 8-pole SCIM and the rotor construction did not have the classical design of spider support arms welded onto the shaft since the rotor core was keyed directly onto the shaft but there were eight kidney shaped air ducts in the rotor core. This is analogous to having eight spider support arms in the rotor's magnetic circuit and therefore the magnetic reluctance is changing as a function of time as illustrated for the 4-pole case in Figures 9.1 and 9.2. The first MCSA test was on May 4, 2007 and at that time, the rotor construction was unknown to the end user which is very often the case. The spectrum clearly showed $\pm 2sf$ sidebands around the supply component, which were 37 dB down on $f$ and this is normally indicative of cage winding breaks. After the first test, subsequent MCSA tests over several weeks at constant load, produced virtually identical spectra. The OEM of the motor was contacted and confirmed that the rotor design was as described above and the end user was advised to drop the load by 30–40%. The result was, that the sidebands at $\pm 2sf$ increased by 7 dB relative to the supply component compared to their magnitude at the higher load condition.

This proved that the $\pm 2sf$ sidebands were solely being produced due to the magnetic circuit in the rotor as fully explained in Section 9.1.2, which verifies that a false positive of broken bars can be prevented if the load is changed, when the number of poles equals the number of axial cooling ducts. For this particular design, the rotor core is keyed onto the shaft.

Of course, if there are also broken rotor bars then there is a complex situation due to the two phenomena and the diagnostic strategy proposed by References 9.2 and 9.3 should be applied.

### 9.2.2 MCSA Measurements and Analysis (2007)

The nameplate provided the following data: 3-phase, 4160 V, 385 A, 2238 kW/3000 HP, 60 Hz, 892 r/min SCIM. The rotor had 145 rotor bars. The nominal full-load slip, $s_{FL}$ is 0.0088 (0.88%) based on the rated full-load speed of 892 r/min at 60 Hz. In May 2007, the spectrum from the first MCSA test on this motor operating at 272 A (70.6% of the full-load current) appears in Figure 9.7 which clearly shows $\pm 2sf$ sidebands around the supply component ($f$ at 60 Hz) at an average of

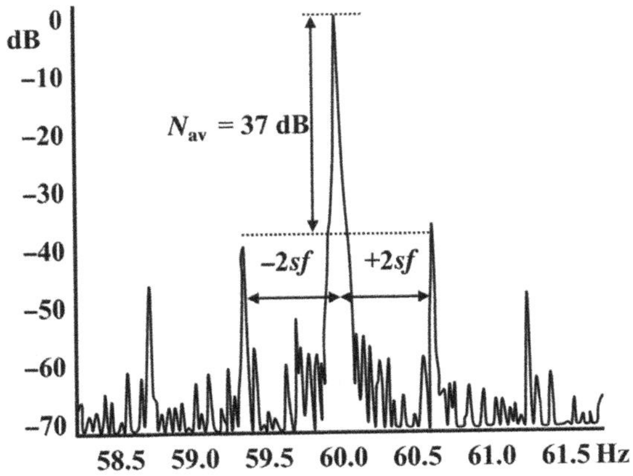


Figure 9.7 Current spectrum, 272 A, operating slip of 0.54%, 10.16 mHz/line.

$N_{av} = 37$ dB down on $f$. The operating slip from the measured $\pm 2sf$ sidebands was 0.0054 (0.54%) and thus the motor was operating at 61% of the full-load slip.

The equivalent broken rotor bar factor ($BB_f$) from equation (4.5) referred to the full-load slip, and using $N_{av} = 37$ dB, an operating slip of 0.54%, a full-load slip of 0.88%, and $R = 145$, gives

$$BB_f = \{2 \times 145/(10^{(37/20)} + 16)\} \times (0.88/0.54) = 7$$

Recall from Chapter 4, the $BB_c$ correction factor due to the motor being on reduced load, in this case, is $s_{FL}/s_{op} = 0.88/0.54$

Based on all the case histories previously presented in Chapters 5 and 6, supported by the theory given in Chapter 4, this equates to a very serious broken rotor bar problem with multiple broken rotor bars. The end user was very concerned about the MCSA result and was advised to drop the load on the motor by 30–40%. Fortunately, this could be achieved with this induced draft (ID) fan unit in a power station.

The spectrum shown in Figure 9.8 is at a current of 167 A (43.3% of the full-load current) and the result clearly shows the sidebands at $\pm 2sf$ have increased by 7 dB (factor of 2.2 in absolute units) to an average $N_{av} = 30$ dB down on the supply component. The operating slip from the $\pm 2sf$ sidebands (0.3 Hz) in Figure 9.8 is now only 0.0025 (0.25%), which is only 28% of the full-load slip.

The equivalent broken rotor bar factor ($BB_f$) from equation (4.5) referred to the full-load slip is now

$$BB_f = \{2 \times 145/(10^{(30/20)} + 16)\} \times (0.89/0.25) = 22$$

The value for $BB_f$ from the light load (slip = 0.25%) is now a factor of three times greater than the $BB_f$ at higher load but this cannot be the case at only 28% of the full-load slip. This quantitative result further confirms that the $\pm 2sf$ sidebands were not due to broken rotor bars. A false positive was prevented by changing the load.

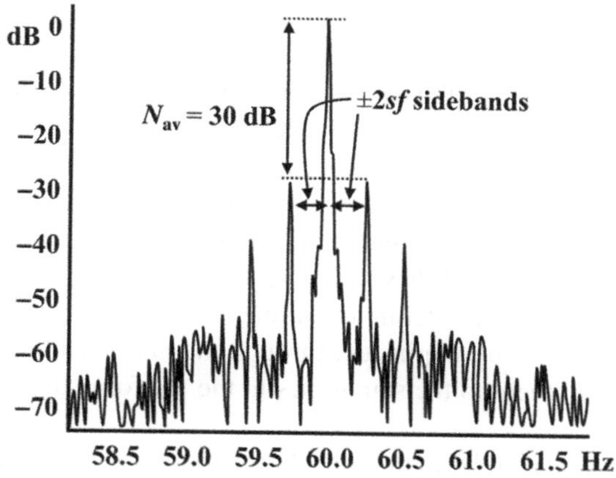


Figure 9.8 Current spectrum, 167 A, operating slip of 0.25%, 10.16 mHz/line.

## 9.3 TWO CASE HISTORIES (2005–2008)—NORMAL AND ABNORMAL PUMPING DYNAMICS IN TWO SCIM SEAWATER LIFT PUMP DRIVE TRAINS

### 9.3.1 Summary and Key Features

It was clearly proven in Chapters 7 and 8 that load fluctuations, gearbox dynamics, and mechanical phenomena from different loads can be reflected back into the rotor of a SCIM and consequential current components can be induced in the stator winding. During routine MCSA surveys to ascertain whether $\pm 2sf$ sidebands exist in the current spectrum from cage winding breaks, it was observed in this case history that abnormal current components existed in the current spectrum from one of the SWLP motors. These components could not be predicted by any formula and it was suspected that they were due to abnormal drive dynamics downstream of the motor. It was therefore impossible to determine the actual physical root cause of the abnormality by current analysis, but since comparisons between nominally identical drive trains could be made there was clearly a problem downstream of the motor, which was producing an abnormal current spectrum.

A subsequent inspection did reveal a faulty caisson (see Section 9.3.2 for full details) in one of the submersible SCIMs, driving a SWLP on the offshore oil production platform. This was a positive byproduct of MCSA but the authors emphasize that comparisons between spectra from nominally identical SCIM drive trains were required for the diagnosis of a problem downstream of this motor. The prediction was that an abnormality downstream of the motor existed but that was the only statement which could be given and therefore the interpretation is not truly diagnostic, although sufficient to ensure the required action was taken.

### 9.3.2 Description of a Seawater Lift Pump (SWLP) Drive Train

A caisson is a solid pipe that goes down into the sea and the opening is at the bottom end only. There are normally "*diver bars*" over the end of the caisson to stop any large objects being drawn into it. The SWLP unit is attached to numerous lengths of standpipe that are lowered down the caisson from the offshore oil production platform, and held by the *"lid"* of the caisson which has a flanged connection with the pipework coming through it, and with a penetration to feed the HV cable through to the HV submersible SCIM. The motor and pump-set are connected to the bottom of the pipework, which, in this case, was approximately 35 m long (115 ft). With this installation there is an anti-fouling unit, which is a cage made of two metals (zinc and copper), which have impressed current supplies to try to prevent marine growths around the seawater intake strainer. The flow of water should be from the bottom of the caisson, up the inside, through the anti-fouling unit and into the pump suction strainer, then up the standpipes to the platform and into the seawater header tank.

If there are any open holes in the side of the caisson, seawater bypasses the anti-fouling assembly, and due to the high erosion rate of seawater being drawn through the hole, the caisson could wear away and become a dropped object risk. If a caisson failed, it would either strike the leg structural members, which have a span around 17 m (56 ft), and are not designed to withstand the impact of a large object (like a caisson hitting them), and if it missed the structural members, it would land on the oil storage "gravity base tank" on the seafloor under this oil production platform. This is a storage facility that holds 500,000 barrels of oil and is not designed to withstand the impact of a large object from above. Consequently, the mechanical condition of the caisson has to be intact to prevent its failure resulting in a very serious safety hazard if it damaged either a platform leg or the oil storage facility.

### 9.3.3 MCSA Measurements and Analysis (2005)

The nameplate of the subsea SCIM was not accessible but the oil company provided the following information: 3-phase, 6600 V, 520 kW/700 HP, 63 A, 60 Hz, 1785 r/min, submersible SCIM. After a strip down, inspection, overhaul, and testing of an identical submersible motor the following additional information was available.

No-load current: 25 A (40% of $I_{FL}$)

No-load p.f.: 0.2 lag

Stator slots: 36

Rotor slots: 46

In these special submersible, HV SCIMs, the cage winding is of the deep bar copper construction and comprises six copper plates at each end of the rotor. The copper plates form an integral part of the rotor core package and the copper bars protrude through the copper plates which are then brazed to the plates to form a cage rotor. There is therefore no overhang between the ends of the core and the short circuiting end rings as is the case with a standard copper fabricated cage design. The construction of this unusual cage rotor is shown in Figure 9.9.

A routine MCSA survey of two of the SWLP motors was carried out in 2005 to assess the operational condition of the cage windings and the current spectra are shown in Figures 9.10 and 9.11 for motors A and B, respectively. The motors

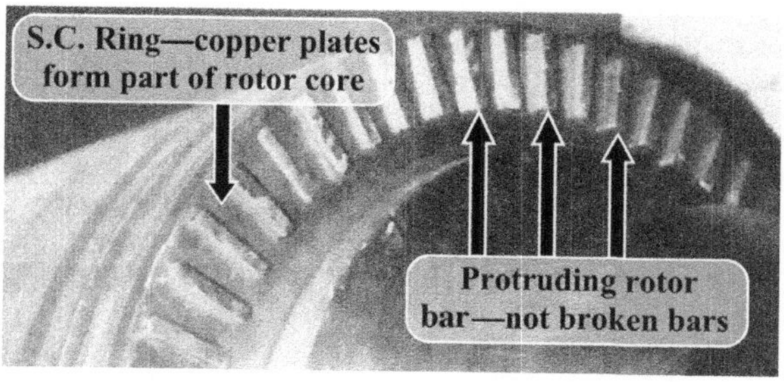


Figure 9.9 Special rotor cage construction for submersible SCIM.

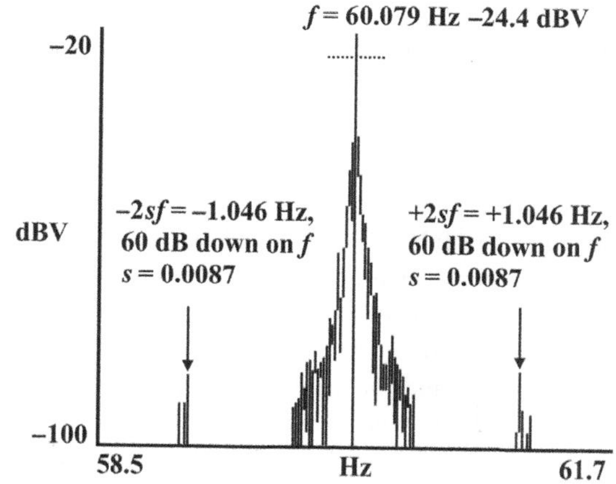


Figure 9.10 2005 motor A, 62 A, current zoom spectrum, 10.16 mHz/line.

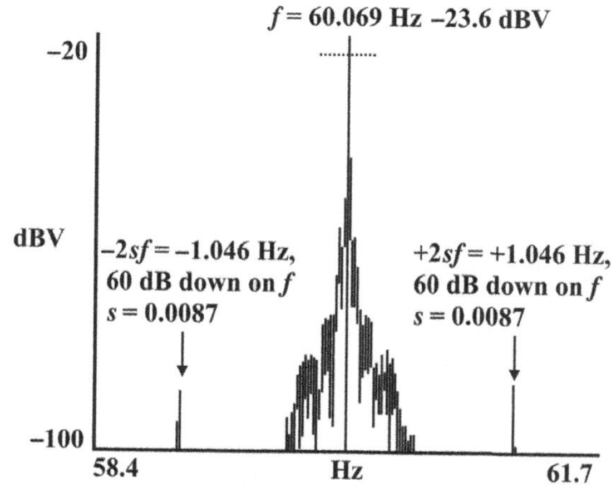


Figure 9.11 2005 motor B, 64 A, current zoom spectrum, 10.16 mHz/line.

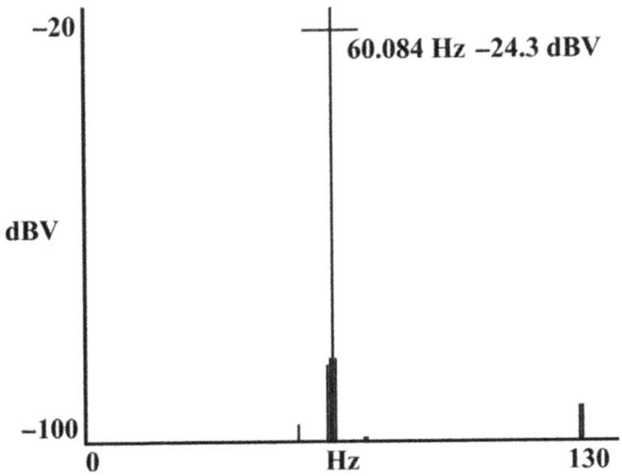


Figure 9.12 2005 motor A, 62 A, current spectrum, 10.16 mHz/line.

were operating at 62 A and 64 A and were considered to be on full-load with a nominal full-load rated speed of 1785 r/min, giving a full-load slip of 0.83% (0.0083) at the nameplate frequency of 60 Hz. The $\pm 2sf$ sidebands at full-load are nominally $\pm 1.0$ Hz around the supply component, $f$. The analogue, *in situ* ammeters indicated no fluctuations in the currents to the two drives.

The results in Figures 9.10 and 9.11 indicate that there are $\pm 2sf$ sidebands but they are 60 dB down on $f$ and the equivalent broken bar factor ($BB_f$) at full-load using equation (4.5) is 0.092 which means that the cage windings are normal. The current spectra in Figures 9.12 and 9.13 show that there are no unusual components from drive dynamics downstream of the submersible motors and this was expected because the motors and pumps were directly coupled and the currents were steady.

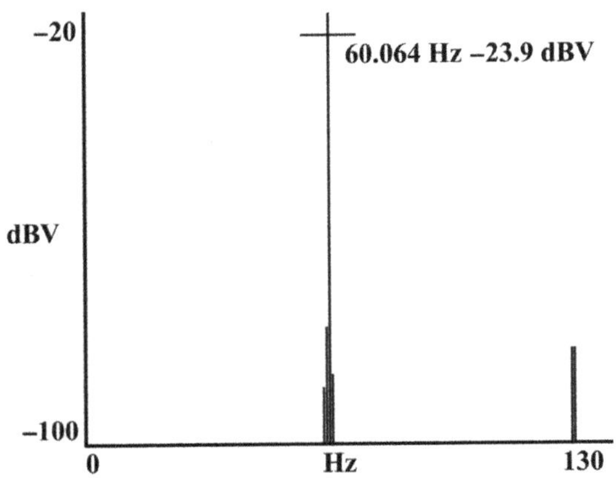


Figure 9.13 2005 motor B, 64 A, current spectrum, 10.16 mHz/line.

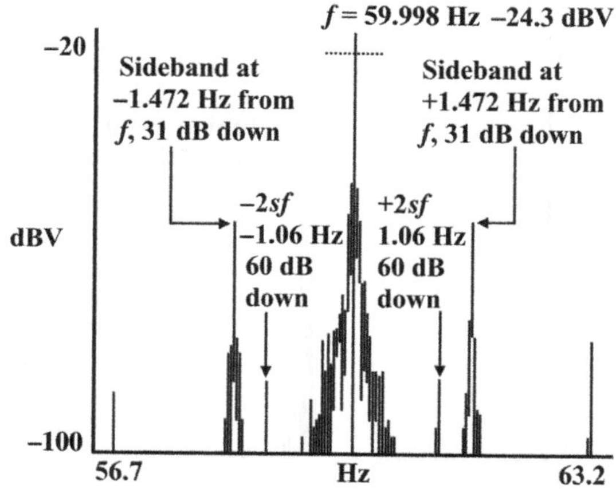


Figure 9.14 2008 motor A, 62–67 A, current zoom spectrum, 10.16 mHz/line.

### 9.3.4 MCSA Measurements and Analysis (2008)

The current spectra recorded in 2008 are shown in Figures 9.14 and 9.15. The analogue ammeter was fluctuating between 62 and 67 A for motor A and the current was steady at 67 A for motor B.

It is now obvious from Figure (9.14) compared to Figure 9.10 that there are additional sidebands at ±1.47 Hz around $f$ which are very prominent at 31 dB down on $f$ but the $\pm 2sf$ sidebands are still present at 60 dB down as they were in 2005 and the cage winding is normal. There has been no change in the spectrum as shown in Figure 9.15 (2008) compared to Figure 9.11 (2005) from the identical SWLP motor B since there are only sidebands at $2sf$ which are 61 dB down on $f$. The current spectra between 0 and 130 Hz from motors A and B are shown in Figures 9.16 and 9.17.

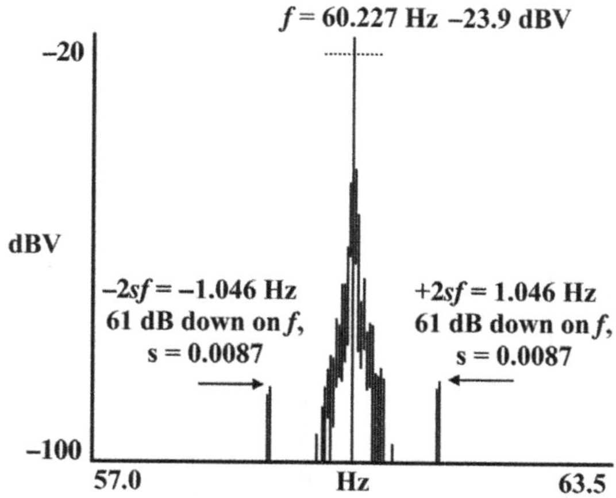


Figure 9.15 2008 motor B, 67 A, current zoom spectrum, 10.16 mHz/line.

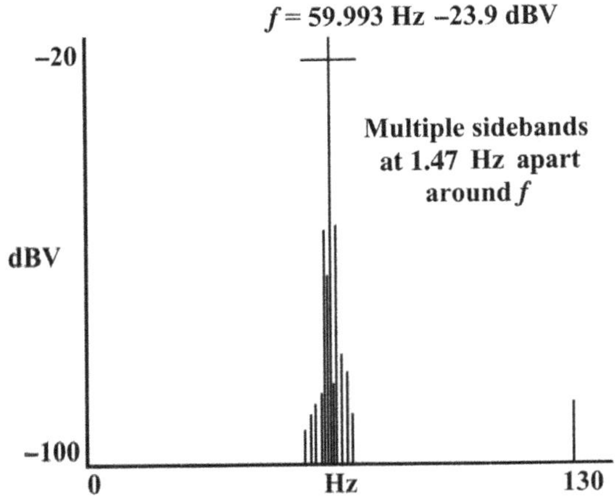


Figure 9.16 2008 motor A, 62–67 A, current spectrum, 10.16 mHz/line.

This shows that there are multiple sidebands spaced ±1.47 Hz apart from SWLP A whereas they do not exist from SWLP B. The fundamental root cause of these sidebands from SWLP A was not from the motor and must therefore be coming from abnormal pumping dynamics downstream of the motor since units A and B were in their normal operating mode. It is not possible to determine the root cause of these components due to abnormal pumping dynamics but it can be stated that they should not exist and there is a problem.

Figure 9.18 shows that a subsea inspection revealed a large banana shaped hole of approximately 650 mm (25.6 inches) × 80 mm (3.15 inches) in the solid caisson. The small holes are part of a cable tray within the caisson. The caisson is covered with crustaceans and other growths. Following the discovery of the holes, the oil company carried out a swage repair, which is a liner pushed down from above, and swaged

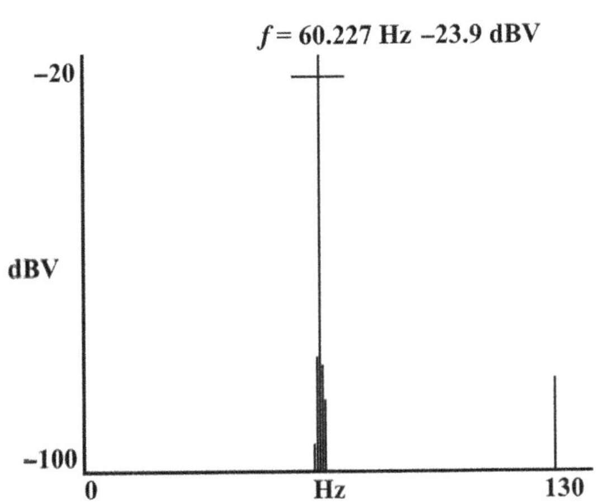


Figure 9.17 2008 motor B, 67 A, current zoom spectrum, 10.16 mHz/line.

Figure 9.18 Photograph of the damaged caisson. Reproduced with permission of EM Diagnostics Ltd.

outward with a special tool to form an interference fit with the existing caisson. The caissons are all made from carbon steel.

## 9.4 MCSA CASE HISTORY (2006–2007) – SLACK AND WORN BELT DRIVES IN TWO SCIM COOLING FAN DRIVES IN A CEMENT FACTORY

### 9.4.1 Summary and Key Features

Routine MCSA surveys of SCIMs in a cement factory, to assess the condition of cage windings, indicated numerous current components being induced due to the drive dynamics downstream of the motors, which were turning cooling fans via belt drives. It was subsequently established that the belts were slack, cracked, and worn and the current spectra from the motors demonstrate the differences between current spectra produced by normal belts and those from worn/slack belts.

### 9.4.2 MCSA Measurements and Analysis

The motor nameplate provided the following data: 3-phase, 3300 V, 132 kW/180 HP, 28 A, 50 Hz, 1489 r/min SCIM. Number of rotor bars provided by the OEM was 38. The full-load slip, $s_{FL} = 0.0073$ (0.73%), and therefore any sidebands at $\pm 2sf$ on full-load would equal $\pm 0.73$ Hz when the supply is 50 Hz.

The motors A and B were operating with input currents of 23 and 24 A, respectively, compared to the full-load current of 28 A. Both motors were operating at virtually the same load at a lower slip than the full-load slip, therefore any sidebands at $\pm 2sf$, will be less than $\pm 0.733$ Hz around $f$. The current spectra in Figures 9.19 and 9.20 recorded in November 2006 for motors A and B indicate that there are no $\pm 2sf$ sidebands around the supply component $f$ and the cage windings in both motors are

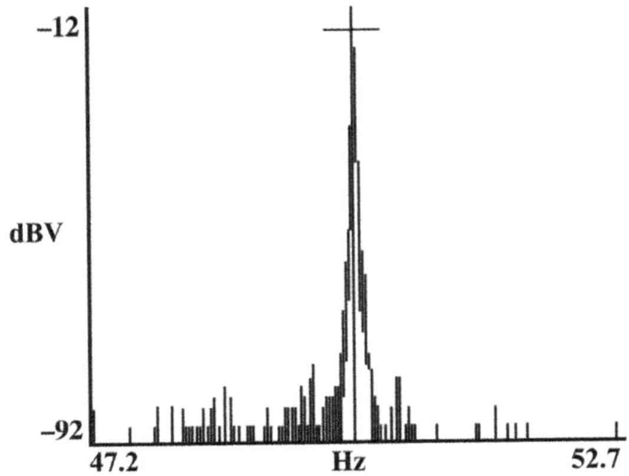


Figure 9.19 Motor A, current zoom spectrum, 10.16 mHz/line.

perfectly normal. It was established that there was a speed reduction of 2.56 from the motor to the fan via the different diameters of the driving and driven pulleys.

Predictions of current components around the supply component, $f$ due to drive dynamics at full-load:

$$\pm f_r = \pm 1489/60 \text{ Hz} = \pm 24.8 \text{ Hz}$$
$$\pm f_n = \pm 24.82/2.56 \text{ Hz} = \pm 9.7 \text{ Hz}$$

where

$f_r$ = rotor speed frequency of the motor at full-load, Hertz

$f_n$ = speed frequency of the driven fan at full-load, Hertz

The current spectra shown in Figures 9.21 and 9.22 indicate that there are numerous components, which are in fact due to the mechanical dynamics from the belts and pulleys being reflected back into the rotor in the SCIMs. There is a distinct

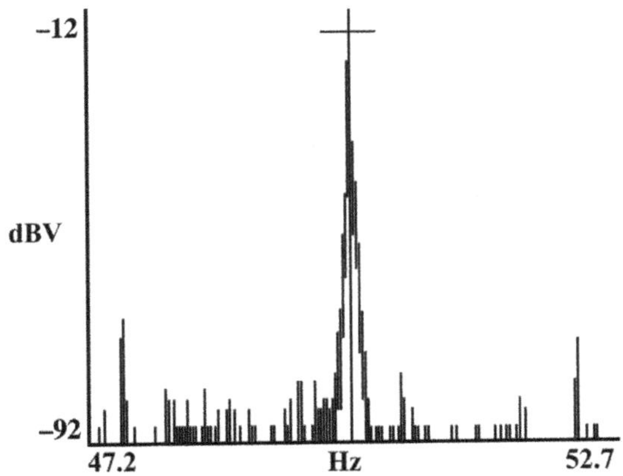


Figure 9.20 Motor B, current zoom spectrum,10.16 mHz/line.

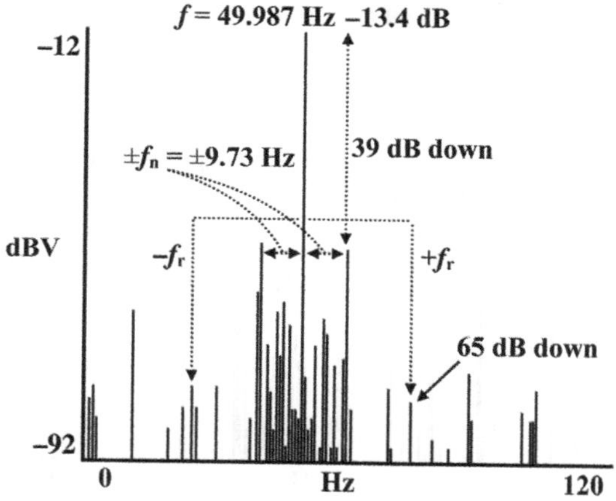


Figure 9.21 Motor drive A, current spectrum, November 2006, 10.16 mHz/line.

difference between the two spectra in that, from drive train A, the magnitude of sidebands at $\pm f_n = \pm 9.73$ Hz around $f$ are 39 dB down on the supply component ($f$) but for drive train B the components at the same frequencies are 52 dB down on $f$, which is 13 dB lower in the latter and equates to a factor 4.46 times lower in absolute units of amperes. In addition, the component at $+f_r$ around $f$ in drive train B is 77 dB down which is four times lower than in drive train A.

Since the frequency components at $\pm f_n$ equate to the speed frequency of the fans it was suspected that the belts in drive train A were worn/cracked and probably loose. An inspection of drive train A, while it was running, indicated that a loud acoustic noise could be heard emanating from the belts in drive train A, a new set were fitted at a later date, when the motor could be shut down. The old belts were slipping and were badly cracked and worn. The current spectrum with a set of new belts fitted to drive train A was recorded in March, 2007, as shown in Figure 9.23 and the components at

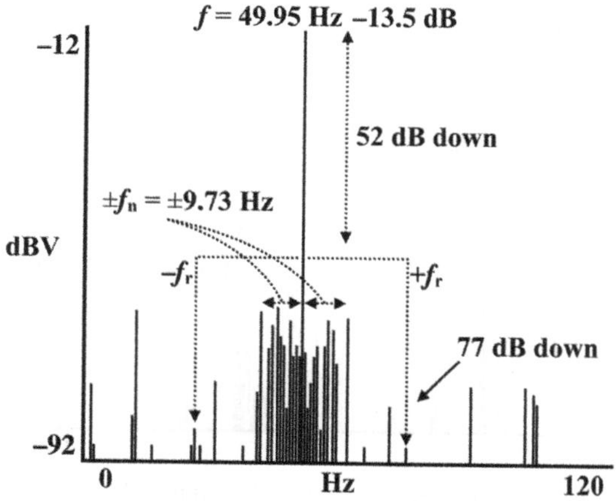


Figure 9.22 Motor drive B, current spectrum November 2006, 10.16 mHz/line.

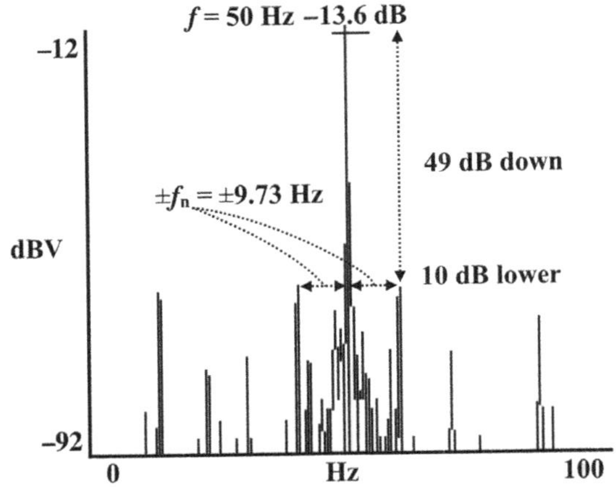


Figure 9.23 Motor drive A, current spectrum, retest in March 2007 after new belts fitted, 10.16 mHz/line.

$\pm f_n$ around $f$ have dropped by 10 dB (factor of 3.16 times smaller) from 39 to 49 dB down on $f$ and the loud noise disappeared. In March 2007, drive train B was MCSA tested again with the same belts as in November 2006 and Figure 9.24 shows that the $\pm f_n$ components have increased by 6 dB (double in absolute units of amperes), the belts were inspected and they were slipping and there was evidence of cracking and abnormal wear.

This case history shows that a mechanical problem such as worn belts and/or slipping belts can be diagnosed via MCSA but comparisons between spectra from identical drive trains are required and also subsequent trending of the spectra, thus this approach is not truly diagnostic as is the case with detecting $\pm 2sf$ sidebands due to broken rotor bars.

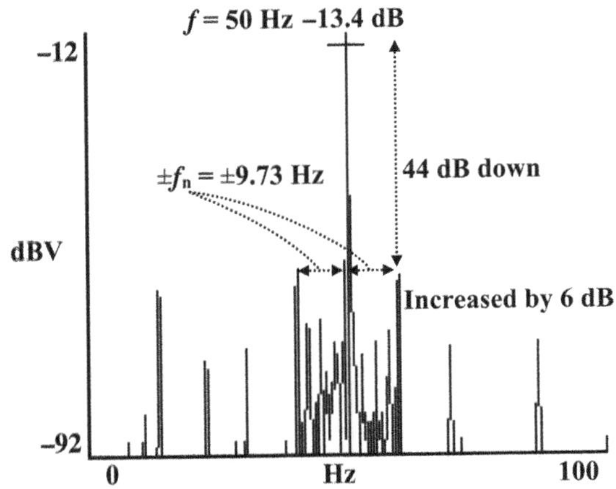


Figure 9.24 Motor drive B, current spectrum, retest in March 2007, same belts as in November 2006, 10.16 mHz/line.

## 9.5 APPLICATION OF MCSA TO INVERTER-FED LV AND HV SCIMs

### 9.5.1 Summary and Key Features

It is generally accepted that SCIMs fed by inverters do not normally experience cage winding breaks because soft starts are applied, thus reducing (as fully discussed in Chapter 3) the mechanical and electromagnetic forces and thermal stresses on the cage winding in comparison to the very high stresses caused by DOL starts. Therefore, there is not a definitive chapter or separate section or focus in this book concerned with applying MCSA to inverter-fed SCIMs.

The authors have had no MCSA industrial case histories of detecting broken rotor bars in inverter-fed SCIMs but that is due to the limited number of HV SCIMs (fed via inverters), which they have MCSA tested but it does not mean that cage winding breaks cannot occur in inverter-fed cage rotors. If, for example, the rotor cage winding is overheated due to, for example, a stall or operating at above full-load for long periods of time and with certain designs of overhangs between the place where the bars leave the rotor core and the joints between the bars to end rings, then it is possible that broken bars could occur in an inverter-fed SCIM. For completeness, two industrial case histories will be presented to demonstrate that MCSA can be applied to LV and HV motors to assess the condition of the cage windings. MCSA results were recorded by Thomson in 1987 and only partially reported in Thomson and Stewart [9.4] in 1988. Additional results are now presented in this section to demonstrate that MCSA can detect broken rotor bars in an inverter-fed induction motor, albeit in a small laboratory-based PWM inverter drive (415 V, 11 kW SCIM).

For the avoidance of doubt, the types and operation of the inverters in the following case histories will not be discussed since that is not the function of this book. Also, information on the inverter types and designs was not provided by the end users during MCSA testing.

### 9.5.2 MCSA Results from Laboratory-Based PWM Inverter-Fed SCIM (1987)

The nameplate data for the SCIM under test was as follows: 3-phase, delta connected, 415 V, 11 kW (14.7 HP), 20.5 A, 50 Hz, 1470 r/min. MCSA measurements were carried out with two aluminum die-cast cage rotors with 28 bars (deep bar design), one with a healthy cage winding and the other with three broken bars via holes drilled at the joints between three bars and an end ring. The tests were carried out in 1987 in the R&D electrical machines laboratories at The Robert Gordon University, Aberdeen, Scotland. The motor was supplied via a PWM inverter.

Figure 9.25 shows the time domain current waveform from the inverter when set to a fundamental base frequency of 50 Hz and there are clearly high frequency components. The current spectrum shown in Figure 9.26 from the healthy cage winding and the motor running on full-load (20.5 A) at a full-load slip of 0.02 (2%) verifies that there are no $\pm 2sf$ sidebands around the supply component, $f$. An identical rotor

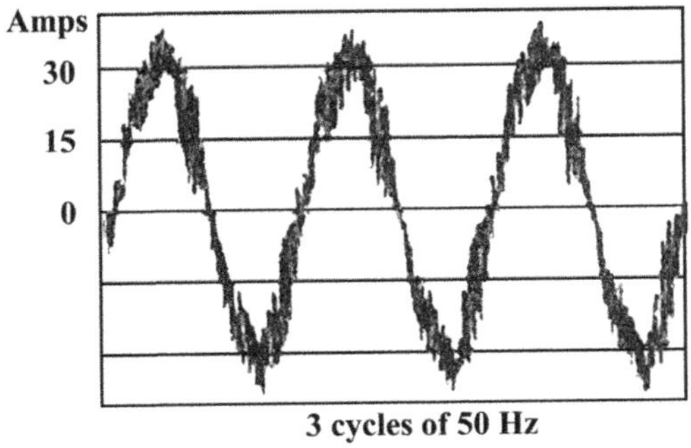


Figure 9.25 Time domain current waveform.

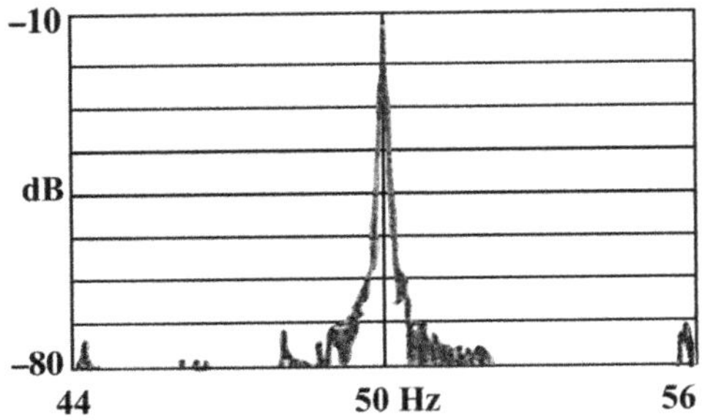


Figure 9.26 Current zoom spectrum, 10.16 mHz/line.

but with three broken rotor bars was then MCSA tested at full-load slip and Figure 9.27 shows the current spectrum and there are $\pm 2sf$ sidebands around $f$ at 38 dB down.

With three broken bars, the fundamental base frequency from the inverter was then dropped to 40 Hz and the motor was operated on a reduced load so that the $\pm 2sf$ sidebands were at $\pm 1.0$ Hz around $f$ as shown in Figure 9.28. Based on the dB difference chart (extract from Table 4.2 is given in Table 9.1), it is clear that broken

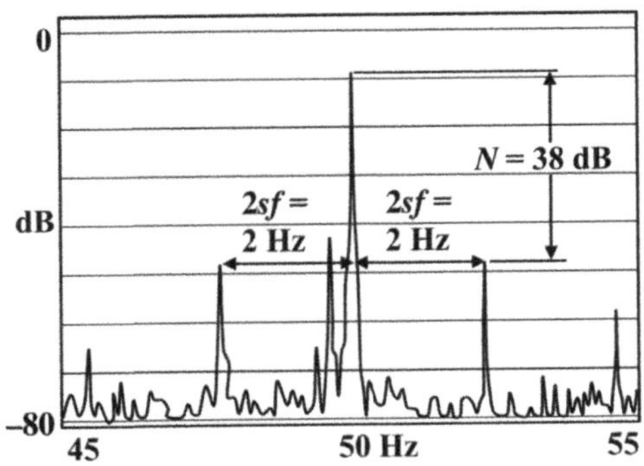


Figure 9.27 Current zoom spectrum, 10.16 mHz/line.

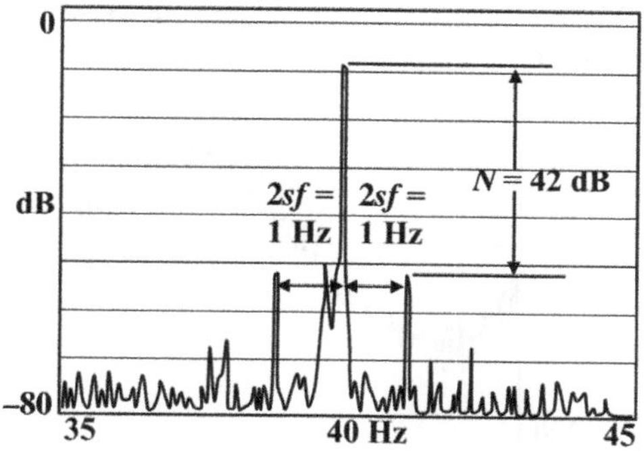


Figure 9.28 Current zoom spectrum, 10.16 mHz/line.

rotor bars can be detected via MCSA when applied to inverter-fed SCIMs operating at different loads and with changes in the fundamental base frequency from the inverter.

### 9.5.3 Case History (2005)—Measurements and Analysis from an LV Inverter-Fed SCIM With No Cage Winding Breaks

The motor nameplate provided the following data: 3-phase, 440 V, 450 kW/600 HP, 710 A, 60 Hz, 1193 r/min SCIM. The full-load slip, $s_{FL} = 0.0058(0.58\%)$ and therefore any sidebands at $\pm 2sf$ on full-load would equal ±0.7 Hz at 60 Hz. The rotor was aluminum die-cast with 74 bars and that is the only information the OEM would provide to the end user. This motor was driving an anti-scale pump on an offshore oil and gas production platform.

The fundamental base frequency from the inverter was 23.4 Hz at the time of the MCSA test and the time domain current waveform (recall the MCSA CT gives an output in volts) is shown in Figure 9.29; it is clear there are high frequency components in the waveform. The current spectrum shown in Figure 9.30 confirms that

**TABLE 9.1 Extract from table 4.2. Broken Rotor Bar Severity *"ESTIMATOR"* Based on the Average dB Difference ($N_{av}$) Between the $\pm 2sf$ Sidebands and the Supply Component**

| $N_{av}$ = dB Difference | **Estimated Condition of Cage Winding** |
|---|---|
| 35–40 dB | A broken rotor bar problem exists. |
| | When $N_{av} = 35$ dB in a 2-pole motor having typically 38, 46, or 50 bars (actual designs of HV, high power SCIMs) then the probability is high that several bars are broken. |
| | When $N_{av} = 40$ dB in a 4-pole motor normally one or two bars are broken. |
| | When $N_{av} = 35$ dB with for example a 36-pole (240 slot) motor a serious broken bar problem with multiple broken bars, would exist. |
| | The actual measured $N_{av}$ values were 38 dB and 42 dB for the experimental SCIM fed from an inverter having three broken bars operating at the nominal full-load slip (supplied at a fundamental of 50 Hz) and a slip of 50% (supplied at a fundamental of 40 Hz) of the nameplate full-load slip, respectively. |

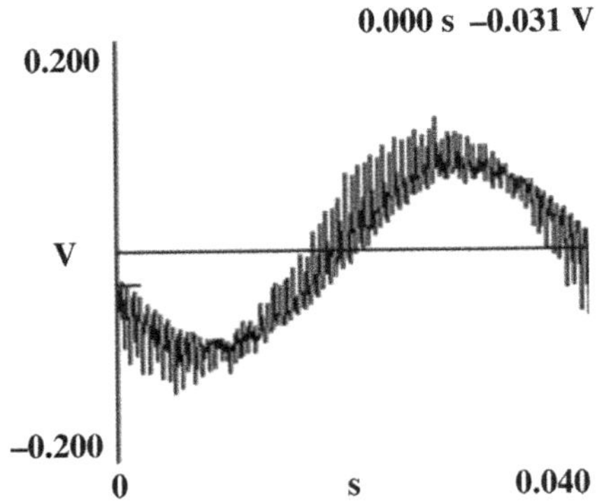


Figure 9.29 Time domain current waveform.

there are no $\pm 2sf$ sidebands around the fundamental supply component and the cage winding is perfectly healthy.

### 9.5.4 Case History (2008)—Measurements and Analysis from a Large, 6300 kW/8445 HP Inverter-Fed SCIM With No Cage Winding Breaks

The motor nameplate provided the following data: 3-phase, 6600 V, 6300 kW/ 8450 HP, 633 A, 50 Hz, 1493 r/min SCIM. The number of rotor bars as provided by the OEM was 82. The full-load slip, $s_{FL} = 0.0047$ (0.47%) and therefore any side-bands at $\pm 2sf$ on full-load would equal ±0.47 Hz on a 50 Hz mains supply.

The cage winding was copper fabricated with rectangular bars inserted into annular grooves in each end ring and the joints were puddle brazed. The rotor cage had 82 bars. This motor was driving a compressor at an LNG onshore gas processing plant. The only information provided on the inverter was that it was a "*perfect harmony design*" which is a proprietary name by the OEM of the inverter.

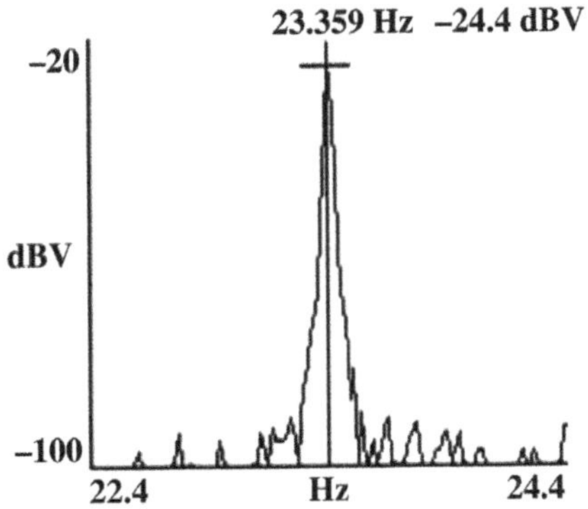


Figure 9.30 Current zoom spectrum, baseband span 0–50 Hz, 12,800 lines, 3.9 mHz/line.

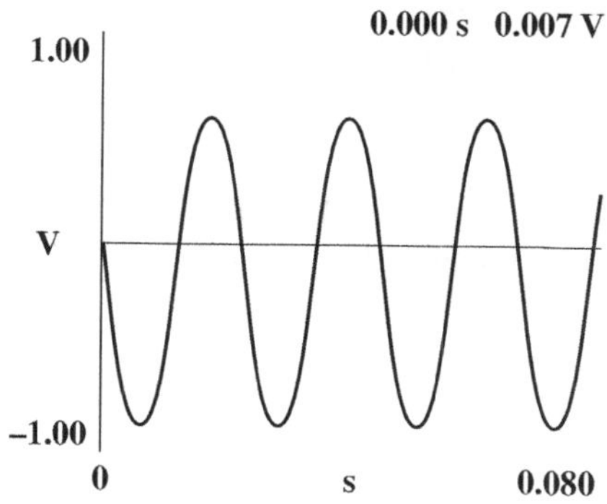


Figure 9.31 Time domain current.

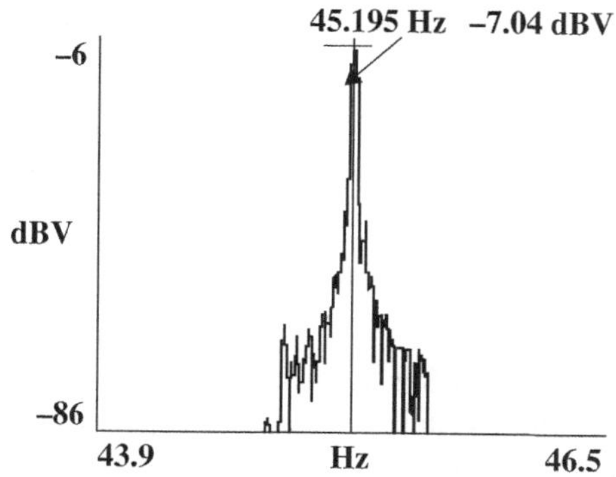


Figure 9.32 Current zoom spectrum, baseband span 0–50 Hz, 12,800 lines, 3.9 mHz/line.

The fundamental base frequency from the inverter was nominally 45 Hz and the operating current was 516 A at the time of the MCSA test. The current to the motor was directly measured via a clip-on CT (1000:1 with a shunt giving 0.1 V/amp rms output) over one of the output supply cables feeding the motor. The measured current to the motor was not filtered by any processing within the inverter; note that the pk-pk volts in Figure 9.31 is 0.73 V (0.516 V rms output from the CT which gives 516 A). The time domain current waveform is shown in Figure 9.31 and the current spectrum shown in Figure 9.32 confirms there are no $\pm 2sf$ sidebands around the fundamental supply component and the cage winding is perfectly healthy.

## 9.6 CASE HISTORY (1990)—ASSESSMENT OF THE MECHANICAL OPERATIONAL CONDITION OF AN ELECTRICAL SUBMERSIBLE PUMP (ESP) DRIVEN BY A SCIM USED IN ARTIFICIAL OIL LIFT

Special ESPs driven by 3-phase SCIMs are used world wide to provide artificial lift in oil wells. Initial R&D by Thomson et al. [9.5] in 1989–1991 verified that MCSA

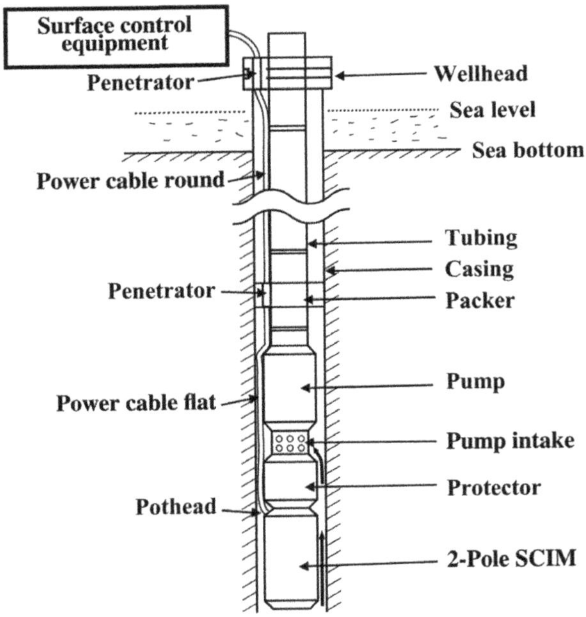


Figure 9.33 General layout of an ESP installation.

could be applied to inverter-fed ESPs via trials in an onshore test well at an OEM's factory in Scotland. In offshore installations they can operate at depths up to 12,000 ft (≅ 3500 m) at temperatures and pressures of 300°F (150°C) and 4000 psi (27 Mpa), respectively. This can result in very short run lives. For example, after 2–3 years in deep and hostile subsea oil wells, ESPs are often pulled for a full inspection and refurbishment and complete failures can also occur within that period. A typical layout of an ESP is shown in Figure 9.33.

If a problem is identified before a catastrophic failure, an outage for that drive train can be planned for the removal of the entire unit and pumping uptake can be redirected to other pumps to avoid a sudden failure of the faulty one (e.g., operate it on reduced load). Vibration monitoring can be difficult to implement due to the depths, temperatures, and pressures at which the ESPs may operate in, for example, oil and gas offshore installations. There is the problem of selecting the optimum position to mount a vibration sensor to obtain meaningful vibration data to diagnose mechanical faults because the mechanical stiffness is a variable between the source of the vibration problem and different mounting positions of an accelerometer. Therefore, the measured vibration is a function of accelerometer position and this is a problem in ESPs, which is evident in the layout of an ESP as is shown in Figure 9.33. MCSA is an alternative or better still, in addition to vibration since the MCSA measurements can be simply taken top-side.

### 9.6.1 MCSA Result

This is a short case history since there are no predictions of current components due to the complex design of an ESP. The induction motors used in ESPs on offshore

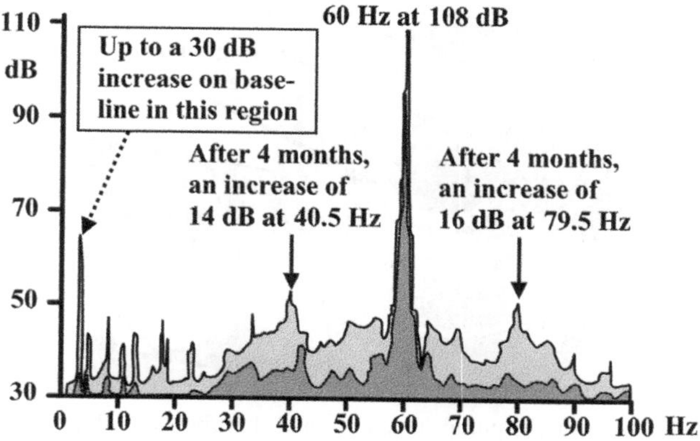


Figure 9.34 Current spectra recorded 4 months apart, baseband span 0–100 Hz, 12,800 line, 7.8 mHz/line.

installations are not conventional since there is a series of cage rotors mounted on a common shaft and each rotor is supported by roller element bearings, hence there are multiple sets of bearings and there are also the main thrust bearings.

This was an offshore field trial with the following objective:

(i) To ascertain whether a purely mechanical problem such as pump wear can cause a clear change in the current spectrum and that MCSA is a viable CM method for ESPs.

Two 3-phase, 4160 V, 2-pole, (2 × 210 HP/2 × 157 kW), 60 Hz SCIMs operating in tandem, were used to drive an ESP in a deep well on a North Sea offshore oil platform, off the coast of Scotland. The ESPs were operating at a depth of 7500 ft (2280 m) in a well deviation of 45°.

Figure 9.34 shows two current spectra taken at different times (4-month span) and the magnitude of the 60 Hz component did not change, since the motor was operating on full-load in both cases. The difference between the two current spectra is significant, for example, there are two distinct components at ±9.5 Hz around the supply component and compared to 4 months earlier they have increased by up to 16 dB (a factor of 6.32 times greater) and at lower frequencies by up to 30 dB (32 times increase) with respect to the baseline spectrum.

The causes of increases at these frequencies are completely unknown but they are certainly not caused by cage winding breaks in the multiple and independent cage windings in this ESP. The operator reported that the pumping efficiency had dropped, which provided supporting evidence that the pump's performance had changed. The offshore operator pulled (this is a major and very costly operation) the ESP based on the drop in efficiency and it has to be stated it was not removed because of the MCSA result since the latter was only an MCSA trial.

The inspection revealed that there was radial wear on the pump stage hubs and also shaft/bushing wear and Figure 9.35 illustrates that observation.

It can be concluded that this field trial indicated there was potential for MCSA to be used for monitoring the health of ESPs. However, it is emphasized that trending of the current spectrum is a necessity to indicate that there is a problem (exact cause

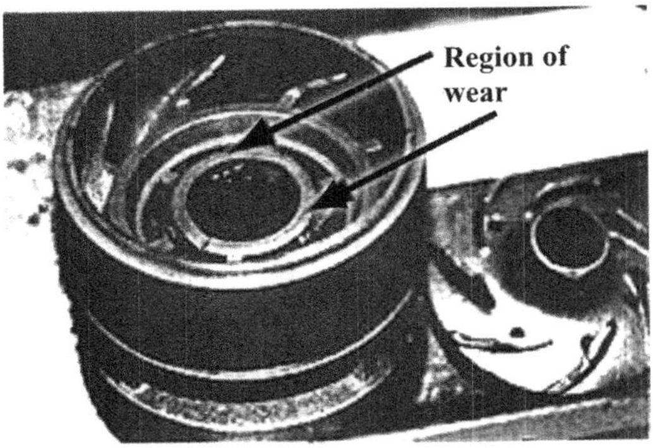


Figure 9.35 Radial wear on the pump stage hubs and shaft/bushing. Reproduced with permission of EM Diagnostics Ltd.

unknown) with the pump, therefore it is not a truly diagnostic analysis to detect the fundamental root cause of the changes in the current spectrum. Note that this case history was in 1990.

In a relatively recent publication (2012), Harihara and Parlos [9.6] have developed a fault detection indicator for "Fault Diagnosis of Centrifugal Pumps using Motor Electrical Signals" which requires three voltages and three currents to be monitored. This publication does include an industrial case history (a field trial) of a boiler feedwater pump driven by a 298 kW/400 HP induction motor (no other details on the motor were given), which indicated that it was feasible to determine a mechanical problem in the motor which was subsequently found to be a bearing fault. This is indeed an encouraging development and the reader is referred to Reference 9.6 for further information. Further publications by Harihara and Parlos [9.7, 9.8] should also be followed up, since they cover the detection of cavitation and impellor cracks in centrifugal pumps by sensing and analyzing three currents and three voltages from the SCIMs driving the pumps.

## 9.7 QUESTIONS

**9.7.1** When a 3-phase SCIM is supplied at constant voltage and frequency, why is the main flux/pole constant?

**9.7.2** If the number of poles ($2p$) equals the number of spider support arms ($S_p$) on the rotor of a SCIM, explain why the magnitude of the supply component is modulated at $2sf$.

**9.7.3** The current displayed on an *in situ* analogue ammeter from a SCIM operating on light load with a rotor having $2p = S_p$ is slowly oscillating at $50 \pm 5$ A; explain to the maintenance manager of the plant, who is a mechanical engineer, why this the case.

**9.7.4** Why do some OEMs manufacturer SCIMs with 4-poles and above with the number of poles equal to the number of spider support arms on the rotor shaft?

**9.7.5** Why are large, 2-pole, HV SCIMs never designed with the number of spider support arms equal to the number of poles on the rotor shaft?

**9.7.6** Although large, HV SCIMs having the number of poles equal to the number of spider support arms do not have a fault because of this particular rotor shaft design, why is this design undesirable?

**9.7.7** Why is the percentage magnitude of the $\pm 2sf$ current components relative to the supply component higher on light load compared to full-load current when $2p = S_p$?

**9.7.8** The owner of a power station has issued a purchase order for your company to carry out MCSA tests on a large, HV, 6.6 kV, 6-pole, 60 Hz SCIM to assess the operational condition of the cage winding. This motor has a rotor shaft design with $2p = S_p$. The motor operates at a constant load which cannot be changed at the time of the MCSA test. The client has knowledge and experience of using MCSA to detect the classical $\pm 2sf$ sidebands around the supply component with other SCIMs, which do not have $2p = S_p$.

Explain to the client that it is not possible to detect broken rotor bars in a cage rotor of a SCIM by monitoring the $\pm 2sf$ sidebands around the supply component with $2p = S_p$ when the load cannot be changed.

However, then explain that it is possible to monitor certain space harmonics, which are only a function of broken bars in SCIMs with $2p = S_p$ and thus convince the client that the MCSA test should proceed. Reference should be made to relatively recent publications [9.2, 9.3].

**9.7.9** Discuss the application of MCSA to detect broken rotor bars in inverter-fed induction motors with respect to its reliability of diagnosis and comment on whether you consider that it is justifiable for MCSA to be applied on an annual survey basis if the first MCSA test on a soft-start, inverter-fed SCIM verified that the cage winding was normal.

**9.7.10** If a fixed number of broken bars exist in an inverter fed SCIM, will the magnitude of the $\pm 2sf$ sidebands be constant at a given load when the frequency is changed.

## REFERENCES


[9.1] W. T. Thomson, "On-Line Current Monitoring—The Influence of Mechanical Loads/Unique Rotor Designs on the Detection of Broken Rotor Bars in Squirrel Cage Induction Motors," ICEM '92, University of Manchester Science and Technology (UMIST), UK, September 1992.

[9.2] C. Yang, T.-J. Kang, S. B. Lee, J.-Y. Yoo, A. Bellini, L. Zarri, and F. Fiilippetti, "Screening of false induction motor fault alarms produced by axial air ducts based on the space-harmonic-induced current components," *IEEE Transactions on Industrial Electronics*, vol. 62, no. 3, March 2015, pp. 1803–1813.

[9.3] S. Lee, J. Hong, S. B. Lee, E. J. Wiedenburg, M. Teska, and H. Kim, "Evaluation of the influence of rotor axial air duct on condition monitoring of induction motors," *IEEE Transactions on Industrial Applications*, vol. 49, no. 5, September/October 2013, pp. 2024–2032.

[9.4] W. T. Thomson and I. D. Stewart, "On-Line Current Monitoring for Fault Diagnosis in Inverter Fed Induction Motors," Proceedings of IEE 3rd International Conference in Power Electronics and Variable Speed Drives, Savoy Place, London, July 1988.

[9.5] W. T. Thomson, I. D. Stewart, and J. May, "On-Line Current Monitoring to Detect Mechanical Faults in Electrical Submersible Pumps," Proceedings of International Conference in Electrical Machines ICEM '90, MIT, Boston, MA, August 1990.

[9.6] P. P. Harihara and A. G. Parlos "Fault Diagnosis of Centrifugal Pumps using Motor Electrical Signals," in *Centrifugal Pumps*, edited by D. Papantonis, Intech, February 2012, ISBN: 978 953-51-0051-5.

[9.7] P. P. Harihara and A. G. Parlos, "Sensorless Detection of Impellor Cracks in Motor Driven Centrifugal Pumps," Proceedings of ASME International Mechanical Engineering Congress and Exposition, Boston, MA, USA, October 31–November 6, 2008, pp. 17–23, ISBN: 9780791848661.

[9.8] P. P. Harihara and A. G. Parlos, "Sensorless Detection and Isolation of Faults in Motor-Pump Systems," Proceedings of ASME International Mechanical Engineering Congress and Exposition, Boston, MA, October 31–November 6, 2008, pp. 43–50, ISBN: 9780791848661.

CHAPTER 10

# MCSA TO ESTIMATE THE OPERATIONAL AIRGAP ECCENTRICITY IN SQUIRREL CAGE INDUCTION MOTORS

*William T. Thomson*

## 10.0 SUMMARY AND INTRODUCTION

The main objective of this chapter is to provide relevant knowledge for the industrial application of MCSA to estimate the operational airgap eccentricity level in SCIMs. An explanation of airgap eccentricity is given at a level that is intended to be suitable for electrical, mechanical, instrumentation, and maintenance engineers who have to apply MCSA in industry. The different types of airgap eccentricity, namely static and dynamic, are defined and their causes discussed as is unbalanced magnetic pull (UMP) between the rotor and stator, because it is a byproduct of airgap eccentricity.

The fundamental concepts relating to the flux waves produced by rotor slotting, which are a function of static airgap eccentricity and the rotating flux waves due to dynamic airgap eccentricity are presented. These lead to the equation for predicting the current signature pattern, which in turn is a function of the combination of static and dynamic eccentricity, namely the total air gap eccentricity. The mathematical derivations are not presented, since they have already been published by Cameron et al. in 1986 [10.1] and this book's central theme is on the application of MCSA as a reference text for industry. However, the industrial case histories should be of interest to academia, since in some cases they open up new possibilities for further research, particularly with respect to large HV SCIMs.

A presentation is given on the signal processing strategy used to identify a truly unique current signature pattern, which is a function of airgap eccentricity in SCIMs. An explanation is given on the interpretation of the current signature required to estimate the operational airgap eccentricity (in percentage bands) with respect to the nominal radial airgap and two industrial case histories are presented in this chapter.

*Current Signature Analysis for Condition Monitoring of Cage Induction Motors: Industrial Application and Case Histories*, First Edition. William T. Thomson and Ian Culbert.

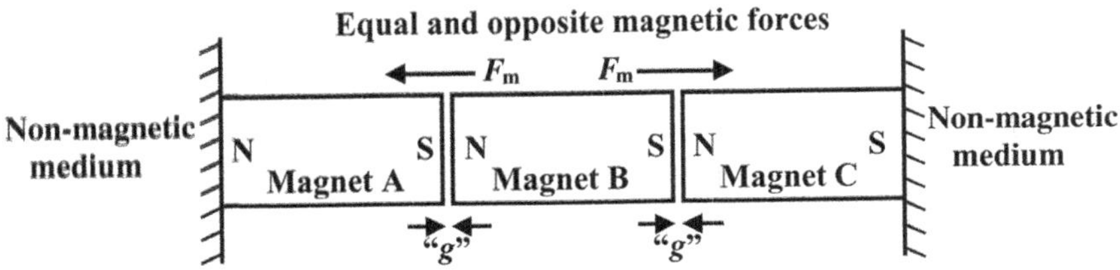


Figure 10.1 Idealized illustration of uniform airgaps "$g$" between magnets.

A further 15 case histories, which assess the operational airgap eccentricity in a diverse range of HV SCIMs, are presented in Chapter 11 and these include successful and unsuccessful cases. Only the key references are cited in this chapter but numerous supplementary references are also listed.

## 10.1 DEFINITION OF AIRGAP ECCENTRICITY

Figure 10.1 gives an illustration of three permanent magnets that have exactly the same magnetic field strength and each magnet has exactly the same cross-sectional area and length and they are positioned exactly opposite to and aligned with each other. If the airgaps "$g$" between each magnet are identical then the magnetic pull between the magnets, A and B and B and C due to the N–S poles are identical but in opposite directions, therefore the resultant pull ($F_m$) on magnet B is zero. This is of course an idealized illustration and not possible to produce in practice since there will always be some tolerance variations in dimensions between manuactured components.

Now consider an induction motor with north and south poles created by the magnetic field produced by the ampere-turns (mmf) from the stator winding as illustrated in Figure 10.2.

In Figure 10.2 it is assumed that the stator bore and circumference of the rotor core are perfectly round and that the center of rotation of the rotor coincides exactly with the center of the stator bore and of the rotor itself. Thus the radial airgap between the stator bore and rotor circumference will be constant both statically and dynamically. The resultant force on the rotor between the N–S poles is zero. Therefore, the airgap eccentricity and what is referred to as the UMP is zero, as was the case in Figure 10.1.

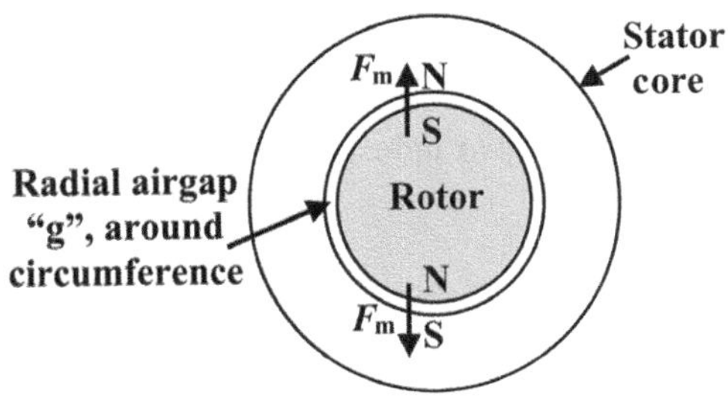


Figure 10.2 Idealized case of constant radial airgap "$g$".

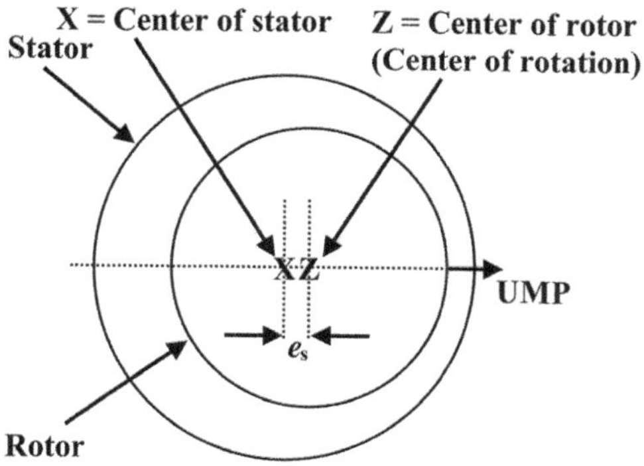


Figure 10.3 Schematic illustration of static airgap eccentricity only.

**Definition of Static Airgap Eccentricity:** This occurs when the central axes of rotation and of the rotor co-align through (Z) but the stator bore central axis through (X) is displaced from that co-alignment. In this case horizontally as shown in Figure 10.3 by $e_s$ the static eccentricity.

**Definition of Dynamic Airgap Eccentricity:** This is when the central axes of rotation and stator bore co-align through (X) but that of the rotor through (Z) is displaced and the UMP rotates as shown in Figure 10.4.

Combination of static and dynamic eccentricity (10.1)

$$g_{(\theta,t)} = g(1 - e_s \cos(\theta) - e_d \cos(\omega t - \theta)) \tag{10.1}$$

where

$g_{(\theta,t)}$ = radial airgap length as a function of static and dynamic airgap eccentricity, millimeter or mils

$\theta$ = angular position in degrees around the circumference

$t$ = time in seconds

$g$ = nominal design value for the radial airgap length, mm/mils

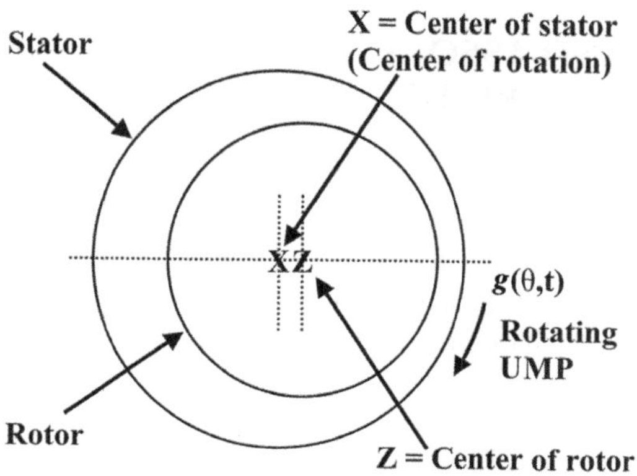


Figure 10.4 Schematic illustration of dynamic airgap eccentricity only.

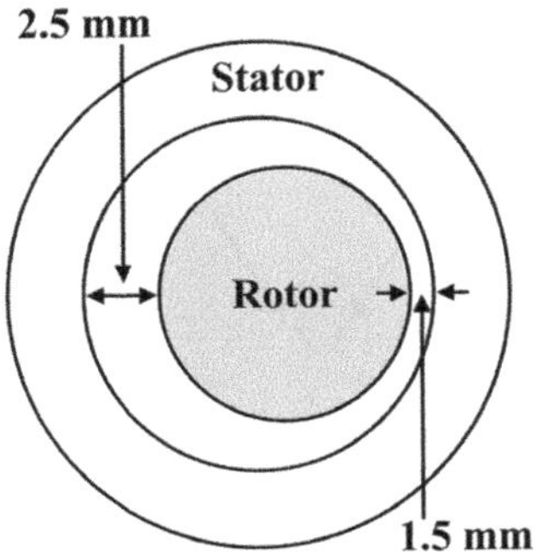


Figure 10.5 Illustration of only static airgap eccentricity.

$e_s$ = level of static eccentricity, mm/mils

$e_d$ = level of dynamic eccentricity, mm/mils

$\omega$ = angular speed in rad/sec

Equation (10.1) describes the combination of static and dynamic airgap eccentricity, which is what exists in practice. For purely static eccentricity, $e_d = 0$, and equation (10.1) becomes

$$g_{(\theta)} = g(1 - e_s \cos(\theta))$$

An illustration of only static airgap eccentricity is shown in Figure 10.5 where the stator bore and rotor are nominally perfectly round but the stator assembly has been moved horizontally to the left.

In industry, it is normal to state the airgap eccentricity as a percentage of the nominal radial airgap, and in this illustration only static eccentricity exists.

$e_s = 0.5/2.0 = 0.25$ or 25% of the nominal airgap of 2 mm (80 mils or thou).

Therefore, the radial airgap is defined as $g = 2 \pm 0.5$ mm or $g = 2$ mm($\pm$25%) or $g = 80$ mils($\pm$25%).

## 10.2 CAUSES AND ASSOCIATED TYPES OF AIRGAP ECCENTRICITY

In practice, it is not feasible to manufacture SCIMs with zero airgap eccentricity because each component part is produced to be within a specific tolerance specification. The OEMs' goal is to achieve concentricity in both the stator bore and outside diameter of the rotor core and to ensure that the center of the rotor and its center of rotation coincide with the center of the stator bore. To achieve this also requires the centers of the end frames (or end bells) and bearing housings to be the center of rotation. This is indeed a challenging specification but the OEMs take great care to achieve as near perfection as is practically possible via stringent quality control and quality assurance procedures.

**(a)** Typical Causes of Static Airgap Eccentricity include:

  **(i)** The bore of the stator core does not have a constant diameter around its circumference and along its axial length and is not satisfying the concentricity tolerance specification set by the OEM.

  **(ii)** The rotor core outside diameter and stator bore diameters are constant and concentric to each other and the center of rotation is the center of the rotor but the complete stator core assembly is off-set by an amount ($e_S$) from the center of rotation, as illustrated in Figure 10.3.

  **(iii)** End bells (or end frames in the United Kingdom) in large HV SCIMs that house the bearings are often doweled to ensure concentricity between the center of the stator bore and the center of rotation of the shaft. Static eccentricity can therefore occur when concentricity is not achieved.

**(b)** Typical Causes of Dynamic Airgap Eccentricity include:

  **(i)** The rotor core and stator bore diameters are constant around their circumference and along their lengths but the center of the rotor does not coincide with the central axis of rotation and center of the stator bore, this was illustrated in Figure 10.4 and the minimum airgap rotates with the rotor.

  **(ii)** The outer diameter of the rotor core is not constant around its circumference or along its full axial length.

In practice, combinations of static and dynamic airgap eccentricity simultaneously exist.

During the manufacture of the rotor, the outside diameter is checked for concentricity along its axial length, which is referred to by the OEMs as the total indicated run out (TIR) and this takes account of the concentricity of the complete shaft-rotor assembly between bearing centers and the bearing journals. The manufacturing tolerance set for the TIR of the rotor core may well vary between OEMs and can also be a function of the design of the motor and subsequent operational conditions. The eccentricity of the rotor core is manufactured to be as low as is practically possible and OEMs and electric motor repair companies take great care to so do. An example of an OEM's typical tolerance specifications for the various parts of a squirrel cage rotor is given in Table 10.1.

The TIR for the rotor core is 0.05 mm (2.5 mils/thou) thus for a 2-pole, 7460 kW/10,000 HP, 60 Hz SCIM with a nominal design airgap of 5 mm (200 mils/thou) the rotor's dynamic airgap eccentricity is only 1%. The bore of the stator core may on occasions have to be bored after the stator is assembled to improve its concentricity, Bonnett and Soukop [10.2]. It is normally the case that the combination of static and dynamic airgap eccentricity should be within ±10% of the radial airgap. The smaller the airgap eccentricity the lower the electromagnetic forces that are transmitted to the stator core, teeth, and HV stator coils.

The portion of HV coils in the stator slots have a semi-conductive coating, often referred to as a corona shield (see Section 12.2, Figure 12.1), which can be mechanically damaged by vibration due to high airgap eccentricity because the coils are in direct contact with the stator teeth along the axial length of the stator core.

**TABLE 10.1 Typical TIR Tolerance Criteria for Parts of a Squirrel Cage Rotor**

**Acceptance TIR (mm)**

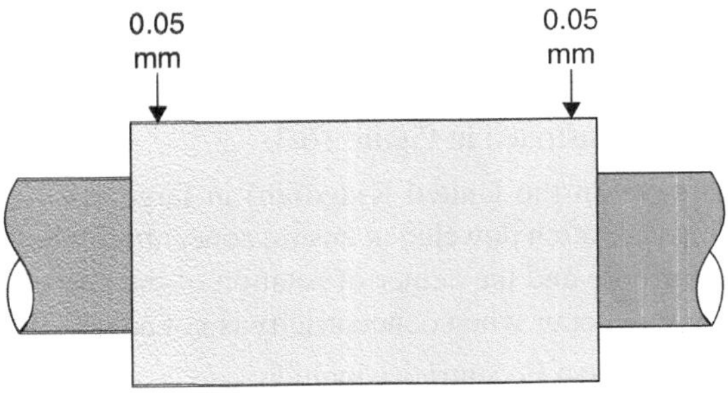


**Rotor core**

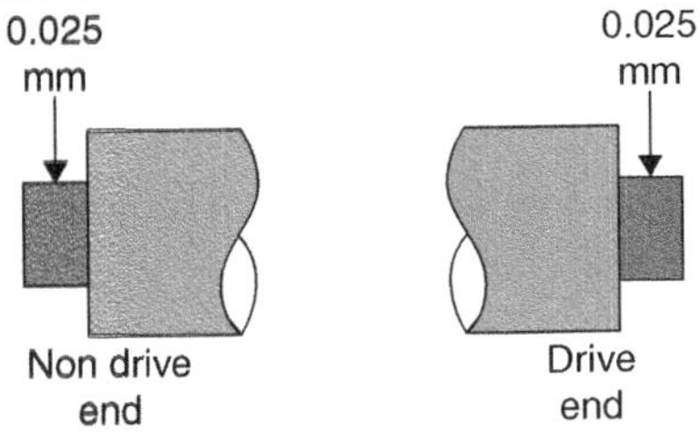


**NDE & DE shaft ends**

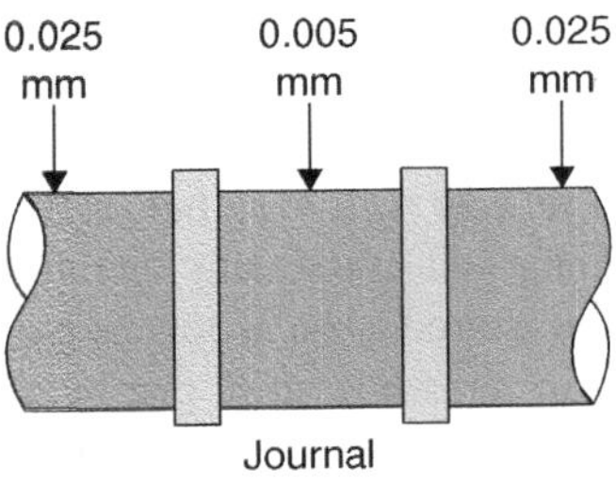


**NDE & DE bearing journals**

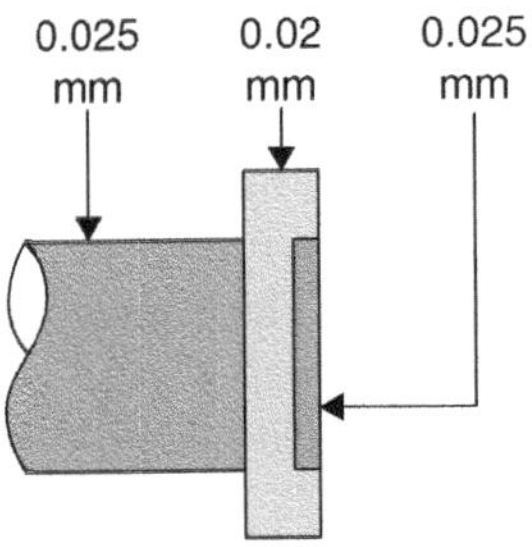


**Solid flanged end**

The dynamic eccentricity is kept to a minimum to limit the UMP which rotates at the speed of the rotor and is transmitted to the bearings.

A case history in Section 11.6 is presented which verifies that MCSA diagnosed an operational airgap eccentricity of between 25% and 30% (which is considered to be severe, Reference 10.2) in a large, HV SCIM. On strip down the semi-conductive coatings on all the HV stator coils were found to be mechanically damaged due to high vibration, caused by the high level of airgap eccentricity and the complete stator was rewound. An increase in airgap eccentricity can also be caused by the following operational conditions after the motor is manufactured and during its operational life.

**(i)** Advanced bearing wear, particularly with roller element bearings (dynamic airgap eccentricity).

**(ii)** Mechanical degradation of any shims under the feet of the motor over a period of time due to rust or chemical contamination causing the stator to drop.

The length of radial airgaps are different between the ratings, design, and number of poles in 3-phase SCIMs and samples of actual airgaps are given in Table 10.2

**TABLE 10.2 Typical Radial Airgaps**

| Different OEMs Defined as A, B, C | Power, HP/kW | V | A | Full-Load Speed, r/min | Frequency, Hz | Airgap Length |
|---|---|---|---|---|---|---|
| A | 6700 HP<br>5000 kW | 11,000 | 324 | 3580 (2-pole) | 60 | 6.35 mm<br>250 mils |
| B | 9580 HP<br>7147 kW | 11,000 | 447 | 3564 (2-pole) | 60 | 5 mm<br>200 mils |
| B | 780 HP<br>580 kW | 6600 | 59 | 3580 (2-pole) | 60 | 4 mm<br>160 mils |
| C | 300 HP<br>230 kW | 415 | 370 | 1478 (4-pole) | 50 | 1.8 mm<br>70 mils |
| D | 1600 HP<br>1194 kW | 6600 | 145 | 1784 (4-pole) | 60 | 2.5 mm<br>100 mils |
| B | 322 HP<br>240 kW | 6600 | 28 | 1187 (6-pole) | 60 | 1.5 mm<br>60 mils |
| E | 1940 HP<br>1450 kW | 11,000 | 103 | 742 (8-pole) | 50 | 2.5 mm<br>100 mils |
| E | 4350 HP<br>3245kW | 11,000 | 204 | 592 (10-pole) | 50 | 2.8 mm<br>110 mils |
| F | 670 HP<br>500 kW | 3300 | 120 | 593 (10-pole) | 50 | 1.5 mm<br>60 mils |
| F | 750 HP<br>560 kW | 11,000 | 41 | 593 (10-pole) | 50 | 2.2 mm<br>87 mils |
| D | 320 HP<br>240 kW | 11,000 | 17.5 | 592 (10-pole) | 50 | 1.6 mm<br>63 mils |
| D | 1000 HP<br>746 kW | 11,000 | 51 | 494 (12-pole) | 50 | 2.2 mm<br>87 mils |

With reference to Table 10.2, the airgap at 6.35 mm/250 mils in the 5000 kW/6700 HP, 2-pole, 60 Hz SCIM is 2.5 times greater than the airgap (2.5 mm/ 100 mils/thou) in the 1450 kW/1944 HP 8-pole, 50 Hz SCIM. There are two main reasons for large 2-pole SCIMs having larger radial airgaps than higher pole number motors of the same rating, and these are

**(i)** The centrifugal forces in a 2-pole rotor are much greater than those in higher pole numbers and for the two motors cited above, the 2-pole motor's centrifugal force is 23 times greater than that of the 8-pole motor assuming the same mechanical unbalance in the rotors of both motors.

**(ii)** It is the case that 2-pole SCIMs, particularly those with spider support arms welded to the rotor shaft, will normally pass through the rotor's first critical speed during the run-up sequence, which is when the first lateral bending natural frequency equals the speed of the rotor and resonance occurs. It can be the case that 2-pole rotors in SCIMs are sometimes termed as being flexible by the OEM, reference discussions with Mr. Len Jones, Chief Engineer and Engineering manager, Parsons Peebles Group, Scotland.

The combination of items (i) and (ii) plus the existing airgap eccentricity therefore increases the risk of a rotor-to-stator rub during the run-up sequence.

In many cases after SCIMs are assembled it is not possible to measure the airgaps at the DE and NDE unless special access slots in the end frames (end bells) are provided to allow the insertion of feeler gauges. If such airgap access is not provided the radial airgap is therefore solely dependent on the manufacturing QC and QA procedures set by the OEM. With large 2-pole motors very long feeler gauges are required due to the length of the end winding overhangs. Also, where the feeler gauges are inserted, the stator and rotor core teeth must be clear of any varnish that may exist. The procedure is shown in Figure 10.6.

The airgaps are measured at four fixed reference positions on the stator and the rotor is then turned through 90 degree steps and the measurements are repeated, thus 16 measurements are recorded, the rotor steps can be less, at say 45 degrees, thus giving 32 measurements. This is done at the DE and NDE for each rotor position. It is now appropriate to discuss UMP since it is caused by airgap eccentricity and is the force that can ultimately lead to a rotor-to-stator rub, commonly referred to as "rotor pull-over."

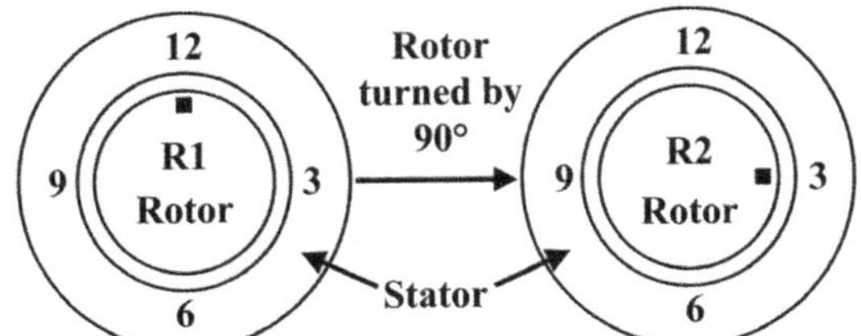


Airgap measured at 12, 3, 6, and 9 positions on stator with rotor in position R1 (12 O'clock)

Airgap measured at 12, 3, 6, and 9 positions on stator with rotor now at position R2 (3 O'clock)

Figure 10.6 Measurement of radial airgap.

## 10.3 UNBALANCED MAGNETIC PULL (UMP) AND ROTOR PULL-OVER

The subject of UMP and airgap eccentricity has been investigated since the early part of the twentieth century, for example, in 1918 by Gray and Pertsch [10.3], and Rosenberg [10.4], and it is worth noting that Rosenberg's paper (available via IEEE Xplore Digital Library) is in fact 45 pages and contains invaluable information. Also, an excellent book *(as quoted by Rosenberg,* [10.4]) by Miles Walker [10.5] in 1925 on the "Specification and Design of Dynamo-Electric Machines" covers UMP and airgap eccentricity. These publications present the fundamental theory, its relevance to designing electrical machines, and formulae for estimating UMP as a function of design parameters and airgap eccentricity. These have not changed and were used by many designers and OEMs in the early years and are still relevant in the twenty-first century.

An excellent paper by Bonnett and Soukop [10.2] in 1986 on "Rotor Failures in Squirrel Cage Induction Motors" discusses UMP and makes specific reference to Richard L. Nailen's [10.6] paper in 1966. This gives an unambiguous and very good description of the rotor pull-over process, which leads to a rub between the rotor and stator and that description follows:

> *"The pull-over process begins with a decrease of the airgap on one side while it increases at the opposite side. In an alternating magnetic field, the result of decreasing the airgap is a greater force of attraction across the smaller gap. The reluctance, which is the opposition to the passage of flux in the magnetic flux path, is reduced due to the smaller airgap. The magnetizing current can generate more flux across the smaller airgap, thus leading to greater pull, and at the same time, the airgap is being increased in the opposite direction of the machine. The reluctance is greater across the larger airgap so the flux and magnetic side pull are reduced. The greater pull on the side of the smaller airgap tends to move the rotor in that direction, making the gap even smaller. This process may continue until the gap becomes zero and the rotor comes into contact with the stator core and pull-over has occurred. However, if some airgap eccentricity exists, as it always does, then why does this pull-over not occur more often? The answer is that the rotor movement is restrained by the mechanical stiffness of the rotor."*

Figures 10.7 and 10.8 show examples of rotor-to-stator rubs, which occurred in these brand new motors and the pull-over shown in Figure 10.9 occurred after a repair in which part of the stator core pack was replaced. Prior to these failures, MCSA had not been applied at commissioning or during any test run at the OEMs' factory or the motor repair company's factory.

With reference to Figure 10.9, the first rotor pull-over was due to a bearing collapse. Part of the stator core pack was replaced during the rewind. The failure in Figure 10.9 occurred after only 10 DOL starts of the repaired motor and was caused by out of tolerance concentricity of the partly replaced stator core pack. MCSA was not applied at re-commissioning, but it should have been.

The potential for rotor pull-over is a function of various variables and it is not the purpose of this chapter to carry out an in-depth study of UMP. A major review in 1963 by Von Kaehne (64 pages and 64 references) on UMP [10.7], presents various

Figure 10.7 3-Phase 3300 V, 670 HP/500 kW, 110 A, 50 Hz, 593 r/min, 10-pole, p.f. = 0.85, efficiency = 93.6% SCIM. Rotor-to-stator rub and a consequential stator failure.

formulae for estimating the magnitude of UMP for different designs and ratings of SCIMs. This review is based upon papers by numerous authors. A list of additional publications [10.29–10.58] on airgap eccentricity and UMP and also on MCSA for monitoring airgap eccentricity is given at the end of this chapter. It is interesting to observe that a relatively recent paper by Dorrell et al. [10.50] in 2014 proposed, and to quote from the conclusions, "*a new U.M.P. factor and U.M.P. gap factor which attempts to quantify the U.M.P. factor to allow a direct comparison of different machines and sizes.*"

This indicates the prediction of UMP has now been studied for over 115 years, which is a rather interesting statistic, bearing in mind, that from a practical perspective

Figure 10.8 3-Phase, 4350 HP/3245 kW, 11 kV, 204 A, 50 Hz, 592 r/min, p.f. = 0.88, efficiency = 95% SCIM. Catastrophic rotor-to-stator rub and stator winding failure in a brand new SCIM, August 2004. A segmental laminated rotor core—radial movement of the rotor core.

Figure 10.9 3-Phase, 11,000 V, 109/86 A, 1610/1122 kW (2158/1500 HP), 594/495 r/min, 50 Hz, star connected, PAM SCIM. Catastrophic rotor-to-stator rub, FD fan motor.

and not an academic one, the manufacturers of induction motors have successfully designed and manufactured reliable induction motors for many years indeed, without problems of regular pull-overs in new induction motors. In fact, rotor pull-over is a relatively rare event.

This chapter is not about calculating UMP since it is focused on providing a basic and practical guide for engineers applying MCSA in industry to estimate the operational level of airgap eccentricity and it is not about the design of induction motors to prevent high UMP and a rotor-to-stator rub. The *key parameters* which govern UMP are described by equation (10.2), which has been known for many years, Gray and Pertsch [10.3], Rosenberg [10.4], Walker [10.5], Friese and Jordan [10.8], and Bradford [10.9].

$$F_{(\mathrm{UMP})} = \left(\pi D L B_{\mathrm{p}}^2 e_{\mathrm{c}}\right)/4\mu_0 \tag{10.2}$$

$F_{(\mathrm{UMP})}$ = force due to UMP as a result of static airgap eccentricity, Newtons

where

$D$ = rotor diameter, millimeter

$L$ = rotor length, millimeter

$B_{\mathrm{p}}$ = peak flux density, Tesla

$e_{\mathrm{c}}$ = absolute airgap eccentricity level, millimeter

$\mu_0 = 4\pi \times 10^{-7}$ H/m

For a given airgap eccentricity and flux density, the UMP increases as D or L increases and in general as the power rating increases then the DL product will increase for a given pole number. If, for example, the flux density increases from 1.0 to 1.2 T then the UMP will increase by 44% for a given eccentricity, provided the motor is not operating in the saturated region of the B–H curve, [10.4–10.6]. As the airgap eccentricity increases for a given motor design, the restraining force that essentially prevents a rotor-to-stator rub is provided by the inherent mechanical stiffness of the rotor assembly. The rotor shaft, has the lowest diameter in the rotor assembly and it must be of a sufficient diameter and mechanical stiffness to prevent a rotor-to-stator rub with a normal level of airgap eccentricity during starting and running conditions, plus a margin of safety to allow for any increase in the operational airgap eccentricity. The mechanical stiffness in the lateral direction of a round shaft is a function of the second moment of area:

$$I_{\mathrm{M}} = \pi D^4/64\,(\mathrm{m}^4) \tag{10.3}$$

Since the stiffness is proportional to the shaft diameter raised to the fourth power, if the shaft diameter is increased, for a given shaft length between the bearing supports, the stiffness can be greatly increased. For example, if the shaft diameter is increased by 20% then the stiffness increases by a factor of 2.0. Finally, the risk of pull-over is normally at its highest during the start-up sequence of a SCIM, when the ampere-turns from the stator winding is at its highest and it is also worth noting that parallel paths in each phase of the stator winding have been shown to reduce UMP, Bonnett and Soukop [10.2] and Yang [10.9]. If a SCIM had a rotor-to-stator rub due to pull-over, which caused a stator winding failure and if the stator had no parallel paths, it would be worthwhile to consider a stator rewind with parallel paths in each phase winding of the stator, whose action should reduce the UMP.

## 10.4 CURRENT SIGNATURE PATTERN DUE TO AIRGAP ECCENTRICITY

### 10.4.1 Predictor Equations

As reported by Morrell [10.11] in 1940 and Alger [10.12] in 1954, it has been known for many years that rotor slotting, due to magnetic reluctance variations caused by rotor slot openings, results in high frequency components in the main airgap flux waveform of a SCIM. This phenomenon is often referred to as the "*rotor slotting ripple*" which can induce emfs in the stator winding and consequential current components can flow in the stator. An illustration of the high frequency rotor slotting ripple in the current waveform is shown in Figure 10.10.

These flux waves are rotating with respect to the stationary stator winding and their frequencies are given by the following equation, Alger [10.12]:

$$f_{\mathrm{rs}} = f[(R/p)(1-s) \pm n_{\omega\mathrm{s}}] \tag{10.4}$$

where

$f_{\mathrm{rs}}$ = rotor slot passing frequencies in Hz, a series of frequency components spaced twice the supply frequency apart

$f$ = supply frequency

$R$ = number of rotor slots

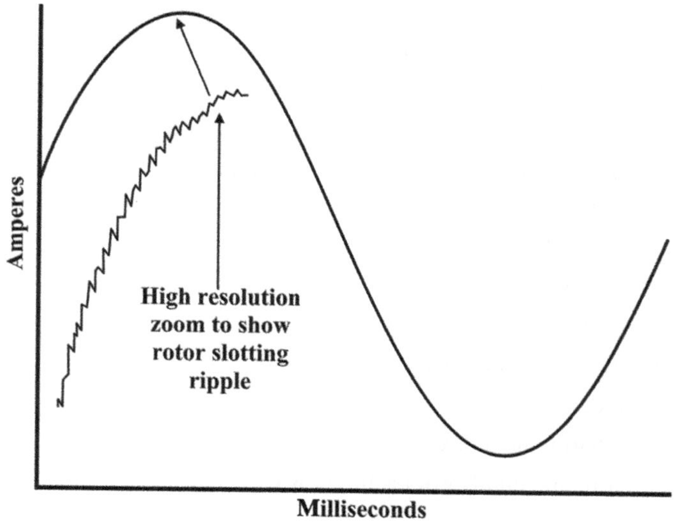


Figure 10.10 Idealized illustration of rotor slotting ripple in current–time domain waveform.

$p$ = pole-pairs

$s$ = operational slip

$n_{\omega s}$ = integers for the fundamental time domain mmf ($n_{\omega s} = +1$), and its odd harmonics, 3, 5,...

These flux waves are normal and inherent to a 3-phase SCIM and are not due to a fault. The magnitude (in dB or amperes) of the flux components due to rotor slotting is a function of airgap eccentricity but a crucial fact is that their magnitudes are a function of numerous other factors such as

- Rotor slot design
- Rotor bar skew
- Flux density
- Magnetic saturation
- Magneto motive force (mmf) distribution
- Operating load

During their studies on the effect of airgap eccentricity on acoustic noise from SCIMs, in 1981 Yang [10.10] and Ellison and Yang [10.14], derived an equation for the frequencies of the eccentricity flux waves as a function of rotor slotting taking account of magnetic saturation and dynamic eccentricity, given by

$$f_{ec} = f[(R/p)(1-s) \pm n_{\omega s} \pm 2n_{sa}p] \pm n_d f(1-s)/p \quad (10.5)$$

$$m = R \pm S \pm n_s \pm n_d \pm 2n_{sa}p \pm n_{\theta s}p \quad (10.6)$$

Equation (10.5) is in fact the combination of the classical rotor slot passing frequencies (RSPFs) ($f_{rs}$) from equation (10.4) plus the rotational speed frequency of the rotor, $f_r = f\,(1-s)\,/p$, with $n_d = 1$.

$f_{ec}$ = the frequencies of the flux waves, which are characteristic of the combination of static and dynamic airgap eccentricities which exist simultaneously in practice, Hertz

$R$ = number of rotor slots

$S$ = number of stator slots

$n_s$ = first order static eccentricity integer = 1.0

$n_d$ = first order dynamic eccentricity integer = 1.0

$n_{sa}$ = saturation index

$n_{\theta s}$ = stator space harmonic index

$m$ = pole-pair harmonic number of each flux wave

For the application of MCSA, a simplified version of the equation derived by Yang in Reference 10.10 was used by Cameron et al. [10.1] in 1986 to identify current components, which were a function of the combination of static and dynamic airgap eccentricity:

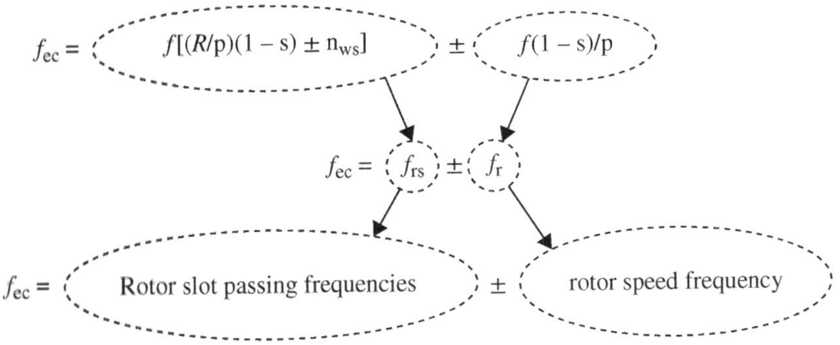


(10.7)

Equation (10.7) neglects saturation since it is a second order effect ($n_{sa} = 0$), but includes the first order dynamic eccentricity index, $n_d = 1$. The emfs and corresponding current components predicted by equation (10.7) *can only exist* if the pole-pairs of each particular flux wave (from equation 10.6) has a corresponding harmonic pole-pair in the 3-phase stator winding, given by

$$\text{Odd harmonics } m_o = p\,(6c \pm 1) \tag{10.8}$$

$$\text{Even harmonics } m_e = p\,(6c \pm 2) \tag{10.9}$$

$p$ = pole-pairs

$c = 0, 1, 2, 3, 4, 5, \ldots$

If there is no pole-pair compatibility, then an emf cannot be induced in the stator winding and a current will not flow at that frequency. Toliyat et al. [10.15] and Nandi et al. [10.16] considered the requirement for pole-pair compatibility and verified the theory via controlled tests using a small power 3-phase SCIM in a university laboratory where design details of the test motor were available for the purpose of research work. Even when pole-pair compatibility does exist and corresponding currents flow in the stator, their magnitude (in dB) is a function of the harmonic pole-pair stator winding factors, namely the coil span and distribution factors.

The practical reality is that for SCIMs operating in industry it is impossible to obtain design details of the stator winding from the OEMs. The MCSA test has to be carried out with the minimum of information (nameplate data) but to predict the current signature pattern due to airgap eccentricity requires the number of rotor slots and even such basic information can be difficult to obtain from OEMs.

An MCSA industrial case history is presented in Section 11.9 where there was incompatibility of pole-pairs between certain (not all) flux waves given by equation (10.7) and the harmonic pole-pairs of the stator winding given by equation (10.6) and an analysis is presented which successfully diagnosed an airgap eccentricity problem.

### 10.4.2 Example of Signal Processing Strategy

It is now appropriate to go through a step-by-step educational illustration to assist the users of MCSA to identify the current components, which are a function of the combination of static and dynamic airgap eccentricity. This work inherently covers the signal processing strategy and the presentation is deliberately detailed for the benefit of industrial engineers who apply MCSA.

Motor nameplate details: 3-phase, 6600 V, 60 Hz, 3600 kW/4830 HP, 376 A, 3564 r/min, p.f. 0.88, efficiency 95.2%, star connected SCIM. Number of rotor slots is 46, single cage winding, copper fabricated rotor.

**Step 1.0**
Calculate the series of RSPFs given by

$$f_{rs} = f[(R/p)(1 - s) \pm n_{\omega s}] \tag{10.4}$$

The information on the nameplate is all that is available but at least the nominal full-slip can be calculated but note it must be borne in mind that the full-load speed on the nameplate may differ by several r/min compared to the actual value when the motor is operating at the rated full-load current.

$$s_{FL} = (3600 - 3564)/3600 = 0.01$$

$$R = 46, \ f = 60 \text{ Hz}, \ p = 1.0$$

These values are used in equation (10.4) and with $n_{\omega s} = 1, 3, 5$, which corresponds to the fundamental, third, and fifth time harmonics of the mmf (ampere-turns) distribution from the stator winding. This gives the normal and inherent RSPF components for this particular motor at nominally full-load.

$f_{rs(+1)} = 2792$ Hz, for $+n_{\omega s} = +1$
$f_{rs(-1)} = 2672$ Hz, for $-n_{\omega s} = -1$
$f_{rs(+3)} = 2912$ Hz, for $+n_{\omega s} = +3$

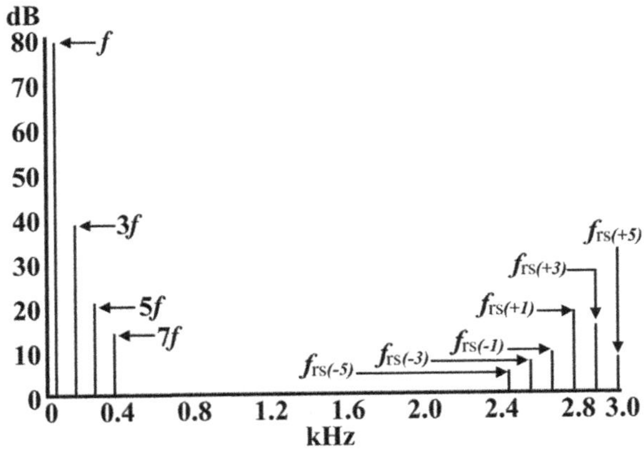


Figure 10.11 Idealized current spectrum indicating the spectrum between 0 and 3000 Hz.

$f_{rs(-3)} = 2552$ Hz, for $-n_{\omega s} = -3$
$f_{rs(+5)} = 3032$ Hz, for $+n_{\omega s} = +5$
$f_{rs(-5)} = 2432$ Hz, for $-n_{\omega s} = -5$
$f_{rs(+1)}$ is often referred to as the principal RSPF.

The calculated flux, emf, and current components are separated by twice the supply frequency (2*f*) and none of them is due to a fault. An idealized current spectrum for this example is shown in Figure 10.11.

The magnitude of RSPF current components can typically be greater than 60 dB down (a factor of 1000 times smaller) on the main supply component as will be verified in the case histories and are a function of the motor's design and operating load on the SCIM as discussed in Section 10.4.1. Recall that 40, 60, and 80 dB equates to factors of 100, 1000, and 10,000 in absolute units and the idealized spectrum in Figure 10.11 illustrates that fact, the third, fifth, seventh,.... harmonics of the supply frequency (*f*) are typically more than 40 dB down (or 1% ) on the supply component *f*.

Clearly the supply frequency component dominates the spectrum and in practice this component is filtered out so that the components from equation (10.5) can be detected. The baseband frequency for the first stage spectrum analysis is typically set from 70 Hz up to the RSPF given in equation (10.4) which equates to $+n_{\omega s}$ equal to 3 and in this case at the full-load slip and for this illustration the upper frequency limit is set to 3000 Hz. It is recommended that the minimum number of spectral lines for a 2-pole SCIM is 12,800 but ideally 25,600 lines is desirable, this gives a frequency resolution of 0.25 Hz/line and 0.125 Hz/line, respectively.

It has to be pointed out at this stage that certain condition monitoring companies, particularly those whose expertise is in vibration monitoring, often consider an MCSA test as merely an add on. These companies tend to lack a thorough understanding of the fundamental theory and operation of SCIMs. They seem to believe that they can simply monitor the current components given by equation (10.4) and if the magnitude of these components increases, then their contention is that the airgap eccentricity has increased but that belief is fundamentally incorrect. In fact, the authors have read reports by vibration condition monitoring companies who have stated that if $f_{rs(-1)}$ and $f_{rs(+3)}$ exist there is an airgap eccentricity problem. Yet another misconception.

**Step 2**

Equation (10.7) is used to predict the components in the current signature pattern which are due to the combination of static and dynamic airgap eccentricity (i.e., the total airgap eccentricity) and recall they cannot be separated in practice via MCSA.

Calculate $f_r = f\,(1-s)\,/p$ at full-load slip, with $n_d = 1$ in equation (10.7)

$$f_r = \frac{60\,(1-0.01)}{1.0} = 59.4\ \text{Hz}$$

or from the nominal full-load speed on the nameplate,

$$f_r = 3564/60 = 59.4\ \text{Hz}$$

Consequently, the signature pattern is that each RSPF $f_{rs}$ may have $\pm f_r$ components around it, and at a nominal full-load slip of 0.01 these are as follows:

$$f_{rs(+1)} \pm f_r = 2792 \pm 59.4\ \text{Hz}$$

$$f_{rs(-1)} \pm f_r = 2672 \pm 59.4\ \text{Hz}$$

$$f_{rs(+3)} \pm f_r = 2912 \pm 59.4\ \text{Hz}$$

$$f_{rs(-3)} \pm f_r = 2552 \pm 59.4\ \text{Hz}$$

$$f_{rs(+5)} \pm f_r = 3032 \pm 59.4\ \text{Hz}$$

$$f_{rs(-5)} \pm f_r = 2432 \pm 59.4\ \text{Hz}$$

Note: For the avoidance of doubt it is certainly not acceptable to round down $f_r = 59.4$ to 59 Hz, recall that 0.1 Hz equates to 6 r/min.

**Step 3**

Initially select the largest (in dB) RSPF component which is normally $f_{rs(+1)}$ because this is a function of the fundamental component of the mmf waveform from the stator winding. The 60 Hz component is filtered out of the spectrum shown in Figure 10.11 and the spectrum is then auto-scaled to obtain an 80 dB dynamic range of the filtered spectrum. A zoom spectrum analysis is then displayed around $f_{rs(+1)}$.

In this case, since it is an idealized case with the motor on full-load slip, search for $\pm f_r$ components around $f_{rs(+1)}$ at ±59.4Hz. For this example, this search reveals the current signature pattern, which is used to detect airgap eccentricity problems and the idealized zoomed current spectrum is shown in Figure 10.12.

In this idealized case the $\pm f_r$ components are an arithmetic* average of $N_{ec} = (N_1 + N_2)/2$ equal to 42 dB down on the principal RSPF component, $f_{rs(+1)}$.

*Note: The use of arithmetic dB average was validated in Chapter 4.

Table 10.3 is an interpretation of the current signature due to the combination of static and dynamic airgap eccentricity, and normally gives sensible estimations of the operational level (as a percentage band) of air gap eccentricity. It is based

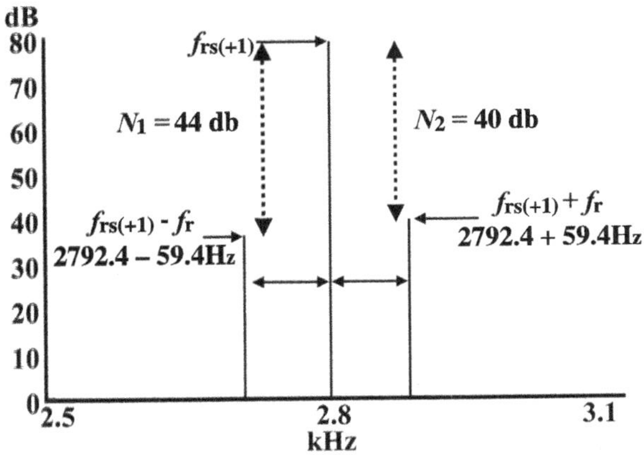


Figure 10.12 Idealized current zoom spectrum for the airgap eccentricity signature.

**TABLE 10.3 Estimate of Operational Airgap Eccentricity from Current Signature Pattern**

| $N_{ec}$ Average dB difference between $f_{rs(+1)}$ and $\pm f_r$ components | **Estimate of "*operational*" airgap eccentricity that is combination of static and dynamic** |
|---|---|
| $N_{ec} \geq 40$ dB | Airgap eccentricity is within 5% of the radial airgap |
| $N_{ec}$ = 35–40 dB | Airgap eccentricity is within 10% of the radial airgap |
| $N_{ec}$ = 25–35 dB | Airgap eccentricity is ≥10% of radial airgap but at $N_{ec}$ = 25 dB it is typically around 15% of the radial airgap |
| $N_{ec}$ = 15–25 dB | Airgap eccentricity is ≥15% of the radial airgap and the closer $N_{ec}$ is to 15 dB it is typically 20% of the radial airgap |
| $N_{ec} \leq 15$ dB | Airgap eccentricity is normally >20% of the radial airgap and this is considered to be a severe level. Suggested actions:<br>**(i)** In older SCIMs check the condition of any shims under the base of the motor to establish that they have not rusted or even disintegrated, which would cause the stator assembly to drop.<br>**(ii)** Check the bearings.<br>**(iii)** If possible (i.e., if access is available via the end frames) check the airgaps at the DE and NDE. If not, the motor will need to be removed from site for concentricity checks on the stator bore, end frame (end bell) spigots, and TIR of the rotor, etc.<br>Secondary effects<br>• High electromagnetic forces and vibration being transmitted to the stator core and teeth, windings, and rotor and bearings.<br>• Possible mechanical damage to the corona shields in HV coils due to high vibration of stator teeth and core.<br>• Possible pull-over due to high UMP during start up. |

on publications by Cameron and Thomson et al. [10.1] and Thomson et al. [10.17–10.27] plus the results from MCSA tests (during the past 34 years by Thomson) on thousands of SCIMs operating in industry. Furthermore, its accuracy and efficacy is demonstrated in the industrial case histories presented in Chapter 11.

### 10.4.3 Two Case Histories (1985)—Assessment of Operational Airgap Eccentricity in HV SCIMs Driving Large FD Fans

This was part of a power utility funded research and development project which commenced in 1982 and it was known that an FD fan motor had stator core ovality but by how much was unknown by the power station, since the motors were commissioned in 1962. The objective was to evaluate the research (via an on-site trial) carried out between 1982 and 1985 before the power utility would provide further research funding to Thomson (author).

The available nameplate information on the FD fan motors was as follows: 3-phase, 11,000 V, 1200 kW/1600 HP, 75 A, 50 Hz, 492 r/min (12-pole), p.f. = 0.88, efficiency = 96% and the rotor had 112 slots. Two FD fan motors, still coupled to the fans, were made available for MCSA testing but only during a very lightly loaded run. The design value of the radial airgap was unknown and could not be obtained from the OEM.

**COMMENTARY ON FREQUENCY RESOLUTION**

For the avoidance of any doubt and since this is the first case history on MCSA for estimating the operational level of airgap eccentricity, the following points should be noted.

1. The first main objective is to accurately identify the frequency components that are a function of the total airgap eccentricity.
2. Therefore, it is not appropriate to round up certain predicted or measured frequencies (e.g., $f_r$) to whole numbers or one decimal place otherwise incorrect components can be selected and this will be evident in the industrial case histories in Chapter 11.
3. If the predicted or measured frequencies are quoted to three decimal places then round up/down to two decimal places when predicting or measuring $f_r$ Hz, which is the rotational speed frequency of the rotor.
4. As an example, the nominal full-load speed for this motor is 492 r/min and is therefore the starting point and there is no other option but to use that value.
5. The nominal, full-load rotational speed frequency is therefore $f_r = N_r/60 = 8.2$ Hz—this should not be rounded down to 8 Hz, since that would give a speed of 480 r/min, which is 12 r/min lower than the rated value.

6. These motors were running on very light load and their speeds were very close to the synchronous speed of 500 r/min, the rotation speed frequency is taken to be $f_r = 500/60 = 8.33$ Hz or $499/60 = 8.32$ Hz.
7. Either of these can be used but neither of them can be rounded down to 8 Hz, since that would give a no-load speed = 480 r/min, which is less than the full-load nameplate speed when the motors are operating on no-load! From the foregoing, the frequency resolution required for MCSA is very clear.
8. With respect to the RSPFs these are much higher than $f_r$ and rounding up or down to one decimal place is acceptable, if so desired.

The first step was to calculate the current components from equation (10.5) using the available nameplate data and the number of rotor slots.

$$f_{ec} = f[(R/p)(1-s) \pm n_{\omega s}] \pm n_d f(1-s)/p = 50[(112/6) \pm 1] \pm 50/6$$

$n_d = 1$, $s =$ approximately zero on light load, $p = 6$, $R = 112$

$$f_{ec} = f_{rs(+1)} \pm f_r = 983 \pm 8.33 \text{ Hz}$$

The principal RSPF $f_{rs(+1)}$ was selected. The measured zoom current spectra for motors A and B are shown in Figures 10.13 and 10.14, respectively.

Component (i) $= f_{rs(+1)} = 978.5$ Hz at 91 dB

Component (ii) $= f_{rs(+1)} - f_r = 970.2$ Hz at 47 dB

Component (iii) $= f_{rs(+1)} + f_r = 986.6$ Hz at 46 dB

From the spectrum in Figure 10.13 for motor A, the average dB difference $N_{ec}$ between $f_{rs(+1)}$ and the $\pm f_r$ components is 46.5 dB and Table 10.3 gives an operational airgap eccentricity of less than 5% of the radial airgap for this low speed SCIM.

During the MCSA tests the power station confirmed that the supply frequency was 49.85 Hz and 50.2 Hz for motors A and B, respectively, at the time of testing. The

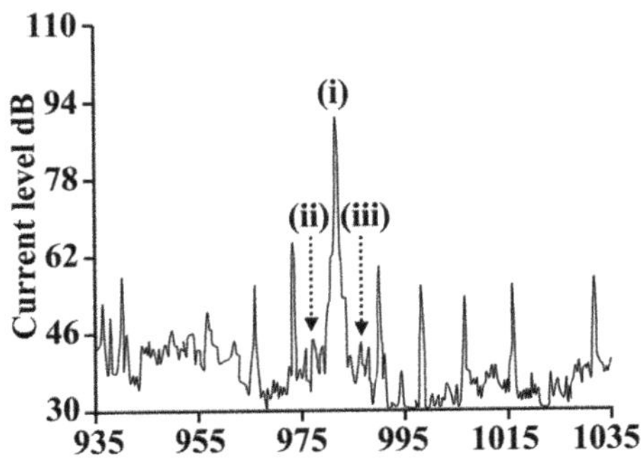


Figure 10.13 Current zoom spectrum around $f_{rs(+1)}$ for motor A, baseband span 70–1200 Hz, 12,800 lines, 88.3 mHz/line. *Source*: Cameron et al. [10.1]. Reproduced with permission of the Institution of Engineering and Technology.

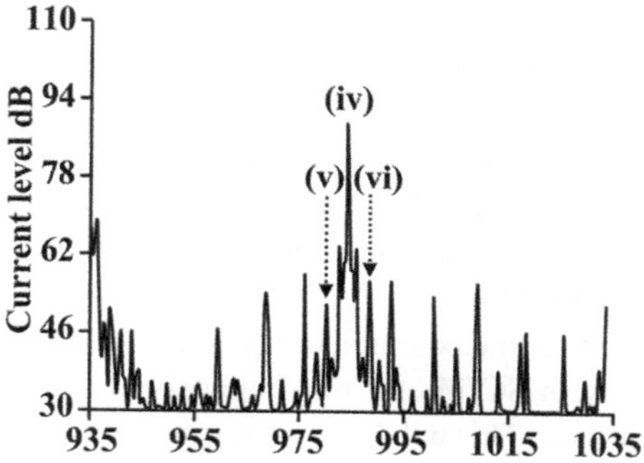


Figure 10.14 Current zoom spectrum around $f_{rs(+1)}$ for motor B, baseband span 70–1200 Hz, 12,800 lines, 88.3 mHz/line. *Source*: Cameron et al. [10.1]. Reproduced with permission of the Institution of Engineering and Technology.

prediction for $f_{rs(+1)}$ was based on the nameplate frequency giving 983 Hz, but the supply frequency when motor A was tested was actually 49.85 Hz giving a prediction of 980 Hz with the slip assumed to be zero but the measured $f_{rs(+1)}$ for motor A was 978.5 Hz, because the slip was not actually zero.

The slip can now be back calculated from equation (10.4) for $f_{rs(+1)}$ using the measured value as 978.5 Hz, and the measured supply frequency, $f$ at 49.85 Hz.

Operating slip,

$$s = 1 - \{((f_{rs(+1)}/f) - 1)(p/R)\} = 1 - \{((978.5/49.85) - 1)(6/112)\}$$

$$s = 0.002(0.2\%)$$

where

Measured $f_{rs(+1)} = 978.5$ Hz

Measured supply frequency, $f = 49.85$ Hz

Pole-pairs, $p = 6$

Number of rotor slots, $R = 112$

The operational slip for motor A was 0.002 (0.2%). It is worth mentioning that certain inverters for speed control of 3-phase SCIMs now monitor the $f_{rs(+1)}$ component to obtain the operational slip and hence speed signal for feedback control rather than having to measure the speed via a separate sensor. This is now referred to as sensor-less speed control of SCIMs, Kiyotake [10.28].

Component (iv) $= f_{rs(+1)} = 985$ Hz at 89 dB

Component (v) $= f_{rs(+1)} - f_r = 976.7$ Hz at 52 dB

Component (vi) $= f_{rs(+1)} + f_r = 993.3$ Hz at 56 dB

From the spectrum in Figure 10.14 for motor B, the average dB difference $N_{ec}$ between $f_{rs(+1)}$ and the $\pm f_r$ components is 35 dB and from Table 10.3 this gives an estimated operational airgap eccentricity of around 10% of the radial airgap since it is on the border of two estimated airgap eccentricity bands. There is clearly a significant difference between the two spectra in that the $\pm f_r$ components around $f_{rs(+1)}$ in motor B are an average of 35 dB down whereas for motor A they are 44.5 dB down. This means that for motor B, the $\pm f_r$ components relative to $f_{rs(+1)}$ are three times greater than the $\pm f_r$ components relative to $f_{rs(+1)}$ for motor A. It is this difference between the current spectra from motors A and B, which is the *key outcome* from these MCSA on-site trials in 1985. Of course, the operational airgap eccentricity in motor B is perfectly acceptable at around 10%.

The case was proven for the potential of MCSA to successfully assess the operational airgap eccentricity in SCIMs for the diagnosis of a severe operational level (above 20–25% of the airgap). This will be demonstrated in the case histories in Chapter 11 but it will also be shown that with certain SCIM drives it is not possible to apply MCSA to assess operational airgap eccentricity levels due to the complex mechanical dynamics of the driven load and therefore false positives of high airgap eccentricity are possible.

## 10.5 QUESTIONS

**10.5.1** Explain the differences between static and dynamic airgap eccentricity and give three causes for each of them.

**10.5.2** Explain what is meant by a TIR measurement of a squirrel cage rotor.

**10.5.3** Why is the tolerance on the concentricity of the rotor core kept to a minimum by the OEMs of SCIMs?

**10.5.4** Discuss the reasons why airgap eccentricity (combination of static and dynamic) should be kept to, for example, $\leq$10% of the mean radial airgap.

**10.5.5** Why do large, HV, 2-pole, 60 Hz SCIMs have larger airgaps compared to higher pole number SCIMs of similar rating and voltage?

**10.5.6** What are the differences between UMP caused by dynamic and static airgap eccentricity? Comment on the consequential secondary effects of high (e.g., >20%) static and high (>20%) dynamic airgap eccentricity.

**10.5.7** Why is it the case that a rotor-to-stator rub (or pull-over) is a rare event in brand new SCIMs? Also explain the possible causes of rotor pull-over which sometimes occurs during the operational life of a large SCIM.

**10.5.8** Explain why rotor slotting in a cage rotor produces rotating flux waves with respect to the stationary stator winding.

**10.5.8.1** Why is there a series of rotor slotting flux components spaced $2f$ apart from equation (10.4), and are these components normal in a 3-phase SCIM or due to a problem within the motor?

**10.5.8.2** The magnitudes of the rotor slotting flux components are completely independent of the load on the motor. Is this statement correct or incorrect; justify your answer.

**10.5.8.3** Why can current components, which are at the same frequencies as the rotor slotting flux components, flow in the stator windings?

**10.5.9** Discuss the physical meaning of equation (10.7) which is used to predict the frequency content of the current signature pattern, which is a function of the combination of static and dynamic airgap eccentricity.

**10.5.9.1** Can static and dynamic eccentricity levels be separated via MCSA? Justify your answer.

**10.5.9.2** What is meant by pole-pair compatibility between the rotor slotting flux components and harmonic pole pairs of the stator winding so that emfs and corresponding currents can actually be induced in the stator winding?

**10.5.10** MCSA was applied to the following motor: 3-phase SCIM, star, 13.8 kV, 6800 kW/9115 HP, 326 A, 3580 r/min, 60 Hz, Class F, p.f. = 0.89, efficiency = 97.2%. Number of rotor slots = 50.

**10.5.10.1** Calculate the rotor slotting current components when the motor is running at its nominal full-load current and speed with the supply frequency at 60 Hz.

**10.5.10.2** Repeat 10.5.10.1 but with the motor running on no-load.

**10.5.10.3** Calculate the rotor speed frequency on full-load and no-load.

**10.5.10.4** State the baseband frequency span you would select to display the rotor slotting components with, $n_{\omega s} = \pm 1$, and $n_{\omega s} = \pm 3$ at full-load slip and no-load slip. What is the number of lines of resolution you would select to display the current spectra? What is the frequency resolution in Hz/line that you have selected?

**10.5.10.5** From your calculations in 10.5.10.1 and 10.5.10.2, state the frequency of the principal rotor slotting current component $f_{rs(+1)}$ for no-load and full-load slips. Propose the zoom spectrum analysis band to display the current signature pattern from equation (10.7) which is

$$f_{ec} = f[(R/p)(1-s) \pm n_{\omega s}] \pm n_d f(1-s)/p \tag{10.7}$$

For $n_d = 1, f_{ec} = f_{rs} \pm f_r$

The zoom spectrum at full-load slip gave the following result:

The dB difference between $f_{rs(+1)}$ and $f_{rs(+1)} + f_r = 18$ dB

The dB difference between $f_{rs(+1)}$ and $f_{rs(+1)} - f_r = 16$ dB

**(i)** Estimate the operational band of airgap eccentricity as a percentage of the nominal radial airgap.

**(ii)** What are your recommendations to the end user?

## REFERENCES


[10.1] J. R. Cameron, W. T. Thomson, and A. B. Dow, "Vibration and current monitoring for detecting airgap eccentricity in large induction motors," *IEE Proceedings*, vol. 133, Part B. no. 3, pp. 155–163.

[10.2] A. H. Bonnett and G. C. Soukop, "Rotor failures in squirrel-cage induction motors," *IEEE Transactions on Industry Applications*, vol. 1A-22, no. 6, 1986, pp. 1165–1173.

[10.3] A. Gray and J. G. Pertsch, "Critical review of the bibliography on unbalanced magnetic pull in dynamo electric machines," *Transactions of the AIEE*, 1918, Part II, vol. 37, pp. 1417–1424.

[10.4] E. Rosenberg, "Magnetic pull in electrical machines," *Transactions of the AIEE*, 1918, Part II, vol. 37 pp. 1425–1460.

[10.5] M. Walker, *Specification and Design of Dynamo-Electric Machines*, 4th Impression, Longmans, Green and Co., London, 1925.

[10.6] R. L. Nailen, *Stop Rotor Troubles Before They Start*, Plant Engineering, December 1966.

[10.7] P. Von Kaehne, *Unbalanced Magnetic Pull in Rotating Electrical Machines*, ERA report, Ref. Z/T142, England, UK, 1963.

[10.8] W. Friese and H. Jordan, Unbalanced magnetic pull in 3-phase A.C. machines," *ETZ-A*, vol. 83, no. 9, 1962. Translation: C. E. Trans. 7836.

[10.9] M. Bradford, "Unbalanced magnetic pull in a 6-pole induction motor," *Proceedings of the IEE*, vol. 115, no. 11, 1968, pp. 1619–1627.

[10.10] S. J. Yang, *Low Noise Electric Motors*, Monographs in Electrical and Electronic Engineering, IEE, Savoy Place, London, 1981.

[10.11] W. J. Morrill, "Harmonic theory of noise in induction motors," *Transactions of the AIEE*, vol. 59, no. 8, August 1940, pp. 474–480.

[10.12] P. L. Alger "Magnetic noise in poly-phase induction motors," *Transactions of the AIEE*, vol. 73, Part IIA, 1954, pp. 118–125.

[10.13] P. L. Alger, *Induction Machines—Their Behavior and Uses*, Gordon and Breach Publishers, 1965.

[10.14] A. J. Ellison and S. J. Yang, "Effects of rotor eccentricity on acoustic noise from induction machines," *Proceedings of the IEE*, vol. 118, pp. 174–184.

[10.15] A. H. Toliyat, S. Nandi, S. Choi, and H. Meshgin-Keik, *Electrical Machines: Modelling, Condition Monitoring, and Fault Diagnosis*, CRC Press, October 2012, pp. 272. ISBN: 9780849370274.

[10.16] S. Nandi, S. Ahmed and A. H. Toliyat, "Selection of rotor slot and other eccentricity related harmonics in a three phase induction motor with different rotor cages," *IEEE Transactions on Energy Conversion*, vol. 16, no. 3, September 2001, pp. 253–260.

[10.17] J. R. Cameron, W. T. Thomson, and A. B. Dow, "On-Line Current Monitoring of Induction Motors—A Method of Calculating the Level of Airgap Eccentricity," Proceedings of IEE EMDA '87 Conference, London, November 1987.

[10.18] D. G. Dorrell, W. T. Thomson, and S. Roach, "Analysis of airgap flux, current and vibration signals as a function of the combination of static and dynamic airgap eccentricity in 3-phase induction motors," *IEEE Transactions on Industry Applications*, vol. 33, no. 1, January/February 1997, pp. 24–34.

[10.19] A. Barbour and W. T. Thomson, "Finite Element Study of Rotor Slot Designs with Respect to Current Monitoring for Detecting Static Airgap Eccentricity in Squirrel-Cage Induction Motors," 32nd IAS Annual meeting, IAS '97, IEEE Industry Applications Conference, New Orleans, LA, October 1997.

[10.20] W. T. Thomson and A. Barbour, "On-line current monitoring and application of a finite element method to predict the level of airgap eccentricity in 3-phase induction motors," *IEEE Transactions on Energy Conversion*, vol. 13, no. 4, December 1998, pp. 347–357 (includes discussion and closure).

[10.21] W. T. Thomson and A. Barbour, "The On-Line Prediction of Airgap Eccentricity Levels in Large (MW Range) 3-Phase Induction Motors," International Conference IEMD '99 on Electric Machines and Drives, Seattle, WA, May 1999.

[10.22] W. T. Thomson, D. Rankin, and D. G. Dorrell, "On-line current monitoring to diagnose airgap eccentricity in large 3-phase induction motors—an industrial case history verifies the predictions," *IEEE Transactions on Energy Conversion*, vol. 14, no. 4, December 1999, pp. 1372–1378.

[10.23] W. T. Thomson and M. Fenger, "Case Histories of Current Signature Analysis to Detect Faults in Induction Motor Drives," Proceedings of IEEE International Conference on Electrical Machines and Drives (IEMDC), University of Wisconsin, Madison, USA, June 2003.

[10.24] W. T. Thomson, "A Review of On-Line Condition Monitoring Techniques for Three-Phase Squirrel-Cage Induction Motors—Past, Present, and Future," IEEE Symposium on Diagnostics for Electrical Machines, Power Electronics and Drives, Gijon, Spain, 1999, pp. 3–18 (opening keynote address).

[10.25] W. T. Thomson, A. Barbour, C. Tassoni, and F. Fillippetti, "An Appraisal of the MMF-Permeance Method and Finite Element Models to Study Static Airgap Eccentricity and Diagnosis in Induction Machines," Proceedings of ICEM '98, Istanbul, 1998.

[10.26] W. T. Thomson and M. Fenger, "Case Histories of Current Signature Analysis to Detect Faults in Induction Motor Drives," Proceedings of IEEE International Conference on Electrical Machines and Drives (IEMDC), University of Wisconsin, Madison, WI, June 2003.

[10.27] W. T. Thomson and I. D. Stewart, "On-Line Current Monitoring for Detecting Airgap Eccentricity and Broken Rotor Bars in Inverter Fed Drives," Proceedings of International Conference in Electrical Machines ICEM '88, Pisa, Italy, September 1988.

[10.28] H. Kiyotake, K. Shinohara, and D. Katoh, "A New Detection Method of Slot Harmonics for Speed Sensorless Vector Control of Induction Motors," IEEE Power Conversion Conference—Nagoya. PCC 2007, pp. 845–851.

[10.29] J. Fisher-Hinnen, *Dynamo Design*, Van Nostrand, pp. 260–265, 1899.

[10.30] F. W. Carter, "Magnetic centering of dynamo electric machinery," *Proceedings of Institution of Civil Engineers (London)*, vol. 187, May 1912, pp. 311–318.

[10.31] R. C. Robinson, "*The calculation of Unbalanced Magnetic Pull in Synchronous and Induction Motors*," EI Engineering, October 1943, pp. 620–623.

[10.32] A. Covo, "Unbalanced magnetic pull in induction motors with eccentric rotor," *AIEE, III, B*, vol. 73, 1954, pp. 1421–1425.

[10.33] S. A. Swann, "The Magnetic Field in a Non-Salient Pole Dynamo-Electric Machine with Eccentric Rotor," Ph.D. Thesis, University of Nottingham, England, 1959.

[10.34] O. K. Gashus, "The Airgap Magnetic Field in Electric Machines with Special Reference to Eccentricity," Ph.D. Thesis, University of Glasgow, Scotland, 1960.

[10.35] S. A. Swann, "Effect of rotor eccentricity on the magnetic field in the airgap of a non-salient pole machine," *Proceedings of the IEE*, vol. 110, no. 5, 1963.

[10.36] K. J. Binns, and M. Dye, "Identification of the principal factors causing unbalanced magnetic pull in cage induction motors," *Proceedings of the IEE*, vol. 120, no. 3, 1973, pp. 349–354.

[10.37] M. T. Wright, D. S. M. Gould, and J. J. Middlemiss, "The Influence of Unbalanced Magnetic Pull on the Critical Speed of Flexible Shaft Induction Machines," Proceedings of the 1st IEE International, Conference on Electrical Machines, Design and Application. London, UK, 1982, pp. 61–64.

[10.38] S. Williamson and M. A. S. Abdel-Magied, "Unbalanced Magnetic Pull in Induction Motors with Asymmetrical Rotor Cages," Proceedings of the 2nd IEE International Conference on Electrical Machines, Design and Application. London, UK, 1985, pp. 218–222, 1989.

[10.39] R. Belmans, A. Vandenput, and W. Geyson, "Influence of unbalanced magnetic pull on the radial stability of flexible-shaft induction machines," *Proceedings of the IEE*, vol. 124, Pt B, no. 2, 1987.

[10.40] S. Williamson, M. A. Mueller, J. F. Eastham, and L. H. Lim, "Transient Unbalanced Magnetic Pull in Change-Pole Induction Motors," Proceedings of the4th IEE Conference on Electrical Machines and Drives Conference.

[10.41] D. G. Dorrell, "U.M.P. in Cage Induction Motors," Ph.D. Thesis, University of Cambridge, England, May 1993.

[10.42] M. J. DeBertoli, S. J. Salon, D. W. Burow, and C. J. Slavik, "Effects of rotor eccentricity and parallel windings on induction machine behavior: a study using finite element analysis," *IEEE Transactions on Magnetics*, vol. 29, no. 2, 1993, pp. 1676–1682.

[10.43] S. J. Salon, D. W. Burow, C. J. Slavik, and M. J. DeBertoli, "Effects of Airgap Ovality on Induction Machine Behavior," International Conference on Electrical Machines (ICEM 96), Vigo, Spain, vol. 1, pp. 401–404.

[10.44] D. G. Dorrell and A. C. Smith, "Calculation of unbalanced magnetic pull in small cage induction motors with eccentric rotors, part 2: experimental investigation," *IEE Proceedings of the Electric Power Applications*, vol. 143, 1996, pp. 202–210.

[10.45] A. Arkkio and O. Lindgren, "Unbalanced magnetic pull in a high speed induction motor with an eccentric rotor," *International Conference on Electrical Machines (ICEM 94)*, vol. 1, pp. 53–58

[10.46] A. J. M. Cardoso and E. S. Saraiva, "Predicting the level of airgap eccentricity in operating three-phase induction motors by park's vector approach," *IEEE Industry Applications Annual Meeting, Conference Record*, vol. 1, 1992, pp. 132–135.

[10.47] M. Drif and A. J. M. Cardoso, "Airgap Eccentricity Fault Diagnosis, in Three-phase Induction Motors by the Complex Apparent Power Signature Analysis," IEEE SPEEDAM, Conference Proceedings, 2006, pp. 61–65.

[10.48] A. J. M. Cardoso and E. S. Saraiva, "Computer aided detection of airgap eccentricity in operating three-phase induction motors by park's vector approach," *IEEE Transactions on Industry Applications*, vol. 9, no. 5, September/October 1993, pp. 897–901.

[10.49] A. Arkkio, M. Antila, K. Pokki, A. Simon, and E. Lantto, "Electromagnetic force on a whirling cage rotor," *IEE Proceedings – Electrical Power Applications*, vol. 147, no. 5, September 2000, pp. 353–360.

[10.50] D. G. Dorrell, "Sources and characteristics of unbalanced magnetic pull in three-phase cage induction motors with axial-varying rotor eccentricity," *IEEE Transactions on Industry Applications*, vol. 47, no. 1, January/February, 2011, pp. 12–24.

[10.51] D. G. Dorrell, J. K. H. Shek, and M.-F. Hsieh, "The Measurement and Indexing of Unbalanced Magnetic Pull in Electrical Machines," IEEE Energy Congress and Exposition (ECCE), 2014, pp. 4827–4834.

[10.52] S. Fruchtenicht, E. Pittius, and H. Seinsch, "A Diagnostic System for Three-Phase Asynchronous Machines," Proceeding of the IEE Conference, EMDA'89, vol. 310, IEE Savoy Place, London, 1989, pp. 163–171.

[10.53] A. Stavrou and J. Penman, "Modelling Dynamic Eccentricity in Smooth Airgap Induction Machines," Proceeding of the ICEM 94, Paris, France, 1994, pp. 238–243.

[10.54] A. Stavrou and J. Penman, "The on-line Quantification of Airgap Eccentricity in Induction Machines," IEEE International Conference on Electrical Machines and Drives, IEMDC, 2001, pp. 864–871

[10.55] H. A. Toliyat, M. S. Arefeen, and A. G. Parlos, "A method for dynamic simulation and detection of airgap eccentricity in induction motors", *IEEE Transactions on Industry Applications*, vol. 32, no. 4, 1996, pp. 910–918.

[10.56] X. Huang and T. G. Habetler, "Detection of mixed Airgap Eccentricity in Closed-loop Drive-Connected Induction Motors," 4th IEEE International Symposium on Diagnostics for Electric Machines, Power Electronics and Drives, SDEMPED, 2003, pp. 312–316.

[10.57] X. Huang, T. G. Habetler, and R. G. Harley, "Detection of rotor eccentricity in closed-loop drive-connected induction motor using an artificial neural network," *IEEE Transactions on Power Electronics*, vol. 22, no. 4, 2007, pp. 1552–1559.

[10.58] J. F. Bangura, R. J. Povineelli, N. A. Demerdash, and R. H. Brown, "Diagnostics of eccentricities and bar/end-ring connector breakages in polyphase induction motors through a combination of time-series data mining and time stepping coupled FE-state-space techniques," *IEEE Transactions on Industry Applications*, vol. 39, no. 4, July/August 2003.

CHAPTER 11

# CASE HISTORIES—SUCCESSFUL AND UNSUCCESSFUL APPLICATION OF MCSA TO ESTIMATE OPERATIONAL AIRGAP ECCENTRICITY IN SCIMS

*William T. Thomson*

## 11.0 SUMMARY AND LIST OF CASE HISTORIES

This chapter reports on 15 case histories (i.e., results from 15 motors), which cover successes and failures of MCSA to assess the operational airgap eccentricity in SCIMs. It is artificial to try to subdivide these case histories into different chapters and to assist the reader in selecting case histories of personal interest. A list of salient data and diagnosis from each case history is as follows:

**Section 11.2:** A centrifugal air compressor is driven by a 340 kW/456 HP, 4-pole SCIM. MCSA diagnosed airgap eccentricity <10%. This is a detailed step-by-step presentation in a coaching style.

**Section 11.3:** Two condensate extraction pumps are driven by 1230 kW/1650 HP, 4-pole SCIMs. MCSA diagnosed airgap eccentricity in motor A (<10%) and B (10–15%). This is a comparison between two nominally identical motors.

**Section 11.4:** Four slow speed cooling water, vertical pumps are driven by 1550 kW/2078 HP, 24-pole SCIMs. One of the motors had developed a severe vibration problem but vibration CM vendors could not identify the cause. MCSA diagnosed an airgap eccentricity of the order of 15% which was 3 times greater than the maximum allowable (5%) by the OEM for this vertical motor.

**Section 11.5:** Two centrifugal pumps are driven by 1400 kW/1876 HP, 8-pole SCIMs. MCSA identified high airgap eccentricity of at least 30% in one of

---

*Current Signature Analysis for Condition Monitoring of Cage Induction Motors: Industrial Application and Case Histories*, First Edition. William T. Thomson and Ian Culbert.

the motors. Vibration CM vendors could not identify the cause of the DE bearing vibration and temperature being at alarm level.

**Section 11.6:** A cooling water pump is driven by a 1200 kW/1608 HP, 4-pole SCIM. MCSA diagnosed 20–25% airgap eccentricity; the consequential vibration on the stator core and teeth caused severe damage to the semi-conductive coatings (often referred to as corona shields) on the coil sides of HV coils in the stator slot portions and the stator was rewound.

**Section 11.7:** A new, inverter-fed, HV, 6300 kW/8450 HP, 4-pole SCIM. During the commissioning of the motor it was run uncoupled but since the high frequency inverter harmonics coincided with RSPFs it was impossible to determine the airgap eccentricity.

**Section 11.8:** A centrifugal gas compressor was driven by an inverter-fed 4500 kW/6030 HP, 4-pole SCIM at 66% $I_{FL}$; the airgap eccentricity $<10\%$.

**Section 11.9:** A boiler primary air fan was driven by a 2238 kW/3000 HP, 6-pole SCIM. Interpretation of the current spectrum was complex since a pole-pair compatibility analysis was required (Sections 10.4.1 and 11.9.3) and it was estimated that the airgap eccentricity was 25–30%.

**Section 11.10:** A reciprocating compressor was driven by a 796 kW/1067 HP, 10-pole SCIM. MCSA was unable to detect airgap eccentricity features in the spectrum due to their coincidence with current components caused by the cyclic load.

**Section 11.11:** A centrifugal compressor was driven by a 6714 kW/9000 HP, 4-pole SCIM, but the number of rotor slots was unknown, however, the RSPF current components were found by an iterative signal processing methodology using a selection of different numbers of rotor slots (see Tables 4.3 and 4.4). The airgap eccentricity was less than 10%.

For clarity, a flow chart of the MCSA diagnostic and interpretation strategy is presented in Section 11.1 before the detailed presentation of each case history.

## 11.1 FLOW CHART OF MCSA PROCEDURE TO ESTIMATE OPERATIONAL AIRGAP ECCENTRICITY

Obtain the SCIM's nameplate data before going on-site to carry out an MCSA test. Obtain No. of rotor slots (R) from OEM
If R cannot be obtained, then see case history 11.11

Measure rms current to SCIM

Measure the supply frequency using a dB versus frequency spectrum, use, e.g., 0 to 130 Hz, and 12,800 lines

↓

Calculate $f_{rs(+1)}, f_{rs(-1)}, f_{rs(+3)}, f_{rs(+5)}$, at no-load and full-load

↓

Filter out supply frequency component and measure spectrum: $\{70 \text{ to } (f_{rs(+5)} + 100)\}$Hz

↓

Search for the $f_{rs(+1)}$ frequency component in the spectrum which should be between its value at no-load and full-load

↓

Reverse calculate the operating slip using the measured $f_{rs(+1)}$ and supply frequency. Calculate $f_r = (f(1-s)/p)$ using operating slip and frequency. Check that $f_{rs(+1)}, f_{rs(-1)}, f_{rs(+3)}, f_{rs(+5)}$ are $2f$ apart from each other.

↓

Select the highest RSPF component (in dB), which is normally $f_{rs(+1)}$ and produce a zoom spectrum around $f_{rs(+1)}$ or the highest RSPF (in dB), search for $\pm f_r$ around $f_{rs(+1)}$. If $\pm f_r$ components exist, measure average dB difference ($N_{ec}$) between $\pm f_r$ and $f_{rs(+1)}$

↓

Go to Table 10.3 and use $N_{ec}$ to obtain an estimate of operational airgap eccentricity band.

Recall from equation 10.4:

$$f_{rs} = f[(R/p)(1-s) \pm n_{\omega s}]$$

where

$f_{rs}$ = rotor slot passing frequencies in Hz, a series of frequency components spaced twice the supply frequency apart

$f$ = supply frequency

$R$ = number of rotor slots

$p$ = pole-pairs

$s$ = operational slip

$n_{\omega s}$ = integers for the fundamental time domain mmf ($n_{\omega s} = +1$), and its odd harmonics, 3, 5, …

## 11.2 CASE HISTORY (1989)—LOW LEVEL OF AIRGAP ECCENTRICITY IN A SCIM DRIVING A CENTRIFUGAL AIR COMPRESSOR

### 11.2.1 Summary

This case history gives a detailed presentation to illustrate that attention to detail is essential when carrying out MCSA to analyze the current to estimate the operational airgap eccentricity.

Motor data: 3-phase, 3300 V, 340 kW/456 HP, 74 A, 60 Hz, 1789 r/min SCIM. The number of rotor slots is 50 and the rotor winding is a single cage design. The motor was driving a centrifugal air compressor on an offshore oil production platform. The nominal design airgap was unavailable from the end user or OEM. The analysis estimated that the motor had a normal airgap eccentricity of certainly less than ±10% of the radial airgap.

### 11.2.2 MCSA Diagnosis

The operating current was 70 A, which is very close to the full-load current of 74 A, and therefore the full-load slip can be used for the initial predictions.

The current spectrum in Figure 11.1 confirms that the operating supply frequency during the MCSA tests was 59.9 Hz and not 60 Hz. This means that the synchronous speed of the rotating magnetic field produced by the stator winding was 1797 r/min and not 1800 r/min.

The RSPF components $f_{rs(+1)}, f_{rs(-1)}, f_{rs(+3)}$ are calculated at full-load and no-load using equation (10.4) and the measured supply frequency of 59.9 Hz.

$$f_{rs} = f[(R/p)(1 - s) \pm n_{\omega s}]$$

Slip on no-load is taken to be zero.

Nominal full-load slip, $s_{FL} = (1800 - 1789)/1800 = 0.0061$ or (0.61%).

where

$f = 59.9$ Hz, $R = 50$, $p = 2$ pole-pairs, $s = 0$ on no-load, and $s = 0.0061$ on full-load

$f_{rs(+1)} = 26f$ and 1548.3 Hz; for $s = 0$ and $s = 0.0061$, respectively, and for $+n_{\omega s} = +1$

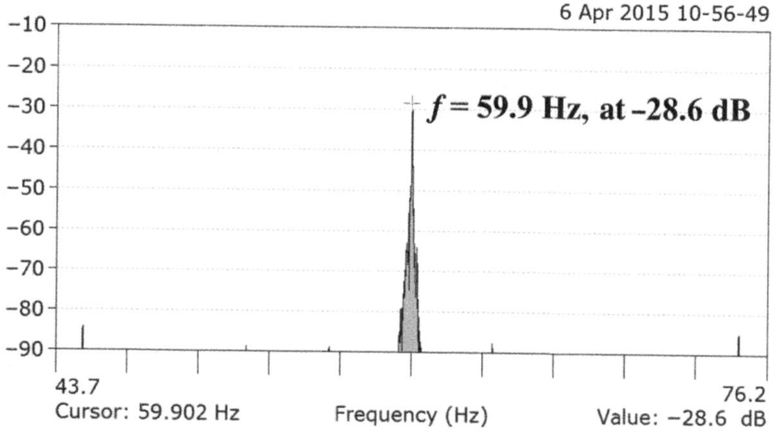


Figure 11.1 Current zoom spectrum, baseband span 0–130 Hz, 12,800 lines, frequency resolution is 10.16 mHz/line.

$f_{rs(-1)} = 24f$ and 1428.5 Hz; for $s = 0$ and $s = 0.0061$, respectively, and for $-n_{\omega s} = -1$

$f_{rs(+3)} = 28f$ and 1668 Hz; for $s = 0$ and $s = 0.0061$, respectively, and for $+n_{\omega s} = +3$

$f_{rs(-3)} = 22f$ and 1308.5 Hz; for $s = 0$ and $s = 0.0061$, respectively, and for $-n_{\omega s} = -3$

Note that $f_{rs(+1)}$ is often referred to as the principal RSPF.

The next step is to select the baseband frequency span but "with the supply frequency component filtered out" and for this case it is taken to be 70–1800 Hz.

It is important to identify the harmonics of the supply so that they can be disregarded with respect to the airgap eccentricity analysis. Figure 11.2 is with the motor operating at 70 A and gives the classical odd harmonics up to the 29th of the supply frequency (59.9 Hz) with the highest in dB being the 5th harmonic at 46 dB down on $f$ which is 0.5% of the supply frequency current component. Figure 11.3 shows a component at 1548.4 Hz which is the same as the predicted $f_{rs(+1)}$ component at 1548.3 Hz for an operational full-load slip of 0.0061.

Recall from section 10.4.3 on frequency resolution and accuracy of presentation it is perfectly valid and very often necessary in MCSA to quote the frequency up to two decimal places, particularly when measuring the supply frequency and $f_r$ via high resolution zoom spectrum analysis.

The next step is to carry out a zoom spectrum around $f_{rs(+1)}$, and this is shown in Figure 11.4, which verifies that another two of the inherent and normal RSPF current components exist and also confirms that there is a difference of $2f$ (119.8 Hz) between each of these components with respect to $f_{rs(+1)}$.

For completeness, the frequencies and magnitudes of the rotor slot passing frequency current components are as follows:

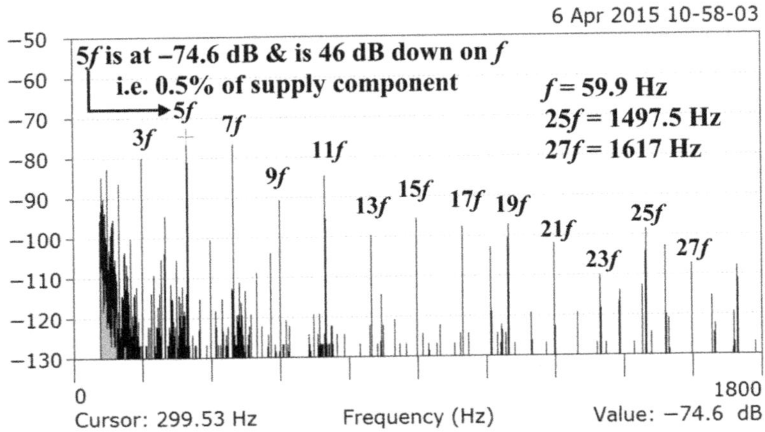


Figure 11.2 Current spectrum, baseband span 70–1800 Hz, 12,800 spectral lines, frequency resolution is 0.135 Hz/line.

$f_{rs(+1)} = 1548.4$ Hz, for $+n_{\omega s} = +1$, at −103 dB
$f_{rs(-1)} = 1428.6$ Hz, for $-n_{\omega s} = -1$, at −114 dB
$f_{rs(+3)} = 1668$ Hz, for $+n_{\omega s} = +3$, at −116 dB

The supply component $f$ is at −28.6 dB and it is interesting to note the following:

$f_{rs(+1)}$ is 74.6 dB smaller than $f$
$f_{rs(-1)}$ is 83.6 dB smaller than $f$
$f_{rs(+3)}$ is 85.6 dB smaller than $f$

This demonstrates the advantage of the dB scale since very small components can be identified, which would not be visible in a linear spectrum. As is often the

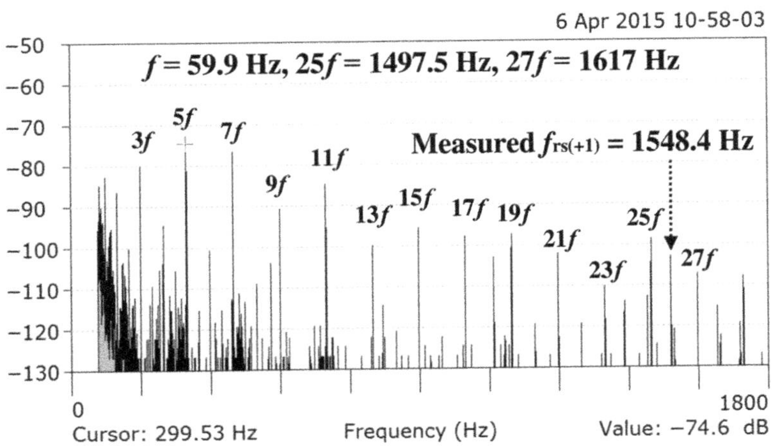


Figure 11.3 Current spectrum, baseband span 70–1800 Hz, 12,800 spectral lines, frequency resolution 0.134 Hz/line, $f_{rs(+1)}$ is identified as 1548.4 Hz, operating current was 70 A.

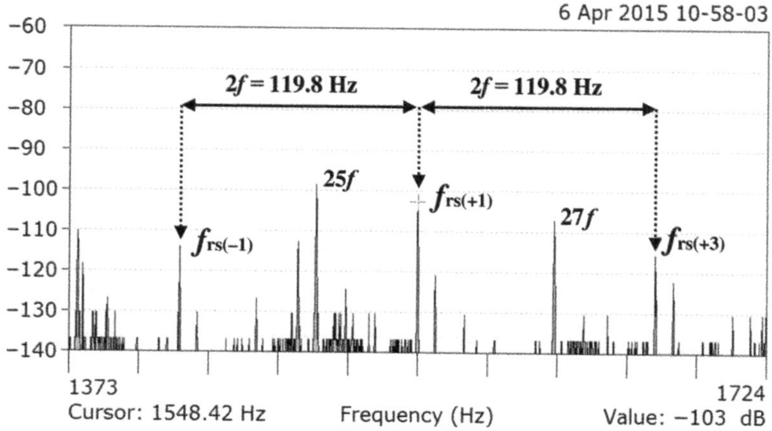


Figure 11.4 Current zoom spectrum showing $f_{rs(+1)}$, $f_{rs(-1)}$, and $f_{rs(+3)}$, spaced $2f$ apart.

case, the principal RSPF component ($f_{rs(+1)}$) has the largest dB magnitude (note $n_{\omega s} = +1$ integer corresponding to the fundamental mmf) and is therefore selected for the final zoom analysis to display the current spectrum, which is a function of the combination of static and dynamic airgap eccentricity. Equation (10.7) gives the equation that predicts the signature pattern, which is a function of airgap eccentricity:

$$f_{ec} = f[(R/p)(1 - s) \pm n_{\omega s}] \pm f(1 - s)/p$$

$$f_{ec} = f_{rs} \pm f_r$$

The speed frequency ($f_r = f(1 - s)/p$) of the rotor can now be calculated, but firstly the actual operating slip must be back calculated from the measured value of the principal rotor slot passing frequency, $f_{rs(+1)} = 1548.4$ Hz

Operating slip,

$$s = 1 - \{((f_{rs(+1)}/f) - 1)(p/R)\} = 1 - \{(1548.42/59.9) - 1)(2/50)\}$$

$$s = 0.006(0.6\%)$$

The rotor speed frequency, $f_r = f(1 - s)/p = 59.9(1 - 0.006)/2 = 29.77$ Hz

The rotor speed = 1786 r/min

Recall that the nominal, full-load nameplate speed was 1789 r/min when supplied at 60 Hz with the motor taking full-load current but the supply frequency was 59.9 Hz, hence the speed at full-load is less than the rated speed. The final zoom current spectrum is shown in Figure 11.5, which has a span of 87 Hz and a frequency resolution per line of 0.135 Hz/line as per the spectra displayed in Figures 11.2 and 11.3.

The interpretation of Figure 11.5 confirms that there are no $\pm f_r$ components around $f_{rs(+1)}$ and therefore the airgap eccentricity is perfectly normal and certainly less than ±10% of the radial airgap.

For completeness, it is interesting to note that an even harmonic at $26f$ exists in Figure 11.5. This requires an explanation, since classical 3-phase SCIM theory

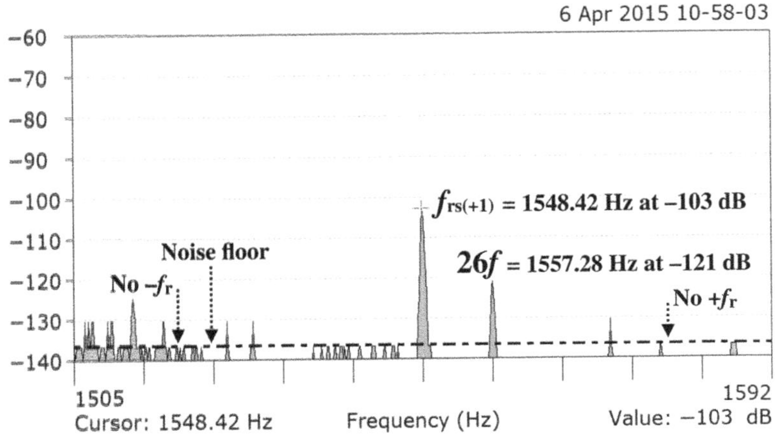


Figure 11.5 Current zoom spectrum around $f_{rs(+1)}$, no $\pm f_r$ components around $f_{rs(+1)}$.

states that no even harmonics can exist in the current to a 3-phase SCIM when it is supplied by a "perfectly balanced" 3-phase voltage supply where each phase is exactly of the same voltage and displaced by exactly 120 degrees in the time domain and also that the axes of the spatially distributed 3-phase windings "are spaced exactly 120 electrical degrees apart." However, perfect symmetry cannot be achieved in practice and with a minuscule deviation away from the 120 degrees, even harmonics of the supply frequency can flow. The key factor is that they would not be detected when the supply frequency and its harmonics are measured and displayed via a linear current spectrum, which would be dominated by the supply frequency and consequential odd harmonics.

With a dB logarithmic spectrum having a dynamic range of 80 dB (a factor of 10,000 in absolute linear units) the $26f$ current component in Figure 11.5 is visible in the filtered spectrum with the supply component removed. Although it is minuscule at 92.5 dB down on the supply component, when the airgap eccentricity analysis is carried out the even harmonic at $26f$ is in the zone of interpretation. Thus all harmonics of the fundamental must be accurately measured and subsequently disregarded in the analysis to prevent any misinterpretation.

### 11.2.3 Illustration of Airgap Eccentricity Analysis During a Lightly Loaded Run and a No-Load Uncoupled Run

As an illustration, assume this motor was operating on very light load (still supplied at 59.9 Hz) with an operational slip of 0.1% compared to the actual full-load slip of 0.6%, thus the operational $f_{rs(+1)}$ component would be 1556 Hz which is only 1.3 Hz smaller than the even harmonic at 1557.3 Hz. Figure 11.5a illustrates the new position (not an actual measurement) of $f_{rs(+1)}$, which is now very close to the even harmonic at $26f$.

Now consider this motor during a no-load, uncoupled run on the offshore platform the slip would be very close to zero and experience has shown that $f_{rs(+1)}$ can virtually coincide with $26f$. To separate these two components requires a much higher

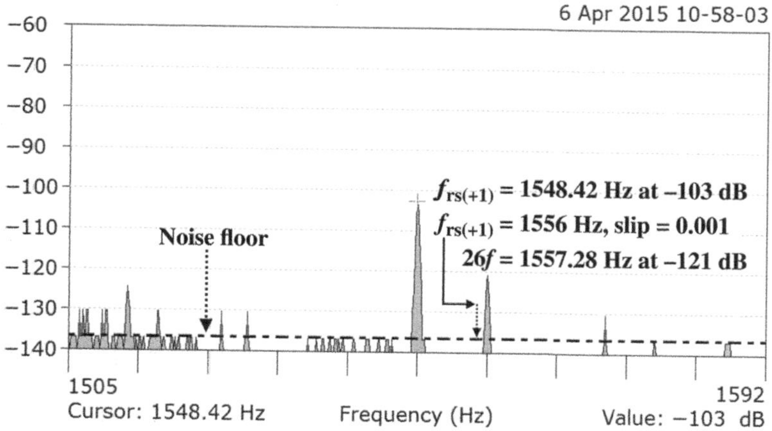


Figure 11.5a An illustration of the position of $f_{rs(+1)}$ at a slip of 0.001 (0.1%).

number of spectral lines, for example, 25,600 or preferably 51,200, which gives 0.0676 and 0.0338 Hz/line, respectively, for this case history with a baseband frequency span from 70 to 1800 Hz. Please refer to Section 4.4 on spectrum analysis of current for further information on signal processing terminology.

Therefore, for this motor, if the difference was only 0.5 Hz between $26f$ and $f_{rs(+1)}$ during a no-load, uncoupled run, if 51,200 spectral lines were available in an MCSA instrument or spectrum analyzer, there will be 15 spectral lines between $26f$ and $f_{rs(+1)}$, which is sufficient to provide separation. With $f_{rs(+1)}$ a difference of 0.5 Hz down from $26f$ gives $f_{rs(+1)}$ equal to 1556.78 Hz, and using equation (10.4) to back calculate the corresponding no-load slip gives a value of 0.0004 (0.04%) which is equivalent to 1796 r/min when the synchronous speed is 1797 r/min (recall the frequency was 59.9 Hz) for this motor. This is a drop of 1 r/min which is typical for a plain bearing SCIM operating during an uncoupled, no-load run.

## 11.3 TWO CASE HISTORIES (2004)—OPERATIONAL AIRGAP ECCENTRICITY IN NOMINALLY IDENTICAL SCIMS DRIVING PUMPS IN A CCGT POWER STATION

### 11.3.1 MCSA Measurements and Diagnosis

Two nominally identical SCIMs were used to drive "Condensate Extraction Pumps" in a Combined Cycle Gas Turbine (CCGT) power station. The need for these pumps is that after the steam has passed through a gas turbine and it has given up all its energy to the turbine, it collapses back into water in the condenser. This water is called "condensate" and the pumps remove the condensate from the condenser and deliver it to the feed system for delivery to the boilers, to be raised back into steam for reuse in the turbine, in what is theoretically a closed system.

The nameplate provided the following data: 3-phase, 3300 V, 1230 kW/1650 HP, 244 A, 50 Hz, 1485 r/min SCIM. The number of rotor

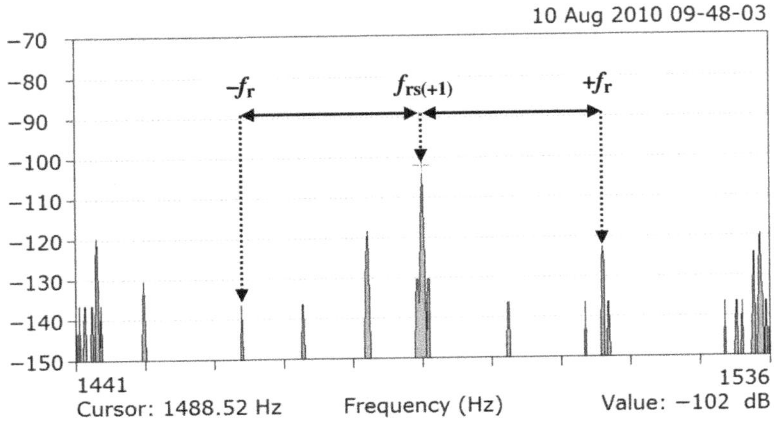


Figure 11.6 Current zoom spectrum, motor A, $f_{rs(+1)}$ is 1488.5 Hz at −102 dB, operating current was 224 A.

slots is 58 and the rotor winding is a single cage design but the design value of the airgap is unknown. The supply frequencies during the testing of motors A and B were 50.044 and 49.97 Hz, respectively, and their operational currents were 224 and 216 A, respectively, compared to a nominal full-load current of 244 A. Therefore, the motors were operating very close to the nominal full-load current and speed, thus the corresponding full-load slip was used to initially calculate the principal rotor slot passing frequency for both motors.

**Motor A:** $f_{rs} = f[(R/p)(1 - s) \pm n_{\omega s}]$
where $f = 50.044$ Hz, $R = 58$, $p = 2$ pole-pairs, $s_{FL} = 0.01$

$$f_{rs(+1)} = 1486.8 \text{ Hz for } +n_{\omega s} = +1$$

**Motor B:** $f_{rs} = f[(R/p)(1 - s) \pm n_{\omega s}]$
where $f = 49.97$ Hz, $R = 58$, $p = 2$ pole-pairs, $s_{FL} = 0.01$

$$f_{rs(+1)} = 1484.6 \text{ Hz for } +n_{\omega s} = +1$$

The zoom current spectra for motors A and B are shown in Figures 11.6 and 11.7, respectively, and the principal RSPF ($f_{rs(+1)}$) components for motors A and B from the spectra are 1488.5 and 1487.7 Hz. The magnitude of the $f_{rs(+1)}$ components in motors A and B is the same at −102 dB and this is expected since the operating amperes differ by only 8 A and as a percentage of the nominal full-load current of 244 A, this is only 3.3%.

Recall that the magnitudes (in dB) of the RSPF flux components are a function of numerous parameters (Section 10.4.1) and one of these is the leakage flux around each of the current carrying rotor bars and as the rotor current increases with load current the magnitude of these flux components and hence the corresponding induced emfs and currents in the stator increase between no-load and full-load.

This fact is subsequently proven in case history 11.6 via Figures 11.33 and 11.34, which present the principal RSPF component at a fixed airgap eccentricity

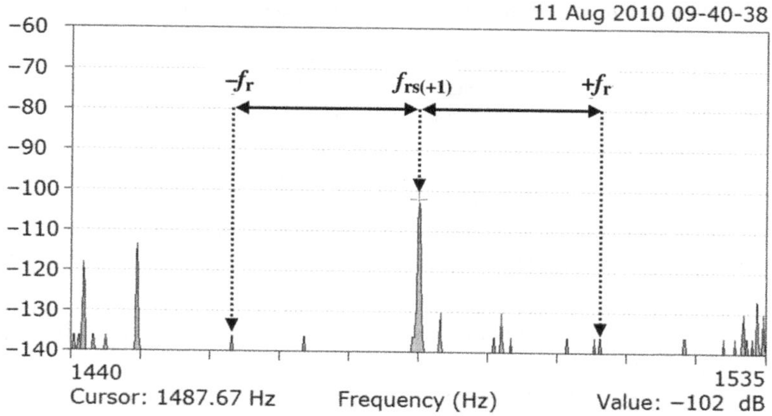


Figure 11.7 Current zoom spectrum, motor B, $f_{rs(+1)}$ is 1487.7 Hz at –102 dB, operating current was 216 A.

with a SCIM on no-load and at 46% $I_{FL}$ and the RSPF component has increased by 10 dB (a factor of 3.16 times larger).

The actual operating slips for motors A and B are back calculated from the measured $f_{rs(+1)}$ components and actual supply frequencies.

Motor A: operating slip, $s = 0.0088$

Motor B: operating slip, $s = 0.0079$

The slips are different even though the operating amperes differ by only 3.3% with respect to the full-load amperes, but note the supply frequencies are different and the synchronous speeds of the rotating magnetic fields from the stator winding are 1501 r/min and 1499 r/min for motors A and B, respectively—a difference of 2 r/min

The rotational speed frequency $f_r = f(1 - s)/p$ for motors A and B at the operating frequencies and slips are

Motor A: $f_r = 24.8$ Hz, rotor speed = 1488 r/min

Motor B: $f_r = 24.788$ Hz (very close to 24.8 Hz), rotor speed = 1488 r/min

Interpretation of the spectra in Figures 11.6 and 11.7 verifies that $\pm f_r$ components exist around each of the $f_{rs(+1)}$ components:

Motor A:

The $+f_r$ component at 24.8 Hz up from $f_{rs(+1)}$ is 21 dB down on $f_{rs(+1)}$

The $-f_r$ component at 24.8 Hz down from $f_{rs(+1)}$ is 34 dB down on $f_{rs(+1)}$

The dB difference between $+f_r$ and $-f_r$ at 13 dB, is too large a difference to use the arithmetic sum to calculate the average dB difference $N_{ec}$ so the true value of $N_{ec}$ is calculated as shown in Section 4.5.3 C, for this case (Figure 11.6).

Convert to absolute differences and take the average, $\{10^{-\left(\frac{21}{20}\right)} + 10^{-\left(\frac{34}{20}\right)}\}/2 = 54.5 \times 10^{-3}$

**EXTRACT FROM TABLE 10.3 Estimate of Operational Airgap Eccentricity from Current Signature Pattern**

| | |
|---|---|
| $N_{ec}$<br>Average dB difference between $f_{rs(+1)}$ and $\pm f_r$ components | Estimate of "operational" airgap eccentricity, i.e., combination of static and dynamic |
| $N_{ec}$ = 35–40 dB | Airgap eccentricity is within 10% of the radial airgap |
| $N_{ec}$ = 25–35 dB<br>MCSA measured $N_{ec}$ = 25 dB | Airgap eccentricity is ≥10% of radial airgap and at $N_{ec}$ = 25 dB, it is typically around 15% of the radial airgap |

Convert back to obtain the average dB difference, $N_{av} = 20\{\log_{10}(1/(54.5 \times 10^{-3})\} = 25$ dB

The true average dB difference between the $\pm f_r$ components and $f_{rs(+1)}$ is $N_{ec}$ = 25 dB.

An extract from Table 10.3 in Section 10.4.2 proposes that this equates to an estimated operational airgap eccentricity of 15%, which is acceptable for an operational airgap eccentricity in a horizontal motor of 30 years old.

Motor B:

The $+f_r$ component at 24.8 Hz up from $f_{rs(+1)}$ is 34 dB down on $f_{rs(+1)}$

The $-f_r$ component at 24.8 Hz down from $f_{rs(+1)}$ is 34 dB down on $f_{rs(+1)}$

The average dB difference between the $\pm f_r$ components and $f_{rs(+1)}$ is $N_{ec}$ = 34 dB

This equates to an estimated operational airgap eccentricity of close to 10% (it is a borderline case between two categories) but is of course acceptable as an operational airgap eccentricity. The analysis indicates the differences in the current spectra between two nominally identical motors, which both have acceptable operational airgap eccentricity levels and these spectra can be trended to identify any increase in the estimated airgap eccentricity.

## 11.4 FOUR CASE HISTORIES (2005)—ABNORMAL LEVEL OF AIRGAP ECCENTRICITY IN A LARGE, LOW SPEED, HV MOTOR DRIVING A COOLING WATER PUMP IN A POWER STATION

### 11.4.1 Summary

The four main cooling water pumps in a CCGT power station are driven by the following four SCIMs: 3-phase, 11 kV, 1550 kW/2078 HP 123 A, 50 Hz, 246 r/min. These motors are referred to as CW1, CW2, CW3, and CW4, they are low speed, large diameter, 24-pole, vertically mounted and were manufactured in 1975. The rotors are single cage with 162 rotor bars and the OEM's airgap design value was 2 mm/80 mils ± (a maximum of 5%). Photos of one of the CW motors on its pump stool are shown in Figures 11.8a and 11.8b.

In 1997, after 22 years in regular operation, a stator winding failed in unit CW1 and the power station subsequently decided to carry out a phased sequence of

(a) (b)

Figure 11.8 (**a**) and (**b**) Photos of CW units. Reproduced with permission of EM Diagnostics Ltd.

inspections, overhauls, and stator rewinds (where necessary) on all CW pump motors. The motor in slot CW3 was overhauled and its stator was rewound by a motor repair company, it was returned to service with no vibration measurements having been taken during the uncoupled and coupled runs at re-commissioning.

There are no permanent vibration monitoring sensors on the bearing housings and it seemed that the motor had no operational problems after the repairs. In 2002, during a routine look, listen, and feel check of the motor in CW3 the maintenance electrician reported to the power station's senior electrical engineer, that while standing on top of the motor, there was "an oscillatory movement best described as being on a ship." Further checks indicated that the drive end bearing (DE) housing was visually deflecting and CW3 was subsequently shut down and put into a standby mode as an emergency spare unit.

Vibration measurements and analyses were carried out during 2003 by three different companies (not the authors of this book) but they all had different ideas as to the possible cause(s) of the problem and suggested the following:

**(i)** Check the drive unit's alignment.

**(ii)** Inspect the bearings.

**(iii)** Carry out visual inspections of the rotor and stator.

**(iv)** Check the airgaps at each end with the motor *in situ* (but this was not possible) as is evident from the photos in Figures 11.8a and 11.8b.

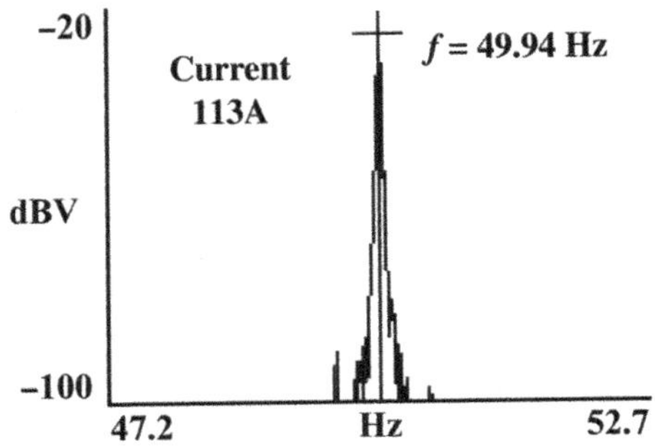


Figure 11.8 CW1 current spectrum.

These can be typical suggestions made by condition monitoring companies who do not have expertise in the design, manufacture, and operation of SCIMs and is what the authors of this book call the "shopping list of possible checks" given to the end user without a definitive diagnosis of the cause of the problem. At the end of 2003, the findings were inconclusive and the fundamental cause of the problem had not been identified.

The power utility wanted to establish the root cause of the problem before embarking on the removal of the CW3 motor for a complete strip down, since it had been rewound and overhauled only 3 years previously. The cause of the symptoms remained unresolved for another 2 years until MCSA which diagnosed an unacceptable operational level of airgap eccentricity of the order of 15% for this vertically mounted SCIM was applied for the first time. The OEM's specification, at the time of design and manufacture allowed a maximum of 5%.

### 11.4.2 MCSA Measurements and Analysis

The current spectra for the four SCIMs are shown in Figures 11.8–11.11, which gives the exact operational frequency applied to each motor during the MCSA measurements.

The full-load current is 123 A and the operating current for CW3 was 115 A. Therefore, the motor was running close enough to full-load to allow the nominal

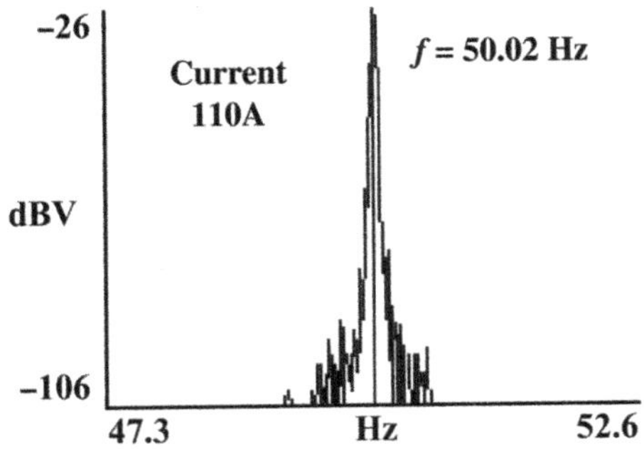


Figure 11.9 CW2 current spectrum.

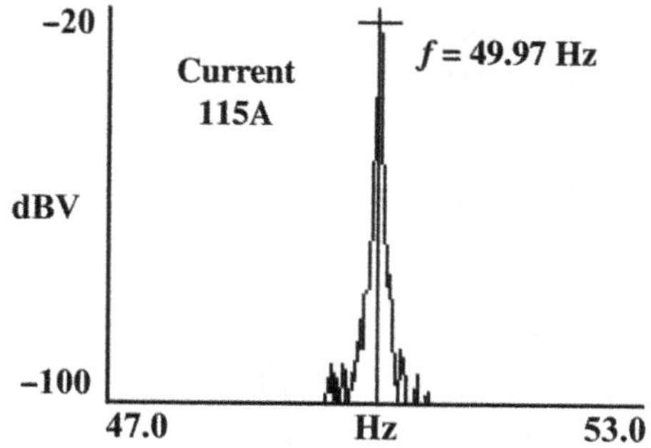


Figure 11.10 CW3 current spectrum.

full-load slip calculated from the nameplate speed and frequency to be used to predict the principal RSPF ($f_{rs(+1)}$) using the measured supply frequency of 49.97 Hz.

$$s_{FL} = (250 - 246)/250 = 0.016(1.6\%)$$

$$f_{rs} = f[(R/p)(1 - s) \pm n_{\omega s}]$$

where $f = 49.97$ Hz, $R = 162$, $p = 12$ pole-pairs, $s_{FL} = 0.016$

$$f_{rs(+1)} = 713.77\,\text{Hz}, \text{ for } + n_{\omega s} = +1$$

The baseband frequency span for the spectrum of the filtered current from the flow chart in Section 11.1 is

$$70 \text{ to } (f_{rs(+5)} + 100) = 70\text{–}915 \text{ Hz}$$

The baseband upper frequency was set to 1000 Hz, which using 12,800 spectral lines gives a frequency resolution of 0.066 Hz/line. The current zoom spectra for each of the four CW pump motors are presented in Figures 11.12–11.15, to identify $f_{rs(+1)}$ from each motor.

As expected, the measured frequency values of $f_{rs(+1)}$ (from 715 to 717 Hz) differ from the predicted value (713.77 Hz), since the operational currents and slips are less than the full-load values and the supply frequencies are also different. The

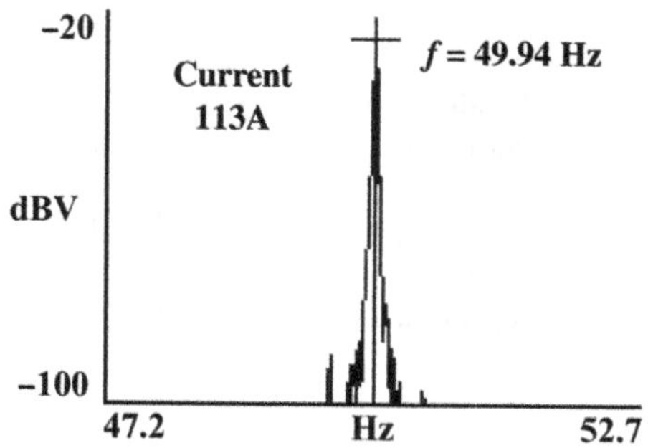


Figure 11.11 CW4 current spectrum.

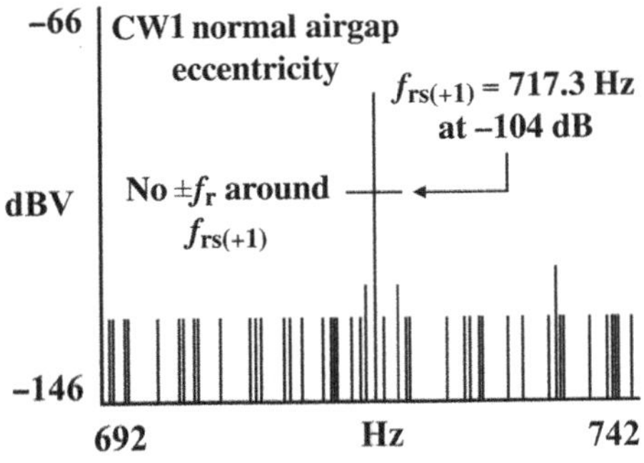


Figure 11.12 CW1 current spectrum

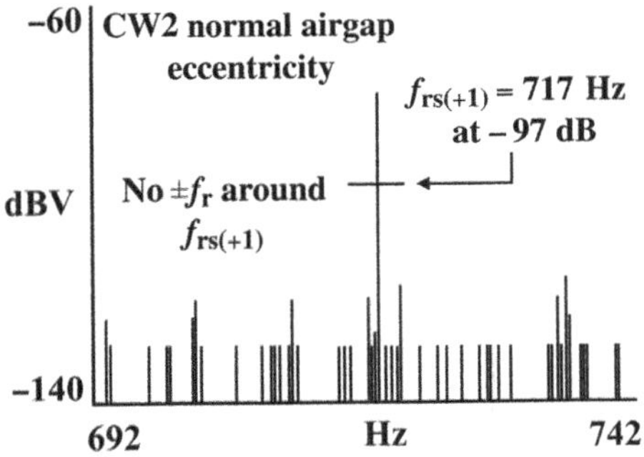


Figure 11.13 CW2 current spectrum.

next step is to reverse calculate the actual operating slip using equation (10.4) and the measured $f_{rs(+1)}$ and supply frequency which are 716.3 and 49.97 Hz, respectively.

Operating slip,

$$s = 1 - \{((f_{rs(+1)}/f) - 1)(p/R)\} = 1 - \{(716.3/49.97) - 1\}(12/162)$$

$$s = 0.0122(1.22\%)$$

The rotor speed frequency, $f_r = f(1 - s)/p = 49.97(1 - 0.0122)/12 =$ 4.113 Hz

The rotor speed = 246.8 r/min

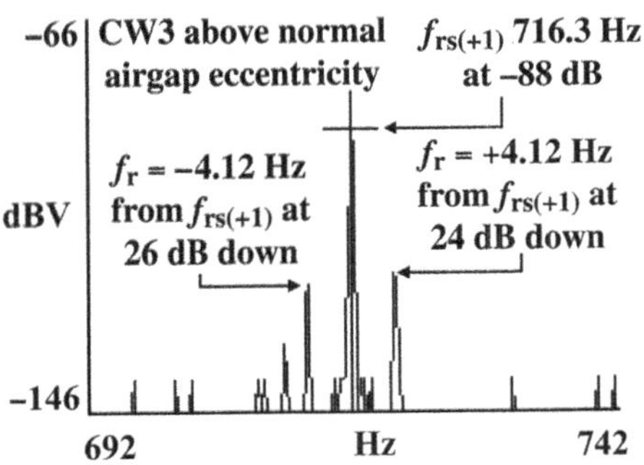


Figure 11.14 CW3 current spectrum.

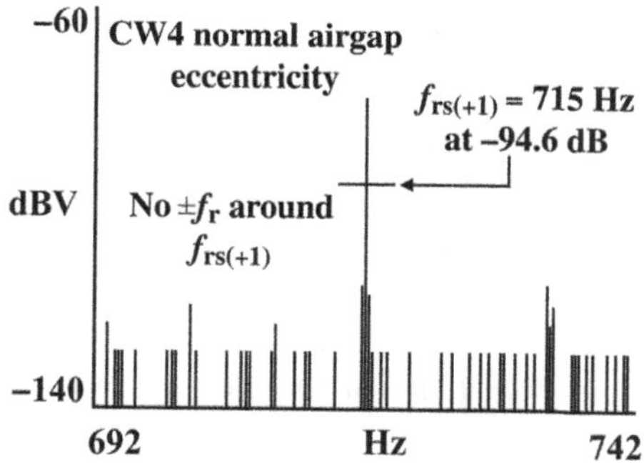


Figure 11.15 CW4 current spectrum.

Note that the nominal full-load speed of 246 r/min at a full-load current of 123 A is 0.016 (1.6%), hence a slip of 1.22% (246.8 r/min) at 115 A is as expected. In fact, MCSA is a very accurate method of measuring the exact slip and speed of the motor. An interpretation of the current zoom spectra in Figures 11.12–11.15 clearly shows that there are no $\pm f_r$ components around $f_{rs(+1)}$ in the CW1, CW2, and CW4 motors but the $\pm f_r$ components are very evident indeed in the spectrum shown in Figure 11.14 for CW3. The average dB difference ($N_{ec}$) between the $f_{rs(+1)}$ and $\pm f_r$ components is 25 dB and an extract from Table 10.3 indicates that the operational airgap eccentricity (combination of static and dynamic) is of the order of 15% of the radial airgap. Recall that the OEM's original specification was that the airgap eccentricity should be $\leq$5%. These motors had been in service for 30 years and have been overhauled by various repair companies during their life time.

The recommendation based on the MCSA result was that the power utility should remove the CW3 motor and return it to the OEM for strip down and a full inspection but it remained on-site as a workable spare for a further 2 years.

### 11.4.3 Airgap Measurements Before and After Modifications and Final MCSA Test

The measured airgaps by the OEM at the DE and NDE before any modifications were as follows:

Note that the actual measurements were recorded in mils (or thou) and not in millimeter.

Figure 11.16 for the DE gives the following:

**AN EXTRACT FROM TABLE 10.3 Estimate of Operational Airgap Eccentricity from Current Signature Pattern**

| $N_{ec}$ Average dB difference between EMBED Equation. DSMT4 $f_{rs(+1)}$ and $\pm f_r$ components | **Estimate of "operational" airgap eccentricity i.e., combination of static and dynamic** |
|---|---|
| $N_{ec}$ = 25–35 dB<br>MCSA measured $N_{ec}$ = 25 dB | Airgap eccentricity is $\geq$10% of radial airgap but at $N_{ec}$ = 25 dB, it is typically around 15% of the radial airgap |

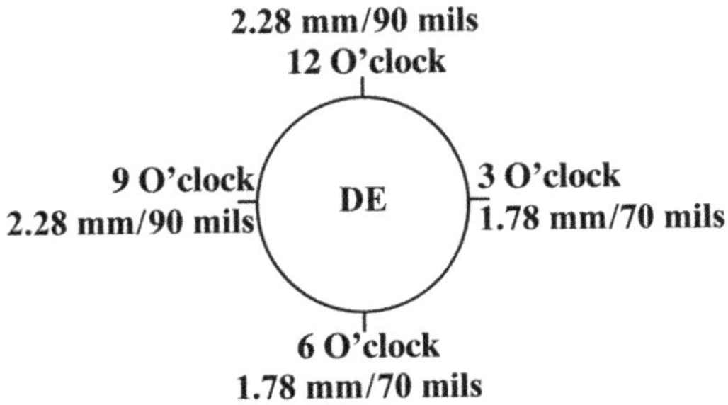


Figure 11.16 DE airgap measurements.

Total gap = 160 mils (4.0 mm)

Mean airgap = 80 mils (2.0 mm)

Actual airgap eccentricity = ±10/80 or 0.254/2.032 = ±12.5% of mean airgap

Therefore, the airgap eccentricity was 7.5% above the maximum design tolerance of 5%.

Figure 11.17 for the NDE gives the following:

Total gap between 3 and 9 O'clock = 150 mils =3.81 mm

Mean airgap = 75 mils (1.905 mm)

Actual airgap eccentricity = ±5/75 or 0.127/1.9 = ±6.67% of mean airgap

Therefore, the airgap eccentricity is 1.67% above the maximum design tolerance of 5%.

The OEM considered the DE airgap eccentricity at 12.5% was unacceptably high for this vertical motor since the design and manufacture specification was for an airgap eccentricity level of ≤5% of the radial airgap. The MCSA result and interpretation estimated that there was an operational airgap eccentricity of the order of around 15%, which is higher than the measured value of 12.5% using feeler gauges. However, it needs to be recognized that due to unbalanced magnetic pull (UMP) forces the actual operational airgap eccentricity can certainly be higher than is measured via feeler gauges when the motor is stationary and at ambient temperature.

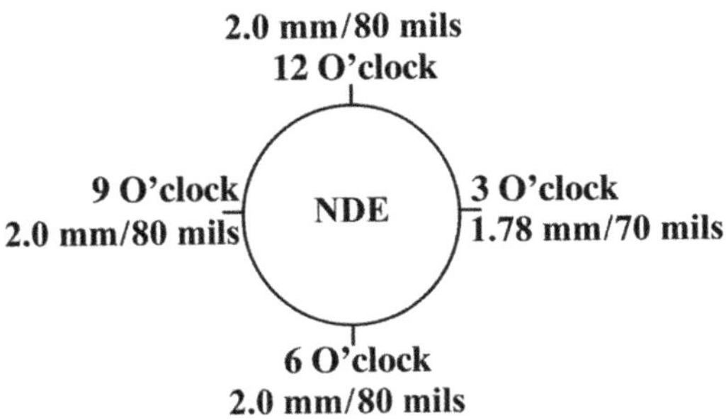


Figure 11.17 NDE airgap measurements.

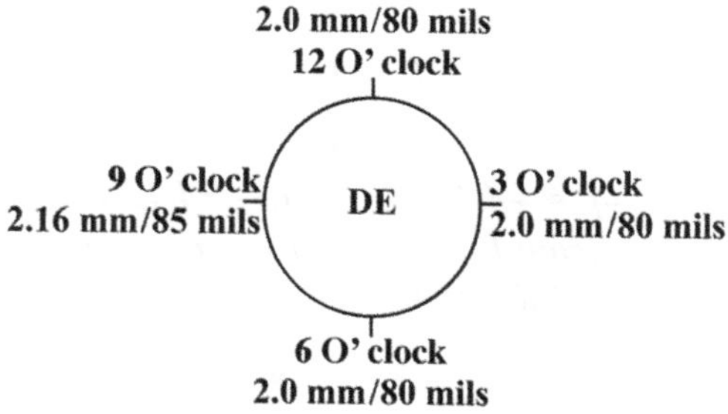


Figure 11.18 DE airgap measurements.

The locating spigot on the DE end frame (or end bell) was machined so that the end shield could be adjusted to reduce the airgap eccentricity. After that adjustment, the following airgaps were achieved as reported in Figures 11.18 and 11.19. The DE end shield was doweled to the motor's frame to ensure correct location in the future.

Figure 11.18 for the DE gives the following:

Total gap = 165 mils (4.2 mm)

Mean airgap = 82.5 mils (2.1 mm)

Actual airgap eccentricity = ±2.5/82.5 = ±3% of mean airgap

This is a reduction of 9.5% compared to the as-received airgap eccentricity in the motor.

Figure 11.19 for the NDE gives the following:

Total gap = 165 mils (4.2 mm)

Mean airgap = 82.5 mils (2.1 mm)

Actual airgap eccentricity = ±2.5/82.5 = ±3% of mean airgap

This is a reduction of 3.67% compared to the as-received airgap eccentricity in the motor.

The current spectrum shown in Figure 11.20 is from CW3 after re-commissioning, which confirms the $\pm f_r$ components have dropped to 41 dB down on $f_{rs(+1)}$ compared to 25 dB down on $f_{rs(+1)}$ as shown in Figure 11.14 when the air-gap eccentricity problem was detected via MCSA. An extract from Table 10.3 states that when $N_{ec}$ is ≥40 dB then the airgap eccentricity is ≤5%, which it is, since the

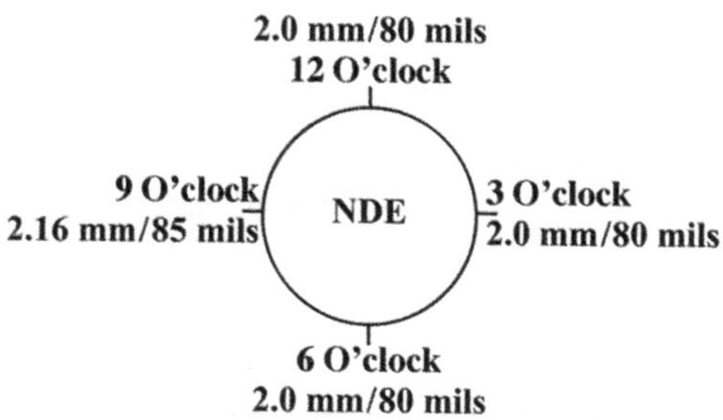


Figure 11.19 DE airgap measurements.

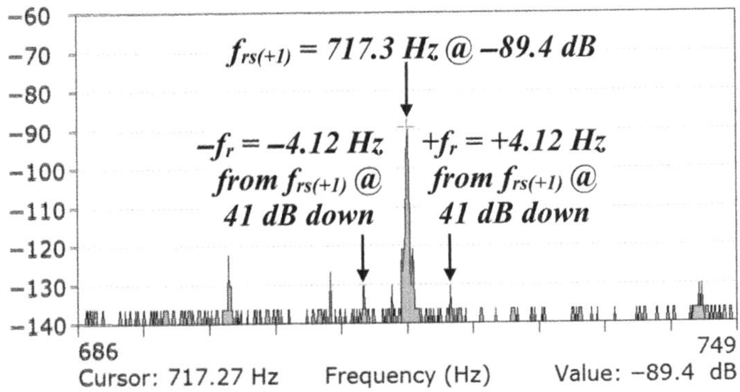


Figure 11.20 Current spectrum after modifications to give 3% airgap eccentricity.

new value after the modifications was 3% and this further confirms the success of MCSA in this case history. Recall that vibration measurements and analysis (by third party vendors (not the authors of this book) at the power station could not identify the root cause of the reported vibration problem, whereas MCSA successfully made that identification.

The motor has continued to operate normally without a repeat of the original problem and the unacceptable motion at the NDE disappeared.

## 11.5 CASE HISTORY (1988)—HIGH LEVEL OF AIRGAP ECCENTRICITY IN AN HV SCIM DRIVING A PUMP IN A LARGE OIL STORAGE TANK FACILITY

### 11.5.1 Summary

Four pumps, which are driven by nominally identical SCIMs, are used to load oil from an oil storage facility onto large tankers in a deep water estuary. The oil storage depot is supplied by a continuous flow of oil via a pipeline from offshore oil production platforms.

Motor data: 3-phase, 11 kV, 1400 kW/1876 HP, 103 A, 742 r/min, 50 Hz, star connected. The number of rotor slots is 62. The nominal design and manufactured radial airgap specified by the OEM was 2 mm/80 mils ± 10%.

**EXTRACT FROM TABLE 10.3 Estimate of Operational Airgap Eccentricity from Current Signature Pattern**

| $N_{ec}$<br>Average dB difference between $f_{rs(+1)}$ and $\pm f_r$ components | **Estimate of "operational" airgap eccentricity i.e., combination of static and dynamic** |
|---|---|
| $N_{ec} \geq 40$ dB<br>MCSA measured $N_{ec} = 41$ dB | Airgap eccentricity is within 5% of the radial airgap |

The pump in unit C had failed and was due for return to service in 20 days, the end user's proposed strategy was to move the motor in drive train C into drive train A so that they had units A, B, and D operational. The faulty motor in unit A could run for up to 1 hour after start up from cold before going into the alarm state for vibration and temperature at the DE bearing. The end user wanted to determine if the problem could be due to broken rotor bars or high airgap eccentricity, before removing the motor from site for a full inspection to establish the root cause of the problem.

The DE bearing in one of the motors was reaching its alarm level settings for bearing temperature and vibration of 85°C/185°F and 127 μm/5 mils pk-pk, respectively. Neither the on-site mechanical engineers nor the incumbent vendor who provided vibration monitoring services could identify the cause of the problem and no additional information on the problem was provided. MCSA results indicated that the fundamental cause was high airgap eccentricity of at least 30% of the radial airgap, which was subsequently verified from off-line measurements of the airgaps at the DE and NDE in a motor repair shop.

## 11.5.2 MCSA Measurements and Analysis

### *11.5.2.1 Motor A: Broken Rotor Bar Analysis*

The operational current was 91 A compared to the full-load current of 103 A thus the operational slip should be less than the full-load slip. The starting point is to use the full-load slip for the prediction of frequency components due to broken rotor bars or airgap eccentricity at full-load.

$$s_{FL} = 0.0107\,(1.07\%)$$

A current zoom spectrum is shown in Figure 11.21 to ascertain the operational frequency and to diagnose the presence of any sidebands at $\pm 2sf$ around $f$ due to broken rotor bars. There are sidebands at ±0.81 Hz around $f$ but they are 64 dB down and if they are at $\pm 2sf$ around $f$ this gives a slip of 0.0081 (0.81%), which is in line with the operational slip back calculated from one of the measured RSPF components

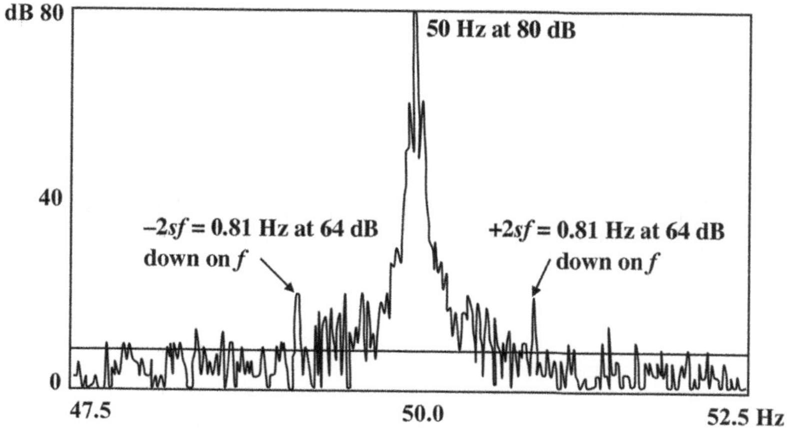


Figure 11.21 Motor A current zoom spectrum.

at 0.0077 (0.77%). There is a negligible difference (of only 0.036%) between the two operational slips from analyses of two different parts of the current spectrum. The $\pm 2sf$ sidebands are 64 dB down on $f$ and therefore the cage winding is considered to be in a perfectly normal condition since the equivalent broken bar factor ($BB_f$) referred to the full-load slip is 0.1.

#### 11.5.2.2 Motor A: Airgap Eccentricity Analysis

The principal RSPF component $f_{rs(+1)}$, was calculated at the full-load slip using equation (10.4) and the measured supply frequency of 50 Hz.

$$f_{rs} = f[(R/p)(1 - s) \pm n_{\omega s}]$$

where $f = 50$ Hz, $R = 62$, $p = 4$ pole-pairs, and $s_{FL} = 0.01$

$$f_{rs(+1)} = 816.7 \text{ Hz}$$

and $f_{rs(+3)}, f_{rs(+5)}, f_{rs(+7)} = 916.7$, 1016.7, and 1116.7 Hz, respectively, since the RSPFs are spaced $2f$ (100 Hz) apart. The initial baseband frequency span was selected as 70 to ($f_{rs(+5)}$, +100) = 1119 Hz, but it was found that this had to be increased to an upper limit of 1200 Hz since in this case history $f_{rs(+1)}$ was not the rotor slot passing frequency with the highest dB value. The current zoom spectrum between 800 and 1200 Hz shown in Figure 11.22 indicates that $f_{rs(+1)}, f_{rs(+3)}, f_{rs(+5)}$, and $f_{rs(+7)}$ are present and $f_{rs(+5)}$ has the highest dB value and is thus selected for the next analysis.

As expected, the frequency of $f_{rs(+5)}$ is slightly higher (1019 Hz) than the predicted value (1016.7 Hz) at the full-load slip since the operational current and slip are actually less than the full-load values. The next step is to reverse calculate the actual operating slip using equation (10.4) and the measured $f_{rs(+5)}$ and supply frequency which are 1019 and 50 Hz, respectively:

Operating slip, $s_o = 1 - \{((f_{rs(+5)}/f) - 5)(p/R)\} = 1 - \{(1019/50) - 5\}$ (4/62)

$$s = 0.0077(0.77\%)$$

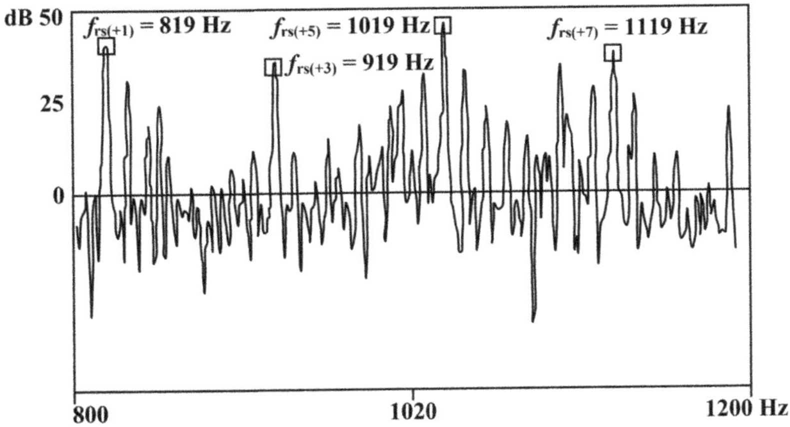


Figure 11.22 Motor A current zoom spectrum to display rotor slot passing frequencies.

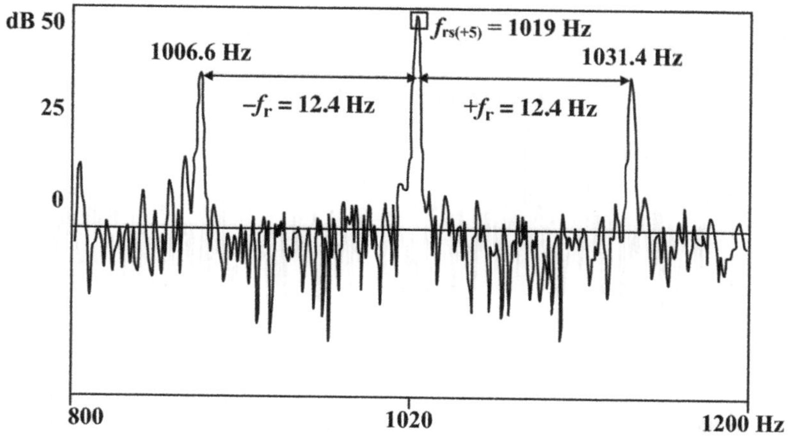


Figure 11.23 Motor A current zoom spectrum to display $\pm f_r$ components around $f_{rs(+5)}$.

The rotor speed frequency, $f_r = f(1-s)/p = 50(1-0.077/4) = 12.4$ Hz
The rotor speed = 744 r/min

To detect the $\pm f_r$ components around $f_{rs(+5)}$, a narrower band zoom spectrum is used as shown in Figure 11.23 and pronounced $\pm f_r$ components exist around $f_{rs(+5)}$ at an average of 10 dB down on $f_{rs(+5)}$. The extract from Table 10.3 indicates a very severe level of operational airgap eccentricity of the order of at least 30%. MCSA therefore identified the fundamental cause of the high temperature and vibration at the DE bearing, which is of course due to electromagnetic forces and a consequentially high level of UMP being transmitted to the motor's bearings. It is expected that the high airgap eccentricity will be at the DE due to high vibration and temperature at the DE bearing.

#### 11.5.2.3 Motor B: Airgap Eccentricity Analysis

The current zoom spectrum from the motor in drive train B, which had neither a vibration nor temperature problem, is presented in Figure 11.24, which shows that there is not the characteristic current signature indicative of an airgap eccentricity problem since no $+f_r$ component exists above $f_{rs(+5)}$ and $-f_r$ is 41 dB down on $f_{rs(+5)}$.

**EXTRACT FROM TABLE 10.3 Estimate of Operational Airgap Eccentricity from Current Signature Pattern**

| $N_{ec}$ Average dB difference between $f_{rs(+1)}$ and $\pm f_r$ components | **Estimate of "operational" airgap eccentricity i.e., combination of static and dynamic** |
|---|---|
| $N_{ec} \leq 15$ dB | Airgap eccentricity is normally >20% of the radial airgap and this is considered to be a severe level. |
| $N_{ec} \leq 10$ dB | Airgap eccentricity is normally >30% of the radial airgap and this is considered to be a very severe level. |
| MCSA measured $N_{ec} = 10$ dB | |

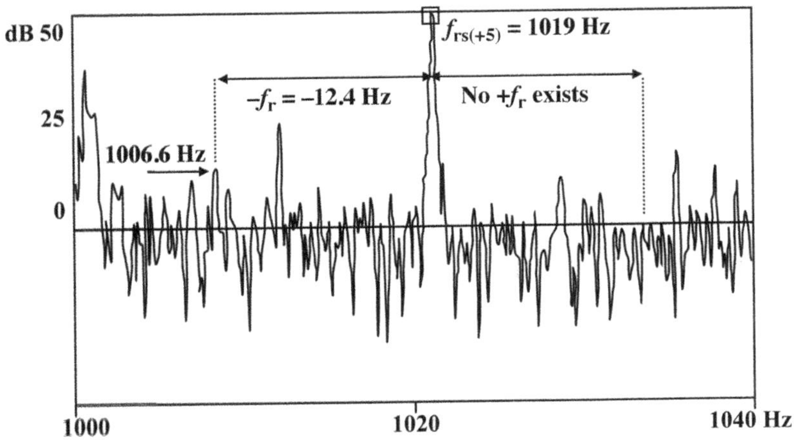


Figure 11.24 Motor B current zoom spectrum to display $\pm f_r$ components around $f_{rs(+5)}$.

Also a comparison between the current spectra in Figure 11.23 from motor A and that of Figure 11.24 from motor B clearly shows the distinct differences.

Based on the MCSA diagnosis of a severe level of operational airgap eccentricity in the motor in drive train A it was taken out of service and subsequently sent to the OEM for checks on the airgaps at the DE and NDE. A good motor from drive train C (which had a failed pump) was subsequently moved into drive train A to replace the problem motor.

#### 11.5.2.4 Motor A: Measurement of Airgaps

At the time the motor was manufactured the OEM specified that the motor had a radial airgap length of 2.0 mm/ 80 mils ±10%. The rotor was manufactured with a total run out (TIR) of 0.05 mm/2 mils giving a dynamic eccentricity of only 2%. The suspect airgaps were measured via feeler gauges at the manufacturer's workshop with the motor at room temperature and the results are shown in Figures 11.25 and 11.26.

This gives an operational airgap eccentricity at the DE of 30% (severe) and at the NDE of 15% (acceptable for a running motor), respectively. It should be appreciated that although the off-line checks gave an airgap eccentricity of 30% at the DE, this could be higher when the motor is running due to UMP forces. It is worth noting

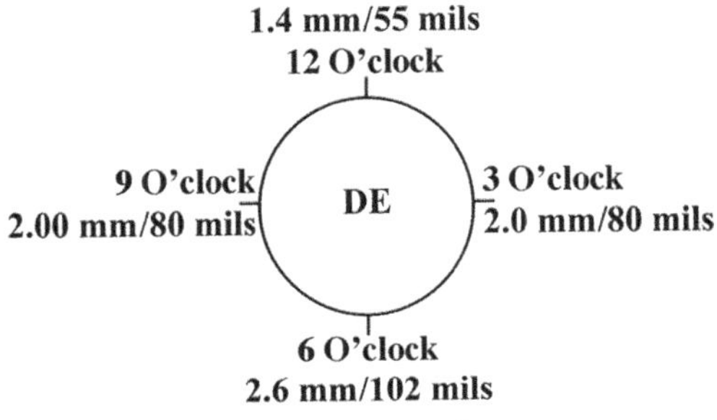


Figure 11.25 DE airgaps.

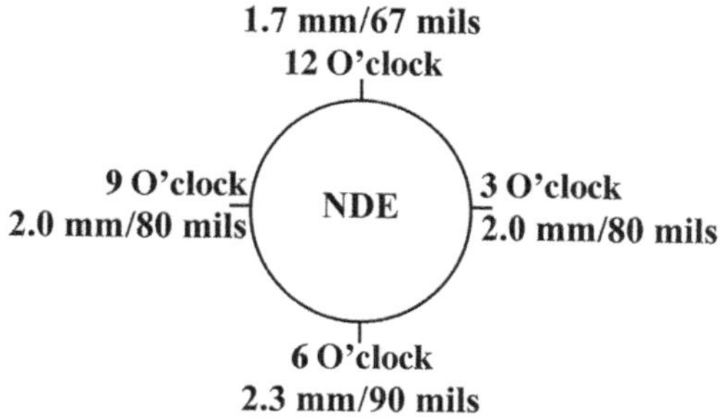


Figure 11.26 NDE airgaps.

that the high airgap eccentricity was at the DE and it was the DE bearing that had high vibration and temperature levels. The rotor was removed and the TIR value was 0.05 mm/2 mils as per the original run out, thus the problem was high static airgap eccentricity. The airgaps at the 9 and 3 O'clock are even at 2 mm and the high static airgap eccentricity is in the vertical direction, which is directly in line with the DE bearing, hence the magnetic pull is acting directly on the rotor and the DE bearing.

It was established that the end frame at the DE had not been securely doweled during the last strip down by a motor repair company. This was rectified and the OEM reported that off-line airgap measurements via feeler gauges confirmed that the airgaps at the DE and NDE at the 12, 3, 6, and 9 O'clock positions were now 2 mm/80 mils ±10%. The motor was reinstalled into drive train A and the high bearing temperature and vibration problem no longer existed since the values at the DE bearing end were now 50 μm/2 mils pk-pk and 65°C/149°F which further proves that MCSA had identified the root cause of the initial problem.

A repeat MCSA test on the repaired motor A was carried out and the current zoom spectrum is shown in Figure 11.27.

There are still $\pm f_r$ components but they are now an average of $N_{ec} = 25$ dB down on $f_{rs(+5)}$ which equates to an operational airgap eccentricity of the order of 15% as shown in the extract from Table 10.3. This is acceptable for a running SCIM,

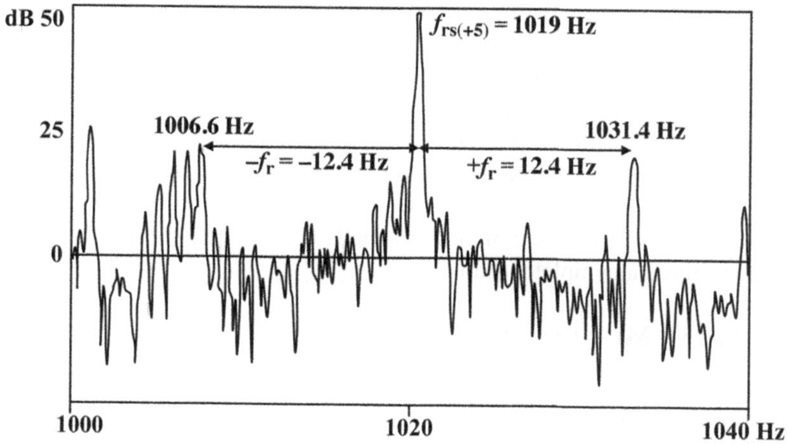


Figure 11.27 Motor A current zoom spectrum after airgap eccentricity was reduced to 10% with motor at ambient temperature and stationary.

**EXTRACT FROM TABLE 10.3 Estimate of Operational Airgap Eccentricity from Current Signature Pattern**

| $N_{ec}$ Average dB difference between $f_{rs(+1)}$ and $\pm f_r$ components | **Estimate of "operational" airgap eccentricity i.e., combination of static and dynamic** |
|---|---|
| $N_{ec}$ = 25–35 dB | Airgap eccentricity is ≥10% of radial airgap but at $N_{ec}$ = 25 dB it is typically around 15% of the airgap |
| MCSA measured | |
| $N_{ec}$ = 25 dB | |

although it is 5% higher than the off-line feeler gauge measurements. It is a fact that when a SCIM is running and has reached its stable operating temperature its operational airgap eccentricity can certainly differ from that obtained by off-line airgap measurements from a limited number of feeler gauge measurement positions when the motor is stationary and at ambient temperature.

The motor has continued to operate at the normal operating temperature and vibration levels for the DE bearing and no repeat of the original problem has since been reported.

## 11.6 CASE HISTORY (2001)—HIGH AIRGAP ECCENTRICITY IN A COOLING WATER PUMP MOTOR THAT CAUSED SEVERE MECHANICAL DAMAGE TO HV STATOR COILS

### 11.6.1 Summary

This case history reports on the MCSA diagnosis of a high level (20 to 25%) of operational airgap eccentricity in a cooling water pump motor on an offshore oil production platform.

Motor data: 3-phase, SCIM, 6.6 kV, 1200 kW/1610 HP, 130 A, 60 Hz, 1764 r/min, 0.86 p.f. The rotor has 50 rotor slots with a copper fabricated cage winding, standard sash bar with the bars brazed onto the outer periphery of the end rings. The OEM stated the design airgap was 2 mm/80 mils ± 5% (OEM's maximum allowable).

MCSA was carried out on this motor as part of an annual survey of all HV motors on the offshore oil production platform. There were no reports of any abnormal vibration being measured on the bearing housings and no displacement probes were fitted in the housings to measure the shaft displacement and its orbital profile. Due to the high airgap eccentricity, the electromagnetic forces and subsequent vibration on the stator core and teeth caused severe mechanical damage to the semi-conductive coatings of all the coil sides of the HV stator coils in the slot portions. The motor required a complete stator rewind.

### 11.6.2 MCSA Measurements and Analysis

The operating current was 106 A compared to the full-load current of 130 A, hence the operational slip was less than the full-load slip. The current zoom spectrum of Figure 11.28 gives the operational supply frequency as 60.064 Hz.

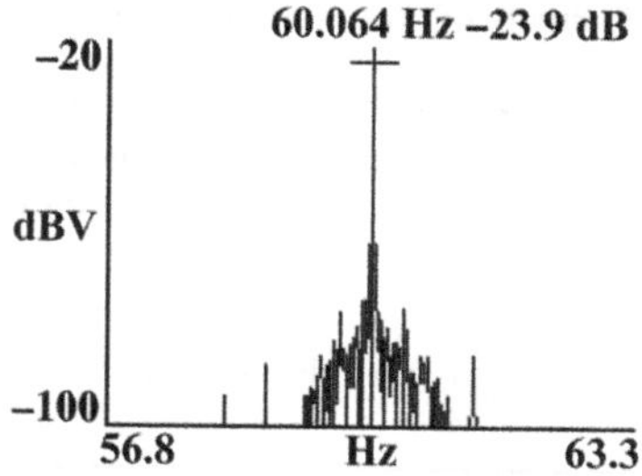


Figure 11.28 Current zoom spectrum to measure the supply frequency at 60.064 Hz; frequency resolution of 0.01 Hz/line.

Using the measured supply frequency and the full-load slip, an initial prediction of the principal rotor slot passing frequency component $f_{rs(+1)}$ is as follows:

$$f_{rs(+1)} = f[(R/p)(1-s) \pm n_{\omega s}]$$

where $f = 60.064$ Hz, $R = 50$, $p =$ pole-pairs $= 2$, $s_{FL} = 0.02$, and $+n_{\omega s} = +1$ $f_{rs(+1)} = 1531.6$ Hz, and $n_{\omega s}$ at $-1$, $+3$, and $+5$ gives $f_{rs(-1)}, f_{rs(+3)}, f_{rs(+5)}$ at 1411.5, 1651.7, and 1771.8 Hz, respectively, since the RSPFs are spaced $2f$ (120.13 Hz) apart. The baseband frequency span was set at 70 to ($f_{rs(+5)} + 100$) Hz, and with 12,800 spectral lines this gives a frequency resolution of 0.14 Hz/line.

The spectrum in Figure 11.29 shows the $f_{rs(+1)}$ component is 1539.3 Hz, which is higher than the predicted full-load value since the operational slip is less than the full-load values. The next step is to reverse calculate the actual operating slip using equation (10.4) and the measured $f_{rs(+1)}$ and supply frequency which are 1539.3 and 60.064 Hz, respectively.

Operating slip,

$$s_o = 1 - \{(f_{rs(+1)}/f) - 1)(p/R)\} = 1 - \{(1539.3/60.064) - 1\}(2/50)$$

$$s = 0.0149(1.49\%)$$

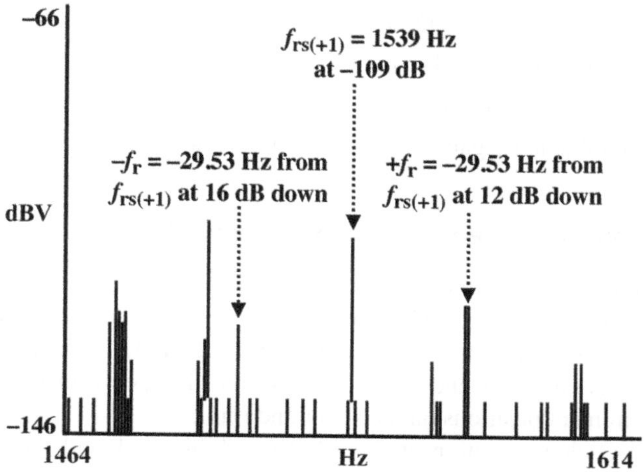


Figure 11.29 Current zoom spectrum to detect $f_{rs(+1)}$ and $\pm f_r$ components.

**EXTRACT FROM TABLE 10.3 Estimate of Operational Airgap Eccentricity from Current Signature Pattern**

| $N_{ec}$ Average dB difference between $f_{rs(+1)}$ and $\pm f_r$ components | **Estimate of "operational" airgap eccentricity i.e., combination of static and dynamic** |
|---|---|
| $N_{ec} \leq 15$ dB MCSA measured $N_{ec} = 14$ dB | Airgap eccentricity is normally >20% of the radial airgap and this is considered to be a severe level. |

The rotor speed frequency, $f_r = f\,(1 - s)\,/p = 60.064\,(1 - 0.0149)\,/2 = 29.58$ Hz

The rotor speed, $N_r = 1775$ r/min

The measured $\pm f_r$ components are $\pm 29.53$ Hz around $f_{rs(+1)}$ as shown in Figure 11.29 compared to the predicted $\pm f_r$ value of $\pm 29.58$ Hz, which is a difference of only 0.05 Hz and this verifies the operational slip and airgap eccentricity analysis. The $\pm f_r$ components around $f_{rs(+1)}$ are an average of $N_{ec}$ equal to 14 dB down on $f_{rs(+1)}$. The extract from Table 10.3 indicates a severe level of operational airgap eccentricity of the order of 20–25%. The motor was removed at the first opportunity and returned to the OEM for off-line measurements of the airgaps and a full inspection.

### 11.6.3 Inspection: Bearing Clearances & White Metal Shells

The bearing clearances were checked by the OEM, who confirmed that the normal clearance for this sleeve bearing alignment was between 0.14 mm/5.5 mils and 0.19 mm/7.5 mils. The measured value indicated the clearance was 0.03 mm/1.2 mils above 0.19 mm/7.5 mils and this was considered to be an acceptable difference. The bearing shells were inspected and there were no signs of any abnormal wear, so that had not contributed to the problem in the motor.

**Nominal Design Values of Stator and Rotor**

Stator bore diameter $D_s = 450$ mm/17.72 inches

Rotor core diameter $D_r = 446$ mm/17.56 inches

Thus giving a nominal radial airgap ($g$) of 2 mm or 80 mils/thou.

Thus a total diametric gap of 4 mm/160 mils/thou.

### 11.6.4 Airgap Measurements in the Received Motor

For the airgap measurements, the OEM selected fixed positions at 10.00, 14.00, 16.00, and 20.00 as shown in Table 11.1 which were fixed on the stator core and the rotor was turned through sequential steps of 90 degrees, starting at the 12 O'clock reference for the rotor to then get the 3, 6, and 9 O'clock positions. Excess varnish was removed from the measurement positions on the stator and rotor.

At the 3 O'clock position at the DE from Table 11.1 and the stator fixed positions of 10.00 and 16.00 the airgap was 2 mm/79 mils and 1.25 mm/49 mils, respectively, thus giving a total airgap of 3.25 mm/128 mils and a mean gap of 1.625 mm/64 mm compared to the design value of 2 mm/79 mils. The maximum

**TABLE 11.1 Airgap Measurements Via Feeler Gauges in the As-Received Motor**

| DE | NDE |
|---|---|
| 75* 1.9” 1.95” 77*<br>53* 1.35” 1.25” 49*<br>12 O’clock | 79* 2.0” 2.0” 79*<br>49* 1.25” 1.25” 49*<br>12 O’clock |
| 77* 1.95” 1.9” 75*<br>53* 1.35” 1.25” 49*<br>9 O’clock | 79* 2.0” 1.9” 75*<br>52* 1.3” 1.25” 49*<br>9 O’clock |
| 73* 1.85” 1.9” 75*<br>53* 1.35” 1.25” 49*<br>6 O’clock | 79* 2.0” 1.9” 75*<br>49* 1.25” 1.2” 47*<br>6 O’clock |
| 79* 2.0” 1.95” 77*<br>53* 1.3” 1.25” 49*<br>3 O’clock | 77* 1.95” 1.95” 77*<br>49* 1.25” 1.2” 47*<br>3 O’clock |

Note: is in mm and * is in mils or thou

airgap eccentricity is $e_c = 0.375/1.625 = 23\%$. Likewise, at the NDE at the 3 O’clock position the airgap eccentricity is 24%. This is 19% above the OEM’s original specification of a maximum allowable airgap eccentricity of 5%, that is, $(2 \pm 0.1)$ mm or $(80 \pm 4)$ mils/thou.

It should be borne in mind that the measured airgap eccentricity of 24% was via feeler gauges at a limited number of positions and at ambient temperature and there are no UMP forces during these off-line measurements. The MCSA prediction for the operational airgap eccentricity was 20–25% with the motor running on load and with operational UMP forces. The operational airgap eccentricity can certainly be higher than that given by feeler gauge measurements. The reason for the undersized airgap was either an out of tolerance stator bore or rotor diameter.

### 11.6.5 Total Indicated Run Out (TIR) and Modifications to the Rotor

The measured rotor diameter ($D_r$) was 446.57 mm (17.58 inches), which is 0.57 mm/22 mils above the design value of 446 mm (17.72 inches). If the stator bore were perfectly round to produce a constant stator diameter ($D_s$) of 450 mm and the other build tolerances were correct, then the nominal airgap would be

$g = 1.715$ mm/67 mils compared to the design value of 2 mm/80 mils.

The OEM measured the TIR of the rotor core along the rotor's diameter as being 0.15 mm/6 mils, giving a dynamic eccentricity of

$d_e = 0.15/1.715 = 8.75\%$

The TIR of a SCIM rotor of this rating is typically set up to a maximum of 0.05 mm/2 mils/thou as discussed in Section 10.2 and presented in Table 10.1. The actual level of rotor core run out was unacceptable and so the rotor was machined (skimmed) and was re-balanced to G0.4 (a low level) as defined in ISO 1940-1:2003 [11.1]. The design value of the rotor diameter ($D_r$) is 446 mm (17.56 inches). The new rotor diameter ($D_r$ measured) was 445.98 mm compared to 446.57 mm (17.58 inches) prior to skimming the rotor.

The new maximum rotor run out on the rotor core was measured as 0.02 mm (0.8 mils) compared to 0.15 mm (60 mils) prior to skimming the rotor. The OEM then carried out concentricity checks on the stator assembly and bearing centers and modifications were carried out by machining the spigots on the end shields and then doweling them to the main frame so that the end frames (end bells) were secure and could always be located at the same position. The airgaps at the DE and NDE were checked after all the modifications and the results from the feeler gauge measurements are presented in Table 11.2.

At the 3 and 9 O'clock positions at the DE and NDE respectively, the maximum eccentricity was reduced to 1.33%.

### 11.6.6 Mechanical Damage to Stator Coils

When the motor was initially stripped down for the rotor TIR measurements, skilled personnel, who tap tested the stator winding slot wedges found that at least 80% were not tight in the slots. The stator core and teeth were being subjected to much higher than normal electromagnetic forces and vibration from the UMP caused by high airgap eccentricity, due to the combination of static and dynamic airgap eccentricity. Recall that dynamic eccentricity was at least 8.75%.

A number of coils were removed as shown in Figure 11.30, and Figure 11.31 is a view of five coils from a line end group of coils and a coil side from the star end. Both sets of coils clearly show mechanical damage to the line and star end coils thus confirming the fundamental root cause of this damage to the HV coils was mechanical damage due to the high airgap eccentricity. It is well known that the star end coils of a star connected 6600 V SCIM have a very low electrical stress compared to the line end coils, Stone et al. [11.2], thus the damage on the star connected coil is purely due to mechanical damage whereas the line end coils have experienced enhanced levels of PD activity due to mechanical degradation of the semi-conductive coatings on the portions of the HV coils in the stator slots. The stator was completely rewound.

**TABLE 11.2 Airgap Measurements after Modifications**

| DE | NDE |
| --- | --- |
| 75* 1.9" 1.85" 73*<br>73* 1.85" 1.9" 75*<br>12 O'clock | 75* 1.9" 1.85" 73*<br>73* 1.85" 1.9" 75*<br>12 O'clock |
| 75* 1.9" 1.9" 75*<br>73* 1.85" 1.85" 73*<br>9 O'clock | 75* 1.9" 1.9" 75*<br>73* 1.85" 1.85" 73*<br>9 O'clock |
| 73* 1.85" 1.9" 75*<br>73* 1.85" 1.9" 75*<br>6 O'clock | 75* 1.9" 1.9" 75*<br>73* 1.85" 1.85" 73*<br>6 O'clock |
| 75* 1.9" 1.85" 73*<br>73* 1.85" 1.9" 75*<br>3 O'clock | 73* 1.85" 1.9" 75*<br>73* 1.85" 1.9" 75*<br>3 O'clock |

**Note: is in mm and * is in mils or thou**

Figure 11.30 Stator core assembly showing coils have been removed from the slots.

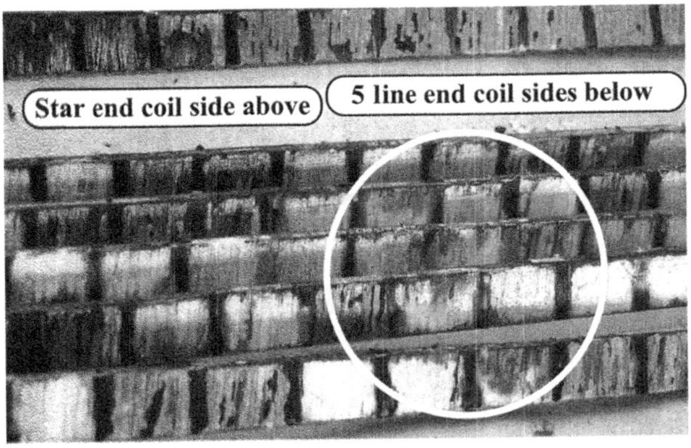


Figure 11.31 Samples of severely damaged semi-conductive coatings due to mechanical forces and vibration enhanced by consequential slot discharges of the line end coils.

### 11.6.7 MCSA Measurements and Analysis on Refurbished Motor

The current spectrum in Figure 11.32 was for the no-load, uncoupled run to give an accurate measurement of the supply frequency, which was 59.962 Hz. The no-load current was 16 A.

The principal rotor slot passing frequency is given by

$$f_{rs(+1)} = f[(R/p)(1 - s) \pm n_{\omega s}]$$

where

$f = 59.962$ Hz, $R = 50$, $p =$ pole-pairs $= 2$, the slip $s =$ approximately zero during an uncoupled run, and $n_{\omega s} = +1$

$$f_{rs(+1)} = 1559 \text{ Hz}$$

The predicted rotor slot passing current component $f_{rs(+1)} = 1559$ Hz and the measured spectrum in Figure 11.33 gives 1557.4 Hz, which is a difference of only 0.1%.

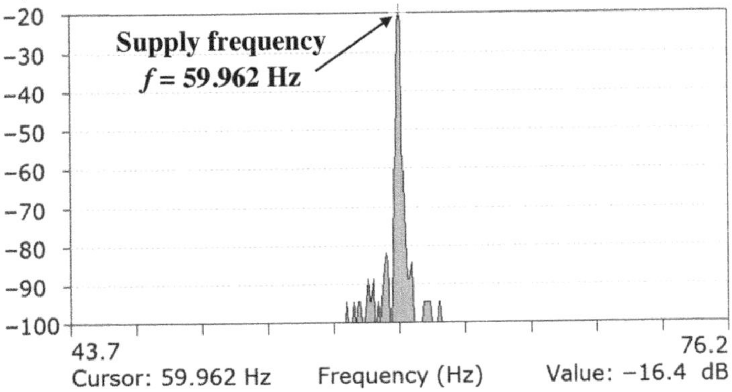


Figure 11.32 Current zoom spectrum to measure supply frequency.

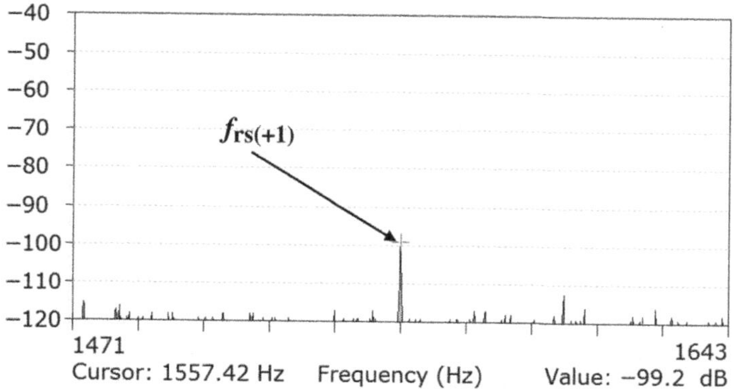


Figure 11.33 Current zoom spectrum to measure $f_{rs(+1)}$ and to identify any $\pm f_r$ components around $f_{rs(+1)}$, uncoupled, no-load run.

The current signature pattern in Figure 11.33 indicates that there are no $\pm f_r$ components around $f_{rs(+1)}$ during the uncoupled run hence the airgap eccentricity is now certainly less than 5%. Figure 11.34 shows the spectrum at a load current of 60 A (Note: The motor repair company only had a limited load test facility) and this also confirms there are no $\pm f_r$ components around $f_{rs(+1)}$. The current spectra in Figures 11.33 and 11.34 confirm that the operational airgap eccentricity is below 5% and as is known the measured airgap eccentricity via feeler gauges was reduced to only 1.3%.

A comparison between Figures 11.34 and 11.29 clearly demonstrates that the former has no $\pm f_r$ components around $f_{rs(+1)}$ due to the very low airgap eccentricity (1.3%) whereas the latter had $\pm f_r$ components at an average of 14 dB down on $f_{rs(+1)}$. This clearly proves that the original MCSA diagnosis of a high level of airgap eccentricity between 20% and 25% was certainly valid.

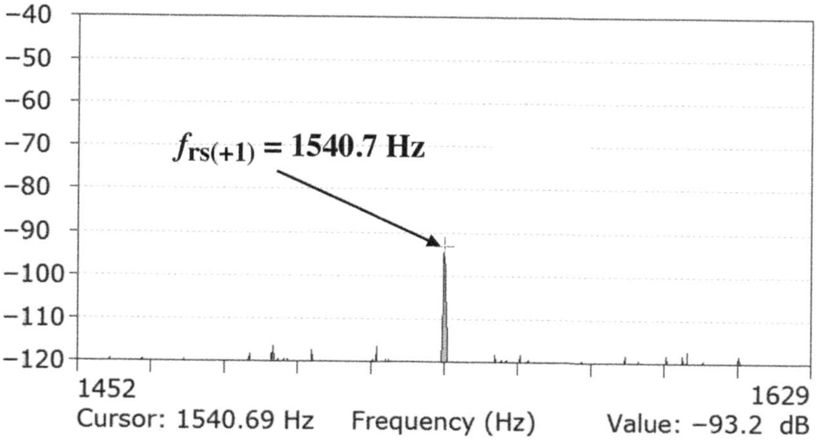


Figure 11.34 Current zoom spectrum to measure $f_{rs(+1)}$ and to identify any $\pm f_r$ components around $f_{rs(+1)}$, coupled run at 60 A (46% full-load current).

For completeness, the operating slips at no-load and with the motor operating on load at 60 A are 0.003 and 0.012, respectively, which were reverse calculated from the measured $f_{rs(+1)}$ components in Figures 11.33 and 11.34 in the usual manner.

It is interesting to note that the magnitude of $f_{rs(+1)}$ between Figures 11.33 and 11.34 has increased by 6 dB (a doubling) between no-load (slip = 0.003) and a load current of 60 A (0.012), thus proving that the magnitude (in dB) of this RSPF current component is indeed a function of load with a fixed level of airgap eccentricity as previously discussed and predicted in Section 10.4.1 in Chapter 10. The increase in the current component at $f_{rs(+1)}$ is therefore due to an increase in the rotor bar currents between a slip of 0.003 and 0.012 (a factor of 4). The consequential increase in leakage flux around each current carrying rotor bar causes an increase in the slot ripple flux at $f_{rs(+1)}$ and hence the induced emf and current in the stator winding.

## 11.7 CASE HISTORY (2008)—UNSUCCESSFUL APPLICATION OF MCSA APPLIED TO A LARGE (6300 kW), INVERTER-FED, 6600 V SCIM DURING A NO-LOAD RUN TO ASSESS ITS OPERATIONAL AIRGAP ECCENTRICITY

### 11.7.1 Summary

This case history reports on an MCSA test of a brand new inverter-fed, HV SCIM to assess its operational airgap eccentricity during a no-load, uncoupled run. This motor drives a centrifugal compressor in an onshore LNG processing plant. The motor data is as follows: 3-phase, 6600 V, 6300 kW/8445 HP, 633 A, 50 Hz, 1493 r/min SCIM. The number of rotor slots is 82. The design value of the airgap was unavailable from the OEM. The only information available on the inverter was that it was a "perfect harmony design" which is the OEM's proprietary name for the product. The end user required a baseline current spectrum to assess the operational airgap eccentricity during the commissioning of the motor.

The results will verify that during a no-load uncoupled run it was not possible to assess the operational airgap eccentricity since there was coincidence between the rotor slot passing frequencies and the supply harmonics from the inverter.

### 11.7.2 Unsuccessful MCSA Measurements and Analysis—Uncoupled Run

The motor was run uncoupled with the inverter frequency set to 50 Hz and the time domain current signal (converted to voltage) is presented in Figure 11.35a. The current was sensed via a 1000:1 clip on CT with an in-built shunt of 1.0 Ω, which was installed around one of the supply phase cables (2 cables/phase) from the inverter to the SCIM. Although there was a signal proportional to the motor's current within the inverter circuitry it was unknown as to how it had been processed. From Figure 11.35a the peak volts is 0.083 V across the shunt of the CT and converted to current is 117 A (rms), taking into account there are 2 cables/phase.

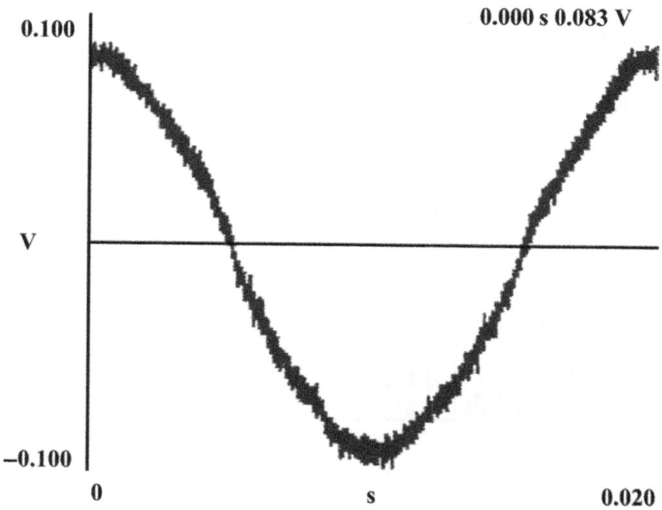


Figure 11.35a Time domain current waveform to the motor, note the shunt across the CT gives a voltage.

This is a good quality current waveform from the "perfect harmony" inverter and as expected there are high frequency ripples and corresponding frequencies in the waveform.

The supply frequency was 50.034 Hz as shown in Figure 11.35b and the principal rotor slot passing frequency is given by

$$f_{rs(+1)} = f[(R/p)(1 - s) \pm n_{\omega s}]$$

where

$f$ = 50.034 Hz, $R$ = 82, $p$ = pole-pairs = 2, the slip $s$ = approximately zero during an uncoupled run, and $+n_{\omega s} = +1$

$$f_{rs(+1)} = 2101.4 \text{ Hz}$$

and $f_{rs(-1)}, f_{rs(+3)}, f_{rs(+5)}$ = 2001.4, 2201.5, and 2301.6 Hz, respectively, since the rotor slot passing frequencies are spaced $2f$ (100.1 Hz) apart.

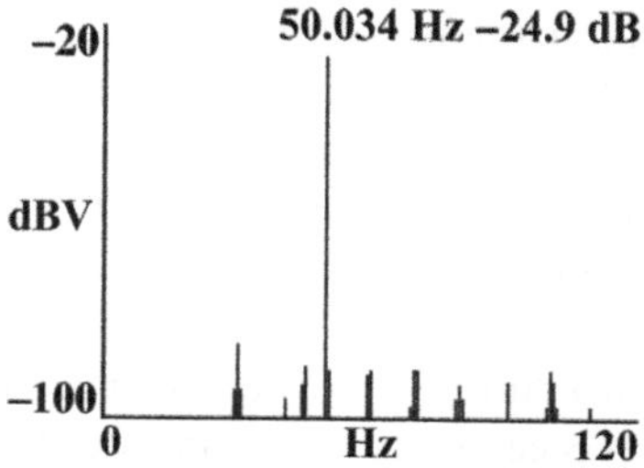


Figure 11.35b Current spectrum, baseband frequency 0—120 Hz, 12,800 spectral lines, frequency resolution 9.38 mHz/line.

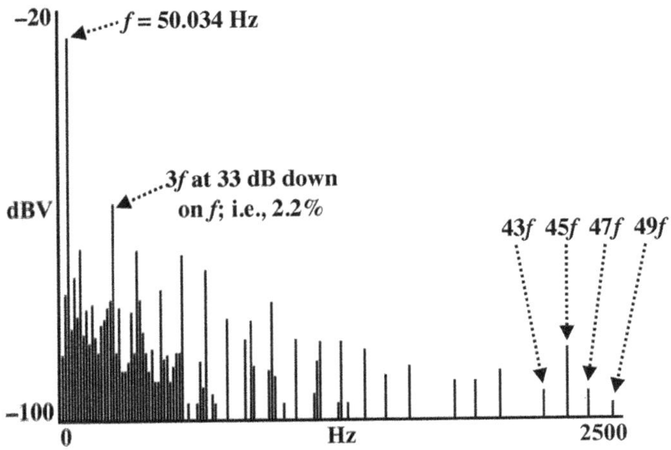


Figure 11.36 Current spectrum 0–2500 Hz to show the higher harmonics relative to the inverter base frequency of 50.034 Hz.

Since the motor was inverter-fed the first step was to measure the unfiltered spectrum up to 2500 Hz to ascertain whether high frequency harmonics of the inverter base frequency could be observed in the region of the RSPFs. This is not normally the case in the unfiltered current spectrum from a mains-fed SCIM since the high frequency components are normally only visible in the filtered current signal after the supply frequency $f$ is removed, as has been demonstrated in the previous six case histories in this chapter.

Figure 11.36 shows that high frequency harmonics at the 43rd, 45th, 47th, and 49th of $f$ can be observed in the unfiltered spectrum between 0 and 2500 Hz. These are due to the inverter supply and are in the spectral zone where the RSPFs exist. As expected, there is no evidence of any rotor slot passing frequencies in the unfiltered spectrum, which means that these high frequency harmonics from the inverter are considerably larger than the RSPFs and any $\pm f_r$ components around them when the filtered spectrum is analyzed. This of course makes the airgap eccentricity analysis more challenging.

For the filtered spectrum, the baseband frequency span was set at 70 to $(f_{rs(+5)} + 100)$ Hz, but in this case the upper limit was set to 2500 Hz, which using 12,800 spectral lines gives a frequency resolution of 0.19 Hz/line. Figure 11.37 presents a zoom spectrum between the 43rd and 45th inverter supply harmonic to illustrate the coincidence between the 44th harmonic and $f_{rs(+3)}$, please refer to case history 11.2 and Sections 11.2.1 and 11.2.2 for the explanation and proof, which confirmed that even harmonics of the supply can be observed using an 80 dB logarithmic spectrum. The 43rd and 45th harmonics from the inverter are 10 and 12 dB higher (3.16 and 4 times higher, respectively) than the component at 2201.6 Hz which is coincident with $f_{rs(+3)}$ and $44f$.

The analysis verifies that during a no-load uncoupled run it was not possible to assess the operational airgap eccentricity of this inverter-fed, HV SCIM since there was coincidence between the rotor slot passing frequencies and even supply harmonics from the inverter.

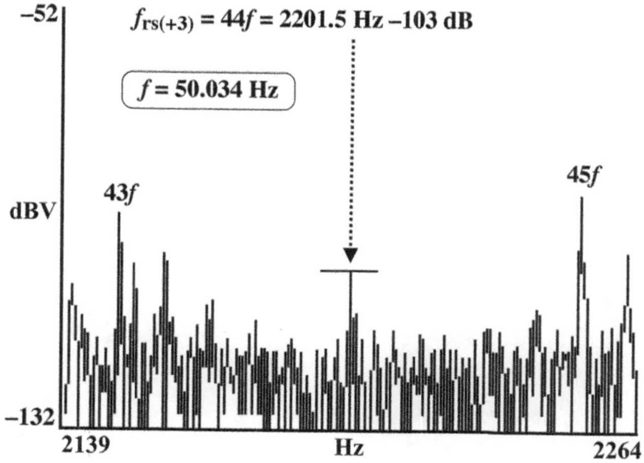


Figure 11.37 Current zoom spectrum.

## 11.8 CASE HISTORY (2008)—SUCCESSFUL APPLICATION OF MCSA APPLIED TO A LARGE (4500 kW), INVERTER-FED, 3300 V SCIM TO ASSESS ITS OPERATIONAL AIRGAP ECCENTRICITY

### 11.8.1 Summary

This case history verifies that the operational airgap eccentricity can be assessed in a large, inverter-fed, HV SCIM via MCSA when the motor is on load. The motor drives a centrifugal compressor on an offshore oil and gas production platform.

Nameplate information: 3-phase 3.3 kV, 4500 kW/6032 HP, 914A, 50.3 Hz, 1500 r/min., 0.89 p.f., 96.8%. The number of rotor slots is 94 but the nominal design value of the radial airgap is unknown. A photograph of the vertically mounted motor is shown in Figure 11.38.

### 11.8.2 MCSA Measurements and Analysis

Based on the nameplate data the full-load slip is

$$s_{FL} = (1509 - 1500)/1509 = 0.006\,(0.6\%)$$

However, the full-load slip of this motor when supplied at a base frequency of 31.8 Hz is unknown and the OEM did not provide the end user with the full-load slips at different operating frequencies and corresponding full-load currents. The motor was operating on a reduced load current of 600 A (with a supply of 31.8 Hz) compared to the nameplate full-load current of 914 A at its rated frequency (50.3 Hz). Therefore, at an inverter supply frequency of 31.8 Hz as confirmed from Figure 11.39, the only option available is to calculate the RSPFs at no-load (assume the slip is zero).

Figure 11.38 Motor being MCSA tested.

The supply frequency was 31.8 Hz and the principal rotor slot passing frequency is given by

$$f_{rs(+1)} = f[(R/p)(1-s) \pm n_{\omega s}]$$

where $f$ = 31.8 Hz, $R$ = 94, $p$ = pole-pairs = 2, the slip $s$ = approximately zero during no-load, and $+n_{\omega s} = +1$, therefore $f_{rs(+1)}$ = 1526.4 Hz and $f_{rs(-1)}$, $f_{rs(+3)}$, $f_{rs(+5)}$ = 1462.8, 1590, and 1653.6 Hz, respectively, since the RSPFs are spaced $2f$ (63.6 Hz) apart.

The baseband frequency span for the filtered spectrum was set at 70 to ($f_{rs(+5)}$ + 100) Hz, and in this case the upper limit was set to 1800 Hz, which with 12,800 spectral lines gives a frequency resolution of 0.135 Hz/line.

Figure 11.40 presents the spectrum between 70 and 1800 Hz and as expected there are numerous harmonics (odd and even) due to the inverter supply. On no-load,

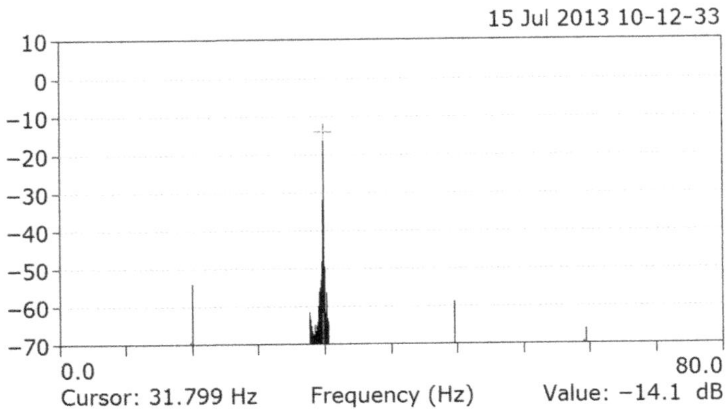


Figure 11.39 Current spectrum, 12,800 lines, 0–80 Hz, 6.25 mHz/line.

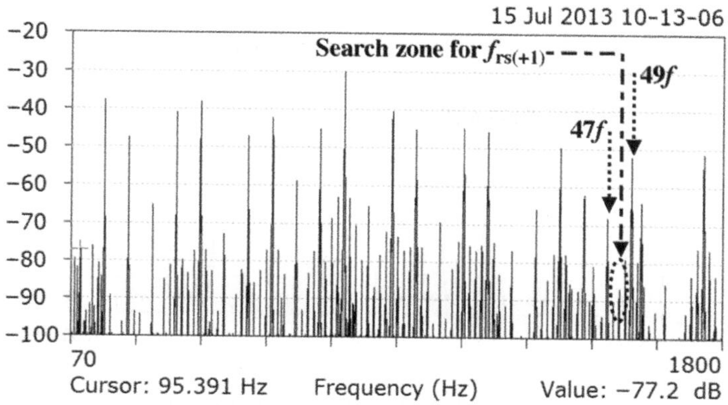


Figure 11.40 Current spectrum, 12,800 lines, 70–1800 Hz.

$f_{rs(+1)}$ coincides with the 48th harmonic of the inverter's base frequency but of course the motor is on-load and operating at a reduced load current therefore by deduction $f_{rs(+1)}$ will be slightly less than the 48th harmonic and that is the region of the spectrum to carry out a zoom spectrum analysis and search for the principal RSPF.

With reference to Figure 11.40 the 47th and 49th harmonics of $f$ (31.8 Hz) are highlighted and are 1494.6 and 1558.2 Hz, respectively, with the 48th harmonic being 1526.4 Hz which coincides with $f_{rs(+1)}$ on no-load.

In the first instance the zoom spectrum is presented in Figure 11.41 without any annotation of the components to provide the reader with a perspective of the measured spectrum prior to interpretation. As a self-learning exercise the reader is encouraged to first identify the actual frequencies of the odd and even harmonics and then deduce the rotor slot passing frequencies in Figure 11.41. This can be done since the span is 187 Hz, therefore each main division is 18.7 Hz and by using a ruler with 0.5 mm or 1/64th of an inch scale, a conversion scale can be easily produced in Hz/mm or Hz/(1/64th of an inch). Simply measure the distances of the components in mm or 1/64ths from the starting frequency of 1428 Hz and then apply the conversion

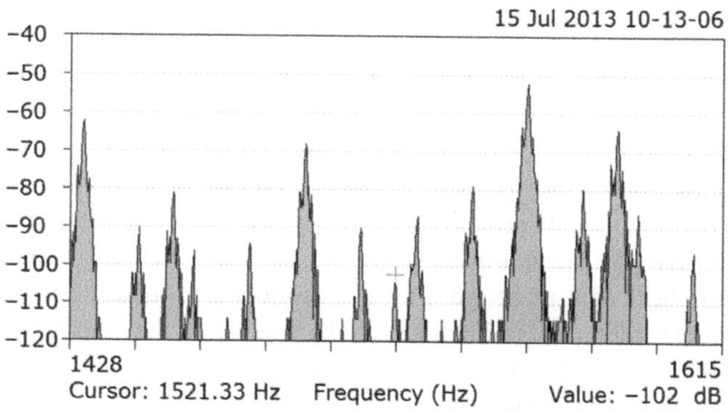


Figure 11.41 Current zoom spectrum, 12,800 lines in baseband span of 70–1800 Hz.

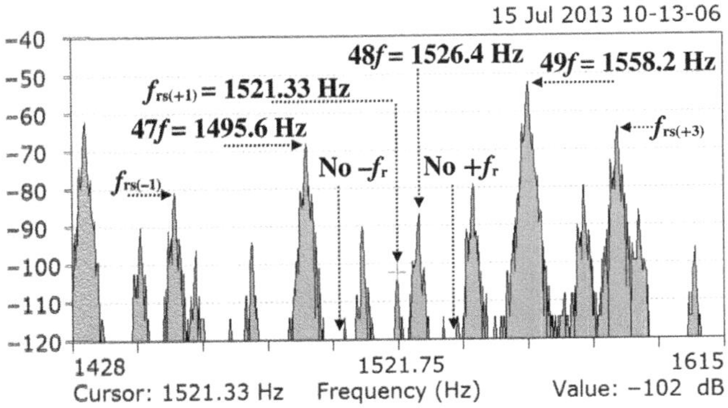


Figure 11.42 Annotated current zoom spectrum.

scaling to get the frequency of each component. It is surprising how accurate this is in practice. This approach can be applied to any frequency spectrum (provided the spectrum has a suitable resolution of Hz/line) and the reader can check spectra in reports from consultants and contractors who provide condition monitoring reports.

The zoom spectrum is applied between 1428 and 1615 Hz and the annotated spectrum in Figure 11.42 shows the following rotor slot passing frequencies at an operating current of 600 A and they are spaced $2f$ (63.6 Hz) apart.

$$f_{rs(+1)} = 1521.33\ \text{Hz}$$
$$f_{rs(-1)} = 1457.73\ \text{Hz}$$
$$f_{rs(+3)} = 1584.93\ \text{Hz}$$

The operational slip can be reverse calculated from the measured RSPFs, using $f_{rs(+1)} = 1521.33$ Hz, the operating slip at 600 A is as follows:

Operating slip, $s_o = 1 - \{(f_{rs(+1)}/f) - 1)(p/R)\} = 1 - \{(1521.33/31.8) - 1\}$ (2/94)

$$s = 0.0034\ (0.34\%)$$

The rotor speed frequency, $f_r = f\,(1 - s)\,/p = 31.8\,(1 - 0.0034)\,/2 =$ 15.846 Hz

The rotor speed, $N_r = 950.76$ r/min (recall $N_s = 954$ r/min)

An examination of the spectrum shown in Figure 11.42 verifies there are no $\pm f_r$ (±15.85 Hz) components around $f_{rs(+1)}, f_{rs(-1)}$, or $f_{rs(+3)}$ and therefore the airgap eccentricity is perfectly normal and certainly less than ±10% of the radial airgap. The reader can check that there are no $\pm f_r$ components around the three rotor slot passing frequencies identified in Figure 11.42 by producing the scaling factor for this spectrum using a ruler with 0.5 mm or 1/64th of an inch scale and then a conversion factor can be produced in Hz/mm or Hz/(1/64th of an inch) as per the approach for the interpretation of Figure 11.41.

## 11.9 CASE HISTORY (2007)—ADVANCED MCSA INTERPRETATION OF CURRENT SPECTRA WAS REQUIRED TO VERIFY HIGH AIRGAP ECCENTRICITY IN AN HV SCIM DRIVING A PRIMARY AIR (PA) FAN IN A POWER STATION

### 11.9.1 Summary

A coal fired power station utilizes four SCIMs to drive boiler primary air (PA) fans and the relevant motor data for MCSA is as follows:3-phase, 4160 V, 2238 kW/3000 HP, 371 A, 60 Hz, 1184 r/min. The number of rotor and stator slots as provided by the end user is 133 and 108, respectively.

It was reported by the power station in January 2006, that the four PA fan motors were producing high vibration and that it seemed to be electromagnetic in origin and they had a suspicion that the problem was high airgap eccentricity. However, no quantitative data with respect to where the vibration was measured, no information on the vibration quantity that was measured (e.g., displacement, velocity, or acceleration) and no samples of vibration spectra, were provided.

The authors of this book were contacted to interpret MCSA data recorded by power station personnel using a commercial MCSA instrument because expert interpretation was required to assess the operational airgap eccentricity due to the complexity of the recorded current spectra. The in-depth MCSA analysis of one of the motors predicted that the operational airgap eccentricity was of the order of 25–30%.

Unfortunately, the power station did not receive details from the motor repair shop on the measured airgaps before and after the rotor position was adjusted and this suggested that the problem was an unacceptable level of dynamic airgap eccentricity and this was borne out by the spectrum analysis.

The authors did contact the power station during the writing (May 2014 to April 2016) of this book but no quantitative information was available. This illustrates the difficulty in obtaining information when trying to close the loop between an MCSA diagnosis of high airgap eccentricity in large induction motors and what is actually found on strip down and the corrective action taken. Unless of course the end user carries out an RCFA and instructs the motor repair company with a definitive scope of work.

### 11.9.2 MCSA Measurements and Analysis

The supply frequency was 60.02 Hz during the MCSA tests and the motor's operating current was 249 A (67% of full-load current). Since the nameplate data is the only information available the RSPF components are calculated using the full-load and no-load slips. This then gives the lower and upper values, within which the measured RSPF components will exist, since the operational slip at 67% full-load current is unknown.

$$f_{\text{rs}} = f[(R/p)(1 - s) \pm n_{\omega s}]$$

Slip on no-load is taken to be zero.

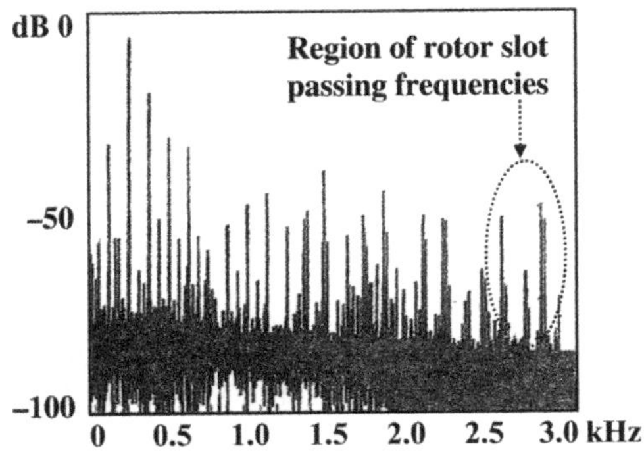


Figure 11.43 Filtered current spectrum, 60 Hz removed, baseband span 70–3000 Hz, 12,800 lines, 0.23 Hz/line.

Nominal full-load slip, $s_{FL} = (1200 - 1184)/1200 = 0.0133$ or (1.33%) where

$f = 60.02$ Hz, $R = 133$, $p = 3$ pole-pairs.

For $s = 0$ on no-load and $s = 0.0133$ on full-load

$f_{rs(+1)} = 2720.9$ Hz and 2685.5 Hz; $+n_{\omega s} = +1$
$f_{rs(-1)} = 2600.8$ Hz and 2565.5 Hz; $-n_{\omega s} = -1$
$f_{rs(+3)} = 2840.9$ Hz and 2805.5 Hz; $+n_{\omega s} = +3$
$f_{rs(-3)} = 2480.8$ Hz and 2445.5 Hz; $-n_{\omega s} = -3$
$f_{rs(+5)} = 2960.9$ Hz and 2925.5 Hz; $+n_{\omega s} = +5$

The next step is to select the baseband frequency span and for this case it is taken to be 70–3000 Hz. Figure 11.43 is the unfiltered spectrum and the region of the RSPFs is identified.

Figure 11.44 shows the current zoom spectrum and the RSPF at $f_{rs(+3)}$ is 2818 Hz which at −54 dB has the highest dB magnitude of the RSPF components and as in the previous case histories it is the RSPF component with the highest dB value, which is selected for the interpretation and assessment of operational airgap eccentricity. The speed frequency ($\pm f_r$) of the rotor is now calculated but firstly the operating slip must be back calculated from the measured value of $f_{rs(+3)}$.

Operating slip, $s_o = 1 - \{((f_{rs(+3)}/f) - 3)(p/R)\} = 1 - \{(2818/60.02) - 1)(3/133)\}$

$$s = 0.00862(0.862\%)$$

The rotor speed frequency, $f_r = f(1 - s)/p = 60.02(1 - 0.00862)/3 = 19.83$ Hz

The rotor speed = 1190 r/min at 67% of full-load current

The full-load speed at nominally full-load current = 1184 r/min

Examination of Figure 11.44 indicates there is a component at $-f_r$ below $f_{rs(+3)}$ which is in fact 3 dB higher but there is no component at $+f_r$ above $f_{rs(+3)}$. This was an interesting result, and was a first for the authors between 1982 and 2006 and still is in 2016. This required further investigation via a pole-pair compatibility analysis.

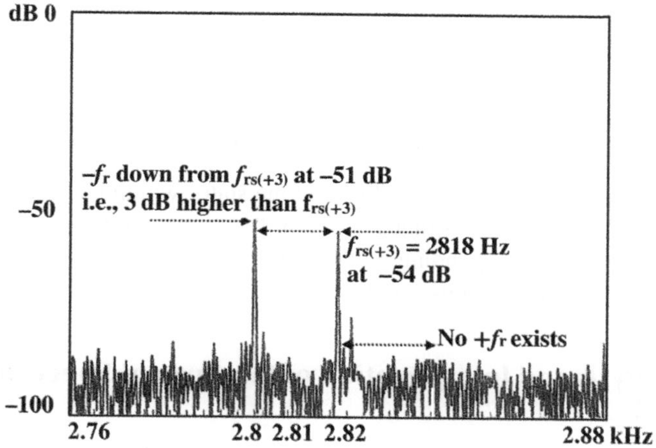


Figure 11.44 Filtered current zoom spectrum to identify $f_{rs(+3)}$ and any $\pm f_r$ components.

### 11.9.3 Pole-Pair Compatibility Analysis

Recall from Section 10.4.1 on airgap eccentricity predictor equations that the following applies:

The harmonic pole-pairs in a 3-phase stator winding are governed by the following and well-known equations, Alger [11.3], Say [11.4], Vickers [11.5]:

$$\text{Odd harmonics } m_o = p(6c \pm 1) \tag{11.1}$$

$$\text{Even harmonics } m_e = p(6c \pm 2) \tag{11.2}$$

where $p$ = pole-pairs; $c = 0, 1, 2, 3, 4, 5$

Let $m_{ps}$ = combination of odd and even harmonic pole-pairs

For this motor with three pole-pairs

$m_{ps} = 3, 6, 12, 15, 21, 24, 30, 33, 39, 42, 48, 51, 57, 60, 66, 69, 75, 78, 84,\ldots$ and so on

From Chapter 10, the RSPFs flux waves and the waves at $\pm f_r$ around the RSPFs are given by

$$f_{ec} = f[(R/p)(1-s) \pm n_{\omega s}] \pm f(1-s)/p$$

$$f_{ec} = f[(R/p)(1-s) \pm n_{\omega s}] \pm f_r \tag{11.3}$$

$f_{ec}$ = the frequencies of the flux waves characteristic of the combination of static and dynamic airgap eccentricity which exist simultaneously in practice.

$R$ = number of rotor slots = 133

$S$ = number of stator slots = 108

$n_s$ = first order static eccentricity integer = 1.0

$n_d$ = first order dynamic eccentricity integer = 1.0

$n$ = any integer

$n_{\theta s}$ = space harmonic index

From Section 10.4, each of these flux waves has a corresponding number of pole-pairs given by:

$$m = R \pm S \pm n_s \pm n_d \pm n_{\theta s}p \quad (11.4)$$

$m$ = pole- pair harmonic number of each flux wave

For emfs (and currents) to be induced in the stator winding at the frequencies given by equation (11.3) requires that a pole-pair ($m$) corresponding to the particular flux wave must have a corresponding pole-pair ($m_{ps}$) from the stator winding:

Pole-pair compatibility must exist: $m = m_{ps}$

### 11.9.4 Analysis and Interpretation of the Current Spectrum in Figure 11.44

A sample pole-pair compatibility analysis is now given for the interpretation of Figure 11.44. The highest (in dB) RSPF current component of the various RSPFs was at $f_{rs(+3)} = 2818$ Hz. The corresponding flux wave that induced an emf to drive this current has a corresponding number of pole-pairs given by

$$m = R \pm S \pm n_s \pm n_d \pm n_{\theta s}p = 133 - 108 - 1.0 + 9 = 33$$

where

$n_{\theta s} = 3$, and $+n_{\omega s} = 3$ in the equation for $f_{rs(+3)}$

$p = 3$

Applying the principle of superposition

$n_s = 1.0$ for first order static airgap eccentricity

$n_d$ = zero

There is first order compatibility with $m = m_{ps} = 33$

This confirms why there is a current component at $f_{rs(+3)}$ in Figure 11.44

Now consider $f_{rs(+3)} + f_r$

$$m = R \pm S \pm n_s \pm n_d \pm n_{\theta s}p = 133 - 108 + 1.0 + 9 = 35$$

where

$n_{\theta s} = 3$, and $+n_{\omega s} = 3$ in the equation for $f_{rs(+3)}$

$p = 3$

Applying the principle of superposition

$n_d = +1$, corresponding to the $+f_r$ component above $f_{rs(+3)}$

$n_s = 0$

$m = 35$ and there is no pole-pair compatible with a pole-pair from $m_{ps}$

$$m \neq m_{ps}$$

The measured current spectrum in Figure 11.44 confirms the analysis that there is no current component at $f_{rs(+3)} + f_r$.

Now consider $f_{rs(+3)} - f_r$

$$m = R \pm S \pm n_s \pm n_d \pm n_{\theta s} p = 133 - 108 - 1.0 + 9 = 33$$

where

$n_{\theta s} = 3$, and $+n_{\omega s} = 3$ in the equation for $f_{rs(+3)}$

$p = 3$

Applying the principle of superposition

$n_d = -1$, corresponding to the $-f_r$ component below $f_{rs(+3)}$

$n_s = 0$

$m = 33$ and there is indeed a pole-pair compatible with a pole-pair from $m_{ps}$

$m = m_{ps} = 33$

The measured current spectrum in Figure 11.44 confirms that there is a current component at $f_{rs(+3)} - f_r$ and the above analysis is verified.

The assessment of the operational airgap eccentricity has to be based on the fact that the $f_{rs(+3)} - f_r$ component is 3 dB higher than $f_{rs(+3)}$. It was therefore concluded that there was a serious airgap eccentricity problem of the order of at least 25–30% and that this was likely to be a dynamic airgap eccentricity problem since the dB magnitude of $f_{rs(+3)} - f_r > f_{rs(+3)}$. The end user in 2015 could only confirm the following from their records on the repair, and to quote:

> *"The motor was sent to a motor repair service shop and the air gaps were found to be well outside the normal ±10% tolerance and the position of the rotor was adjusted to correct this problem. The motor could not be load tested in the service shop, but when it was put back into service its vibration was found to be significantly reduced to acceptable levels."*

The fact that the records state that *"the rotor was adjusted to correct the problem"* does suggest that the problem was an unacceptable level of dynamic airgap eccentricity. The end user was satisfied with the diagnosis and there have been no further vibration problems with this motor or the other three motors since the positions of the rotors in these motors were also adjusted to give airgap eccentricity levels of ≤10%.

## 11.10 CASE HISTORY (1990)—UNSUCCESSFUL MCSA CASE HISTORY TO ASSESS OPERATIONAL AIRGAP ECCENTRICITY IN AN HV SCIM DRIVING A SLOW SPEED RECIPROCATING COMPRESSOR

### 11.10.1 Summary

This case history shows that MCSA was unable to assess the airgap eccentricity in a SCIM driving a reciprocating compressor. The inherent mechanical dynamics of the compressor produces a pulsating load which is normally damped by a flywheel between the compressor and the motor. This reduces but does not normally eliminate the pulsations transferred to the cage rotor of the induction motor.

The current signature pattern, used to identify an airgap eccentricity problem presented in the theory in Chapter 10 and subsequently verified in previous case histories in this chapter, is in fact swamped by the inherent pulsating dynamics from the reciprocating compressor, which produce a current pattern, which makes that caused by airgap eccentricity problems virtually indistinguishable. To ascertain whether there is truly an inherent airgap eccentricity problem within the motor requires an uncoupled run and in this case, with a 20-pole, 128 slot rotor, this would be possible because the high frequency supply harmonics do not coincide with any of the rotor slot passing frequencies. Identification of the signature pattern from an airgap eccentricity problem by this method would certainly be feasible, since there would be no coincidence between any of the rotor RSPF current components and harmonics of the supply, when operating at a no-load slip of zero on a 20-pole, 128 slot rotor motor.

However, this motor was operating in an industrial plant, not in a research laboratory environment and the end user would certainly not simply shutdown an operational and strategic compressor drive train to provide for a no-load uncoupled run of an HV SCIM for a routine MCSA test. It would need to be arranged at a planned outage and even then the end user will not be keen to split the coupling. The old adage of *"leave well alone"* prevails in many cases in industry.

## 11.10.2 MCSA Measurements and Analysis

The motor's relevant nameplate data was as follows:

3-phase, 6600 V, 796 kW/1067 HP, 126 A, 60 Hz, 355 r/min (i.e., 20-poles). The OEM confirmed the number of rotor slots was 128. The nominal full-load slip using the nameplate data is $s_{FL} = 0.0139$ (1.39%)

The current spectrum in Figure 11.45 gives the supply frequency as 59.74 Hz, thus the actual synchronous speed ($N_s$) is 358.4 r/min and not 360 r/min at the nameplate frequency of 60 Hz. The motor was operating at 105 A (83.3% of full-load

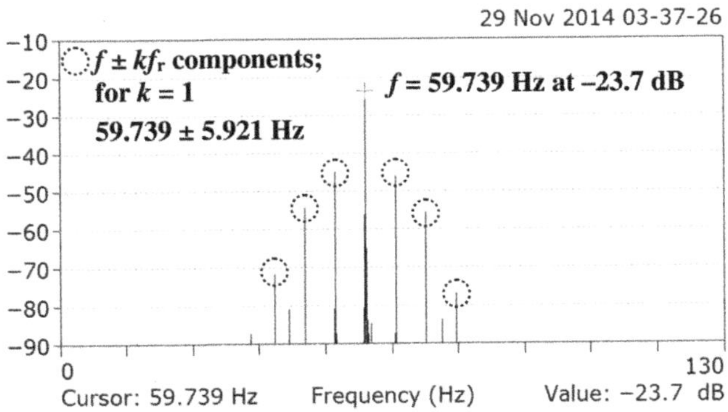


Figure 11.45 Current spectrum, 12,800 line, 0–130 Hz, frequency resolution 0.01 Hz/line.

current) and the current spectrum is very typical of that from a SCIM driving a reciprocating compressor. As expected, the spectrum clearly shows that current components exist at the following frequencies:

$$f \pm kf_r \text{ Hz} \tag{11.5}$$

where $k = 1, 2, 3$.

These components with $k = 1, f_r = 5.92$ Hz are the first order $\pm f_r$ components, which are an average of 20 dB down on the supply component, recall 20 dB equates to a factor of 10, which is 10% of the operating current (105 A) and the *in situ* ammeter clearly showed a pronounced swing of around 10 A.

From equation (11.5) and the spectrum in Figure 11.45, the rotational speed frequency of the rotor $f_r$ is 5.92 Hz ($N_r$ = 355.3 r/min). The synchronous speed is 358.4 r/min at the operational frequency, thus giving an operational slip, $s$ = 0.00865 (0.865%). The predicted RSPFs, $f_{rs(+1)}$ and $f_{rs(-1)}$ at the operating slip are as follows:

$$f_{rs} = f[(R/p)(1 - s) \pm n_{\omega s}]$$

where $f = 57.74$ Hz, $R = 128$, $p = 10$ pole-pairs, $s = 0.00865$

$$f_{rs(+1)} = 817.8 \text{ Hz; for s} = 0.00865 \text{ and } n_{\omega s} = +1$$

$$f_{rs(-1)} = 698.32 \text{ Hz; for s} = 0.00865 \text{ and } n_{\omega s} = -1$$

The accuracy required for measurement of frequency is again very evident in the current spectra shown in Figures 11.46 to 11.48 due to the very closely spaced components.

A baseband span of 70 to 1200 Hz was selected to produce the filtered current spectrum, which is shown in Figure 11.46. Before searching for the predicted RSPF components in Figure 11.46, which is deliberately not annotated at this stage, it is always very good practice to step back and observe key features in this spectrum, particularly in this case, due to the influence of the rotor dynamics produced by the reciprocating compressor.

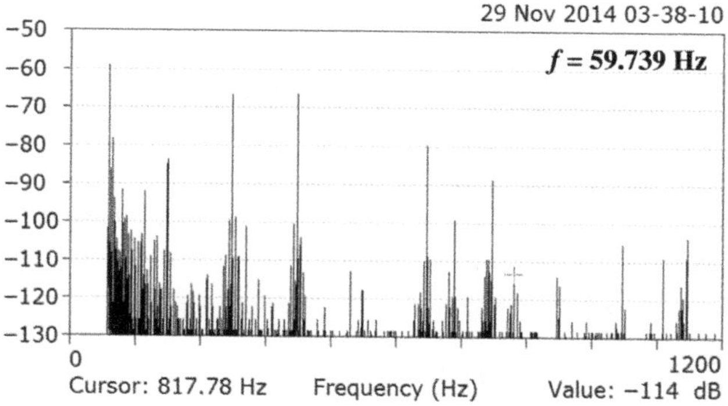


Figure 11.46 Filtered current spectrum, 70 to 1200 Hz, 12,800 lines, 0.883 Hz/line.

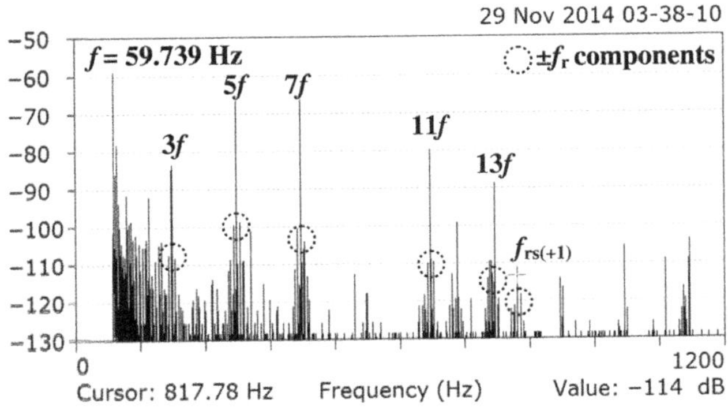


Figure 11.47 Filtered current spectrum, 70 to 1200 Hz, 12,800 lines, 0.883 Hz/line.

From Figure 11.46, it is clear that there are numerous components, which have sidebands around them, and it can be termed a *"very busy spectrum"* and one that a stand-alone MCSA instrument would have great difficulty in automatically analyzing without external human expertise. The annotated spectra are given in Figures 11.47 and 11.48 and the supply harmonics have $\pm f_r$ components around them and the principal RSPF components (Figure 11.48) at $f_{rs(+1)}$ and $f_{rs(-1)}$ also have $\pm f_r$ components around them.

From Figures 11.45 to 11.48 there are $\pm f_r$ sideband components around the following components:

**(i)** $f$ the supply component,

**(ii)** supply harmonics at $3f, 5f, 7f, 11f,$ and$13f$

**(iii)** $f_{rs(+1)}$ and $f_{rs(-1)}$

Conclusions:

**(a)** The $\pm f_r$ sidebands around all these different components are in fact due to the cyclic pulsations from the mechanical dynamics of the reciprocating compressor and are *"polluting"* the spectrum with respect to identifying the current signature pattern, which is truly a function of airgap eccentricity caused by the SCIM and not the external load.

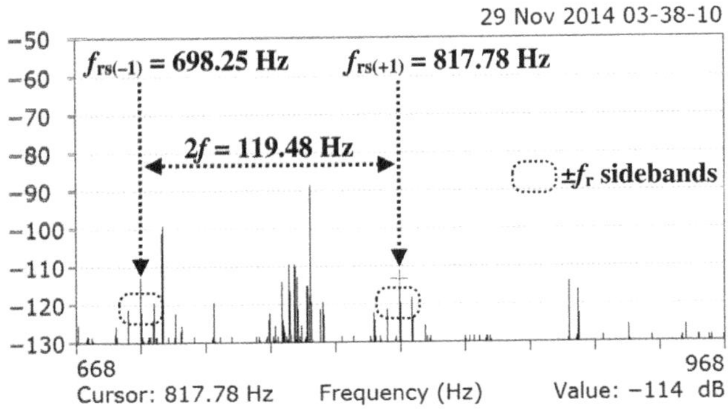


Figure 11.48 Current spectrum to display RSPF components.

**(b)** It is therefore not possible to separate the signature pattern due to an airgap eccentricity problem from the frequencies produced by the normal compressor dynamics.

## 11.11 CASE HISTORY (2002)—PREDICT NUMBER OF ROTOR SLOTS AND ASSESSMENT OF OPERATIONAL AIRGAP ECCENTRICITY IN A LARGE 6600 V, 6714 kW/9000 HP SCIM DRIVING A CENTRIFUGAL COMPRESSOR

### 11.11.1 Summary—Unknown Number of Rotor Slots

The fact that the number of rotor slots is required to enable a prediction to be made of the current signature pattern, which is a function of airgap eccentricity, is a limitation of MCSA since it has to be recognized that many end users can have considerable difficulty in obtaining the number of rotor slots from an OEM. This is not an inherent limitation of the MCSA diagnostic algorithms used to quantify an airgap eccentricity problem but can be an external constraint. The number of rotor slots is never stamped on the nameplates of SCIMs but it would be very easy to so do.

However, it is possible to overcome this limitation but it requires substantial expertise in MCSA and also knowledge about the design of SCIMs particularly in terms of the ranges of number of rotor slots possibly used in the design of a specific motor. The number of slots can vary dependent on power rating, voltage, number of poles (etc.) and due to the individual ideas and experience of different OEMs, who do use different numbers of slots for SCIMs with nominally identical nameplate data. Clearly, if a method can be developed, which allows the number of slots to be determined, the problem no longer exists and the detection of the RSPF components can proceed in the same way as in the previous cases in this chapter. To determine the number of rotor slots requires many years experience of interpreting current spectra and the accumulation of a large data base of knowledge from MCSA tests on a diverse range of industrial cage induction motors, in which the number of rotor slots was known and unknown. Close study of the information provided in this case history can overcome these requirements.

### 11.11.2 MCSA Measurements and Analysis

Motor data: 3-phase, 6600 V, 6714 kW/9000 HP, 686 A, 60 Hz 1780 r/min, 0.89 p.f., efficiency 96.2%. The number of rotor slots was unknown.

The operating current and supply frequency were 93% of full-load current and 59.9 Hz, respectively. Figure 11.49 gives the spectrum for accurate measurement of the supply frequency.

The first stage is to select an upper estimate for the number of rotor slots in a 6600 V, 4-pole, large SCIM, which using Tables 4.3 and 4.4 in Chapter 4, is 86. Taking the no-load slip as being zero, the upper frequency limit can be calculated up to $f_{rs(+3)} + f_r$.

$$f_{rs(+3)} = 59.9[(86/2)(1-0)+3] = 2755 \text{ Hz}; \text{ for } s = 0 \text{ and } +n_{\omega s} = +3$$

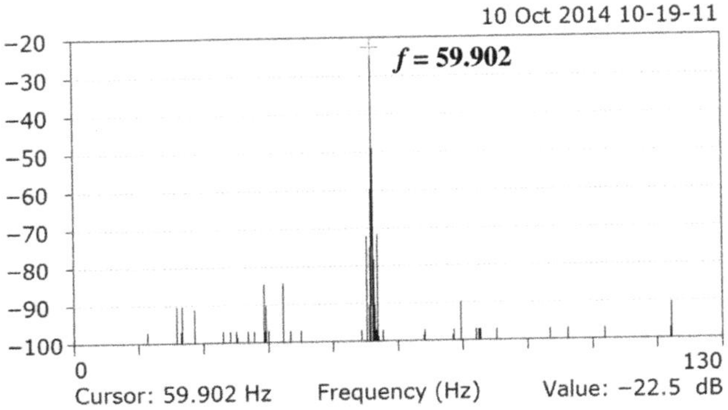


Figure 11.49 Current spectrum 12,800 lines, 0 to 130 Hz, 0.01 Hz/line.

The selected baseband frequency span for the filtered spectrum is set from 70 to 2800 Hz. The filtered spectrum is shown in Figure 11.50 but is not annotated at this stage. The next step is to disregard all odd and even harmonics of the actual operating frequency at 59.9 Hz. As an example, Figure 11.51a shows the odd and even harmonics up to the 20th supply harmonic and Figure 11.51b gives the supply harmonics between the 23rd and 43rd.

It is at this stage that knowledge about the design of SCIMs has to be applied. Consider a 4-pole, 6714 kW/9000 HP 4-pole, 60 Hz, HV SCIM. With reference to Table 4.3 in Chapter 4, it is highly unlikely that the rotor has only 34 slots due to the high power rating and voltage of this motor and it is much more likely to have a rotor with between say 50 and 86 bars (slots).

In fact, due to the voltage and rating, the number of rotor slots would tend to be at the upper end of the band for $R$ between 50 and 86. The next step is to search the spectrum for the RSPF components assuming a band of 50 to 86. Since the operating slip is unknown the only option is to use the full-load slip from the nameplate data.

$$s_{FL} = 0.011\ (1.1\%)$$

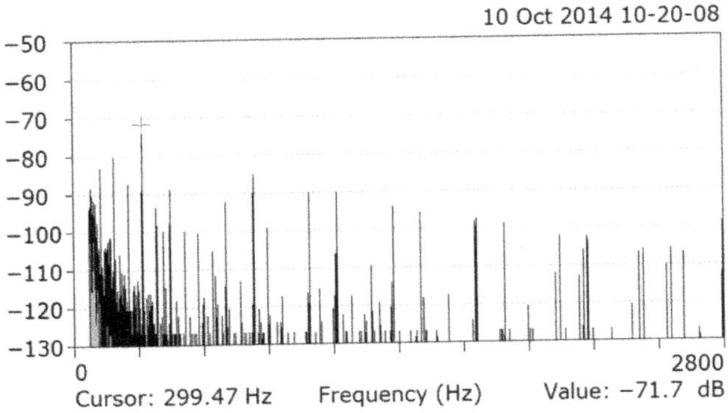


Figure 11.50 Filtered current spectrum 70 to 2800, 12,800 lines, 0.212 Hz/line.

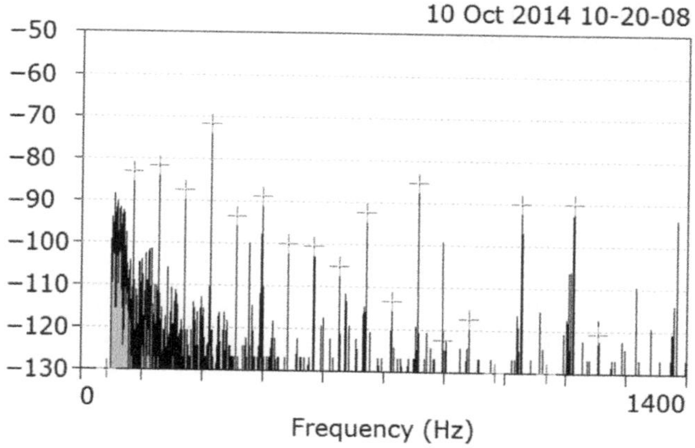


Figure 11.51a Shows the odd and even harmonics up to the 20th supply harmonic.

The search band is therefore selected for the principal rotor slot passing frequency $f_{rs(+1)}$ with $R = 50$ up to $R = 86$ and the predicted principal RSPF at the full-load slip as a function of the number of rotor slots is given in Table 11.3. The measured supply frequency of 59.9 Hz was used to calculate $f_{rs}(+1)$ in this table.

It should be borne in mind that the motor is operating at 93% of its full-load current hence the operating slip will be slightly less than the nominal full-load slip, which means that the predicted principal RSPF will be slightly higher but the supply frequency is 59.9 Hz and not 60 Hz. Examination of Figure 11.52a shows there are no $f_{rs(+1)}$ components close to the corresponding rotor slot values of 50, 52, 58, 62, 66, 74, and 86 hence these values for the number of rotor slots can be disregarded.

However, there are components close to the predicted $f_{rs(+1)}$ components for the corresponding number of rotor slots and these are, from Figure 11.52b, $R = 56$, 60, 68, and 84 hence the range of 11 possible values from Table 11.3 has now been reduced to four possible values for the number of rotor slots and corresponding $f_{rs(+1)}$

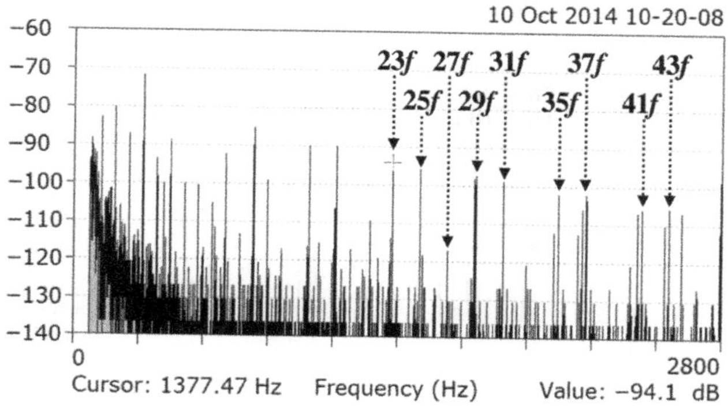


Figure 11.51b This shows the odd and even harmonics from the 23rd up to 43rd supply harmonic.

**TABLE 11.3 Values for $f_{rs(+1)}$ at the Full-Load Slip as a Function of Typical Numbers of Rotor Slots**

| Rotor Slots $R$ | Principal Rotor Slot Passing Frequency $f_{rs(+1)}$ |
|---|---|
| 50 | 1541 |
| 52 | 1600 |
| 56 | 1719 |
| 58 | 1778 |
| 60 | 1837 |
| 62 | 1896 |
| 66 | 2014 |
| 68 | 2074 |
| 74 | 2252 |
| 84 | 2548 |
| 86 | 2607 |

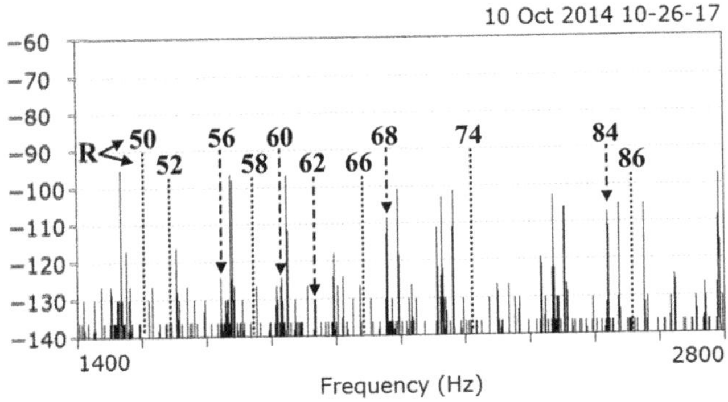


Figure 11.52a Dotted lines give the positions of components which are close to the predicted values for $f_{rs(+1)}$ at the full-load slip as a function of $R$ from table 11.3.

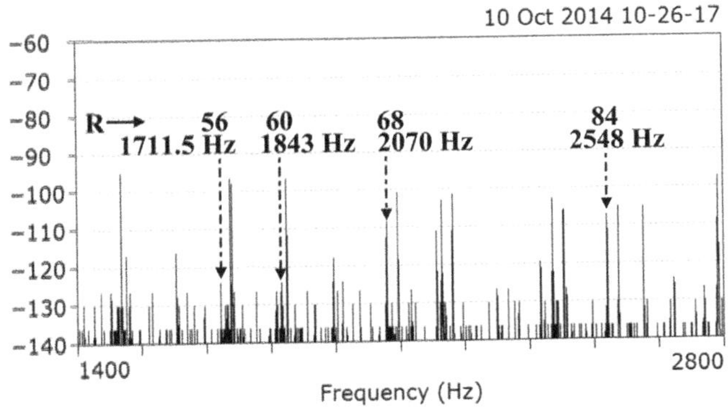


Figure 11.52b Dotted lines give the measured components for $f_{rs(+1)}$ at the full-load slip as a function of $R$ from table 11.1.

values. Figure 11.52b now shows the possibilities for $f_{rs(+1)}$ and its corresponding number of rotor bars.

**R = 56, 1711.5** Hz

The measured component very close to the predicted $f_{rs(+1)}$ with $R = 56 =$ 1711.50 Hz, the operating slip is now calculated using this measured frequency and corresponding number of rotor bars of 56 to ascertain whether a sensible slip value is obtained.

Operating slip, $s = 1 - \{((f_{rs(+1)}/f) - 1)(p/R)\} = 1 - \{(1711.5/59.9) - 1)(2/56)\} = 0.0152(1.52\%)$

*"This slip is too high compared to the full-load slip = 1.1%"*

**R = 60, 1843** Hz

Operating slip, $s = 1 - \{((f_{rs}(+1)/f) - 1)(p/R)\} = 1 - \{(1843/59.9) - 1)(2/60)\} = 0.0083(0.83\%)$

The operating current is 93% of full-load current and thus the operating slip will be greater than 0.83% which is 76% of the full-load slip.

*"For this reason the rotor does not have 60 rotor slots"*

**R = 68, 2070** Hz

Operating slip, $s = 1 - \{((f_{rs}(+1)/f) - 1)(p/R)\} = 1 - \{(2070/59.9) - 1)(2/68)\} = 0.013(1.3\%)$

The operating current is 93% of full-load current and the operating slip would normally be slightly less than the full-load slip, it is not in this case and the operating slip is higher at 1.3% compared to the full-load slip of 1.1%.

*"For this reason, it is very unlikely that the rotor has 68 rotor slots"*

**R = 84, 2548** Hz

Operating slip, $s = 1 - \{((f_{rs}(+1)/f) - 1)(p/R)\} = 1 - \{(2548/59.9) - 1)(2/84)\} = 0.011\,(1.1\%)$

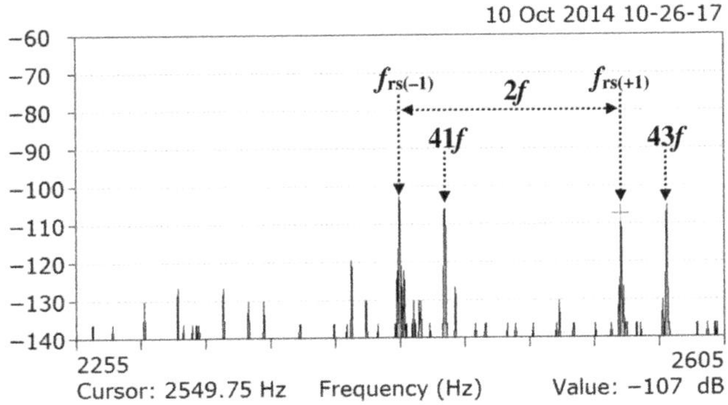


Figure 11.53 Current zoom to display ($f_{rs(-1)}$) and ($f_{rs(+1)}$).

A second test is carried out to detect the inherent rotor slot passing frequency at $-2f$ down (i.e., ($f_{rs(-1)}$) from the principle RSPF ($f_{rs(+1)}$) and indeed there is at 2428 Hz as shown in Figure 11.53.

"*For these reasons the proposed number of rotor slots is 84*"

The principle RSPF = 2548 Hz and the operating slip is 0.011(1.1%).

A final zoom spectrum is presented in Figure 11.54 to ascertain if the characteristic signature pattern of ($f_{rs(+1)}$) $\pm f_r$ exists which is the combination of static and dynamic airgap eccentricity.

There are no $\pm f_r$ components around $f_{rs(+1)}$ and the operational airgap eccentricity is less than 10%. This case history gives an advanced interpretation strategy to predict the number of rotor slots. When assessing the operational condition of a cage rotor and $R$ is unknown the above strategy can also be applied to obtain $R$ to calculate the broken rotor bar factor $BB_f$ rather than only using the dB difference guidance chart thus removing some of the uncertainty which the use of the dB difference guide can entail.

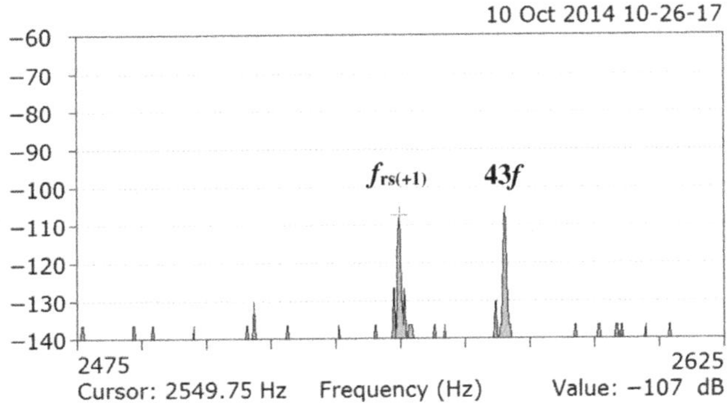


Figure 11.54 Current zoom to display $f_{rs(+1)}$ and any $\pm f_r$ components around it.

## 11.12 QUESTIONS

**11.12.1** Why is it important to accurately measure the supply frequency to a SCIM as part of the strategy to subsequently diagnose a unique current signature pattern that is indicative of the combination of static and dynamic airgap eccentricity?

**11.12.2** Interpret the rotor slot passing frequency equation given by

$$f_{rs} = f[(R/p)(1 - s) \pm n_{\omega s}]$$

What is the major obstacle to being able to calculate the RSPF components? There is only one rotor slot passing frequency—is this statement true or false and give the reasons for your answer.

**11.12.3** A 3-phase SCIM rated at 11,000 V, 3730 kW/5000 HP, 225A, 60 Hz, 3570 r/min, 0.9 p.f., efficiency 96.5% has 52 rotor slots and it is operating at its nominal full-load current and speed.

**(i)** Predict the rotor slot passing frequencies in the current spectrum

**(ii)** Why is it necessary to filter out the supply frequency component?

**(iii)** Select the baseband frequency span to identify the RSPFs up to $f_{rs(+5)}$

**(iv)** How many spectral lines should be selected for the spectrum analysis?

**(v)** If the average dB difference ($N_{ec}$) between the $\pm f_r$ components and $f_{rs(+1)}$ is 16 dB, what is your estimate of the operational airgap eccentricity and recommendation to the end user?

**11.12.4** The motor in question 11.12.3 was uncoupled from the mechanical load. For this motor, it is possible to assess the operational airgap eccentricity with even and odd harmonics of the supply in the dB current spectrum—is this statement true or false, justify your answer.

**11.12.5** The airgap measurements shown in Figures 11.55 and 11.56 were taken via feeler gauges after MCSA estimated an operational airgap eccentricity of 25–30%.

**(i)** State the airgap eccentricity at the DE and NDE from the feeler gauge measurements.

**(ii)** Explain why feeler gauge measurements of the airgap at 20°C/68°F with the rotor stationary and the motor not energized can underestimate the actual operational airgap eccentricity.

**11.12.6** Explain the relevance of the requirement for pole-pair compatibility between the RSPFs $\pm f_r$ flux components and harmonic pole-pairs of the winding distribution to be able to identify the current spectrum that is a function of the combination of static and dynamic airgap eccentricity.

**11.12.7** A 3-phase 12-pole SCIM supplied at 60 Hz, has 112 rotor slots, and operates at a full-load rated speed of 595 r/min. If there are odd and even harmonics in the vicinity of the RSPF components and also close to the $\pm f_r$ components around the RSPF, is it feasible to identify the current signature due to abnormal airgap eccentricity when the motor is running uncoupled from the mechanical load? There is pole-pair compatibility between the RSPFs $\pm f_r$ flux components and harmonic pole-pairs of the winding distribution.

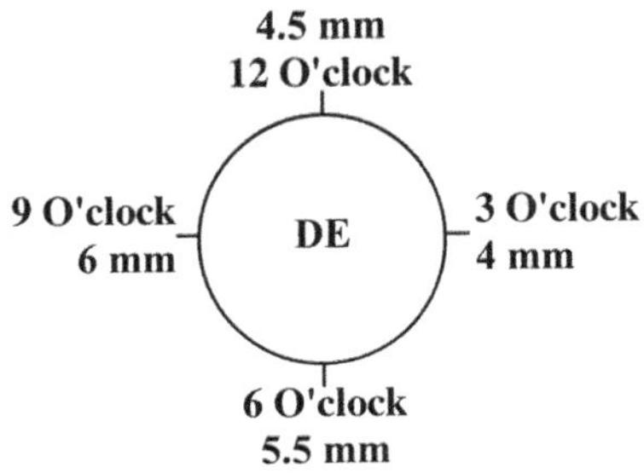


Figure 11.55 Airgap measurements via feeler gauges.

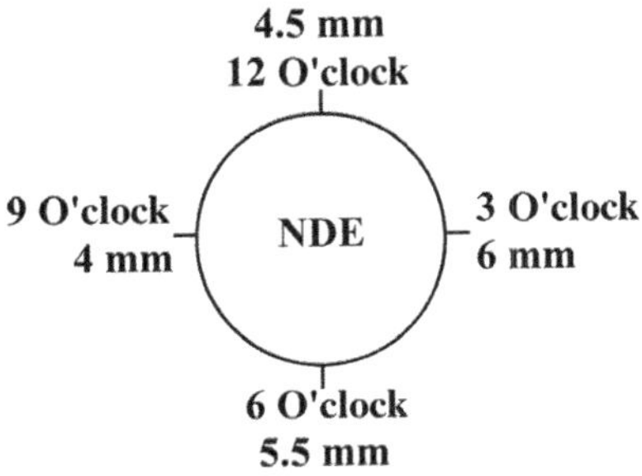


Figure 11.56 Airgap measurements via feeler gauges.

**11.12.8** Why is it not normally possible to identify the current signature that is truly coming from an airgap eccentricity problem within a SCIM when it is driving a reciprocating compressor on full-load?

**11.12.9** A 3-phase SCIM, rated at 11,000 V, 7460 kW/10,000 HP, 464 A, 50 Hz, 1485 r/min drives an air compressor. The rotor has 80 slots and the motor was operating at 454 A.

(i) The current spectra shown in Figures 11.57, 11.58 and 11.59 were recorded. Analyze the current spectra and identify the operational frequency of the

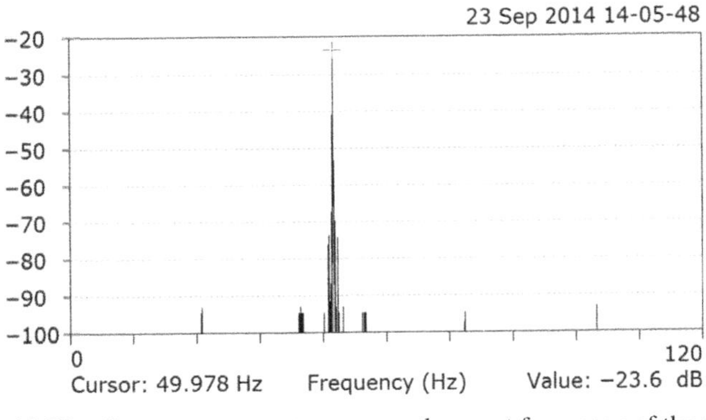


Figure 11.57 Current spectrum to measure the exact frequency of the supply.

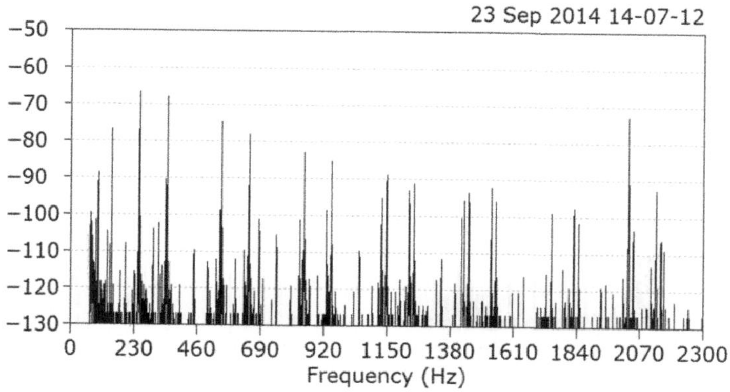


Figure 11.58 Current spectrum, 70–2300 Hz.

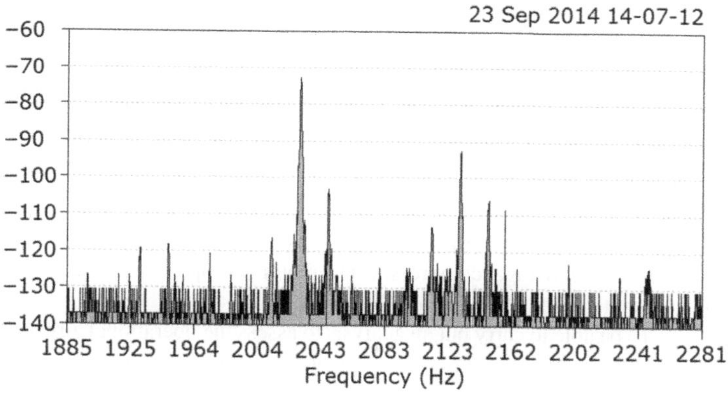


Figure 11.59 Current zoom spectrum.

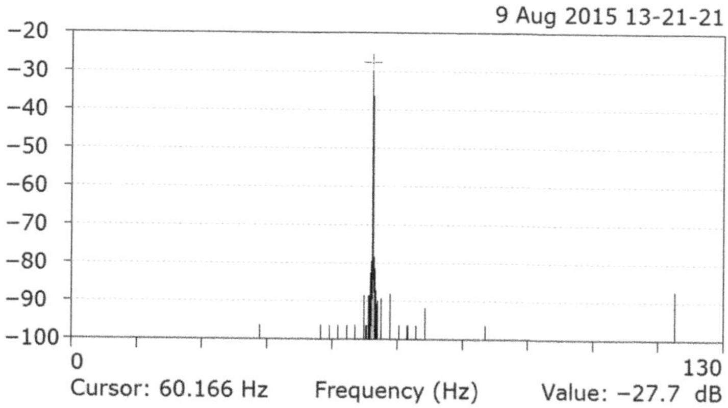


Figure 11.60 Current spectrum to measure the exact frequency of the supply.

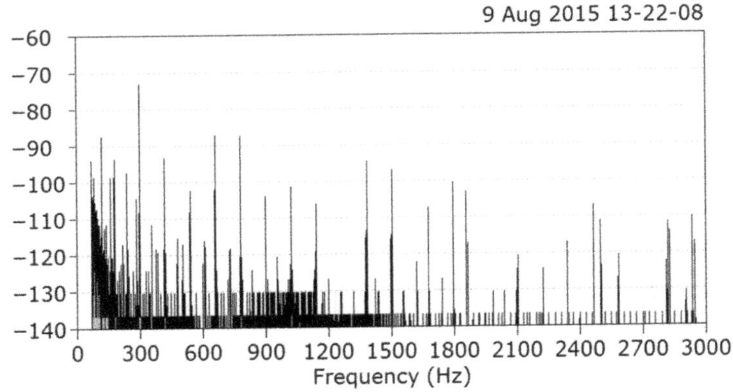


Figure 11.61 Current spectrum, 70–3000 Hz.

principal RSPF ($f_{rs(+1)}$) and whether there are any $\pm f_r$ components around the RSPF ($f_{rs(+1)}$) component.

Although you do not have access to the real-time data that produced the current spectra in Figures 11.57–11.59, by simply using a scale of Hz/mm or Hz/(1/64th of an inch) which you have to determine using a ruler applied to the spectra, it is perfectly possible to answer item (i) above?

**11.12.10** A 3-phase SCIM, rated at 6600 V, 1500 kW/2011 HP, 153 A, 60 Hz, 1785 r/min drives an HP centrifugal compressor. The motor was operating at 92 A. The number of rotor slots is unknown.

To be able to carry out an analysis to assess the operational airgap eccentricity requires the number of rotor slots, analyze the spectra shown in Figures 11.60–11.62 and determine the principal RSPF component, the number of rotor slots, operational slip, and then estimate the operational airgap eccentricity.

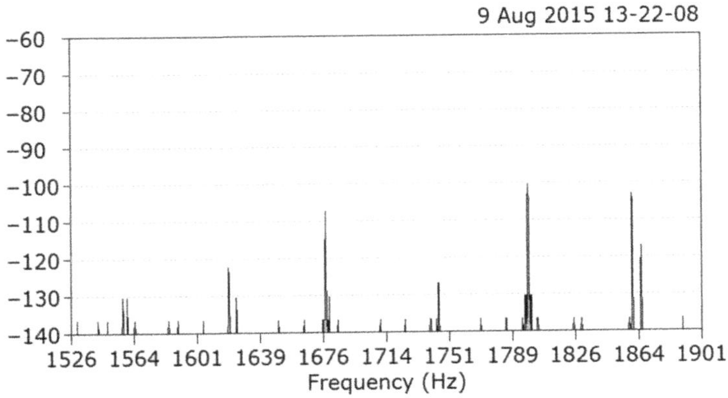


Figure 11.62 Current zoom spectrum.

## REFERENCES

[11.1] ISO 1940-1:2003.
[11.2] G. C. Stone, E. A. Butler, I. Culbert, and H. Dhiranai, *Electrical Insulation for Rotating Machines—Design, Evaluation, Testing and Repair*, Wiley-IEEE Press, 2004, ISBN 0-471-44506-1.
[11.3] P. L. Alger, *Induction Machines—Their Behavior and Uses*, 2nd edition, Gordon and Breach Science Publications Inc., 3rd printing with additions by OPA Amsterdam in 1995.
[11.4] M. G. Say, *Alternating Current Machines*, 4th edition, ELBS and Pitman Publishing, 1976.
[11.5] H. Vickers, *The Induction Motor*, Sir Isaac Pitman and Sons Ltd., London (1st edition, 1924; 2nd edition, 1953).

CHAPTER 12

# CRITICAL APPRAISAL OF MCSA TO DIAGNOSE SHORT CIRCUITED TURNS IN LV AND HV STATOR WINDINGS AND FAULTS IN ROLLER ELEMENT BEARINGS IN SCIMS

*William T. Thomson*

## 12.1 SUMMARY

Examples of research papers on MCSA to diagnose shorted turns in stator windings of LV SCIMs can be found in the following publications: Thomson et al. [12.1–12.3], Stavrou et al. [12.4], Prabhakar et al. [12.5] and a very comprehensive "Bibliography on Induction Motors Faults Detection and Diagnosis" by Benbouzid [12.6] also includes relevant papers on this topic.

A shorted turn or turns in a coil of an HV* or an LV* motor is a serious fault since it will inevitably lead to a phase-to-ground or a phase-to-phase fault and a stator rewind. The time to failure is normally very short indeed, of the order of minutes (and in some cases in a matter of seconds) rather than hours but it can be longer under certain conditions [12.5]. However, it is worth quoting Nailen [12.7, 12.8], a very well-respected electrical machine designer and consultant in the United States and Engineering Editor of *The Electrical Apparatus Magazine* who very aptly stated:

> *"Short circuits in coils in a stator winding cannot be trended since they either exist or they do not, and if present, will soon escalate into a phase to ground fault."*

The authors of this book agree with that hypothesis. Literature surveys indicate that industrial case histories do not seem to have been published on any results

*See Appendix 13A.

*Current Signature Analysis for Condition Monitoring of Cage Induction Motors: Industrial Application and Case Histories*, First Edition. William T. Thomson and Ian Culbert.

obtained from applying MCSA in industry to actually detect shorted turns, and to automatically switch off the power supply before the normal phase-to-earth, phase-to-phase, current unbalance, or negative sequence protection trips are activated in, for example, an HV SCIM. This verifies that end users have not been whatsoever receptive to this application of MCSA and the reasons are indeed obvious and self-explanatory.

With respect to actual times to failure of the stator winding in an LV SCIM due to a shorted turn or turns, Thomson [12.2] did include experimental results from a direct short circuit across turns in one coil, which caused two stator coils to be severely damaged and required the random wound stator to be rewound. This occurred even-though the motor protection did not trip due to a phase-to-earth or phase-to-phase fault. In this instance, nothing was gained by the early detection of shorted turns since the stator still required a rewind. For completeness, these results will be presented in Section 12.3.

A sample of papers on MCSA to diagnose faults in roller element bearings in LV SCIMs is as follows: Steele et al. [12.9], Schoen et al. [12.10], and Knight et al. [12.11]. However, literature surveys indicate that here again actual industrial case histories do not seem to have been reported and this lack of evidence suggests that end users are not receptive to MCSA for detecting bearing faults in industrial SCIMs. Large HV SCIMs use plain bearings, and therefore MCSA is certainly not applicable to detect problems, such as bearing wear, mechanical rotor unbalance, misalignment, and oil whirl/whip in these bearings. The measurement of displacement of the rotating shaft relative to the fixed bearing shells via displacement probes in the bearing housings and supported by shaft orbital plots is a well-proven vibration analysis technique that can detect these problems. This clearly comes under the engineering discipline for mechanical engineers and vibration specialists.

Roller element bearings are predominately used in LV SCIMs and provided the bearing housings can be accessed (or close to), then vibration monitoring is universally applied to detect bearing defects via vibration spectrum analysis (VSA). This is also the responsibility of mechanical and maintenance engineers and vibration specialists and not electrical engineers. A book by Tavner et al. [12.12] has a chapter of 33 pages on vibration monitoring for electrical machines and two excellent books by Taylor [12.3, 12.14] contain numerous industrial case histories on the detection of defects in roller element bearings via VSA. It is a commonsense deduction that mechanical engineers will not switch to MCSA or use MCSA in addition to or instead of vibration monitoring to diagnose faults in roller element bearings in SCIMs, when vibration analysis has a very long and proven track record in industry. It is only when roller element bearings/bearing housings cannot be accessed to mount accelerometers, that MCSA could be used as an alternative option to provide some information that may prevent a catastrophic bearing failure. For example, it has been proposed by Salon et al. [12.15], Gong et al. [12.16], Williamson et al. [12.17], and Zhang et al. [12.18] that due to the cost of fitting numerous accelerometers to a large number of induction generators in a wind farm, the application of MCSA is a much easier and cheaper option to install. As reported in [12.15–12.18], there is the potential that MCSA could possibly assess the condition of roller element bearings and/or gear boxes before the occurrence of a catastrophic failure in a wind generator drive train. However, much more work is required via extensive on-site trials and the reporting of actual industrial case histories of successful and unsuccessful detection of faults in bearings and gearboxes.

MCSA research pioneered (in the 1980s) by Haynes and Kryter [12.19] and Haynes [12.20] for a very specific application was carried out at the Oak Ridge National Laboratory in the United States. This particular application was to check the operational condition of remotely operated, small power (fractional horse power) SCIM drives, which operated valves in nuclear power plants. They proved that gearbox defects and bearing problems could be identified in these drive units, which have a very closely coupled mechanical interaction between the gearbox and/or bearings and fractional horse power SCIMs. However, this very sensitive mechanical interaction is not necessarily the case with much larger SCIM drive trains, which is the main focus of this book.

## 12.2 SHORTED TURNS IN HV STATOR WINDING COILS

Figures 12.1 and 12.2 show the typical construction of HV stator winding coils and the reader is referred to Stone et al. [12.21] for further information. Figure 12.3 shows the cross-section of an 11 kV coil. The term semi-conductive coating bears no relationship whatsoever with semiconductors in the transistor sense and the reader is referred to page 24 of Stone et al.'s book [12.21] for a full explanation of the function of semi-conductive coatings on the coil sides of HV coils in the slot region. The function of this coating is to prevent discharges in the slot region between the surface of the main wall insulation of the HV coil in the slot and the stator core.

(i) **GW** is the ground wall or main wall insulation.

(ii) **ITI** is the inter-turn insulation.

(iii) Example, within a coil, of voids that should not exist.

Since the electrical stress across the inter-turn insulation is very low in comparison to the electrical stress between the line end coils of an HV stator winding and earth and it is well known that partial discharge (PD) monitoring cannot detect the

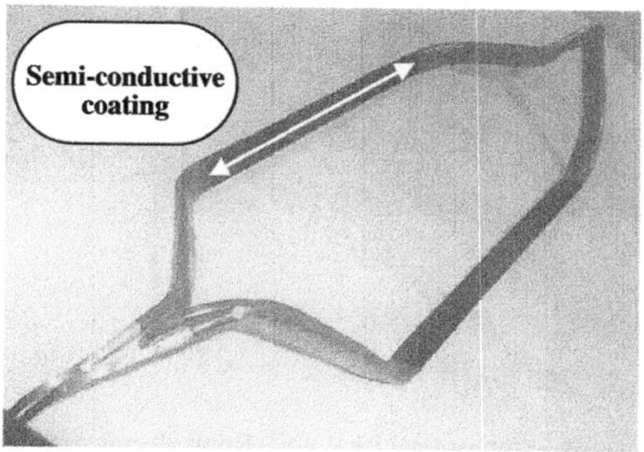


Figure 12.1 Example of HV stator coil. Reproduced with permission of EM Diagnostics Ltd.

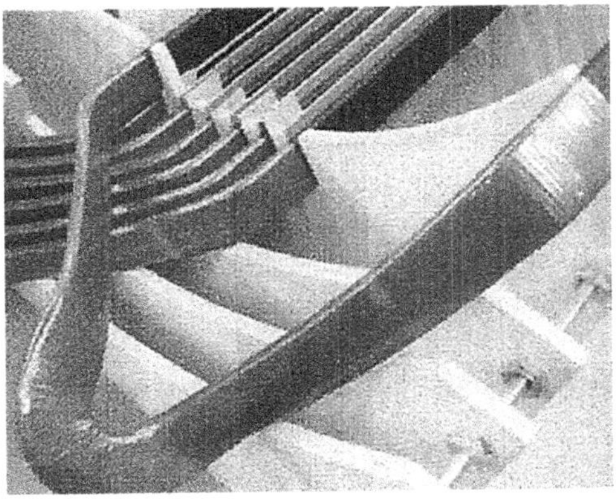

Figure 12.2 Example of adjacent HV stator coils to show the end winding region. Reproduced with permission of EM Diagnostics Ltd.

inception of inter-turn insulation degradation since no PD signals are generated due to the low voltage between turns. To illustrate this fact, an example of the volts/turn in a 6.6 kV stator winding is now presented.

Motor nameplate: 3-phase SCIM, 6.6 kV, 8250 kW/11,060 HP, 869 A, 60 Hz, 1762 r/min, p.f. = 0.87, efficiency = 95.5%. The star connected, stator winding is as follows:

**(i)** Double layer winding in 72 slots (number of coils equals number of slots),

**(ii)** 10 turns/coil, 240 turns/phase,

**(iii)** Four parallel paths (therefore, 60 turns/parallel path across 3180 V rms).

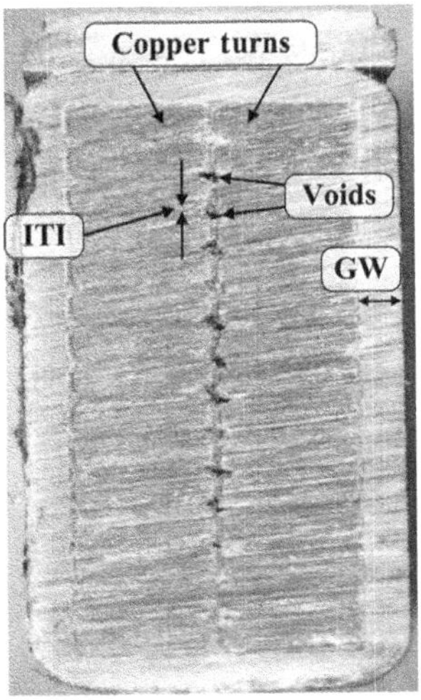


Figure 12.3 Sample cross-section of an 11 kV coil. Reproduced with permission of EM Diagnostics Ltd.

The line voltage is 6.6 kV, which gives a phase voltage of 3810 V and with four parallel paths per phase, this gives 63.5 V/turn (89 V peak). In contrast, there is 3810 V (5388 V peak) between the copper conductors and earth at a line end coil, thus the ground wall insulation is much thicker than the inter-turn insulation as shown in Figure 12.3. In this sample the GW insulation is approximately 11 times the thickness of the inter-turn insulation. This example demonstrates that the electrical stress on the inter-turn insulation is very low in comparison to the stress on the ground wall insulation.

### 12.2.1 Causes of Shorted Turns in HV Stator Windings of SCIMs

**(i)** Transient voltage surges due to lightning strikes or vacuum circuit breakers causing transient voltage surges at the terminals of HV SCIMs can break down the inter-turn insulation. This causes short circuits between turns and very high currents flow, which subsequently cause breakdown of the ground wall insulation. A stator winding failure then occurs due to a phase-to-phase or phase-to-earth fault (if the shorted turns are close to the core or actually in the core region). This failure occurs in a relatively short period of time (it can be a matter of minutes after the shorted turn/turns occurs since very large currents flow).

**(ii)** DOL starts cause mechanical movement of the coils in the end winding region and if repetitive, sequential starts not in accordance with the OEM's specification are a regular occurrence, this can eventually lead to mechanical breakdown of the inter-turn insulation and shorted turns occur.

An example of the catastrophic damage caused by shorted turns in a 6.6 kV stator winding (motor data and an example of volts/turn given in Section 12.2) is shown in Figure 12.4, which was due to item 12.2.1 (ii) and occurred during repeated

Figure 12.4 Example of failure due to shorted turns in a 6.6 kV SCIM.

DOL starts. Installing a permanent on-line MCSA instrumentation system to detect the shorted turn/s before failure would be completely futile.

## 12.3 DETECTION OF SHORTED TURNS VIA MCSA UNDER CONTROLLED EXPERIMENTAL CONDITIONS

An academic exercise was carried out by Thomson [12.2] in 1988 to demonstrate that MCSA could detect a shorted turn/turns in an LV stator winding but it was not intended that this application of MCSA would become a practical reality in industry.

Only a sample of results is presented to demonstrate that the input current spectrum changes due to a shorted turn/turns in an LV stator winding, when the short circuit current is limited by an external resistor, as other researchers have done Stavrou et al. [12.4]. However, Thomson [12.2] went that stage further and introduced a direct short circuit between turns to cause severe damage to the stator and thus provide an indicator of the time it took for this to happen in an LV stator winding.

It is well known that the following equation Thomson et al. [12.3] and Stavrou et al. [12.4] can predict certain frequency components in the airgap flux waveform when a shorted turn occurs and thus emfs are induced in the stator winding that drive current components at these frequencies in the stator:

$$f_{st} = f[(n/p)(1 - s) \pm k] \tag{12.1}$$

where

$f_{st}$ = current components, Hertz
$f$ = supply frequency, Hertz
$n = 1, 2, 3, \ldots$
$k = 1, 3, 5, \ldots$
$p$ = pole-pairs
$s$ = slip

A 11 kW, 415 V, 20.5 A, 1440 r/min, 50 Hz, delta connected SCIM was used for the investigation under controlled laboratory conditions at the Robert Gordon University, Scotland. The stator had 36 slots and a three-tier concentric winding with random wound coils. The current spectrum with no shorted turns is shown in Figure 12.5 when the motor was supplied at rated volts, on no-load other than turning a dynamometer, so the operating slip in equation (12.1) is taken to be zero. Where A = 25 Hz, B = 75 Hz, C = 100 Hz, and D = 150 Hz in Figures 12.5–12.8 and 12.10.

The current spectrum with no shorted turns is shown in Figure 12.5 on no-load. There are components at 25 and 75 Hz (A and B, respectively) due to disturbances to the airgap field from the drive train (coupling plus dynamometer) dynamics downstream of the SCIM as already proven in Section 8.11 in chapter 8. The component at 100 Hz (C) is 60 dB down on $f$, (i.e., 1000 times smaller) and is due to the fact that the stator winding distribution cannot be perfectly symmetrical and the supply voltage was not perfectly balanced. As expected there is a third harmonic component

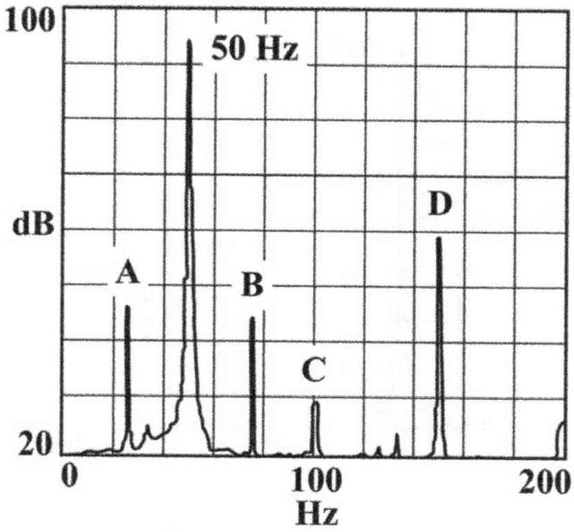


Figure 12.5 R-phase, current spectrum, no shorted turn. *Source*: Thomson (2001) [12.2]. Reproduced with permission of IEEE.

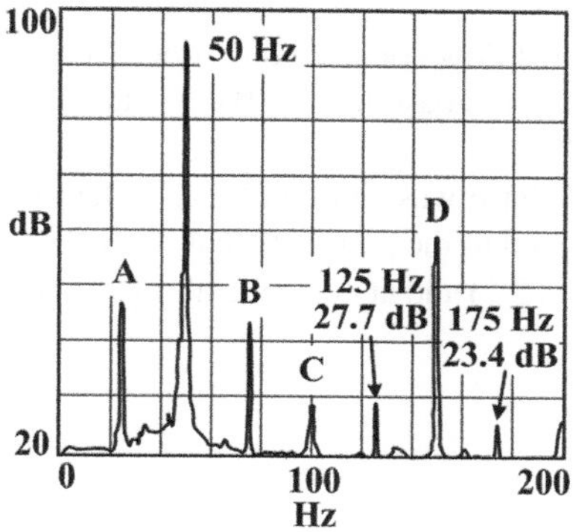


Figure 12.6 R-phase, current spectrum with the short circuit current limited to 3.6 times the full-load phase current. *Source*: Thomson (2001) [12.2]. Reproduced with permission of IEEE.

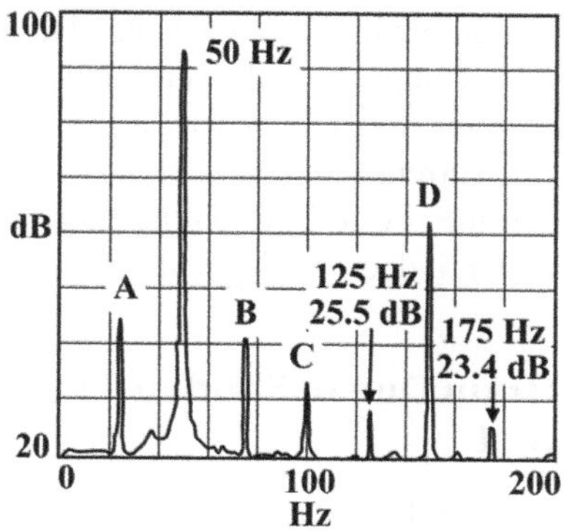


Figure 12.7 Y-phase, current spectrum with the short circuit current limited to 3.6 times the full-load phase current. *Source*: Thomson (2001) [12.2]. Reproduced with permission of IEEE.

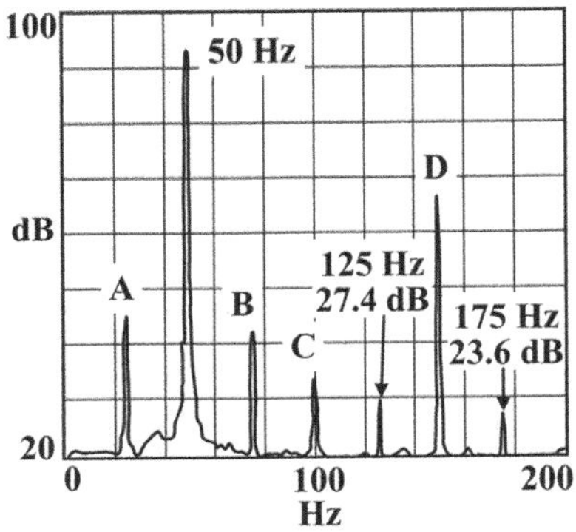


Figure 12.8 B-phase, current spectrum with the short circuit current limited to 3.6 times the full-load phase current. *Source*: Thomson (2001) [12.2]. Reproduced with permission of IEEE.

at 150 Hz (D) which is 34 dB down on $f$ (i.e., 51 times smaller). There are no components at 125 and 175 Hz and the baseline of 20 dB is taken as the reference for these components.

One turn in the red phase was shunted by an external resistor and the current was limited to 42 A compared to the phase current of 11.83 A to avoid initial damage to the coil. From equation (12.1) and at no-load, taking the slip as approximately zero.

For $k = 1, n = 3$,

$$f_{st} = 25 \text{ and } 125 \text{ Hz}$$

For $k = 1, n = 5$,

$$f_{st} = 75 \text{ and } 175 \text{ Hz}$$

Examination of Figure 12.6 shows that due to the shorted turn, there are new components at 125 and 175 Hz at 8 and 5 dB (i.e., factors of 2.5 and 1.8) higher, respectively, compared to the case with no stator faults, but there has been virtually no change in the 25 and 75 Hz components. The current spectra in Figures 12.7 and 12.8 for the Y- and B-phases show the same pattern as the R-phase where the shorted turn exists, thus only one line current needs to be measured and analyzed. Although this result is artificial, due to the severe limit on the short circuit current, which in practice would be much greater, it does verify the theoretical predictions.

### 12.3.1 Current Spectrum Due to Shorted Turns—No Limit on Short Circuit Current

A coil in the red phase was selected and a piece of copper was soldered onto an indeterminate number of adjacent turns on the end winding of the coil after some of the varnish had been removed. The number of turns shorted in the coil was not the main focus of this test, because it was to show what can happen in practice with the main objectives being to observe the current spectrum, in particular, how long the motor could be kept running. The end winding of the coil could be observed during the experiment since the test was carried out on the motor shown in the test rig in Figure 4.2 which was specifically designed in a modular format for research projects. It was inevitably going to be a very short time between switch on and severe damage to the coil, therefore the input line current was recorded on tape for subsequent spectrum analysis. The rated voltage was applied with the motor on no-load and within 4 minutes the following was observed:

- **(i)** One minute after switch on, the shorted turns were obviously overheating.
- **(ii)** The varnish quickly turned black and smoke was emitted from the faulty coil.
- **(iii)** The end winding tape ignited.
- **(iv)** The original copper short circuit was fused on to the end winding.
- **(v)** The coil concentric with and adjacent to the coil with the shorted turns was also severely damaged.

The damaged coils are shown in Figure 12.9 and the motor was manually tripped after 4 minutes but the motor protection had not tripped on a phase-to-earth or phase-to-phase fault.

A comparison between the current spectra in Figures 12.7 and 12.10 with no shorted turns and direct shorts, respectively, is as follows:

- **(i)** There is a negligible change in the input supply current component at 50 Hz; this is as expected since the short circuit current is circulating within the short circuited loop.
- **(ii)** The greatest change in the current spectrum is that the components at 125 and 175 Hz, due to shorted turns are now 56.4 and 47.3 dB, respectively, with respect to the baseline of 20 dB which is an increase of 36.4 and 17.3 dB (i.e., 66 and 7.4 times in absolute amperes, respectively).
- **(iii)** Clearly MCSA can diagnose shorted turns but the most important fact was that severe damage to two adjacent coils occurred within 4 minutes requiring the complete stator to be rewound.

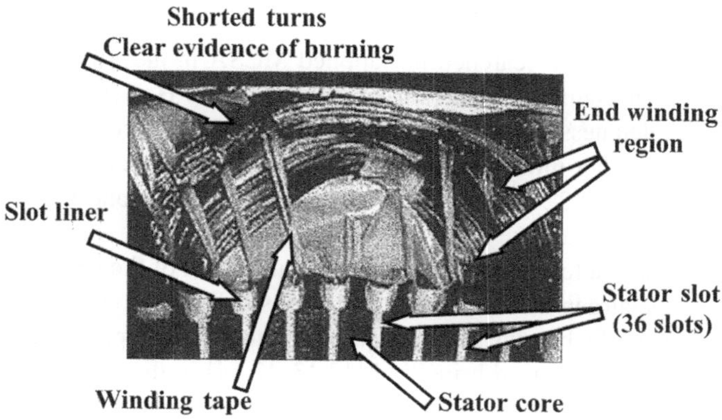


Figure 12.9 Damaged coils due to direct shorted turns in one coil.

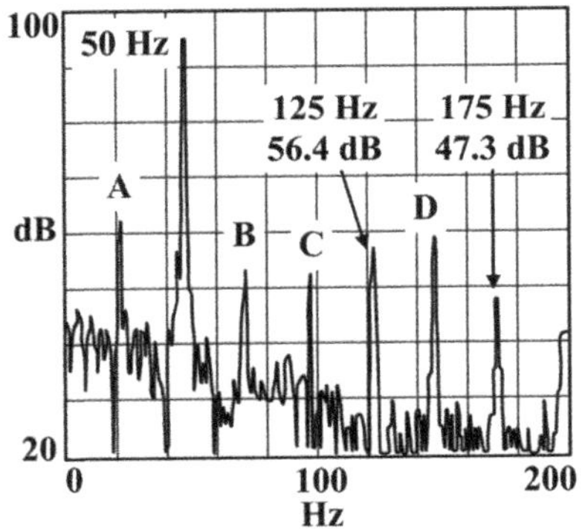


Figure 12.10 Line current spectrum with direct short circuited turns in a coil.

The following points are some of the reasons why end users are not receptive to MCSA monitoring to detect shorted turns in LV stator windings:

**(i)** The MCSA instrumentation has to be permanently on-line which can be costly and has to be justified as proving to be of real value to the end user. In the authors' view such a CM strategy is neither technically sound nor cost effective and is in fact a non-starter.

**(ii)** A broad range of designs of LV and HV SCIMs, supplied by different OEMs, operate in industry and the stator windings are certainly not the same hence the volts/turn varies for each design. For a given number of shorted turns, the magnitudes of the current components will differ, therefore making the setting of protection thresholds to trip the motor via MCSA, before it is tripped out via the normal and reliable phase-to-earth or phase-to-phase protection, to say the least, is a practical impossibility.

## 12.4 DETECTION OF DEFECTS IN ROLLER ELEMENT BEARINGS VIA MCSA

The authors have neither researched nor applied MCSA in industry to detect bearing defects in roller element bearings used in LV SCIMs. However, this section is included for completeness and the reader should refer to References 12.6, 12.9–12.12, 19, and 12.20 for further information and, in particular, more recent publications by Harihara and Parlos [12.22–12.24] which present encouraging and new results and also included an industrial case history.

Severe bearing defects can cause small radial motion of the rotor thus causing disturbances to the airgap flux waveform and very small variations to the radial airgap. As the severity of a bearing fault increases due to, for example, severe and deep pitting or spalling, or advanced brinelling [12.13, 12.14] in the outer or inner races the greater that radial movement of the rotor becomes and eventually the bearing can collapse and a rotor to stator rub occurs. Of course, classical vibration analysis using

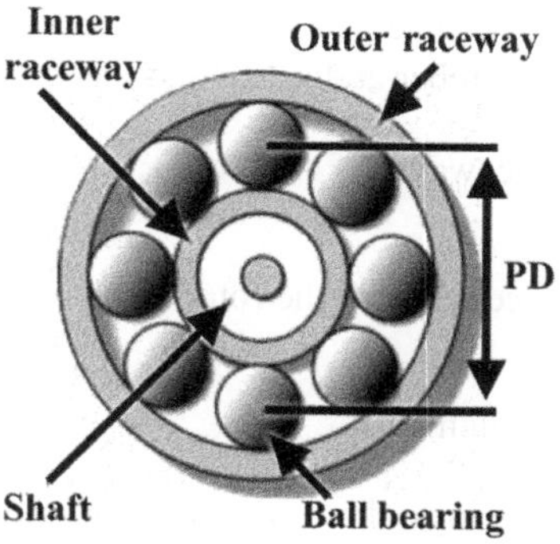


Figure 12.11 Schematic diagram to identify the component parts of a ball bearing.

accelerometers on the bearing housings can detect the early inception of these problems to prevent a catastrophic failure. The flux components at specific frequencies caused by bearing defects induce emfs in the stator windings that drive currents at the following frequencies [12.10]:

$$f_b = f \pm kf_v \tag{12.2}$$

where

$f_b$ = induced stator current components due to bearing faults, Hertz

$f$ = supply frequency, Hz

$k = 1, 2, 3, \ldots\ldots,$ harmonic number

$f_v$ = classical vibrational components due to bearing faults, Hz

The classical vibrational components ($f_v$) due to bearing faults are a function of the bearing dimensions as shown in Figures 12.11 and 12.12, however, information

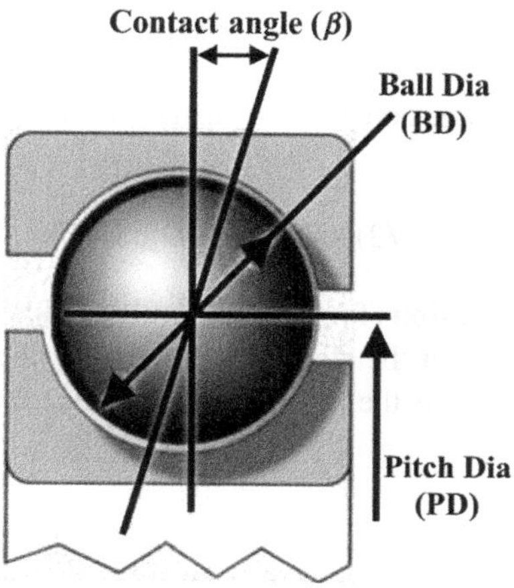


Figure 12.12 Illustration of the contact angle, ball diameter and pitch diameter of a ball bearing.

on the World Wide Web can be accessed to predict the bearing fault frequencies by simply inserting the bearing number and speed and the values for $f_v$ are produced.

$\beta$ = contact angle on raceways

PD = pitch diameter

$f_r$ = rotational speed frequency of the rotor

BD = ball diameter

$n_e$ = number of rolling elements

With radially loaded bearings, the contact areas of the balls and raceways carry the largest loads, hence fatigue failures normally involve these components. The ball spin frequency is produced by the rotation of each ball about its own center. The frequency components due to an inner or outer race defect are generated when each ball passes over a defect. This occurs $n$ times during a complete revolution of the raceway, where $n$ is the number of rolling elements.

**Inner and outer race defects** [12.4]: It is very well known that defects in the inner and outer races produce vibrational frequency components at

$$f_i = (n_e/2)f_r[1 + (\mathrm{BD}/\mathrm{PD})\cos(\pi)] \quad (12.3)$$

$$f_o = (n_e/2)f_r[1 - (\mathrm{BD}/\mathrm{PD})\cos(\pi)] \quad (12.4)$$

where

$f_i$ = vibrational component due to an inner race defect

$f_o$ = vibrational component due to an outer race defect

**Roller element defect** [12.4]: A defective ball will make contact with both the inner and outer races during each revolution hence the ball defect frequency will be double the spin frequency.

$$f_b = (\mathrm{PD}/\mathrm{BD})f_r[1 - ((\mathrm{BD}/\mathrm{PD})\cos(\pi))^2] \quad (12.5)$$

**Cage fault** [12.4]**:** A cage defect results in the following frequency component:

$$f_c = (f_r/2)[1 - (\mathrm{BD}/\mathrm{PD})\cos(\pi)] \quad (12.6)$$

**MCSA predictor equations for bearing faults:** With respect to current components that can be induced in the stator winding due to bearing defects in roller element bearings used in SCIMs the following equations are used, [12.10]:

$$f_{ci} = f \pm kf_i \quad (12.7)$$

$$f_{co} = f \pm kf_o \quad (12.8)$$

$$f_{cb} = f \pm kf_b \quad (12.9)$$

$$f_{cc} = f \pm kf_c \quad (12.10)$$

where

$f_{ci}$ = current components due to an inner race defect
$f_{co}$ = current components due to an outer race defect
$f_{cb}$ = current components due to a ball bearing defect
$f_{cc}$ = current components due to a cage defect

Equations (12.5) and (12.8) can be used to predict the current components due to bearing defects. The reasons why end users have not been receptive to MCSA for the assessment of the operational condition of roller element bearings in SCIMs when bearing vibration monitoring can be applied has already been discussed.

However, in a relatively recent (2012) publication, Harihara and Parlos [12.22] developed a fault detection indicator for "Fault diagnosis of centrifugal pumps using motor electrical signals" but it does require three voltages and three currents to be monitored. This paper did present an industrial case history (via a field trial) of a boiler feed-water pump driven by a 298 kW/400 HP SCIM (no other details on the motor were given, but there should have been.) This indicated that it was possible to determine a mechanical problem in the motor and it was subsequently found to be a bearing fault. This is indeed an encouraging development and the reader is referred to Reference 12.22 for further information. Additional publications by Harihara and Parlos [12.23, 12.24] should also be followed up which cover the detection of cavitation and impellor cracks in centrifugal pumps by sensing and analyzing three currents and three voltages from the SCIMs driving the pumps.

## 12.5 QUESTIONS

**12.5.1** Why is the inter-turn insulation much thinner than the ground wall insulation in HV stator windings?

**12.5.2** Why is it the case that the inception of inter-turn insulation degradation between turns in, for example, an 11 kV stator winding does not emit partial discharges?

**12.5.3** Discuss the causes of sudden breakdown of inter-turn insulation followed by shorted turns in the line coils of 6.6 kV and above, DOL-started SCIMs.

**12.5.4** When shorted turns in a line end coil of an HV winding occur, why is the time to failure of the stator very short indeed?

**12.5.5** State the equation used to predict the current components in the supply current spectrum caused by shorted turns. Are these components unique to shorted turns?

**12.5.6** Why are industrial end users not receptive to using MCSA to diagnose shorted turns in HV stator windings?

**12.5.7** The time between a shorted turn/s occurring and a stator winding failure in LV SCIMs can be longer than that of an HV stator winding but why is it that the end users are still not receptive to MCSA for detecting shorted turns in LV stator windings?

**12.5.8** Why is MCSA not popular with industrial end users to detect faults in roller element bearings?

**12.5.9** Why is it a futile task to try to convince mechanical engineers and vibration experts to use MCSA to detect defects in roller element bearings in SCIMs when vibration sensors can be placed on the bearing housings?

**12.5.10** Where could MCSA be appropriate to try to detect faults in roller element bearings? Give the reasons for your answer.

## REFERENCES

[12.1] W. T. Thomson and C. McRae, "On-Line Current Monitoring to Detect Inter-Turn Winding Faults in Induction Motors," 24th Universities Power Engineering Conference, Queen's University, Belfast, September 1989.

[12.2] W. T. Thomson "On-Line MCSA to Diagnose Shorted Turns in Low Voltage Stator Windings of 3-Phase Induction Motors Prior to Failure," IEEE, PES & IAS IEMDC, MIT, Boston, June 2001.

[12.3] W. T. Thomson and D. Morrison, "On -Line Diagnosis of Stator Shorted Turns in Mains and Inverter Fed Low Voltage Induction Motors," Proceedings of IEE International Conference on Power Electronics and Machines and Drives (PEMD), University of Bath, UK, April 2002.

[12.4] A. Stavrou, H. G. Sedding, and J. Penman, "Current monitoring for detecting inter-turn short circuits in induction motors," *IEEE Transactions on Energy Conversion*, vol. 16, no. 1, March 2001, pp. 32–37.

[12.5] P. Neti, M. R. Shah, K. Younsi, J. Krahn, J. Y. Zhou, and D. Whitefield, "Motor Current Signature Analysis During Accelerated Life Testing of Form Wound Induction Motors," IEEE Power and Energy Society Meeting, 2011, pp. 106–109.

[12.6] M. E. H. Benbouzid, "Bibliography on induction motors faults detection and diagnosis," *IEEE Transactions on Energy Conversion*, vol. 14, no. 4, December 1999, pp. 1065–1074.

[12.7] R. Nailen, "How smart is the intelligent motor," *Electrical Apparatus*, February 2000, pp. 36–37.

[12.8] R. Nailen, "Are those new motor maintenance tests really that great?," *Electrical Apparatus*, February 2000, pp. 31–35.

[12.9] M. E. Steele, R. A. Ashen, and L. G. Knight, "An Electrical Method for Condition Monitoring of Motors," International Conference on Electrical Machines—Design and Applications, July 1982, no. 213, pp. 213–235.

[12.10] R. R. Schoen, T. G. Habetler, F. Kamran, and R. G. Bartheld, "Motor bearing damage detection using stator current monitoring," *IEEE Transactions on Industry Applications*, vol. 31, no. 6, November/December 1995, pp. 1274–1279.

[12.11] A. M. Knight and S. P. Bertani, "Mechanical fault detection in a medium-sized induction motor using stator current monitoring," *IEEE Transactions on Energy Conversion*, vol. 20, no. 4, December 2005, pp. 753–760.

[12.12] P. Tavner, L. Ran, J. Penman, and H. Sedding, *Condition Monitoring of Rotating Electrical Machines*, IET Power and Energy Series 56, Publication, The Institution of Engineering and Technology, London, UK, 2008, ISBN: 978-0-86341-739-9.

[12.13] J. I. Taylor, *The Vibration Analysis Handbook: A Practical Guide for Solving Rotating Machinery Problems*, 2nd edition, VCI 2003, ISBN: 0-9640517-2-09.

[12.14] J. I. Taylor, *The Bearing Analysis Handbook*, 2nd edition, VCI 2003, ISBN: 0-96-40517-3-7.

[12.15] S. Salon and K. Sivasubramaniam, "*Monitoring and Diagnostic Solutions for Wind Generators*," IEEE Power and Energy Society General Meeting, 2011, pp. 1–4, ISBN: 978-1-4577-1000-1.

[12.16] X. Gong and W. Qiao, "Bearing Fault Detection for Direct-Drive Wind Turbines via Stator Current Spectrum Analysis," IEEE Energy Conversion Congress and Exposition, 2011, pp. 313–318.

[12.17] S. Williamson and S. Djurovic, "Origins of Stator Current Spectra in DIGs with finding faults and excitation Asymmetries," IEEE International Conference Electrical Machines and Drives Conference (IEMDC), 2009, pp. 563–571.

[12.18] P. Zhang and P. Neti, "Detection of gearbox defects using electrical signature analysis for doubly fed wind generators," *IEEE Transactions on Industry Applications*, vol. 21, no. 3, May/June 2015, pp. 2195–2200.

[12.19] H. D. Haynes and R. C. Kryter, "Condition monitoring of machinery using motor current signature analysis," *Sound and Vibration Magazine*, September 1989.

[12.20] H. D. Haynes, "Application of signature analysis for determining the operational readiness of motor-operated valves under blowdown test conditions," *Journal of Nuclear Engineering and Design*, no. 118, 1990, pp. 399–408.

[12.21] G. C. Stone, E. A. Butler, I. Culbert, and H. Dhiranai, *Electrical Insulation for Rotating Machines – Design, Evaluation, Testing and Repair*, Wiley-IEEE PRESS, 2004, ISBN: 0-471-44506-1.

[12.22] P. P. Harihara and A. G. Parlos, "*Fault diagnosis of centrifugal pumps using motor electrical signals*," in *Centrifugal Pumps*, edited by D. Papantonis, February 2012, ISBN: 978 953-51-0051-5.

[12.23] P. P. Harihara and A. G. Parlos, "Sensorless Detection of Impellor Cracks in Motor Driven Centrifugal Pumps," Proceedings of the ASME International Mechanical Engineering Congress and Exposition, Boston, MA, October 31–November 6, 2008, pp. 17–23, ISBN: 9780791848661.

[12.24] P. P. Harihara and A. G. Parlos, "Sensorless Detection and Isolation of Faults in Motor-Pump Systems," Proceedings of the ASME International Mechanical Engineering Congress and Exposition, Boston, MA, October 31–November 6, 2008, pp. 43–50, ISBN: 9780791848661.

CHAPTER 13

# APPRAISAL OF MCSA INCLUDING LESSONS LEARNED VIA INDUSTRIAL CASE HISTORIES

*William T. Thomson*

## 13.1 SUMMARY OF MCSA IN INDUSTRY TO DIAGNOSE CAGE WINDING BREAKS

This chapter starts with a flow chart for the application of MCSA to assess the operational condition of cage rotors in SCIMs operating in industry and is for the benefit of engineers who actually use the technology. An appraisal is then presented which summarizes the strengths, weaknesses, external constraints, and in particular, the lessons learned by the author of this chapter during the application of MCSA in industry for 34 years. Suggestions for research and development work on MCSA are presented in Section 13.10.

## 13.2 FLOW CHART FOR MEASUREMENT AND ANALYSIS OF CURRENT TO DIAGNOSE CAGE WINDING BREAKS

**Information gathering**

- Obtain nameplate data.
- Contact OEM and try to obtain number of rotor bars. Are end ring retaining rings fitted?
- Is it a single or double cage rotor?

*Current Signature Analysis for Condition Monitoring of Cage Induction Motors: Industrial Application and Case Histories*, First Edition. William T. Thomson and Ian Culbert.

**Alert: Classical MCSA cannot assess operational condition of outer starting cage of double cage rotor**

- Records on number of DOL starts? Or is it a soft start via an auto-transformer or inverter?
- Have there been any malfunctions in the driven load during start-ups or steady state?
- Any records of previous broken rotor bars in the motor under test or in an identical one?
- If previous MCSA results exist, check spectra, conclusions, and recommendations.
- Obtain information on driven mechanical load—refer to Section 13.3.

↓

**Alert: Nameplate data may not match operational performance**

- The actual speed may not be the same as it is on the nameplate when the full-load rated current is being taken by the motor but this is the only data available in practice.

↓

**Ascertain the available access to measure current while motor is running?**

- HV motor: Can access be provided via the LV instrumentation panel to clip on an MCSA CT to the secondary side of the CT feeding the display ammeter?
- LV motor: What access is available to measure the current via an MCSA clip-on CT?

↓

**Typical operational current**

- What is the expected load on the motor when it is to be MCSA tested?
- If it is on very light load, then MCSA cannot assess the condition of the cage winding.

↓

**Commonsense alert: Stop, look and listen**

- Visit the motor to be MCSA tested "**before**" any MCSA Measurements.
- Have informal discussions with "shop floor" operation and maintenance personnel to obtain historical knowledge on the motor to be tested.

- Listen to acoustic noise from the motor, walk around it to ascertain whether there is a low frequency modulated hum; if possible, listen to an identical motor and compare.

↓

**Commonsense recommendation**

- Before taking any MCSA measurements, check the *in situ* ammeter.
- Record its value and if it is an analogue display, check whether it is steady, swinging very slowly (time it) about a mean position or is it fluctuating in a random nature?

↓

**Set spectrum resolution on MCSA instrument or spectrum analyzer**

- Calculate nominal full-load slip from nameplate data; set spectrum resolution; dB scale of at least an 80 dB dynamic range, base-band span 0–130 Hz, 12,800 lines, and line resolution 10.16 mHz/line

↓

**Alert: Commonsense**

- Do not assume the end user has correctly identified/clipped on the MCSA CT to the correct secondary side of the CT feeding the *in situ* ammeter.
- Cross check the overall rms current from the MCSA instrument with the *in situ* ammeter.

↓

**Interpretation of spectrum to identify $\pm 2sf$ sidebands around $f$ based on "only" having the nameplate data**

- MCSA cannot distinguish between broken rotor bars, a broken end ring, or a high number of inferior bar to end ring joints.
- Measure and record the exact supply frequency from the spectrum.
- Is the operating current of the order of the nominal full-load current, for example, ±5%?
- If yes, search for sidebands at the predicted $\pm 2s_{FL}f$ around $f$ the supply frequency.
- Reverse calculate the nominal $s_{FL}$ from measured $2sf$ and hence operational speed.

- Provided the **nominal nameplate speed is correct** and the slip calculated from the $2sf$ sidebands is equal to the full-load, **rated speed (±2 r/min)**, then the only option is to proceed on the basis you have detected $\pm 2sf$ sidebands.
- Are there higher order sidebands of $\pm 2sf$, at $\pm 4sf$ and/or at $\pm 6sf$ around $f$.

↓

**Alert and lesson learned:**
**The majority of industrial SCIMs operate on reduced load and slip**

- If there are sidebands inboard of the nominal full-load $\pm 2sf$ sidebands, then calculate the slip/s from these sidebands.
- It is here that induction motor knowledge and judgement is required as to whether the slip at the operational current is a sensible value, when only the nameplate data is available.
- However, if the number of rotor bars is known, the operating slip can be obtained as per case histories 6.1, 7.6, 8.5 etc.

↓

**Estimate the operational condition of the cage winding using the dB difference chart (Table 4.2) when the number of rotor bars is unknown***

*The number of rotor bars (slots) can be predicted as is verified in case history 11.11

- Record the dB differences between each of the $\pm 2sf$ sidebands and $f$.
- Normally use $N_{av}$ = average dB difference between the two $\pm 2sf$ sidebands and $f$.
- But when the $-2sf$ sideband is >2 dB compared to $+2sf$ sideband then only the dB difference ($N$) between $-2sf$ and $f$ is used and replaces $N_{av}$ in Table 4.2.
- But NOTE when the $+2sf$ is >2 dB compared to $-2sf$ then $N_{av}$ is used in Table 4.2.

↓

**Number of rotor bars $R$ is known:**
**Estimate operational condition of cage winding**

- If the motor is operating on full-load current and slip, then use equation (4.5) to get equivalent broken factor ($BB_f$) or the broken bar index ($n$) from equation (4.6).
- If motor is operating below full-load slip (minimum of 35%* $s_{FL}$) then use $BB_c$ which is a correction factor to refer $BB_{fs}$ at the operating slip to the full-load slip to give $BB_f$.

- If dB difference between each of the $\pm 2sf$ sidebands with respect to each other is 2 dB and less, use average dB difference $N_{av}$ between the sidebands and supply component ($f$) in equation (4.5) to obtain the equivalent broken rotor bar factor, $BB_f$ or use equation (4.6) to get ($n$).
- When the magnitude in dB of the lower sideband ($-2sf$) is greater than 2 dB compared to the upper sideband then the $-2sf$ component should be used to obtain the dB difference $N$ between the lower sideband and the supply component for equation (4.5).
- When the upper $+2sf$ sideband is greater than 2 dB compared to the lower $-2sf$ sideband then equation (4.6) should be used to estimate the broken bar index ($n$).
- If $BB_f$ or $n$ is 0.5 and greater, it has been shown that the cage winding normally has a defect(s) and the larger $BB_f$ is, the more serious the cage defect.

↓

**Alert and lessons learned**

- Equation (4.5) normally underestimates the severity of the cage winding defects.
- A threshold for $BB_f \geq 0.5$ has been proven to be valid for detecting cage winding breaks in practice, as shown in Chapters 5 and 6.
- Exceptions to the rule occur—case history 5.8; $R = 155$ bars and equation (4.5) predicted 3.7, cage rotor had four clean cut broken bars. This near equality is expected since the number of broken bars was <<< than total number of bars.

## 13.3 MCSA TO DIAGNOSE BROKEN ROTOR BARS IN SCIMS DRIVING STEADY LOADS

MCSA provides the end users with a non-invasive, condition monitoring technique to assess the operational condition of cage windings in SCIMs. However, it is important to indicate the ideal conditions for the successful application of MCSA and also essential to also identify external constraints, weaknesses, and lessons learned when applying MCSA in industry to detect cage winding breaks.

### 13.3.1 Accuracy of Nameplate Data

Ideally, the nameplate data on a SCIM should be correct so that the prediction of the full-load slip and the frequencies of the $\pm 2sf$ sidebands around the supply component at full-load are valid since this is the initial starting point for MCSA.

**External Constraints and Lessons Learned**

**(i)** The nameplate data is the only independent information available for the application of MCSA in industry, unfortunately, the full-load speed at the frequency and full-load current specified on the nameplate, may not be correct and the speed can differ by several r/min, for example, by as much as 5 r/min, Nailen [13.1]. The authors of this book, having MCSA tested thousands of SCIMs during the past 34 years, have also observed that the nominal nameplate speed may not match the actual speed when the input current is the rated full-load current.

In fact, Thomson has experienced a difference as high as 9 r/min (an extreme case) between the nameplate speed and the actual operating speed at the stated full-load current in a 13.8 kV, 6800 kW/9115 HP, 326 A, 60 Hz, 3570 r/min, 2-pole SCIM (see case history 6.1 which proves this large discrepancy).

**(ii)** The full-load slip can be calculated from the nominal nameplate data but when SCIMs are operating on a reduced load, as they very often are in industry, the speed increases and therefore the operating slip is unknown. Ideally, the engineer applying MCSA would like to have access to the current versus speed curve from no-load up to full-load, so that the operating slip at different load currents can be estimated and hence the frequency of the operational $\pm 2sf$ sidebands. In reality, the current versus speed (or slip) curve of a SCIM is normally not available at an industrial plant where MCSA is to be applied. Experience has shown that OEMs will not normally release this information to condition monitoring companies applying MCSA.

**Overcoming this Constraint**

If the number of rotor bars is known, the principal rotor slot passing frequency (RSPF) component ($f_{rs(+1)}$) can be predicted using the nominal full-load slip in equation (10.4) and if the motor is on reduced load the principal RSPF will be higher in frequency than predicted since the slip is lower. The current spectrum is then subsequently searched and the principal RSPF component can normally always be found, its frequency can then be used to reverse calculate the actual operating slip. This was clearly demonstrated in a number of case histories, such as in 6.1, 7.6 and 8.5.

## 13.4 NUMBER OF ROTOR BARS, EXTERNAL CONSTRAINTS, AND LESSONS LEARNED

**(i)** Ideally, the number of rotor bars should be obtained from the OEM to enable a reasonable estimate to be made of the condition of the cage winding via the equivalent broken rotor bar factor ($BB_f$) from equation (4.5) when the motor is operating at full-load slip.

**(ii)** When the motor is operating on reduced load (lower slip) and if the number of rotor bars is known an estimate of the condition of the cage winding can be made since a broken bar correction factor ($BB_c$) from equation (4.7) can be applied, which modifies the estimated equivalent broken rotor bar factor ($BB_{fs}$) at the lower slip such that it reflects the cage condition at full load. However, there is a restriction on using the $BB_c$ in that it is not applicable when the motor is operating at a slip below 35% of the full-load slip since the current in the bars is too low for the correction factor to be applied.

**External Constraints and Lessons Learned**

**(i)** Unfortunately, it can be difficult and sometimes impossible to obtain the number of rotor bars from the OEM. Without the number of rotor bars the only* option is to use the dB difference ($N_{av}$) between the $\pm 2sf$ sidebands and the supply component ($f$) as a means of estimating the condition of the cage winding. *Recall that the number of rotor bars (slots) can be predicted as is verified in case history 11.11.

**(ii)** However, the use of the dB difference ($N_{av}$) chart in Table 4.2 has to be treated with caution since the motor should be operating at a slip of at least 85% of the full-load slip and this is certainly not always the case.

It has been stated several times in preceding chapters that the end user will not normally disturb the plant's process demand at the time of an MCSA test to obtain a high enough load.

**(iii)** Using the dB difference ($N_{av}$) to estimate the severity of cage winding breaks cannot be generally applied to all pole numbers used in SCIMs. For example, an average dB difference ($N_{av}$) of 50 dB between the $\pm 2sf$ sidebands and the supply component in a 2-pole motor and also in a SCIM with a higher pole number and a higher number of rotor bars does not give the same indicator of the condition of the cage winding. This was clearly demonstrated in the graphs in Figures 4.22, 4.23, and 4.24, and also in the case histories presented in Chapter 5.

## 13.5 EFFECT OF END RING RETAINING RINGS (ERRS) ON DIAGNOSIS OF BROKEN ROTOR BARS

In large HV SCIMs, particularly in high speed 2-pole machines, it is often the case that end ring retaining rings (ERRs), which cover the bar to end ring joints, are fitted and it is essential to know when this is the case. Ideally, the actual design of the ERRs should also be known for any recommendation for action when $\pm 2sf$ sidebands are found around the supply component.

### 13.5.1 External Constraint, Disadvantage/Weakness, and Lessons Learned

**(i)** Unfortunately, it can be very difficult to retrospectively obtain confirmation from the end user or an OEM that a cage rotor has ERRs.

**(ii)** If an ERR, which covers the bar to end ring joints is fitted and broken rotor bars lift and touch the conducting ERR, an alternative conducting path for current is provided. Clearly the magnitude of the $\pm 2sf$ sidebands do not then reflect the true condition of the cage winding as was verified in the case history in Section 6.1.

**(iii)** If the ERRs cover the bars between the ends of the rotor core and the joints, then the cage design is fault tolerant to broken rotor bars crashing into the stator winding, as shown in the case history in 6.2. However, if MCSA indicates that there are $\pm 2sf$ sidebands at a level, which indicates cage winding breaks, but the rotor has ERRs that are fault tolerant and that fact is unknown, a premature removal of the rotor can occur, which was in fact the case in Section 6.2.

## 13.6 MCSA APPLIED TO SCIMS DRIVING COMPLEX MECHANICAL PLANT, LESSONS LEARNED, AND RECOMMENDATIONS

(i) There is a diverse range of case histories in Chapters, 7, 8, and 9 which cover the interpretation of current spectra from SCIMs and are a function of complex mechanical dynamics downstream of the motor. These dynamics often make it difficult and in some cases impossible to identify the current signature patterns due to broken rotor bars or airgap eccentricity. The challenge is to clearly separate current components due to normal mechanical dynamics in the drive train which are being reflected back into the SCIM, from those due to broken rotor bars or unacceptable levels of airgap eccentricity. The reader is referred to these case histories since it is pointless regurgitating what already exists in preceding chapters. There are also case histories in Chapter 9, which verify that it is possible to determine abnormal mechanical dynamics/problem downstream of a SCIM.

(ii) It is highly recommended that an engineer applying MCSA to a SCIM drive train obtains information about the mechanical load dynamics downstream of the motor. This includes, for example, the following items (iii)–(vi).

(iii) The number of stages and ratios in slow speed gearboxes driving crushers, and the number of rollers/crushing elements in the crushing process.

(iv) Likewise, for slow speed gearboxes driving conveyors, the conveyor's pay load characteristics are also required.

(v) This type of information is required to predict the current components due to mechanical load dynamics being reflected back into the SCIM otherwise false positives of broken rotor bars can occur.

(vi) With respect to determining abnormal mechanical load dynamics in, for example, a pumping process, or a belt-driven mechanical load, it is essential that comparisons can be made between a healthy SCIM drive train and other identical drive trains, as was the case in a number of cases in Chapter 9. This allows comparisons to be made between complex spectra to determine if a problem exists downstream of the motor and is not a problem within the SCIM.

## 13.7 DOUBLE CAGE ROTORS—CLASSICAL MCSA CAN ONLY DETECT CAGE WINDING BREAKS IN INNER RUN WINDING

Classical MCSA, which is the subject of this book cannot diagnose the condition of the outer cage winding in a double cage rotor since the current in the outer cage is negligible during steady state load conditions. It has been shown by previous publications as referenced [13.2, 13.3, 13.4] in Chapter 7 that an analysis of the transient current during start up can diagnose cage winding breaks but there are no commercially available instruments, which industry can use in practice. This was discussed in Section 7.1.1 and the following points are reiterated:

(i) SCIMs are normally always already running to deliver the production process during a planned MCSA test and end users are very reluctant indeed to disturb the production process to carry out a start for a transient analysis of the current.

(ii) The majority of large and strategic HV induction motors are single cage and that is where there is by far the largest application and demand for classical MCSA.

## 13.8 MCSA TO DIAGNOSE OPERATIONAL LEVELS OF AIRGAP ECCENTRICITY IN SCIMS

This section starts with a repeat of the flow chart presented in Section 11.1 and is a natural follow on from the main flow chart in Section 13.1.

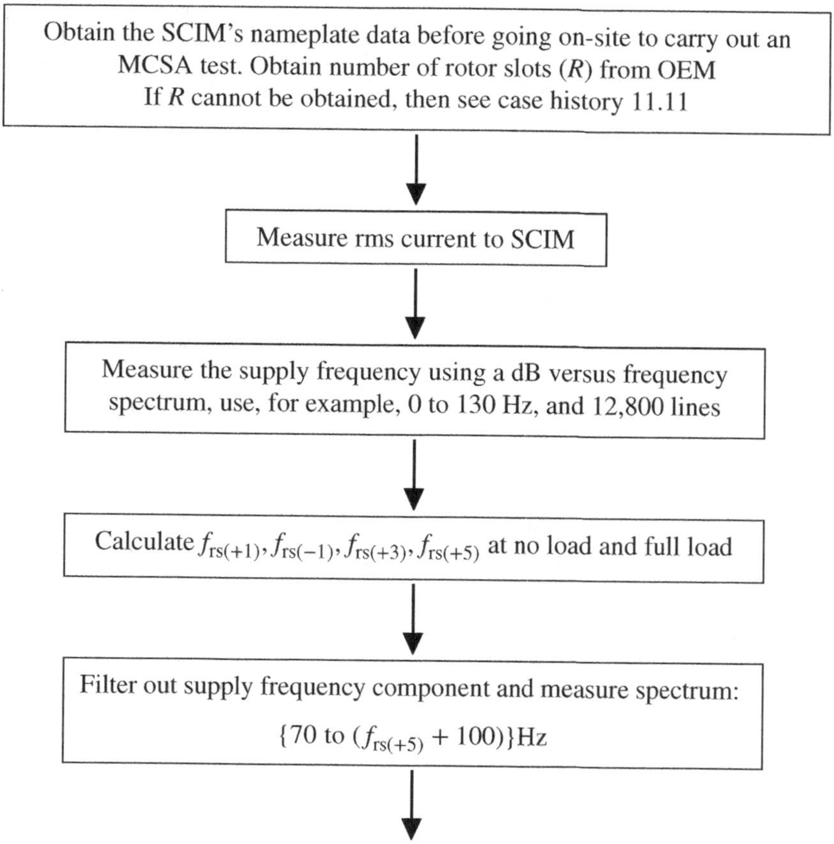

Search for the $f_{rs(+1)}$ component in the spectrum which should be between the value at no-load and full-load

↓

Reverse calculate the operating slip using the measured $f_{rs(+1)}$ and supply frequency. Calculate $f_r = (f(1-s)/p)$ using operating slip and frequency. Check that $f_{rs(+1)}, f_{rs(-1)}, f_{rs(+3)}$, and $f_{rs(+5)}$ are $2f$ apart from each other.

↓

Select the highest RSPF component (in dB), which is normally $f_{rs(+1)}$ and produce a zoom spectrum around $f_{rs(+1)}$ or the highest RSPF (in dB), search for $\pm f_r$ around $f_{rs(+1)}$. If $\pm f_r$ components exist, measure average dB difference ($N_{ec}$) between $\pm f_r$ and $f_{rs(+1)}$

↓

Go to Table 10.3 and use $N_{ec}$ to obtain an estimate of operational airgap eccentricity band.

The industrial case histories in Chapter 11 verify that it is possible to detect unacceptable levels of operational airgap eccentricity in SCIMs but not in all cases. The ideal conditions and external constraints/weaknesses and lessons learned for the application of MCSA in industry to estimate the operational level of airgap eccentricity are as follows.

### 13.8.1 Rotor Slots and Estimate of Airgap Eccentricity, External Constraints/Weaknesses, Lessons Learned, and Possible Solutions

The number of rotor slots is ideally required to identify the current signature pattern, which is unique to the combination of static and dynamic airgap eccentricity.

**(i)** If the number of rotor slots is unknown then standalone MCSA instruments currently in the market place are unable to detect the current signature pattern, which is a function of airgap eccentricity.

**(ii)** However, this lack of information can be overcome by experts in SCIMs and MCSA and this was demonstrated in considerable detail in the case history in Section 11.11 since the number of rotor slots was determined. The information in that case history should assist engineers who use MCSA to apply that knowledge and determine the number of rotor slots.

**(iii)** When MCSA is applied to determine the airgap eccentricity in a SCIM during a no-load uncoupled run and the RSPF ($f_{rs(+1)}$) current component coincides, with a supply harmonic, then MCSA cannot assess the airgap eccentricity. This really cannot be overcome as was shown in the case history in Section 11.7.

**(iv)** MCSA cannot normally identify the signature pattern due to airgap eccentricity when a SCIM is driving a reciprocating compressor. This is due to the coincidence between certain components (i.e., the $\pm f_r$ components around $f_{rs(+1)}$) from the compressor's pulsating characteristics, which coincide with components that are indicative of the airgap eccentricity current signature pattern. This really cannot be overcome as was shown in the case history in Section 11.10.

## 13.9 RECOMMENDATIONS TO END USERS

**(i)** If end users intend applying MCSA then at the invite to tender (ITT) stage given to OEMs for the supply of, for example, large HV SCIMs with copper fabricated cage windings, it should be specified that the OEM must provide the number of rotor bars and also whether ERRs are fitted which can prevent broken rotor bars crashing into the stator winding.

**(ii)** It is often the case that it is the supplier of the pump, compressor, or fan drive train that issues the ITT to OEMs of SCIMs and such information in Section 13.9.(i) above is normally not requested. It is up to end users to insist that this is included in the ITT.

**(iii)** When a motor is sent to a motor repair workshop the repairer should be requested to record the number of rotor bars and whether ERRs are fitted and to include that information in their repair report to the client.

**(iv)** Nameplates on SCIMs do not include the number of rotor bars and therefore the end users should now insist that OEMs stamp the number of rotor bars on the nameplate since condition monitoring via MCSA has now been applied in industry for at least 36 years. The OEMs should know this but still none of the OEMs throughout the world nor the motor standards committees have taken this on board for serious consideration and implementation.

**(v)** It is the authors' view, stamping the number of rotor bars on the nameplate is not releasing highly sensitive IP knowledge by an OEM and this should be emphasized by the end user, after all, the number of rotor bars can be counted at the manufacturing stage by the client during a stage progress visit to the OEM.

**(vi)** MCSA can be used to provide a "*bench mark current spectrum*" to assess (estimate) the operational condition of the cage winding and airgap eccentricity in brand new SCIMs or in motors with repaired cage windings during a full-load heat run so that the end user can refer to the data when applying MCSA as part of the plant's condition monitoring regime. However, this would not be part of a Factory Acceptance Test unless there was a formal contractual agreement to so do.

**(vii)** When a new or repaired HV SCIM is commissioned on-site it is very wise to carry out MCSA tests to obtain base-line results to compare with MCSA results from a full-load heat run as in item (iv) and for future trending of MCSA data.

**(viii)** For ease of applying MCSA in industry, a dedicated CT for each motor to be tested should be permanently installed and the output accessed at the front panel of LV instrumentation cubicles for HV motors and on the starter cubicle for LV motors.

**(ix)** Finally, always include a "Stop, Look, and Listen" policy as information gathering to support MCSA testing.

## 13.10 SUGGESTED RESEARCH AND DEVELOPMENT PROJECTS

### 13.10.1 Fundamental Research on MCSA to Diagnose Broken Rotor Bars in Large 2-Pole SCIMs with End Ring Retaining Rings

To apply, for example, finite element modelling to large HV SCIMs (e.g., above 5000 HP/3730 kW, 2-pole) to predict the magnitude of $\pm 2sf$ sidebands as a function of the following:

**(a)** High resistance joints as a variable between the bars to end rings.

**(b)** Broken rotor bars at different operating load currents with single cage rotor designs and no ERRs.

**(c)** Single cage rotor designs with ERRs in which broken bars have lifted and are touching (i.e., a variable high resistance connection) an ERR thus providing an alternative path for bar currents to flow.

This research work would require electrical, electromagnetic, and mechanical dynamic modelling of the actual drive train to predict the frequency and magnitude of the $\pm 2sf$ sidebands. An electrical–electromagnetic model will only predict the lower $-2sf$ sideband and this is insufficient since both the twice slip frequency sidebands occur in practice. A major challenge for such a research project is to obtain cooperation, which will inevitably be required, via an exclusive, non-disclosure agreement between academia and an OEM for the release of the motor's design details. Inevitably, any publications could not include an OEM's design details of a large SCIM but the results would normally be available for publication without the inclusion of design details. Such a project also requires the cooperation of end users for access to take MCSA measurements from large HV SCIMs.

### 13.10.2 Fundamental Research to Predict the Performance Characteristics for a Large (e.g., 1000 kW/1340 HP and Upward), HV SCIM as a Function of Broken Rotor Bars and the Output Power and Torque Demanded by a Driven Load Such as a Centrifugal Pump

This study is proposed as a result of a case history presented in Section 5.5, since the SCIM was still operating with 20 completely broken rotor bars in a 46 bar cage rotor. The operating slip increased from 0.55% with no broken bars to a slip of 0.75% with 20 broken bars when the pump was demanding full-load power and torque but the production and operating divisions on the offshore oil production platform were still running the motor since it was crucial for oil and gas production. This involves the prediction of the quantities given in the following list (not exclusive) as the number of broken bars increases, and the centrifugal pump is still demanding its full-load torque, speed, and power:

**(i)** Input power and current, power factor, and efficiency.

**(ii)** Rotor bar currents and rotor current.

**(iii)** Power output versus speed curve of the motor.

**(iv)** Torque versus speed curve.

**(v)** Uncoupled run-up time of the motor as a function of broken bars.

**(vi)** Run-up time of motor when driving the pump with open and closed valve settings as a function of the number of broken bars.

This will again require full design details of a large HV SCIM and the torque–speed characteristic of, for example, a centrifugal pump and its associated characteristics. Inertias of the motor and pump will also be required from the OEMs' of the motor and pump.

The first major hurdle is to obtain design data for a large HV SCIM from an OEM and that may well be impossible, since very strict non-disclosure and contractual agreements would need to be put in place, which would prohibit the publication of any results. Academia may well not be interested in a project with such a limitation. Therefore, a *"second best study,"* with the same objectives, could be carried out using a small power 3-phase SCIM since it may be easier to obtain design data from an OEM for a small SCIM, which is mass produced.

Of course a small power 3-phase induction motor can be easily purchased at a very low cost and very carefully (and completely) dissected; an electric motor repair shop's assistance should be obtained. The stator and rotor core and slot dimensions (etc.), stator winding configuration design details can all be determined but the composition of materials used in the motor has to be established, such as grade of magnetic steel, composition of rotor bar material (etc.), but samples from the dissection can be measured by chemical analysis, etc. This can all be done since Thomson (author) did so in the early 1980s when he carried out many experimental tests—see Chapter 4 for information on his test rigs, etc.

A purpose-built test rig is really essential with a small power 3-phase SCIM driving a centrifugal pump in a controlled laboratory environment for proper comparisons between performance predictions and actual measurements as a function of broken rotor bars, the pump demand and run-up times, etc.

### 13.10.3 Development and Application Type Projects

**(a)** Predict the number of rotor bars.

It is well known that it can be difficult to obtain the number of rotor bars, consequently industry need MCSA instruments, which can automatically determine their number. A development type project which is closer to the market place and industry is therefore justified. The catalyst for such a project could be based on the information presented in Tables 4.3 and 4.4, which is knowledge on the range of the number of rotor bars, which are typically used in SCIMs with different pole numbers and ratings, and also the signal processing strategy proposed and verified in the industrial case history in Section 11.11.

**(b)** Signal processing strategy for MCSA applied to crusher and conveyor drive trains

Develop signal processing strategies for inclusion in MCSA instruments to cater to random fluctuations in load from SCIMs driving crushers and conveyors via slow speed gearboxes, which prevent MCSA from identifying the $\pm 2sf$ sidebands in these drives. Case histories were presented on this limitation of MCSA in Chapter 8 which can be used as a catalyst for the project.

## REFERENCES

[13.1] R. Nailen, "How Smart is the Intelligent Motor," *Electrical Apparatus*, February 2000, pp. 36–37.

[13.2] S. Elder, J. F. Watson, and W. T. Thomson, "Fault Detection in Induction Motors as a Result of Transient Analysis," Proceedings of 4th IEE International Conference in Electrical Machines and Drives (EMDA'89), London, September 1989.

[13.3] S. Elder, "The Monitoring of Induction Motor Starting Transients with a View to Early Fault Detection," Ph.D. Thesis, School of Engineering, The Robert Gordon University, Scotland, March 13, 1992.

[13.4] J. Pons-Linares, V. Climente-Alarcon, F. Vedreno-Santos, J. Antonio-Daviu, and M. Riera-Guasp, "Electrical Machine Diagnosis Techniques via Transient Current Analysis," 38th Annual Conference on IEEE Industrial Electronics Society, 2012, pp. 3893–3900.

## APPENDIX 13.A COMMENTARY ON INTERPRETATION OF LV AND HV USED IN SCIMS

The voltages used by OEMs, repair companies and end users to define LV (low voltage) and HV (high voltage) applied to SCIMs, is to say the least, a quagmire of variations and contradictions. This book is not concerned with an appraisal of the variations used to define LV and HV SCIMs and the following has been used in the text:

LV: SCIMs below 690 V

HV: SCIMs 690 V up to 13.8 kV

# LIST OF EQUATIONS

## CHAPTER 1

$$e = -N_{\mathrm{T}}\frac{d\phi}{dt} \tag{1.1}$$

$$f = N_{\mathrm{s}}p/60 \tag{1.2}$$

$$S = (N_{\mathrm{s}} - N_{\mathrm{r}})/N_{\mathrm{s}} \tag{1.3}$$

$$f_2 = sf \tag{1.4}$$

$$N_{\mathrm{fr}} = sN_{\mathrm{s}} \tag{1.5}$$

$$N_{\mathrm{r}} = \left[\left(-\frac{dN_{\mathrm{r}}}{dT}\right)T\right] + N_{\mathrm{s}} \tag{1.6}$$

$$T = \left[\frac{(N_{\mathrm{S}} - N_{\mathrm{r}})}{(N_{\mathrm{S}} - N_{\mathrm{rFL}})}\right]T_{\mathrm{FL}} \tag{1.6a}$$

$$\frac{T_{\mathrm{FL}}}{T} = \frac{s_{\mathrm{FL}}}{s_{\mathrm{o}}} \tag{1.7}$$

$$t = (J/T_{\mathrm{ma}})(w_{\mathrm{r2}} - w_{\mathrm{r1}}) \tag{1.8}$$

$$\Delta t = \frac{(WK^2M + WK^2DE) \times \Delta S}{308 \times (T_{\mathrm{a}})} \tag{1.9}$$

## CHAPTER 3

$$\Delta t = (WK^2M + WK^2DE \times \Delta S)/308 \times T_{\mathrm{a}} \tag{3.1}$$

$$\Delta t = J\,(dwr)\,/T_{\mathrm{av}} \tag{3.2}$$

*Current Signature Analysis for Condition Monitoring of Cage Induction Motors: Industrial Application and Case Histories*, First Edition. William T. Thomson and Ian Culbert.

$$(WK^2) = A\left[\frac{HP^{0.95}}{(RPM/1000)^{2.4}}\right] - 0.0685\left[\frac{HP^{1.5}}{(RPM/1000)^{1.8}}\right] \tag{3.3}$$

## CHAPTER 4

$$f_{\mathrm{sb}} = f\,(1-2s) \tag{4.1}$$

$$f_{\mathrm{s}} = f \pm 2sf \tag{4.2}$$

$$M_{\mathrm{sb}} = 10^{(-N/20)} \tag{4.3}$$

$$\frac{I_{\mathrm{LSB}}}{I_{\mathrm{p}}} \cong \sin\alpha/2p\,(2\pi-\alpha) \tag{4.4}$$

$$BB_{\mathrm{f}} = 2R/\left(10^{(N/20)} + 2p\right) \tag{4.5}$$

$$I_{\mathrm{n}}/I_{\mathrm{p}} \cong n/R \tag{4.6}$$

$$BB_{\mathrm{c}} = s_{\mathrm{FL}}/s_{\mathrm{op}} \tag{4.7}$$

## CHAPTER 6

$$f_{\mathrm{rs}} = f\left\{\left(\frac{R}{p}\right)(1-s) \pm n_{\omega\mathrm{s}}\right\} \tag{6.1}$$

## CHAPTER 7

$$i'_{2(\mathrm{t})} = \frac{\sqrt{2}\,V\sin(2\pi ft)(N_{\mathrm{S}}-N_{\mathrm{a}}) - \sqrt{2}V\sin(2\pi ft)k\sin(2\pi f_{\mathrm{c}}t)}{(R'_{\mathrm{r}}N_{\mathrm{s}})} \tag{7.1}$$

## CHAPTER 8

$$g_{(\theta,\mathrm{t})} = g(1 - e_{\mathrm{s}}\cos(\theta) - e_{\mathrm{d}}\cos(\omega t - \theta)) \tag{8.1}$$

## CHAPTER 9

$$V = \sqrt{2}\pi f\phi_{\mathrm{p}}T_{\mathrm{ph}}k_{\mathrm{s}}k_{\mathrm{d}} \tag{9.1}$$

$$f_{\mathrm{brb,h}} = (k_2\,(1-s) \pm s)f \tag{9.2}$$

## CHAPTER 10

$$g_{(\theta,t)} = g(1 - e_s\cos(\theta) - e_d\cos(\omega t - \theta)) \tag{10.1}$$

$$F_{(UMP)} = \left(\pi DLB_p^2 e_c\right)/4\mu_0 \tag{10.2}$$

$$I_M = \pi D^4/64(m^4) \tag{10.3}$$

$$f_{rs} = f[(R/p)(1 - s) \pm n_{\omega s}] \tag{10.4}$$

$$f_{ec} = f[(R/p)(1 - s) \pm n_{\omega s} \pm 2n_{sa}p] \pm n_d f(1 - s)/p \tag{10.5}$$

$$m = R \pm S \pm n_s \pm n_d \pm 2n_{sa}p \pm n_{\theta s}p \tag{10.6}$$

$$f_{ec} = f[(R/p)(1 - s) \pm n_{\omega s}] \pm f(1 - s)/p \tag{10.7}$$

$$m_o = p(6c \pm 1) \tag{10.8}$$

$$m_e = p(6c \pm 2) \tag{10.9}$$

## CHAPTER 11

$$f \pm kf_r \text{ Hz} \tag{11.5}$$

## CHAPTER 12

$$f_{st} = f[(n/p)(1 - s) \pm k] \tag{12.1}$$

$$f_b = f \pm kf_v \tag{12.2}$$

$$f_i = (n_e/2)f_r[1 + (BD/PD)\cos(\beta)] \tag{12.3}$$

$$f_o = (n_e/2)f_r[1 - (BD/PD)\cos(\beta)] \tag{12.4}$$

$$f_b = (PD/BD)f_r[1 - ((BD/PD)\cos(\beta))^2] \tag{12.5}$$

$$f_c = (f_r/2)[1 - (BD/PD)\cos(\beta)] \tag{12.6}$$

$$f_{ci} = f \pm kf_i \tag{12.7}$$

$$f_{co} = f \pm kf_o \tag{12.8}$$

$$f_{cb} = f \pm kf_b \tag{12.9}$$

$$f_{cc} = f \pm kf_c \tag{12.10}$$

# INDEX



*Current Signature Analysis for Condition Monitoring of Cage Induction Motors: Industrial Application and Case Histories*, First Edition. William T. Thomson and Ian Culbert.

# IEEE Press Series on Power Engineering

**Series Editor: M. E. El-Hawary,** Dalhousie University, Halifax, Nova Scotia, Canada

The mission of IEEE Press Series on Power Engineering is to publish leading-edge books that cover the broad spectrum of current and forward-looking technologies in this fast-moving area. The series attracts highly acclaimed authors from industry/academia to provide accessible coverage of current and emerging topics in power engineering and allied fields. Our target audience includes the power engineering professional who is interested in enhancing their knowledge and perspective in their areas of interest.

1. *Principles of Electric Machines with Power Electronic Applications, Second Edition*
M. E. El-Hawary

2. *Pulse Width Modulation for Power Converters: Principles and Practice*
D. Grahame Holmes and Thomas Lipo

3. *Analysis of Electric Machinery and Drive Systems, Second Edition*
Paul C. Krause, Oleg Wasynczuk, and Scott D. Sudhoff

4. *Risk Assessment for Power Systems: Models, Methods, and Applications*
Wenyuan Li

5. *Optimization Principles: Practical Applications to the Operations of Markets of the Electric Power Industry*
Narayan S. Rau

6. *Electric Economics: Regulation and Deregulation*
Geoffrey Rothwell and Tomas Gomez

7. *Electric Power Systems: Analysis and Control*
Fabio Saccomanno

8. *Electrical Insulation for Rotating Machines: Design, Evaluation, Aging, Testing, and Repair, Second Edition*
Greg Stone, Edward A. Boulter, Ian Culbert, and Hussein Dhirani

9. *Signal Processing of Power Quality Disturbances*
Math H. J. Bollen and Irene Y. H. Gu

10. *Instantaneous Power Theory and Applications to Power Conditioning*
Hirofumi Akagi, Edson H. Watanabe, and Mauricio Aredes

11. *Maintaining Mission Critical Systems in a 24/7 Environment*
Peter M. Curtis

12. *Elements of Tidal-Electric Engineering*
Robert H. Clark

13. *Handbook of Large Turbo-Generator Operation and Maintenance, Second Edition*
Geoff Klempner and Isidor Kerszenbaum

14. *Introduction to Electrical Power Systems*
Mohamed E. El-Hawary

15. *Modeling and Control of Fuel Cells: Distributed Generation Applications*
M. Hashem Nehrir and Caisheng Wang

16. *Power Distribution System Reliability: Practical Methods and Applications*
Ali A. Chowdhury and Don O. Koval

17. *Introduction to FACTS Controllers: Theory, Modeling, and Applications*
Kalyan K. Sen and Mey Ling Sen

18. *Economic Market Design and Planning for Electric Power Systems*
James Momoh and Lamine Mili

19. *Operation and Control of Electric Energy Processing Systems*
James Momoh and Lamine Mili

20. *Restructured Electric Power Systems: Analysis of Electricity Markets with Equilibrium Models*
Xiao-Ping Zhang

21. *An Introduction to Wavelet Modulated Inverters*
S.A. Saleh and M.A. Rahman

22. *Control of Electric Machine Drive Systems*
Seung-Ki Sul

23. *Probabilistic Transmission System Planning*
Wenyuan Li

24. *Electricity Power Generation: The Changing Dimensions*
Digambar M. Tigare

25. *Electric Distribution Systems*
Abdelhay A. Sallam and Om P. Malik

26. *Practical Lighting Design with LEDs*
Ron Lenk and Carol Lenk

27. *High Voltage and Electrical Insulation Engineering*
Ravindra Arora and Wolfgang Mosch

28. *Maintaining Mission Critical Systems in a 24/7 Environment, Second Edition*
Peter Curtis

29. *Power Conversion and Control of Wind Energy Systems*
Bin Wu, Yongqiang Lang, Navid Zargari, and Samir Kouro

30. *Integration of Distributed Generation in the Power System*
Math H. Bollen and Fainan Hassan

31. *Doubly Fed Induction Machine: Modeling and Control for Wind Energy Generation Applications*
Gonzalo Abad, Jesus Lopez, Miguel Rodrigues, Luis Marroyo, and Grzegorz Iwanski

32. *High Voltage Protection for Telecommunications*
Steven W. Blume

33. *Smart Grid: Fundamentals of Design and Analysis*
James Momoh

34. *Electromechanical Motion Devices, Second Edition*
Paul C. Krause, Oleg Wasynczuk, and Steven D. Pekarek

35. *Electrical Energy Conversion and Transport: An Interactive Computer-Based Approach, Second Edition*
George G. Karady and Keith E. Holbert

36. *ARC Flash Hazard and Analysis and Mitigation*
J. C. Das

37. *Handbook of Electrical Power System Dynamics: Modeling, Stability, and Control*
Mircea Eremia and Mohammad Shahidehpour

38. *Analysis of Electric Machinery and Drive Systems, Third Edition*
Paul Krause, Oleg Wasynczuk, S. D. Sudhoff, and Steven D. Pekarek

39. *Extruded Cables for High-Voltage Direct-Current Transmission: Advances in Research and Development*
Giovanni Mazzanti and Massimo Marzinotto

40. *Power Magnetic Devices: A Multi-Objective Design Approach*
S. D. Sudhoff

41. *Risk Assessment of Power Systems: Models, Methods, and Applications, Second Edition*
Wenyuan Li

42. *Practical Power System Operation*
Ebrahim Vaahedi

43. *The Selection Process of Biomass Materials for the Production of Bio-Fuels and Co-Firing*
Najib Altawell

44. *Electrical Insulation for Rotating Machines: Design, Evaluation, Aging, Testing, and Repair, Second Edition*
Greg C. Stone, Ian Culbert, Edward A. Boulter, and Hussein Dhirani

45. *Principles of Electrical Safety*
Peter E. Sutherland

46. *Advanced Power Electronics Converters: PWM Converters Processing AC Voltages*
Euzeli Cipriano dos Santos Jr. and Edison Roberto Cabral da Silva

47. *Optimization of Power System Operation, Second Edition*
Jizhong Zhu

48. *Power System Harmonics and Passive Filter Designs*
J. C. Das

49. *Digital Control of High-Frequency Switched-Mode Power Converters*
Luca Corradini, Dragan Maksimovic, Paolo Mattavelli, and Regan Zane

50. *Industrial Power Distribution, Second Edition*
Ralph E. Fehr, III

51. *HVDC Grids: For Offshore and Supergrid of the Future*
Dirk Van Hertem, Oriol Gomis-Bellmunt, and Jun Liang

52. *Advanced Solutions in Power Systems: HVDC, FACTS, and Artificial Intelligence*
Mircea Eremia, Chen-Ching Liu, and Abdel-Aty Edris

53. *Power Grid Operation in a Market Environment: Economic Efficiency and Risk Mitigation*
Hong Chen

54. *Current Signature Analysis for Condition Monitoring of Cage Induction Motors: Industrial Application and Case Histories*
William T. Thomson and Ian Culburt